Hydraulic Engineering

John A. Roberson
Professor Emeritus,
Washington State University

John J. Cassidy
Bechtel Civil Inc.

M. Hanif Chaudhry
Washington State University

JOHN WILEY & SONS, INC.
New York • Chichester • Brisbane
Toronto • Singapore

To our wives:
Amy, Alice, and Shamim

Art by Boston Graphics, Inc.

ISBN: 0 471 12510 5

Printed in the Unites States of America
10 9 8 7 6 5 4 3

Contents

Preface

This text is intended for junior or senior level students who have completed a course in basic fluid mechanics. One of the primary objectives of the text is to present each subject clearly and completely enough so that the student will develop a depth of understanding often lacking in more superficial presentations. To accomplish this goal in a reasonable-size text, we had to be selective in the subject matter to cover. Therefore, only the common hydraulic engineering topics that civil engineers are likely to encounter in their professional careers are included; specialized topics such as sediment transport and coastal engineering are omitted. Students who wish to acquire expertise in these specialized areas should do so by taking graduate level courses.

We have tried to enhance the learning process by liberally using photos, line drawings, and examples. Some material that the student was exposed to in his or her basic fluid mechanics course is also included to provide continuity in developments of certain areas, such as open channel flow and flow in closed conduits. Throughout the text, we have tried to present the subject matter so that the student will be able to use it in modern engineering practice. This is done by emphasizing fundamental principles and presenting up-to-date analytical procedures for solving problems. For instance, Chapters 10, 11, and 12 are devoted to solving hydraulic engineering problems using computers. Several computer programs are included to facilitate learning these modern approaches.

This text is based on notes successfully used over several years in the undergraduate hydraulic engineering course at Washington State University. The material has been student tested, and feedback from the students was especially helpful in putting the final touches on the manuscript. In the undergraduate course, Chapters 1 through 8 were always covered, and most topics in those chapters were discussed. Chapters 9 through 12 were selectively covered; at times, only certain sections of the latter chapters were covered depending on the emphasis that each instructor wished to make. Chapters 10, 11, and 12 have also served as parts of several specialized courses. Those chapters could be studied by students who choose to independently pursue the area of computational hydraulics.

We wish to acknowledge the contribution of Dr. Alan F. Babb, who wrote most of Chapter 7 on hydraulic structures.

For typing and other assistance in manuscript preparation, we wish to thank Charlena Grimes, Patricia Holiday, Delores Lehn, and Miriam Meyerson. We also sincerely appreciate the constructive criticism given by Jerry R. Bayless, University of Missouri, Rolla, MO; Rafael L. Bras, Massachusetts Institute of Technology, Cambridge, MA; Jerzy Z. Klimkowski, North Carolina A & T State University, Greensboro, NC; Lewis J. Mathers, Villanova University, Villanova, PA; Ronald E. Nece, University of Washington, Seattle, WA; Margaret S. Peterson, University of Arizona, Tucson, AZ; Hsiang Wang, University of Florida, Gainesville, FL; Steven J. Wright, University of Michigan, Ann Arbor, MI; Ben C. Yen, University of Illinois, Urbana-Champaign, IL in reviewing this text.

John A. Roberson
John J. Cassidy
M. Hanif Chaudhry

Meandering river at the base of the Wrangell
Mountains, Alaska. (Photo by Kevin G. Coulton)

Introduction

1-1   Scope of Hydraulic Engineering

Hydraulic engineering is the application of fluid mechanics and other science and engineering disciplines in the development of structures, projects, and systems involving water resources. One can develop an appreciation for the scope of hydraulic engineering by reviewing the traditional types of hydraulic engineering projects and by taking note of recent developments in the field. By focusing on modern developments, the breadth of hydraulic engineering applications is revealed. We discuss both traditional and modern types of hydraulic engineering under the following headings.

Traditional Projects

The popular notion is that hydraulic engineering generally relates to traditional water projects such as multipurpose dams, irrigation systems, and navigation works. The Kentucky Dam on the Tennessee River, a typical multipurpose dam, develops a head of about 50 ft and has an installed generating capacity of 160,000 kW (13). Navigation locks are included in this and other dams on the river so that boats and barges can navigate in slack water from its mouth to 650 mi upstream. Part of the storage of water behind the dam is used for flood control.

An example of a project developed primarily for irrigation is the Colorado-Big Thompson project, which was completed in the mid-1950s. Because of this project, in which some hydroelectric power is developed, more than 500,000 acres of land are supplied with irrigation water. The overall project includes more than 9 major dams,* 14 smaller dams, and 3 pumping plants with a power demand of 10,000 hp each (15). To convey the water from its source high in the Rocky Mountains northwest of Denver (west of the Continental Divide) to the irrigated land north of Denver (east of the Continental Divide) required constructing over 60 mi of tunnels and canals. These canals and tunnels have flow capacities from 500 to 1500 cfs. Thus, a project designed primarily for a single purpose (irrigation) involves several structures. Besides the physical aspects of the project, much effort and expertise was devoted to the political, legal, and operational problems of acquiring property for the structures and of allocating and selling water.

Some of the earliest U.S. water projects were for navigation. The first major navigation canal, the Erie Canal, extended from Buffalo, New York, on the west to Troy and Albany on the east. At the top, it was 40 ft wide and it was more than 360 mi long. The elevation change over the length of the canal was about 568 ft (17). Completed in 1825 at a cost of about $8 million, within a few years

---

* These dams are over 100 ft high.

it was earning more than $1 million a year from tolls (8). Today, one of the busiest navigation systems is farther south on the Ohio River proper. A series of locks and dams from its mouth to Pittsburgh (about 1000 mi), allow ships and barges to navigate through slack water. Over 20 dams produce this slack water, and locks 110 ft wide and 1200 ft long let some of the largest tows use the waterway. Extensive analyses are required in designing locks like these to ensure that the constructed facility does not produce undesirable waves or strong currents in the locks during filling or emptying. Such waves or currents could cause the barges and ships to bump against the sides of the lock damaging the vessel. By designing the supply conduits so that filling or emptying flow is distributed uniformly throughout the lock, waves and currents are diminished. Further, short filling and emptying times are desirable to increase the traffic using the lock. High velocities are therefore required in the flow passages, and unless proper design and construction are achieved, cavitation and damage may occur. Hydraulic engineering is also required in the design approaches to the lock so that barges and boats can easily enter the lock even when the river discharge is large. Often, physical model studies as well as analytical studies are required to obtain satisfactory solutions to design problems. Thus, extensive hydraulic engineering is needed on even the most common traditional projects.

Recent Developments

HYDROELECTRIC PLANTS   The use of water becomes more important as society becomes more technological. Higher technology demands power. Power has historically been generated by means of hydroelectric and thermal power plants. Initially, hydroelectric power plants were small, generating only a few kilowatts. These small plants provided energy for small communities or small industrial plants. Flow rates were in the range of a few cubic feet per second, and operating conditions were relatively simple. As the need for energy grew, an economy of scale arose, and the power facilities became larger. Whereas early hydroelectric plants operated at constant capacity almost continuously, the new and larger plants frequently operate as peaking plants to supplement base-load power provided by thermal plants and are brought on line only a few hours each day during periods of peak demand. Rapidly starting and stopping large hydroelectric plants introduces transient flows in the water conveyance system. As the plant is brought on line in an alternating-current generating system, the generator must be synchronized with the electrical grid it is connected to, and any tendency of the system to surge must be controlled. To bring the plant on line quickly requires controlling low pressure surges, which could collapse steel penstocks or possibly introduce air into the flow system if water in the forebay drops too quickly. If the plant is shut down suddenly, as during a power failure, large transient pressures are introduced, which if not controlled, could rupture penstocks or damage machinery.

Although none of these operating problems are new to hydroelectric plants, they are complicated by the large size of the new plants and the need to efficiently use the water. The hydraulic features of hydroelectric facilities must be carefully analyzed to ensure proper operation and the most economical design.

THERMAL PLANTS   The increasing need for power throughout the world and rapidly increasing costs for oil after 1973 led to extensive development of both nuclear and fossil fuel power plants. In an attempt to develop an economy of scale, nuclear plants were designed to produce up to 1200 MW of power per unit, whereas coal-fired units grew to sizes over 600 MW each. Cooling water requirements for plants of this size are correspondingly large. Circulating water systems for 1000-MW power plants have maximum capacities ranging from 500 to 1000 cfs. Conduit sizes for these large cooling water systems range from 10 to 14 ft in diameter, and the pumps range to as large as 2000 hp. Because of these large sizes, the hydraulic design of such systems has become extremely important from the standpoint of both economics and reliability. These systems operate 24 hours a day, so reducing energy losses in the system produces notable savings in energy. Because of the very large flow rates in these systems, transient pressures during startup and shutdown must be carefully controlled, or serious damage can be caused to the steam condensers, piping, or other components.

Two types of cooling water systems have been developed for thermal power plants. Once-through systems withdraw water from nearby streams, rivers, lakes, or oceans at rates large enough to provide the necessary cooling with one pass through the steam condensers (Fig. 1-1a). After passing through the steam condensers, the heated water is discharged at a few degrees above the ambient stream temperature. Concerns raised in the 1960s about the effect of this "thermal pollution" on the receiving waters caused people to pay great attention to the hydraulics of thermally stratified flows and the quantity and quality of the receiving waters. Similarly, the discharge of cold waters low in dissolved oxygen content from thermally stratified reservoirs required decisive developments in both the understanding of the principles involved and the analytical capabilities related to the prediction of stratification of reservoirs and the flow produced by selective withdrawals.

If a large body of natural water is not available for a once-through cooling system, cooling can be achieved by constructing large cooling ponds (Fig. 1-1b) or by using wet towers (Fig. 1-1c).

In the latter case, the water from the condensers that is to be cooled is sprayed into the interior of the tower, and as the spray droplets fall to the base of the tower, they are cooled by the air that rises within the tower. The cooled water is collected in a reservoir at the base of the tower and is then recirculated through the condensers of the reactor.

Solutions to problems involving the cooling of condenser water require extensive application of hydraulic engineering, basic fluid mechanics, and thermodynamics. Advancements in design and analysis have made it possible to

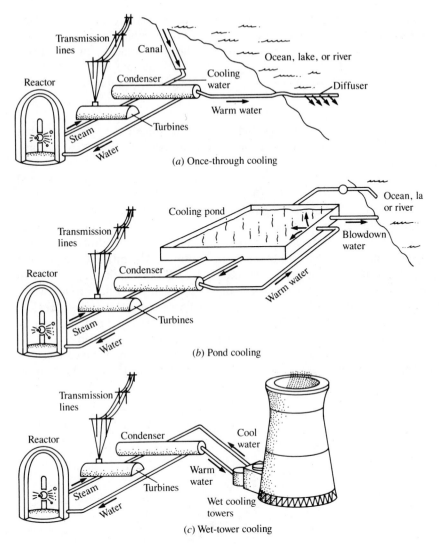

Transmission lines

Canal

Ocean, lake, or river

Reactor

Condenser

Cooling water

Diffuser

Warm water

Steam

Water

Turbines

(a) Once-through cooling

Cooling pond

Ocean, la or river

Transmission lines

Blowdown water

Reactor

Condenser

Warm water

Steam

Water

Turbines

(b) Pond cooling

Transmission lines

Cool water

Reactor

Condenser

Warm water

Steam

Water

Turbines

Wet cooling towers

(c) Wet-tower cooling

Figure 1-1 Three methods of providing cooling water to condenser for a nuclear power plant

successfully design power plants so that environmental hazards associated with them have been minimized.

WASTE STORAGE BEHIND DAMS Other resource developments have also created problems, resulting in the need for better hydraulic and hydrologic analysis of both engineered and natural systems. Increasing numbers of dams have been constructed to contain the tailings from mineral processing plants at mines.

Dams and reservoirs must be developed with virtually every power plant. Some are required to store the bottom ash and SO_2 scrubber sludge, byproducts from the burning of coal, which if released in the environment, could be toxic. Others are required to hold cooling water or to evaporate blowdown, water discharged from the cooling system that because of evaporation has developed a high concentration of salts.

GROUNDWATER DEVELOPMENT   The development of groundwater has proceeded rapidly in the past two decades. Developed around 1920, turbine pumps have greatly stimulated the use of groundwater for agriculture. Figure 1-2 shows the growth of water withdrawals and population for the United States from 1950 to 1980. Total withdrawals for all uses were 450 billion gallons per day (bgd) in 1980, amounting to 2000 gallons per person per day. Of this usage, industrial withdrawals were greatest at 260 bgd. Irrigation ranked second at 150 bgd. However, industrial uses return, generally after treatment, approximately 90% of the water to the streams. Irrigation, on the other hand, because of evapotranspiration, consumes approximately 80% of the diverted water. In many areas, groundwater is pumped for conjunctive use with surface water diversions. In California during 1976–1978, when a record drought occurred, more than 10,000 wells were drilled in the Central Valley to provide supplemental irrigation supplies when water from reservoirs was depleted.

In the 1980s, groundwater pollution has become a great concern in the United States. Pollutants resulting from disposal of industrial waste or from

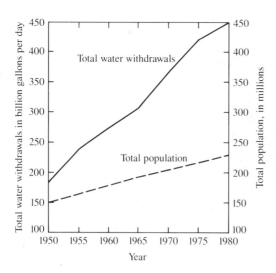

Figure 1-2   Total water withdrawals and population for the United States (10)

agricultural runoff (pesticides, fertilizers, and natural minerals such as salt, selenium, and boron, which are leached from the soil) have caused strong environmental concerns about the quality of the nation's groundwater. Because of the need to understand groundwater movement, advances in analyzing groundwater flow have been achieved. In many cases where pollution has occurred, it has been necessary to install slurry walls (walls constructed across aquifers) to stop the movement of groundwater. In almost all cases, it is necessary to install pumping systems to change the groundwater gradient so that the flow direction of contaminated plumes of groundwater is reversed and a process of removal and treatment can be established. The necessity of designing and placing these structures and well fields to control and remove the contaminated water has required that detailed mathematical modeling be conducted with confidence. The finite-difference and finite-element methods have greatly aided efforts to understand and alleviate these environmental problems.

Modern Developments in Technology

The need to more carefully analyze hydraulic systems, such as cooling-water circulation systems for power plants, large irrigation canals, and sophisticated sewage-treatment plants, has required advancements in the methods of analysis as well as in the experimental testing procedures that must be incorporated in those analyses. Sophisticated mathematical models for transient flows using the method of characteristics were developed to predict transient pressures and to refine the design of control systems and such special devices as surge tanks, pressure relief valves, or compressed air tanks, which are frequently used to control undesirable pressure transients (12).

The need to be more efficient in the management of water has also required better techniques in hydrologic analysis. Foremost in the hydrologic developments of the last decade and a half has been the mathematical watershed model. This numerical modeling of runoff, coupled with the routing of flow in open channels, has made it possible to develop invaluable early warnings of floods through both the direct measurement of streamflow and precipitation and the on-line analysis of these data for quick flood prediction and public warning (2).

The mathematical modeling of thermal discharges and stratified flow has made it possible to predict with reasonable certainty the movement of density-driven flows and the mixing of flows having differential temperatures or concentrations.

All told, the emphasis on large projects and the need to use energy efficiently have lead to many technical advances that allow the hydraulic engineer and the hydrologist to analyze hydraulic and hydrologic systems with increased certainty.

1-2   Historical Perspective

Ancient Hydraulics Works in Egypt and Other Early Civilizations

Early civilizations developed in regions where an abundance of water could be distributed over fairly flat land for irrigation and where a warm climate produced a fast growth of crops. Thus, it is hardly surprising that the earliest remains and accounts of water control were to be found in Egypt, Mesopotamia (Iraq), the Indus Valley (India and Pakistan), and in the Yellow River Valley of China. Egypt is particularly interesting in this regard because the natural features of the Nile River and even the prevailing winds favored the development of a robust civilization. The annual floods over the rich delta land allowed agriculture to flourish even though many people had to move to higher lands during flood season.

To augment the flow of irrigation water during the low flow season, there are signs that one of the early rulers, King Menes (about 3000 B.C.), had a masonry dam built across the Nile near Memphis (about 14 miles upstream from present-day Cairo). This dam was apparently used to divert the river into a canal and, thus, to irrigate part of the adjoining arid lands. As reported by Biswas (1), "the gravity dam seems to have had a maximum height of about 50 ft and a crest length of some 1470 ft."

Egypt was one of the first civilizations to develop an extensive system of river navigation. The Nile traversed the entire length of the country and, in the Delta, divided into seven delta channels, thus providing an extensive system of waterways. A climatic factor favoring the development of water transportation was that the prevailing winds (especially during the summer months) blow from north to south (from the Mediterranean Sea to the Sahara Desert). However, the river flows from south to north so that boatmen used sails to navigate upstream and leisurely drifted downstream (without sails) during the return trip. This mode of transportation is still seen today (see Fig. 1-3).

Civilization in Mesopotamia started about the same time as in Egypt (about 3000 B.C.), and the geography of the two areas is in many ways similar. The Euphrates and Tigris rivers formed a network of channels before finally emptying into the Persian Gulf. Furthermore, the people of the area built many canals for irrigating crops, draining swamps, and water transportation. Early hydraulic "engineering" in this area included developing flood protection works and dam construction (1).

Ancient ruins in the valleys of the Indus River in Asia and the Yellow River in China reveal evidence of water systems developed at least 3000 years ago; however, records of the extent of this development are not as complete as for Egypt and other areas of the Middle East (11).

A different type of ancient hydraulic engineering was developed in Armenia (eastern Turkey) and Persia (Iran) from the seventh to fifth century B.C. Under-

Figure 1-3 Sailboats on the Nile (Photo by
Roger Wood)

ground canals, called qanāts, were dug to intercept groundwater aquifers and
to carry the water from the source areas to cities. Figure 1-4, page 10, shows
the details of this system. Biswas (1) notes the average length of a qanāt was
about 26 mi, and in some places it was as deep as 400 ft. Such ancient water
supply systems, some of which still exist, were truly remarkable.

Roman Water Systems

From about 200 B.C. to 50 A.D., the Romans developed elaborate water-
supply systems throughout their empire. For Rome itself, the usual practice
was to convey water from springs to an aqueduct and then to cisterns throughout the city from which water was delivered to consumers through lead and
baked-clay pipes. Hadas (5) reports that 11 aqueducts supplied Rome with
about 200 million gallons of water daily.

The aqueducts consisted of one or more channels of rectangular cross section and in some locations were supported on spectacular masonry arches. The
channels, which were from 2 to 6 ft in width and from 5 to 8 ft in height (11),
were covered to prevent the water from being contaminated by dust and heated
by the sun. Inspection holes were in channel covers about every 250 ft (11).

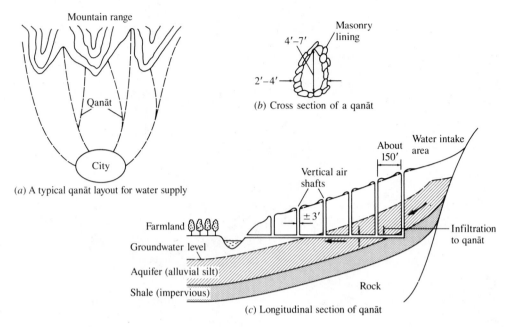

Figure 1-4   Details of qanāt system (1)

The Roman water systems were not limited to the region around Rome. Figure 1-5 shows one of the aqueducts in Spain. This aqueduct was the final stretch of a 60-mi canal that originated in the mountains (5).

Notable Dams Built in This Century

The need for more water resources during this century is the result of a rapidly expanding world population and industrial growth. New machines and methods for manufacturing and placing large quantities of concrete and improved earth-moving equipment provided the means to achieve the rapid growth in major hydropower, irrigation, and flood control projects. In almost all cases, dams are the backbones of these water-resources projects. For hydropower development, a dam is generally needed to develop the head to drive the turbines and to store water to allow power generation. For flood control, dams are used to form reservoirs, which reduce flood peaks by storing the peak flows of flood water. Even a dike constructed to prevent flooding of property near a river is a form of dam. Dams in the United States that are over 15 m high or between 10 m and 15 m high and impound more than 81 acre-ft of water (100,000 m^3) number about 3,000 (9). At the beginning of this century, only 116 of these dams had been built (9).

Figure 1-5 Roman aqueduct in Spain (Photo by
Walter Sanders)

Table 1-1, page 12, lists the world's major dams. The two highest dams are
constructed of earth, which may be surprising to many because we often think
of concrete as the material from which high dams are made. However, since
availability of material and the strength of the foundation dictate the type of
dam to be designed and constructed, earth is often used.

Water Diversion Projects

Some of the world's major water-resource projects divert water from
a drainage basin having a surplus of water to one having a deficiency of water. These projects are usually developed to expand irrigation in the drier
area. Some diversion schemes, however, are developed solely for generating
power. The following two projects involve water diversion.

KEMANO PROJECT   This project is at Kemano, about 400 mi north of
Vancouver, B.C., Canada, and was constructed in the early 1950s. The hydroelectric power generated by the project is used in the electric furnaces of an
aluminum smelter at Kitimat, a deep water port about 50 mi northwest of the
power plant. Alumina from Jamaica is transported by ship to the port at Kitimat
for conversion into aluminum.

Table 1-1 Major Dams of the World*

A. Highest Dams

Name of Dam	Country	Type[†]	Height	
			feet	(meters)
Rogun	USSR	E and R	1099	(335)
Nurek	USSR	E	984	(300)
Grand Dixence	Switzerland	G	935	(285)
Inguri	USSR	A	892	(272)
Vaiont	Italy	A	860	(262)
Chicoasen	Mexico	R	856	(261)
Kishau	India	E and R	830	(253)
Guavio	Colombia	R	820	(250)
Mica	Canada	E and R	804	(245)
Sayano Shushensk	USSR	A and G	804	(245)
Mauvoisin	Switzerland	A	778	(237)
Chivor	Colombia	R	778	(237)
Oroville	USA	E	771	(235)

B. Greatest Volume

Name of Dam	Country	Type[†]	Volume	
			$10^6 \times yd^3$	$(10^6 \times m^3)$
Chapeton	Argentina	E (tailings)	379	(290)
New Cornelia	USA	E (tailings)	273	(209)
Tarbela	Pakistan	E	160	(122)
Fort Peck	USA	E	126	(96)
Guri	Venezuela	E and G and R	99	(76)

C. Greatest hydropower

Name of Dam	Country	Type[†]	Installed Capacity	
			$10^6 \times hp$	(MW)
Grand Coulee	USA	G	8717	(6500)
Sayano Shushensk	USSR	A and G	8583	(6400)
Krasnoyarsk	USSR	G	8046	(6000)
Churchill Falls	Canada	E	7007	(5225)
Bratsk	USSR	E and G	6035	(4500)

* As reported in *International Water Power and Dam Construction*, July 1985.
† Abbreviations are as follows: E = earthfill, R = rockfill, G = gravity, A = Arch.

The main hydraulic engineering features of the Kemano project are:

1. An eastward flowing river, the Nechako River, was dammed to form a large reservoir (water surface elevation = 2800 ft). The rockfill dam, Kenney Dam, is at the east end of the reservoir and is 325 ft high.
2. A 25-ft diameter tunnel was driven through a mountain from the west end

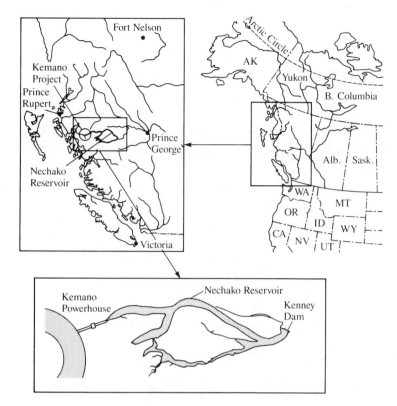

Figure 1-6   Location of Kemano project

of the reservoir to the powerhouse about 200 ft above sea level. The power-house itself is a chamber 1100 ft long, 80 ft wide, and 138 ft high excavated in rock 1000 ft below ground surface.
3. Another tunnel carries the flow from the turbines to the Pacific Ocean, a short distance from the powerhouse.
4. Impulse turbines under a head of 2590 ft have the capacity to generate 140,000 hp each.

Figure 1-6 shows the location of the project, and Fig. 1-7, page 14, shows a plan and elevation of the penstocks and tunnel that deliver the water from the reservoir to the powerhouse. The project's designers chose to have an under-ground power-house because the ground surface profile was rugged; the rock was sound; and surface structures would be subject to snow avalanches, rock slides, and forest fires. The tunnel was designed to carry a flow of about 4500 cfs and is unlined except in stretches (about 35% of the total length) where poor quality of rock required a concrete lining.

Because this aluminum plant and power plant were in such a remote region of Canada, an entire city was developed for the workers and their families. The

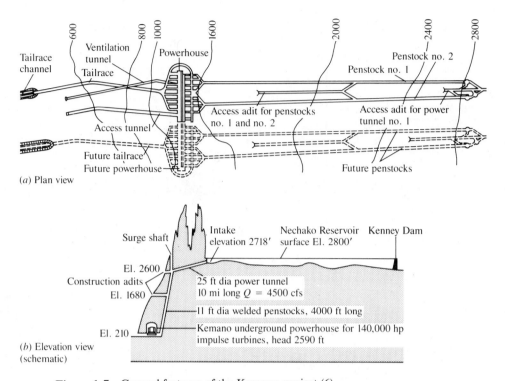

Figure 1-7   General features of the Kemano project (6)

Kemano project is one of the most unusual hydropower projects in the world. [For more information about this project, see the articles in *Civil Engineering* magazine (6) and (7).]

CENTRAL VALLEY PROJECT IN CALIFORNIA   In this project, water is collected during the winter in reservoirs behind several dams and then released for irrigation during the growing season. Figure 1-8 shows the areas served by this project.

The Central Valley is about 500 mi long and has an average width of about 50 mi. One of the two principal rivers in the valley is the Sacramento, which rises in the north end of the valley and flows southward to finally discharge into San Francisco Bay. The other main river is the San Joaquin, which has its source in the Sierra Nevada mountains and flows northwesterly to join the Sacramento a short distance upstream from San Francisco Bay. Together, the two rivers convey enough water to satisfy the irrigation needs of both the northern (Sacramento Valley) and the southern (San Joaquin Valley) parts of the valley; however, the natural flow of the Sacramento River supplies more water than is needed in that valley, and the San Joaquin Valley needs more than is available from the San Joaquin River. Therefore, an elaborate system

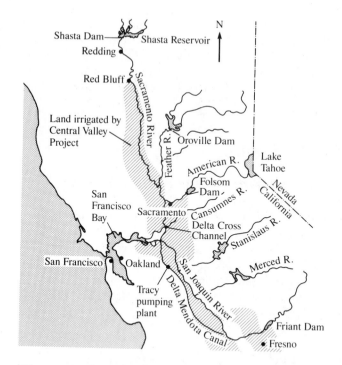

Figure 1-8 Central Valley project

of dams, reservoirs, canals, and pumping stations was designed to transfer water from the north to areas in the south.

The reservoir behind Shasta Dam on the Sacramento River stores the winter flows. During the growing season, water, released through turbines to the Sacramento River, flows south until, about 40 mi upstream from the mouth of the river, some of the river water is diverted into the Delta Cross Channel. This channel discharges water into a slough that conveys the water for about 50 mi to the Tracy Pumping Plant. At the Tracy Pumping Plant, the water is lifted 197 ft to the upper end of the Delta-Mendota Canal. The Tracy Pumping Plant consists of six 22,500-hp pumps, each capable of discharging 767 cfs of water. The energy to run these pumps is supplied from the power generated at Shasta Dam. The Delta-Mendota Canal carries water south along the west side of the San Joaquin Valley for irrigation along its 113 mi-length. This diversion of water from the Sacramento River Basin to the San Joaquin Valley is the main part of the Federal Central Valley Project of the U.S. Bureau of Reclamation; however, developments by the state of California, the U.S. Army Corps of Engineers, and private conservation districts are also major in scope when considered individually. For example, the reservoirs of Oroville Dam, Folsom Dam, New Don Pedro Dam, and Friant Dam all store winter runoff, which is released for irrigation in the valley. The heights of these dams are 735 ft, 340 ft, 600 ft, and 319 ft, respectively, and their combined reservoir capacity is over 6 million acre ft.

The Central Valley Project brings irrigation water to more than 3 million acres, and more than 1 million kW of power is generated from the system's power plants. The project was started in about 1940 and by 1970, was essentially completed.

1-3   Trends for the Future

Water Use in Thermal Power Plants

As already noted, the cooling water systems required for modern thermal power plants are major components of these plants. The trend is for more thermal plants to use coal as fuel. Therefore, more attention will be given to economical ways of delivering and processing coal for these plants. Further, research and development will likely focus more on the transport of coal through slurry pipelines. Thus, hydraulic engineering will continue to be an important aspect of thermal-energy programs.

Small Scale Hydroelectric Development

From about 1940 to 1970, almost all the large-scale hydroelectric sites in the United States were developed. Today, any undeveloped large-scale sites are usually unavailable for development because of public pressure about environmental or recreational concerns. However, with the increased cost of fuel for thermal plants, small-scale hydropower plants became competitive, and small hydraulic turbines are now being manufactured in sizes not readily available a decade ago. Ironically, these small turbines were common 60 to 70 years ago. Most of the potential sites are in the Pacific Northwest, where about 270,000 MW of potential hydropower exists (14). Other areas of the continental United States that have hydropower potential include the midcontinent region (Colorado, Montana, Oklahoma, Wyoming) with about 27,000 MW, the Southeast with about 23,000 MW, and the Northeast with about 8000 MW (14).

One major difference between developing a large- and small-scale project is that most of the equipment such as turbines and valves for the small plants can be bought "off the shelf." However, engineers must still analyze streamflow records to determine the installed capacity of the plant and to assess the economics of the project. Small dams and penstocks are needed to divert and convey the water to the turbines. A number of small-scale hydropower projects will probably be designed and built in the foreseeable future; however, the demand will fluctuate with the price of fuels for the competing plants and with the economy of the nation and the world.

Conflicts in Water Development

As our nation's population increases so does our demand for food, energy, and recreation. Thus, more land is put into irrigation, more water is used for energy production, and more water is used for recreational purposes. These demands on the water environment are, and will continue to be, in conflict with each other. Examples of these conflicts follow:

1. Water diverted for irrigation often leaves too little water downstream from the diversion to support fish and other aquatic life.
2. Drainage from irrigated land often pollutes the river or lake it discharges into because of high levels of sediment, leached elements, or chemicals.
3. Operation of hydroelectric plants often produces fluctuations in streamflow that adversely affect aquatic life and sometimes produces undesirable if not dangerous swimming and boating conditions.
4. Heated effluents from thermal power plants often adversely affect aquatic life.
5. Construction of dams hinders if not destroys anadromous fish runs.
6. Drawdown of reservoirs for flood control often decreases the amount of power that could otherwise be developed and produces unsightly, unpleasant conditions around the reservoir.
7. Release of water from a reservoir for instream flow requirements (to sustain healthy aquatic life, recreation, and pollution control) is often at the expense of efficient hydropower production and irrigation use.

Because of these and other conflicts, engineers and scientists are devoting more attention to understanding the environment and to developing more efficient uses of water. For example, serious problems still exist in designing fish passage facilities at dams so that anadromous fish can traverse the structures without killing a sizable portion of the run. This is especially so of the downstream migration of young fish. Thus, continued research and development is needed in this and other areas of water resources.

Improvements in Engineering

The scientific development of hydraulic and hydrologic engineering has been in many ways forced by the development of natural resources and the design and construction of engineered systems. As projects become larger and more sophisticated, better hydraulic machinery and more sophisticated controls are required. These improvements in turn require better knowledge of hydraulic behavior. Reliability in a project is tremendously important in a large nuclear power plant where, for example, generated revenue may amount to more than $3 million a day. Loss of production due to failure of a hydraulic element can be economically far more important than the actual equipment cost. Thus, for these facilities, improvements in design from better understanding and improved analytical capabilities are vitally important. Further, the design and construction of these facilities tends to foster improved technical developments.

Concern about safety has also increased in the past decade. More than ever before, it is important to improve analyses of flood levels at coastal sites, on sites flooded by rivers, and at locations where a dam failure could cause a great hazard. The ability to analyze the migration of contaminants in both surface water and groundwater also becomes more important as hazardous wastes enter the environment.

Most of these improved methods of analyses involve the application of numerical methods using digital computers. Because many of the problems also involve more than the flow of water itself, multidisciplinary approaches are needed. For example, the migration of certain contaminants requires knowledge about and means of modeling the dispersion and possible change of the contaminant with both time and space. Such a change may be chemical or biological and may be affected by heat-flow considerations. Thus, sophisticated models involving these aspects of the problem must be linked to the basic flow model. There is an increasing need for hydraulic engineers who have the ability and mathematical skills to generate and use sophisticated computational models of this type.

The role of the hydraulic engineer and the engineering hydrologist has increased in importance. Although the modern science of water-resources engineering must be built on the basic elements of hydraulics, hydrology, and fluid mechanics, the requirement for better and more detailed analyses demands engineers with ability in the mathematical and physical sciences as well.

REFERENCES

1. Biswas, A.K. *History of Hydrology*. North Holland Publishing, Amsterdam, 1970.
2. Burnash, R.J.C., and T.M. Twedt. "Event Reporting Instrumentation for Real-Time Flash Flood Warnings." Proceedings of the Conference of Flash Floods: Hydrometeorological Aspects, American Meteorological Society, May 1978.
3. Casson, L. *Ancient Egypt*. Time, Inc., New York, 1965.
4. Giusti, E.V., and E.L. Meyer. "Water Consumption by Nuclear Power Plants and Some Hydrologic Implication." U.S. Geological Survey, Circular 745. Arlington, Virginia, 1977.
5. Hadas, M. *Great Ages of Man*, Vol. One, Imperial Rome. Time, Inc., New York, 1965.
6. Huber, W.G. "Alcan-British Columbia Hydro Project — Tunnels and Underground Penstocks." *Civil Engineering* magazine, vol. 23, no. 2 (February 1953).
7. Huber, W.G. "Alcan-British Columbia Power Project Under Construction." *Civil Engineering* magazine, vol. 22, no. 11 (November 1952).
8. Hulbert, A.B. *Historic Highways of America, The Great American Canals, Vol. II (Erie Canal)*. Arthur H. Clark, Cleveland, Ohio, 1904.

9. Jansen, R.B. *Dams and Public Safety.* U.S. Bureau of Reclamation, U.S. Government Printing Office, Washington, D.C., 1980.
10. Kelly, D. "Estimated Use of Water in the United States, 1983." U.S. Geological Survey, Circular 876. Reston, Virginia, 1983.
11. Rouse, H., and S. Ince. *History of Hydraulics.* Iowa Institute of Hydraulic Research, Iowa City, 1957.
12. Streeter, V.L., and B. Wylie. *Fluid Transients.* FEB Press, Ann Arbor, Michigan, 1982.
13. Todd, David K. *The Water Encyclopedia.* The Water Information Center, Inc., Port Washington, New York, 1970.
14. U.S. Army Corps of Engineers. "Preliminary Inventory of Hydropower Resources." U.S. Army Corps of Engineers National Hydroelectric Power Resources Study, Ft. Belvoir, Virginia, Institute for Water Resources, 1979.
15. U.S. Bureau of Reclamation. *Colorado-Big Thompson Project.* U.S. Government Printing Office, Washington, D.C., 1957.
16. U.S. Bureau of Reclamation. "San Luis Unit," vol. 1 (History, General Description and Geology). U.S. Department of the Interior, 1974.
17. Waggoner, M.S. *The Long Haul West.* G.P. Putnam Sons, New York, 1958.

2

August 8, 1980 photograph of two hurricanes.
Hurricane Isis at left is off the southern tip of Baja,
California, while Hurricane Allen is centered in the
Gulf of Mexico. The photograph was taken from an
altitude of 22,000 mi by GOES (Geostationary
Operational Environmental Satellite. Courtesy of
U.S. National Oceanic and Atmospheric
Administration)

Hydrology

2-1   General Considerations

Hydrology can broadly be defined as the technical field encompassing the occurrence of water *on*, *above*, or *within* the earth. However, in this text, as is generally common practice in engineering, hydrology is considered to deal primarily with surface-water hydrology, or water on the surface of the earth. Hydrogeology will be assumed to deal with the flow and occurrence of groundwater, and meteorology with the occurrence of water in the atmosphere. Because engineers have recognized that the design of major projects must consider unusual storms as design events, hydrometeorology has evolved in the last 30 years as a separate science dealing with the generation of rainfall by major storms including local effects of topography.

The science of hydrology has made major advancements during the twentieth century. Its major developments have occurred partly because of increased collection of data for precipitation, streamflow, and groundwater. Since 1960, developments in electronic instrumentation, satellite communication, and computerized analysis have made data not only easier to collect but also easier to analyze, tabulate, store, and publish. Improved analytical capabilities alone have greatly increased the mathematical sophistication of hydrologic analysis.

2-2   Project Needs

Designers of every project must consider the hazards and benefits of precipitation and streamflow. Since the earliest times, the human race has dwelled near streams or rivers because they provided a source of food, an avenue of transportation, and a water supply for domestic needs. Through trial and error, human beings learned of the hazards due to flooding and drought. As long as people were transient, little was lost by moving everything out of the path of a flood. Drought caused migrations in search of more dependable water supplies and may have contributed to the loss of entire civilizations, such as the one that flourished along the Tigris and Euphrates rivers well before the birth of Christ (3).

Today's comprehensive projects, such as major bridges, airports, factories, power plants, buildings, roads, hydroelectric plants, and irrigation projects, cannot readily be moved to avoid flood hazards, and water supplies that prove inadequate may bankrupt large-investment projects. Thus, it is important to be able to predict the stage of flooding in a river; the depth of rainfall to be expected at a site; or the volume of water a river can provide in a day, a month, or a year. To establish the floor elevation of an industrial facility to be built near a river, choosing a design flood and being able to calculate a maximum water-surface elevation associated with that flood are necessary. If the plant is constructed above that elevation, it will be safe against the design flood and lesser events.

Larger floods will produce higher flood elevations and, as a result, damages. Projecting those damages and assessing their economic impact on a project should be a vital part of selecting the design flood.

A simple irrigation project may involve pumping water from a river directly to a sprinkler system, which in turn provides water for crops. Since developing land for irrigation is expensive, it is critical to know how much water can be pumped from the river during a dry year. How dry is a dry year? Designing for the lowest river flow rate recorded implies that the project cannot function at the design condition for years in which the flow rate is lower. It is important in economic planning for a project to assess the losses that will occur during periods that are drier than the design condition. Associating those losses with their probability of occurrence provides the information needed to determine the project's probable income.

Thus, in hydrology, it is necessary to quantitatively assess streamflow and rainfall and, through statistical analysis, to estimate the probability of occurrence of larger and smaller events. The following sections will deal with each of these requirements.

2-3   Hydrologic Cycle

Hydrologic engineering differs from hydrology primarily in that an engineering application is implied. Thus, engineering considerations deal mostly with estimating, predicting, or forecasting precipitation or streamflow. By contrast, scientific hydrology deals primarily with the basic physical laws governing the elements of hydrology. Certainly, engineering hydrology must be fully aware of the scientific advancements in hydrology to properly apply those advancements in engineering practice. This chapter emphasizes the engineering applications of hydrology while attempting to use a scientific basis.

Hydrology deals intimately with the complex natural phenomena that govern our weather and climate. It is equally difficult to predict rainstorms and the resulting flood flows or extreme droughts and the resulting low flows. All the energy that drives the physical processes of hydrology comes from the sun. Differential warming and cooling of the atmosphere produces both large- and small-scale atmospheric motions. Since earth and water absorb energy at different rates and cool at different rates, the presence of land further complicates atmospheric movement and the occurrence of precipitation. Therefore, weather at any point on the earth exhibits characteristics that appear to be random but actually contain some cyclical components. Figure 2-1 shows the average monthly temperature for Glenwood Springs, Colorado, for the period 1959 through 1960. Figure 2-2 shows the average monthly precipitation for the same period. Note that temperature shows a regular seasonal variation with a 12-month period, but for precipitation, short-term fluctuations tend to mask any evidence of a regular period. Although the qualitative aspects of hydrology are

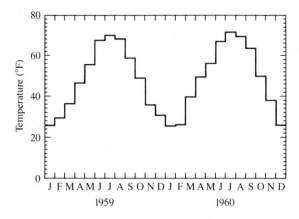

Figure 2-1 Average monthly temperature at
Glenwood Springs, Colorado, 1959–1960

well understood, limitations of the current state of knowledge result in considerable uncertainty about quantitative analysis. Just as weatherforecasters cannot predict the occurrence of rain with 100% accuracy, so are engineering hydrologists' predictions uncertain. Because of this inherent uncertainty, hydrologists must learn to assess the quality of data and to use it reasonably.

The total volume of water on, below, and above the surface of the earth is a constant. Its movement through various phases, known as the hydrologic cycle, is shown in Fig. 2-3. In general, the total cycle involves evaporation from bodies of water, movement of water vapor through the atmosphere, precipitation, infiltration and runoff, groundwater flow, and streamflow, which com-

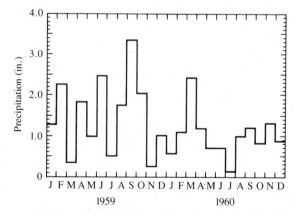

Figure 2-2 Average monthly precipitation at
Glenwood Springs, Colorado, 1959–1960

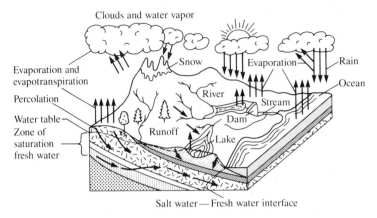

Figure 2-3 Schematic illustration of the hydrologic cycle

pletes the cycle. The volume of water in any given state or zone continuously varies, but the total volume remains constant.

Precipitation is produced when moisture-laden atmosphere is lifted, cooling the air and condensing some of the moisture. The lifting can be produced by local heating, producing the intense local thunderstorms that frequently occur on summer afternoons. Lifting with resultant precipitation also occurs when moving air passes over a mountain range, as shown in Fig. 2-3. The latter, called *orographic* lifting, results in zones of high annual precipitation on the windward side of mountains. For example, rainfall on the ocean side of the Olympic Mountain range along the western coast of the United States increases with elevation, reaching as much as 200 in. per year. On the lee (east) side of the mountains, the air descends and is warmed, increasing its capacity to hold moisture. Thus, central Washington, east·of the Cascade Mountains, has a normal annual rainfall that averages less than 10 in.

The individual processes of the hydrologic cycle are actually complex, and today, neither analytical nor physical models are available to precisely describe them. Because of the complexity of the hydrologic cycle, it is often difficult for students to grasp the essence of it. To help you understand the basic principles of hydrology, a simplified conceptual model is shown in Fig. 2-4. The process shown in Fig. 2-3 is detailed and complex, whereas the conceptual model shown in Fig. 2-4 is simplified and, hence, amenable to approximate analytical simulation.

We will trace the hydrologic cycle beginning with precipitation. As precipitation occurs, part of it is intercepted by trees, grass, stones, or other cover, preventing it from reaching the earth. Part of this intercepted volume is retained and eventually evaporates. Other parts, such as snow on trees, will eventually fall to earth. The part that evaporates is referred to as *interception.* Of the rainfall that reaches the earth, a part is stored in depressions (*depression storage*) and is prevented from running off. Molecular forces and gravity cause water in contact with the earth to be *infiltrated*, or drawn into the openings between soil particles. If the ground is not frozen, *infiltration* begins immediately. When the rainfall intensity is greater than the infiltration rate and the volume of rainfall is enough to more than fill all depression storage, surface runoff begins.

Surface runoff initially flows over the ground surface as sheetflow, eventually concentrating in rivulets and finally in streams or rivers. Infiltration continues as the surface runoff progresses. If precipitation over the drainage basin were uniform, the rate of streamflow would always increase in the downstream direction.

As we mentioned, infiltration begins as soon as precipitation reaches the soil. The rate of infiltration is governed both by the amount and rate at which precipitation reaches the soil and by the ease with which water can penetrate the surface of the soil and move through the interstices between soil particles. Thus, water penetrates loosely packed soils easily. Because small interstitial openings between soil grains require large driving heads, water moves through sands much more readily than through clays.

A given soil can absorb and hold, through forces of molecular attraction, a specific volume of water known as the *field capacity*. Once this field capacity is filled, water moves directly through the soil. In that case, the ability for movement through the soil can limit the rate of infiltration. The volume of water infiltrating, beyond that required to fill the field capacity, moves downward through the soil as *deep percolation* and eventually reaches the *groundwater zone* (the *zone of saturation*). Water moves down gradient through the ground-

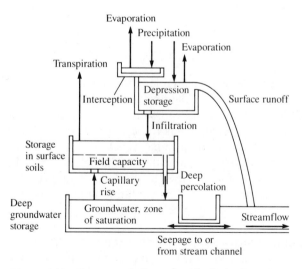

Figure 2-4   Conceptual diagram of hydrologic cycle

water zone eventually flowing into streams, and during periods of zero surface runoff, provides a *baseflow* for the stream. Water may leave the groundwater zone to enter the *root zone* by capillary action. The movement of water through the groundwater zone is much slower than the movement of surface runoff.

That part of precipitation greater in volume than the combined interception, depression-storage, evaporation, and infiltration volumes becomes surface runoff and is known as the *rainfall excess*.

The volume and rate of surface runoff is of great interest in engineering design and water-resources planning. Many different approaches have therefore evolved to estimate runoff rate or volume as a result of a given amount of precipitation. The description of the hydrologic cycle briefly points out the complex considerations affecting surface runoff. A small drainage area, such as a parking lot, may have uniform characteristics (particularly if it is paved) such as depression storage, slope, and texture of the soil surface. However, a large natural basin can be expected to exhibit great areal variations in vegetal cover, soil type, slope and land use. Thus, simple rainfall-runoff relationships such as the rational equation

$$Q_p = CIA \tag{2-1}$$

can be confidently applied only to small drainage areas. In Eq. (2-1), Q_p is peak rate of runoff, I is intensity of rainfall, A is the drainage area, and C is a dimensionless empirical coefficient that is primarily dependent on texture and permeability of the area's surface. The units of Q_p are a volume rate of flow dependent on the units used for A and I. Table 2-9, page 67, lists values of C applicable for various surfaces. We will apply this and other rainfall-runoff methods later.

2-4   Statistics and Probability

In many ways, hydrologic variables resemble random phenomena. Figure 2-5 shows the annual precipitation for Columbia, Missouri. No apparent trend exists in the data. Other characteristics can show a definite trend. For example, temperature in a northern climate tends to be warm in summer and cold in winter. In other areas, such as northern India, a great deal of rain may fall during the monsoon season and virtually none during the remainder of the year. Apparent trends are exhibited, for example, in the temperature data of Fig. 2-1, which shows a cyclical trend, and the precipitation data of Fig. 2-2, which appears to show decreasing monthly precipitation with time although fluctuation is profound.

Extremes in these natural occurrences, their annual averages, and almost any other measure of magnitude vary in ways that appear to be random and can often be considered random. Techniques of probability and statistics are used to analyze these random events. In this section, we discuss these statistical

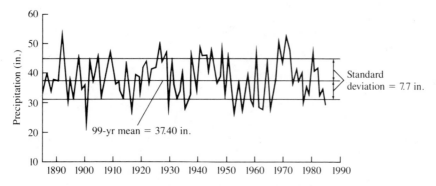

Figure 2-5 Annual precipitation at Columbia, Missouri 1887–1986

tools but neglect most mathematical derivations. You will likely have seen these tools in other contexts.*

The concept of probability is paramount in the field of hydrology. It is convenient to relate probability to a familiar procedure such as tossing a coin, throwing dice, or drawing numbered balls from a hat, all of which are random procedures (if the game is honest). The numerical value of probability ranges from 0 to 1, with 1 being absolute certainty that the event will happen and 0 being absolute certainty that it will not. We will now set forth several relationships regarding probability and then illustrate each with an example.

I. The probability of obtaining either outcome A or B, with A and B independent and mutually exclusive, is the sum of the probability of obtaining each. Thus,

$$P(A \text{ or } B) = P(A) + P(B) \tag{2-2}$$

where $P(A \text{ or } B)$ = probability of obtaining either A or B
$P(A)$ = probability of obtaining A
$P(B)$ = probability of obtaining B

EXAMPLE 2-1 If a roulette wheel contained only 50 numbers, and if either of the numbers 4 or 15 were winners, what would be the probability of winning?

SOLUTION

$$\text{Probability of winning} = P(4) + P(15) = \frac{1}{50} + \frac{1}{50} = \frac{1}{25} = 0.04 = 4\%$$

∎

---

* The serious student can find the theoretical developments in most textbooks on statistics or in textbooks dealing primarily with the statistical methods in hydrologic analysis (8).

II. The probability of obtaining both outcome A and outcome B with A and B independent is the product of the probabilities of obtaining either A or B. Thus,

$$P(A \text{ and } B) = P(A) \cdot P(B) \tag{2-3}$$

In this case, either A or B must be obtained first and then the other.

E X A M P L E 2 - 2   What is the probability of throwing two sixes with a pair of dice? The probability of obtaining one 6 is 1/6.

S O L U T I O N   The probability of obtaining two sixes is

$$P = \frac{1}{6} \cdot \frac{1}{6} = \frac{1}{36} = 0.028 = 2.8\%$$ ∎

III. The probability P of having exactly K occurrences in n trials is

$$P(K \text{ in } n) = \frac{n!}{K!(n-K)!} \, p^K (1-p)^{n-K} \tag{2-4}$$

where $n!$ represents factorial of n, and p is the probability of success in any one attempt.

E X A M P L E 2 - 3   What is the probability of throwing three sixes in eight throws using only one of a pair of dice?

S O L U T I O N   The probability of throwing a six with only one of the dice is 1/6. Thus,

$$P(3 \text{ in } 8) = \frac{8!}{3!5!} \left(\frac{1}{6}\right)^3 \left(1 - \frac{1}{6}\right)^{8-3}$$

or $$P(3 \text{ in } 8) = \frac{6 \cdot 7 \cdot 8}{1 \cdot 2 \cdot 3} \left(\frac{1}{6}\right)^3 \left(\frac{5}{6}\right)^5 = 0.104 = 10.4\%$$ ∎

Probability in hydrology is conveniently referred to in terms of the average *return period* of a particular event. Thus, a rainstorm with an average return period of 100 yr has a probability of occurring or being exceeded once in 100 yr or $1/100 = 0.01$. By average return period, we mean the average number of years between occurrences of this magnitude. Return period T is thus related to probability by the equation

$$P = \frac{1}{T} \tag{2-5}$$

You should understand that the return period or recurrence interval T implies strictly an average number of years between events that equal or exceed the magnitude of an event. The flood event with a return period of 100 yr ($P = 0.01$), thus, has a 0.01 or 1% chance of occurring or being exceeded this year or any other year. Likewise, a 50-yr flood has a 0.02 (2%) probability of being equalled or exceeded in any year.

Equation (2-4) can be used to investigate some questions of interest in hydrology.

EXAMPLE 2-4   What is the probability that exactly three 100-yr floods will occur during the 50-yr expected life of a particular highway bridge?

SOLUTION   For the 100-yr flood: $p = 1/100 = 0.01$. In Eq. (2-4), n is now the 50-yr design life. Thus,

$$P(3 \text{ in } 50) = \frac{50!}{3!(50 - 3)!} (0.01)^3(1 - 0.01)^{47}$$

or $P(3 \text{ in } 50) = 0.012 = 1.2\%$

Thus, if the bridge is designed to be safe during floods up to the 100-yr event (the *design flood*), there is a 1.2% chance that the 100-yr flood will be equalled or exceeded exactly three times during *the design life of the bridge*.   ■

A project is normally designed to be safe for floods of less than or equal to some magnitude, known as the *design flood*. The question posed in Example 2-4 is not the one of real interest in hydrologic design. The important question is: What is the probability that the design flood will be equalled or exceeded at least once during the design life of the project? We can compute the answer with the following equation, which follows directly from Eq. (2-4):

$$P(\text{failure in } n \text{ years}) = 1 - (1 - p)^n \tag{2-6}$$

EXAMPLE 2-5   What is the probability that a 100-yr design flood will be exceeded at least once in a 50-yr project life?

SOLUTION   Using Eq. (2-5):

$$P = \frac{1}{100} = 0.01$$

Using Eq. (2-6):

$$P = 1 - (1 - 0.01)^{50} = 0.39 = 39\%$$

Thus, if an engineer designs a dam to be safe during floods up to a 100-yr event, there is a 39% chance that this design flood will be exceeded during a 50-yr design life. If the dam fails during a flood greater than the 100-yr event, there is a 39% chance that the *dam will fail during its design life* as a result of flooding.

■

Example 2-5 briefly presents the concept of probability of failure. It is interesting to compute a set of design events that must be designed for in order to realize a given risk (probability of failure) during the planned project life. Table 2-1 contains values computed using Eq. (2-6) for the required return period of a design event if that design event is not to be exceeded by more than a chosen probability (risk) in a given project life.

Table 2-1 shows that if the permissible risk of failure is 1%, a project with a 50-yr design life must be designed for an event that has an average recurrence interval of 4975 yr. To find this value we substitute the risk 0.01 into Eq. (2-6) to obtain.

$$0.01 = 1 - (1 - p)^{50}$$

and applying the reciprocal yields

$$T = 4975 \text{ yrs}$$

For reasons of comparison, it is interesting to note that large dams and nuclear power plants are designed to be safe during a probable maximum flood, which is thought to have an exceedance probability in the order of 10^{-6} (see Tables 2-14 and 2-15, pages 88 and 93).

The design life of a project is primarily an economic consideration. At the end of its design life, a project would presumably have zero worth. The actual life of most structures is often much longer than their design life. However, the cost and benefits accrued during the design life balance, and in theory, the project could be abandoned at the end of its design life without loss. This practice

Table 2-1 Required Design Return Interval as a Function of Acceptable Risk

Acceptable Risk of Failure (P)	Planned Useful Project Life (n) (yr)			
	1	25	50	100
0.01	100	2487	4975	9950
0.25	4	87	174	348
0.39	2.6	51	100	203
0.50	2	37	73	145
0.75	1.3	19	37	73
0.99	1.01	6	11	22

is much more common for buildings than for major structures such as hydroelectric plants, bridges, or dams.

Flood Frequency Analysis

The tools of statistics and probability are particularly useful in hydrology. The data on precipitation kept by the U.S. National Weather Service and those on streamflow kept by the U.S. Geological Survey are examples of enormous databases that are of little use unless the data are refined, transformed, and summarized by statistical analysis.

Mean daily, mean monthly, and mean annual values are only three parameters that are continually computed as data are collected. Moreover, information is frequently desired on the return period of particular flows and on the magnitude of a flow that is equalled or exceeded with a given probability (a design flood).

In making a frequency analysis of floods, the annual peak flow is selected for each year of the recorded flows of the stream of interest. The series of flows as shown in Table 2-2 was measured on the Ok Ma River in Papua, New Guinea.

Table 2-2 Peak Flows Above 250 m^3/s for the Ok Ma River in Papua, New Guinea

Date	Peak Flow Rate (m^3/s)	Date	Peak Flow Rate (m^3/s)
11/11/76	253	7/10	353
11/25	296	7/23	560
12/3	353	7/29	393
1/14/77	250	8/1	436
1/30	366	8/5	287
2/4	436	10/21/80	408
2/28	290	1/21/81	303
4/8	505	2/15	290
4/15	273	4/19	387
4/25	560	6/2	285
5/4	273	9/1	260
6/4	766	9/11	509
6/8	380	11/18	327
6/18	301	12/24	300
6/20	340	1/28/82	254
6/27	400	8/13	276
6/30	736	5/2/83	350
7/6	347		

The series of annual maximum flows, so selected, is called, reasonably enough, an *annual series*. In this particular case, all flows greater than 250 m³/s, and not just the maximum for each year, were listed, which is called a *partial duration series*. To analyze these values statistically, we must compute the following parameters where Q_i is the annual peak flow for year i, and N is the number of years of recorded flows.

$$\text{Mean} = \overline{M} = \frac{\sum\limits_{i=1}^{N} Q_i}{N} \tag{2-7}$$

$$\text{Standard deviation} = S = \sqrt{\frac{\sum\limits_{i=1}^{N} (Q_i - \overline{M})^2}{N-1}} \tag{2-8}$$

$$\text{Skew} = G = \frac{\sum\limits_{i=1}^{N} (Q_i - \overline{M})^3}{N} \tag{2-9}$$

These three parameters are, respectively, measures of the first moment of the flows about the origin, and the second and third moments of the flows about the mean. Figure 2-6 illustrates their properties. The mean $\overline{M}$ is, of course, a measure of the location of the centroid of the distribution of flows.

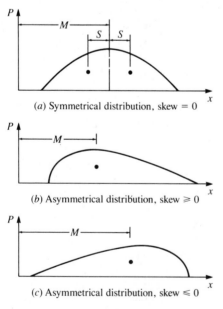

(*a*) Symmetrical distribution, skew = 0

(*b*) Asymmetrical distribution, skew ⩾ 0

(*c*) Asymmetrical distribution, skew ⩽ 0

Figure 2-6   Distributions illustrating statistical parameters

The *standard deviation S* is a measure of the way in which the flows are distributed about the mean, and the *skew G* is a measure of the degree of symmetry of the distribution. Figure 2-6b illustrates a distribution where most values are located to the right side of the mean (a positive skew), whereas Fig. 2-6c illustrates the opposite (a negative skew).

EXAMPLE 2-6 Determine the mean, standard deviation, and skew of the annual peak flows shown in Table 2-2.

SOLUTION In the accompanying table the largest flow for each year has been selected. There are only six years of record, since 1978 and 1979 are missing.

Year	Largest Flow (m³/s)	Order (m)	Apparent Probability (m/N + 1)
1976	353	4	0.57
1977	766	1	0.14
1980	408	3	0.43
1981	509	2	0.28
1982	276	6	0.86
1983	350	5	0.71
	sum = 2662		

Thus, $\bar{M} = 2662/6 = 444$ m³/s. In computing the standard deviation, Eq. (2-8) can be rearranged to give the following equation, which we can compute with a numerical accuracy greater than is inherent in Eq. (2-8).

$$S = \sqrt{\frac{\sum\limits_{i=1}^{N}(Q_i^2) - \left(\sum\limits_{i=1}^{N} Q_i\right)^2 \Big/ N}{N-1}} \qquad (2\text{-}10)$$

$$= \sqrt{\frac{1335586 - 7086244/6}{5}} = 176 \text{ m}^3/\text{s}$$

Instead of calculating the skew directly, a coefficient of skewness is used more frequently and is equal to the skew divided by the cube of the standard deviation. Equation (2-9) can be rearranged to give the following:

$$g = \left[\frac{N^2\left(\sum\limits_{i=1}^{N} Q_i^3\right) - 3N \sum\limits_{i=1}^{N} Q_i \sum\limits_{i=1}^{N} Q_i^2 + 2\left(\sum\limits_{i=1}^{N} Q_i\right)^3}{N(N-1)(N-2)S^3}\right] \qquad (2\text{-}11)$$

$$= \left[\frac{36(757131190) - 3(6)(2662)(1335586) + 2(1.8863582 \times 10^{10})}{6(5)(4)(176)^3}\right]$$

$$= 1.51 \qquad \blacksquare$$

Example 2-6 is a case where the record of flows is very short. A short record of data like this normally does not define a distribution well, and skew particularly is not defined well for a short record. In general, the skew should not be estimated for records much less than 100 yr in length (2). In practice, short records are encountered often, and care must always be used in attempting to extrapolate information calculated from them.

Several theoretical distributions have been used in attempts to fit curves to distributions of peak flows. One such curve is the normal distribution, the symmetrical bell-shaped curve used in many familiar fields. One method used to describe any distribution uses the following equation:

$$Q = \overline{M} + KS \tag{2-12}$$

where $\overline{M}$ and S are the mean and standard deviation as defined in Eqs. (2-7) and (2-8), and K is a factor for which values are given in Table 2-3 for a normal distribution.

A second distribution frequently used for extreme values is the Pearson Type III. Values of K for use in the Pearson Type III distribution are shown in Table 2-4.

The normal distribution is symmetrical about the mean, and thus, the skew and the coefficient of skew are zero for any normal distribution. However, the

Table 2-3 Values of K to be Used in Eq. 2-12 With the Normal Distribution (2)

Exceedance* Probability	K	Exceedance* Probability	K
0.0001	3.719	0.500	0.000
0.0005	3.291	0.550	−0.126
0.001	3.090	0.600	−0.253
0.005	2.576	0.650	−0.385
0.010	2.326	0.700	−0.524
0.025	1.960	0.750	−0.674
0.050	1.645	0.800	−0.842
0.100	1.282	0.850	−1.036
0.150	1.036	0.900	−1.282
0.200	0.842	0.950	−1.645
0.250	0.674	0.975	−1.960
0.300	0.524	0.990	−2.326
0.350	0.385	0.995	−2.576
0.400	0.253	0.999	−3.090
0.450	0.126	0.9995	−3.291
0.500	0.000	0.9999	−3.719

* Exceedance probability is the probability that an event will be equaled or exceeded, and is equal to $1/T$ where T is the return period.

Table 2-4 Values of K to be Used With the Pearson Type III Distribution (2)

Skew Coefficient (g)	Exceedance Probability					
	0.99	**0.90**	**0.50**	**0.10**	**0.02**	**0.01**
3.0	−0.667	−0.660	−0.396	1.180	3.152	4.051
2.5	−0.799	−0.771	−0.360	1.250	3.048	3.845
2.0	−0.990	−0.895	−0.307	1.302	2.912	3.605
1.5	−1.256	−1.018	−0.240	1.333	2.743	3.330
1.2	−1.449	−1.086	−0.195	1.340	2.626	3.149
1.0	−1.588	−1.128	−0.164	1.340	2.542	3.022
0.9	−1.660	−1.147	−0.148	1.339	2.498	2.957
0.8	−1.733	−1.166	−0.132	1.336	2.453	2.891
0.7	−1.806	−1.183	−0.116	1.333	2.407	2.824
0.6	−1.880	−1.200	−0.099	1.328	2.359	2.755
0.5	−1.955	−1.216	−0.083	1.323	2.311	2.686
0.4	−2.029	−1.231	−0.066	1.317	2.261	2.615
0.3	−2.104	−1.245	−0.050	1.309	2.211	2.544
0.2	−2.178	−1.258	−0.033	1.301	2.159	2.472
0.1	−2.252	−1.270	−0.017	1.292	2.107	2.400
0.0	−2.326	−1.282	0.000	1.282	2.054	2.326
−0.1	−2.400	−1.292	0.017	1.270	2.000	2.252
−0.2	−2.472	−1.301	0.033	1.258	1.945	2.178
−0.3	−2.544	−1.309	0.050	1.245	1.890	2.104
−0.4	−2.615	−1.317	0.066	1.231	1.834	2.029
−0.5	−2.686	−1.323	0.083	1.216	1.777	1.955
−0.6	−2.755	−1.328	0.099	1.200	1.720	1.880
−0.7	−2.824	−1.333	0.116	1.183	1.663	1.806
−0.8	−2.891	−1.336	0.132	1.166	1.606	1.733
−0.9	−2.957	−1.339	0.148	1.147	1.549	1.660
−1.0	−3.022	−1.340	0.164	1.128	1.492	1.588
−1.2	−3.149	−1.340	0.195	1.086	1.379	1.449
−1.5	−3.330	−1.333	0.240	1.018	1.217	1.256
−2.0	−3.605	−1.302	0.307	0.895	0.980	0.990
−2.5	−3.845	−1.250	0.360	0.771	0.798	0.799
−3.0	−4.051	−1.180	0.396	0.660	0.666	0.667

Pearson Type III distribution can be used to fit a skewed distribution, and the values of K in Table 2-4 are seen to vary with the value of the skew coefficient. In using the Pearson Type III distribution, logarithms of the discharge are used as the argument while the discharges are used with the normal distribution.

These distributions can be readily demonstrated by example.

EXAMPLE 2-7 Perform a frequency analysis for the annual maximum flows as given in the table for Example 2-6.

SOLUTION In order to perform a frequency analysis it is necessary to select the maximum flow for each year. The table for Example 2-6, page 33, shows the maximum flow for each year as selected from Table 2-2. The third column of the table shows the order number (m) of the flow. The largest flow has order $m = 1$, whereas the smallest has order number $m = 6$. The order numbers are used to calculate the apparent probabilities of occurrence of each annual peak flow. The largest flow appears to have been equalled or exceeded once six years. However, because of the short record, that probability may be erroneous. A relationship used to calculate apparent probability is

$$p = \frac{m}{N + 1} \tag{2-13}$$

where m is the order number, and N is the number of years of record. On the average, this relationship tends to be correct (2).

Applying Eq. (2-13) yields the values shown in column 4. The apparent probabilities computed with Eq. (2-13) are the probabilities that each flow in column 2 will be equalled or exceeded in any year.

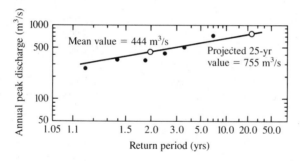

The values shown in column 4 are plotted in the accompanying figure. Probability paper (available commercially) has been used for plotting. ∎

EXAMPLE 2-8 Using the annual flows for the Ok Ma River in New Guinea as given in Table 2-2, and assuming that the flows follow a normal distribution, compute the peak flow that can be equalled or exceeded only once in 25 yr.

SOLUTION In Example 2-6 the mean was calculated as $\overline{M} = 444$ m³/s and the standard deviation S was 176. The mean's return period is approximately 2 years. The value has been plotted in the figure for Example 2-7. For a 25-yr event, the probability of exceedance is

$$P = \frac{1}{25} = 0.04$$

Using Table 2-3, $K = 1.771$ is obtained by interpolation. Thus from Eq. (2-12),

$$Q_{25} = 444 + 1.771(176)$$
$$= 756 \text{ m}^3/\text{s}$$

Note: The skew was assumed to be zero. The mean and the 25-yr value have been plotted in the figure for Example 2-7. The straight line drawn between the mean and the 25-yr values could be used to read apparent peak flows for any given return period.   ■

Risk

As we mentioned, civil engineering design, flood plain delineation, and planning all require the determination of a design flood of a given return interval. The adoption of any particular recurrence interval automatically assumes a particular degree of risk. This degree of risk is not always apparent to engineering planners or designers because many design floods have been fixed far enough in the past that the reasons for those choices are no longer well known. In addition, conditions that reflect hazard, such as population and development, may have changed in the interim. Often, conditions downstream from a dam change drastically with time. Because of reduced frequency and magnitudes of downstream flooding, realized after construction of a dam, a downstream valley that once had few residents and little economic development may experience substantial development and investment after construction of the dam. Thus, at a later date, the area downstream from the dam may need to be reclassified as "high hazard and substantial economic loss," whereas the earlier risk would have been low in both categories.

Extrapolation of Data

Extrapolation of peak floods to return intervals well beyond the period of record is a common requirement because of the shortness of streamflow records on many streams. Extrapolation of these records to a 100-yr flood or larger should be done with much caution and a full appreciation for the uncertainties involved. Figure 2-7 illustrates a historical case. Values in Fig. 2-7 are plotted on special probability paper. The record of annual peak flows from 1901 to 1953 appeared to be readily acceptable for extrapolation to higher return intervals. However, the flood of 1954 is an obvious "outlier" in the statistical series since it plots as a point so far above any smooth curve that might be drawn through the earlier values. Although the apparent recurrence interval of the 1954 flood is only 55 yr, its appearance dramatically changes the series. If either the normal distribution or the Pearson Type III distribution is fit to the data,

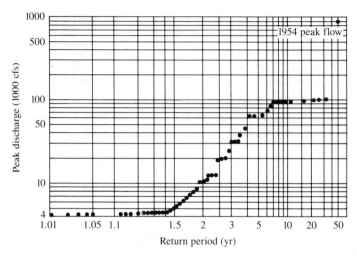

Figure 2-7   Annual maximum peak discharges for
Pecos River near Comstock, Texas, 1901–1954 (16)

the prediction of the 100-yr flood peak, including the 1954 event, will be drastically different from what it will be if the 1954 event were excluded. This particular example should be remembered when fitting a distribution to a short record. Statistical distributions have no recognition of the physical actions that produce the particular events. They are "simple" curves that fit the available historical data reasonably well. One new year of data could easily change the theoretical distribution and drastically change predicted results, particularly if the record is short.

2-5   Precipitation

Precipitation is usually the independent variable in engineering hydrology. Most often we are interested in calculating a rate or volume of runoff that can be expected to occur because of a given amount of rainfall. It now seems terribly obvious that streamflow is the direct result of precipitation, but in 1674 when Pierre Perrault published his book *De L'Origine des Fontaines*, that concept was quite controversial (17). Perrault showed, with the results of his own rainfall measurements, that precipitation was indeed enough to explain the total streamflow of rivers. Measurement of precipitation and streamflow rates provides the basic data on which modern hydrology is based.

Occurrence of Precipitation

Precipitation is quantitatively described in terms of duration (minutes or hours) and depth (inches or millimeters). Intensity is the rate of rainfall (inches

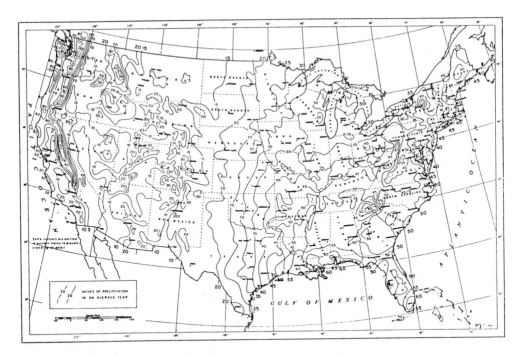

Figure 2-8   Average annual precipitation in the
United States (31) (Enlarged map on page 649)

per hour or millimeters per hour), and can be quite variable throughout the
duration of a storm.

Precipitation occurs as rainfall, sleet, snow, hail, fog, and dew depending
primarily on temperature. Snow falls in the northern climates every winter
almost without fail but only infrequently in southern climates. Precipitation
varies significantly from place to place not only in the amount that falls but
also the seasons during which it falls. Figure 2-8 shows the annual distribution
of precipitation throughout the United States in inches per year. Considerable
variation exists. Generally, more precipitation occurs on the coasts and less in
the higher inland areas. Figure 2-8 has been drawn from the historical measurements of precipitation at approximately 14,000 standard rain gauges maintained
by the U.S. National Weather Service. Further variability exists from year to
year, as can be seen in the data plotted in Fig. 2-2. This timewise and spacewise
variation in precipitation adds considerably to the uncertainty in hydrologic
predictions.

Weather and Meteorology

Temperature and precipitation are the two characteristics of weather
most familiar to all of us. Quantitatively, each is governed by energy given off

by the sun and distribution and absorption of that energy on the earth. All weather, and hence all precipitation, is governed by movement of the air mass surrounding the earth. Motion of that air mass is unsteady and turbulent. The scale of the turbulence is extremely large and, in general, is governed on a grand scale by the size of continents and on a lesser scale by the size of mountains or other topographic features.

In understanding the motion of the earth's atmosphere, two factors are important: atmospheric pressure and the Coriolis effect. Near the earth's surface, air flows into a low-pressure region and out of a high-pressure region (see Fig. 2-9). The Coriolis effect causes the flow of air to bend toward the right as it is initially drawn into a low-pressure area in the northern hemisphere. Thus, a counterclockwise rotation tends to develop around a low-pressure zone and a clockwise rotation around a high-pressure zone in the northern hemisphere, whereas the reverse occurs in the southern hemisphere. Continuity requires an upward flow of air in the middle of a low-pressure region and a downward flow in a high-pressure region. Rising air expands with increasing altitude and decreasing pressure. The expansion causes the air to cool, which in turn lessens its capacity to hold moisture. Moisture condenses forming clouds, and if sufficient moisture is originally contained in the air mass, precipitation occurs. Local low-pressure zones are formed by local heating of the atmosphere and often produce thunderstorms, which generally cover 40 sq mi or less and may cause intense rainfall for short durations, usually less than a few hours.

Large low-pressure regions occur as the result of instabilities in a weather front. Such a large low-pressure zone is called an extratropical cyclone. These intense low-pressure zones may cover areas as small as a few hundred square miles or as large as several thousand square miles. Rainfall produced is generally less intense than that from a local thunderstorm but may last from 24 to 72 hr or even longer.

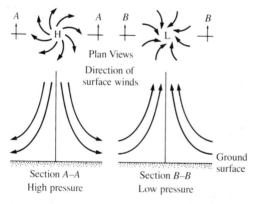

Figure 2-9   Flow of air associated with high- and low-pressure zones in the northern hemisphere

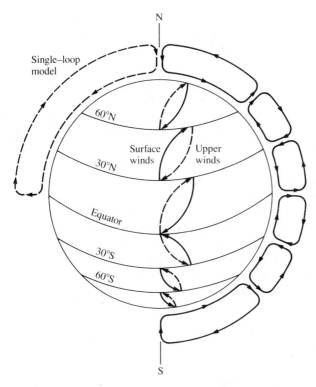

Figure 2-10 Flow patterns of circulation of the earth's atmosphere

Generalities of climate can be discerned from a simplified view of the motion of the atmosphere. Figure 2-10 shows such a simplified picture. Air warmed at the equator rises causing a southerly flow toward the equator in the northern hemisphere. At the North Pole, very cool air descends. Coupling these two simple observations and the principle of continuity, one might expect a southerly flow of air along the surface of the Earth and a northerly flow at higher altitudes (as shown in the single-loop model at the left of Fig. 2-10). However, the rising, cooling air at the equator develops enough momentum to carry it to an elevation above its equilibrium position. Being too high, and thus heavier than its equilibrium weight, it descends again at approximately 30° latitude. Continuity then requires that two other loops form, one between 30°N and 60°N and another between 60°N and 90°N. The position of these loops varies with the seasons in accordance with the elevation of the sun.

High pressure is characteristic of the descending cool air near the poles and at 30°N and 30°S latitude. Low pressure develops where warm air ascends at the equator and at 60°N and 60°S. Thus, one would expect relatively little precipitation to occur near the poles and at 30° latitude. Supporting this conclusion, the world's great deserts are all near 30° latitude, and precipitation near the North Pole averages only about 4 in./yr.

As pointed our earlier, the Coriolis effect causes flows of air to bend to the right in the northern hemisphere (and to the left in the southern hemisphere). Thus, between the equator and 30°N, the flows near the earth's surface are easterly. Between 30°N and 60°N, they are westerly, and between 60°N and the North Pole, the polar easterlies occur (see Fig. 2-10). That most weather patterns in the United States tend to move from west to east is further confirmation of this simple model of atmospheric circulation.

The picture of climate resulting from the simple atmospheric circulation model fails in many places. Because earth and water cool and heat at different rates, and because circulation develops in the oceans, many local differences can be found around the earth. Moreover, atmospheric motion produced by large-scale turbulence and the changing altitude of the sun, respectively, create short-term (daily) and long-term (seasonal) changes.

Monsoon rains are a striking example of seasonal variation. During winter, the land area becomes colder than the ocean, and in some areas such as India this results in rising air masses over the ocean and descending air over the land producing little or no rainfall on the land. During summer, the reverse occurs, and monsoon rainfall occurs over the land.

World's Record Storms

From the standpoint of predicting floods or flood levels, extreme precipitation events are of predominant interest. Extreme precipitation events usually occur because of several simultaneous meteorological occurrences (37). The atmosphere near the earth always contains moisture to a greater or lesser degree. Warm air moving inland from the ocean will generally have a large moisture content because of evaporation from the ocean, whereas cold air descending from aloft in a high-pressure zone will be relatively dry. For a major storm to develop, the atmosphere must contain a large amount of moisture, have a source of energy, and have a liberal inflow of moisture from outside the storm area. Great depths of precipitation occur as a result of a low-pressure zone becoming stalled for long periods with inflow of moisture-laden air. As we mentioned, short-duration storms frequently have high intensities of precipitation, whereas long-duration storms may have lower intensities but great total depths.

Figure 2-11 shows several of the world's greatest rainfall events in terms of depth and duration (13). The short-duration storms on the left of the graph are primarily thunderstorms and can apparently be expected almost anywhere in the tropical, arid, or temperate world. The world's record long-duration storms on the right of the graph, however, have occurred predominantly in areas subject to monsoon rainfall.

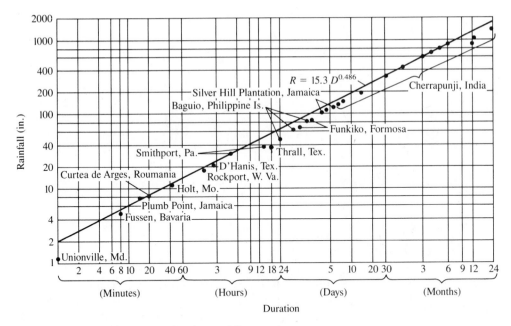

Figure 2-11 Envelope curve for the world's record storms (13)

Rainfall Distribution

To fully describe precipitation quantitatively requires at least two parameters: depth and duration. Alternatively, we could use intensity and duration, since intensity is depth of rainfall divided by duration. Depth is the total amount of precipitation (millimeters or inches) to fall in a given amount of time (*duration*). Precipitation is actually quite variable. Figure 2-12 is a plot of precipitation versus time recorded for actual rainstorms at Pasadena, California, and shows that incremental depth of rainfall (or intensity) varies greatly during a storm. The variability of the precipitation is due to the storm changing both direction and velocity as it moves, the storm cell enlarging and reducing in size with time, and the amount of energy driving the storm changing with time. Precipitation information is normally required to determine runoff and/or streamflow. Thus, quantitative information on precipitation is often vital to hydrologic analysis.

Quantitative information on precipitation is collected by means of a rain gauge. Some rain gauges have recorders or telemeters and provide information on the time rate of rainfall. Other gauges are read only once each day and, thus, provide only data on 24-hr precipitation depths. The amount of rainfall recorded during a period is defined as the "catch" or "depth" of rainfall. The standard rain gauge used in the United States consists of a straight-sided, 8-in.

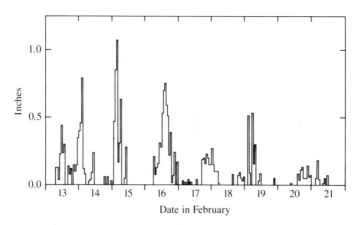

Figure 2-12   Hourly rainfall recorded at Pasadena,
California, during six storms February 13–20, 1980 (4).

diameter tube (30). Many modifications of this gauge have been developed to
collect current data on rainfall as it occurs. Figure 2-13 shows a recording rain
gauge that incorporates a recorder and a radio transmitter that automatically
sends data on rainfall to a central location for analysis and processing. Rain
gauges like this one are also used to provide real-time warnings of possible
flooding.

Because rainfall varies spatially during a storm, the catch of a single rain
gauge may be significantly different from that of another gauge within the same
general area. When several gauges have recorded rainfall on a given drainage
area, the catch of all gauges should be considered in determining the average
depth of rainfall over the drainage area. Three different methods can be used to
compute average depth.

STATION AVERAGE METHOD   Using the station average method, the
catch of all gauges is simply averaged:

$$P_{avg} = \frac{\sum\limits_{i=1}^{N} P_i}{N}$$
(2-14)

where P_{avg} = average precipitation depth over the basin
  P_i = precipitation measured at gauge i
  N = number of gauges

This method is applicable where the drainage area contains a relatively
large number of uniformly distributed gauges. Figure 2-14 shows a typical
drainage area. Some gauges outside the drainage area may be used if they
appear to represent part of the area. The next section describes how, in some

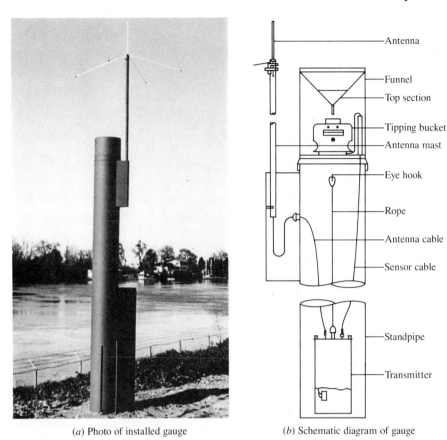

(a) Photo of installed gauge (b) Schematic diagram of gauge

Figure 2-13 Self-reporting rain gauge at Shadun rain station on the Lu River, Hubei, Republic of China. (Courtesy of Sierra-Misco, Inc., Berkeley, California) (a) Photo of installed gauge. (b) Schematic diagram of gauge.

cases, a gauge outside the basin may be representative of some of the area in the basin.

THIESSEN POLYGON METHOD In this method, the precipitation measured by each gauge is assumed to be representative only of the area closest to it. No consideration is given to topography or storm characteristics. To construct the Thiessen polygons, lines are drawn connecting all adjoining gauge locations (the dashed lines in Fig. 2-14). Perpendicular bisectors to each of these (the solid lines in Fig. 2-14) are constructed. The perpendicular bisectors (and the boundaries of the drainage area) then form the portion of the drainage area represented by each gauge. The average precipitation is then computed:

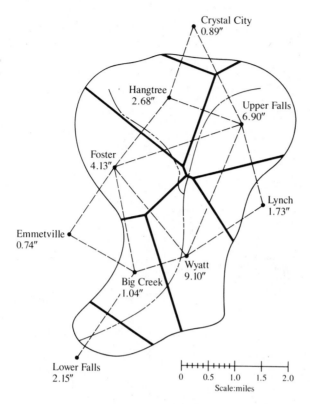

Figure 2-14 Drainage basin showing recorded rainfall for a 24-hour storm and constructed Thiessen polygons

$$P_{avg} = \frac{\sum\limits_{i=1}^{N} P_i \cdot A_i}{\sum\limits_{i=1}^{N} A_i}$$

(2-15)

Where A_i is the area represented by gauge i.

The Thiessen polygon method accounts for nonuniform distribution of stations within the drainage area and also provides guidance on which stations outside the drainage area should be included in calculating the average precipitation. Note that in Fig. 2-14, four stations which are actually outside the drainage area are used to represent rainfall on portions of the drainage area because they are closer to those areas than are any of the other stations within the area.

ISOHYETAL METHOD   The isohyetal method consists of plotting isohyets (contours of equal precipitation) on a map of the drainage area. Figure 2-15

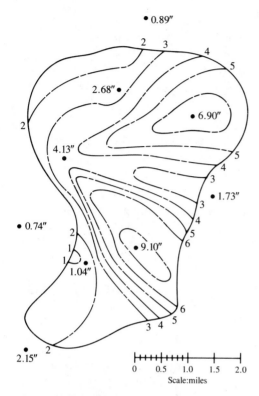

Figure 2-15   Drainage basin and storm shown in Fig.
2-14 with constructed isohyetal lines

illustrates an isohyetal map for the same storm illustrated in Fig. 2-14. Construction of the isohyetal lines is based on the recorded precipitation at the
stations.

In actuality, few drainage areas have enough rain gauges to enable the
isohyets to be constructed using only rainfall data. However, isohyetal lines can
be drawn parallel to ground contour lines where orographic effects appear to be
important. The average precipitation is calculated:

$$P_{\text{avg}} = \frac{\sum\limits_{j=1}^{m} A_j \cdot (P_j + P_{j+1})/2}{\sum\limits_{j=1}^{m} A_j} \tag{2-16}$$

where P_j = the precipitation on isohyet j
  A_j = the area between isohyet j and $j + 1$
  m = the number of intervals between isohyets.

If precipitation depths are interpolated linearly between stations and iso-hyetal lines are then sketched through points of equal precipitation, the calculated average precipitation will be equal to that calculated by the Thiessen polygon method (within the accuracy inherent in the process). Thus, unless there is good reason to believe orographic effects are important, the Thiessen polygon method is adequate.

EXAMPLE 2-9   Using the station average, Thiessen polygon and isohyetal methods, compute the average basin precipitation for the rainfall and drainage basin shown in Figs. 2-14 and 2-15.

SOLUTION   For the station average method, we utilize all stations in Fig. 2-14 and Eq. (2-14). In using the station average method, we have no reason for using data for the four stations outside the drainage area except that those stations are nearby.

$$P_{avg} = \frac{0.89 + 2.68 + 6.90 + 4.13 + 1.73 + 0.74 + 1.04 + 9.10 + 2.15}{9}$$

$$= 3.26 \text{ in.}$$

For the Thiessen polygon method, we must determine the area represented by each station. The construction of the polygons eliminates the Emmetville station from consideration. The accompanying table gives the areas determined for each station. Calculation of the average precipitation for the basin uses Eq. (2-15).

Illustration of Use of Thiessen Polygon Method

(1) Station	(2) Recorded Rainfall Depth P (in.)	(3) Area A Represented by Station (mi^2)	(4) Rainfall Volume (mi^2-in.)
Crystal City	0.89	0.21	0.187
Hangtree	2.68	2.82	7.558
Upper Falls	6.90	3.00	20.700
Foster	4.13	2.64	10.903
Lynch	1.73	1.00	1.730
Emmetville	0.74	0	0
Wyatt	9.10	2.94	26.754
Big Creek	1.04	2.07	2.153
Lower Falls	2.15	0.82	1.763
Totals		15.50	71.748

$$P_{avg} = \frac{71.748}{15.50} = 4.63 \text{ in.}$$

Note: The Thiessen polygon method assigns a very large area to station Wyatt and as a result yields a large average precipitation.

Summary for Use of Isohyetal Method

Rainfall Depth on Isohyet (in.)	Average Rainfall Depth (in.)	Area Between Isohyets (mi²)	Rainfall Volume (mi²-in.)
9.1			
	8.55	0.407	3.480
8.0			
	7.0	1.412	9.884
6.0			
	5.5	0.841 + 1.375 = 2.216	1.219
5.0			
	4.5	0.592 + 1.697 = 2.289	10.300
4.0			
	3.5	3.122	10.927
3.0			
	2.5	2.599 + 0.431 = 3.030	7.575
2.0			
	1.5	2.281	3.422
1.0	1.0	0.05	0.050
6.9			
	6.45	0.693	4.470
6.0			
	Totals	15.500	51.327

For the Isohyetal method, the areas between adjoining isohyets shown in Fig. 2-15 have been determined using a planimeter and are listed in the accompanying table. Using Eq. (2-16), the average precipitation for the basin is

$$P_{avg} = \frac{51.327}{15.50} = 3.31 \text{ in.} \qquad \blacksquare$$

Depth-Duration-Frequency Relations

To develop a design storm for analysis of flooding, it is necessary to have information on the depth of precipitation that can be expected to occur within a given period. It will also be necessary to choose a recurrence interval of storm for use in the design. Information on depth, duration, and frequency of rainfall has been extensively analyzed and published for the United States by the U.S. National Weather Service (35, 36).

Depths of rainfall at a particular point appear to be nearly random events, as is shown in Fig. 2-2, page 23. The variation from year to year effectively hides any long-term variation that may be due to climate. Because of its variation,

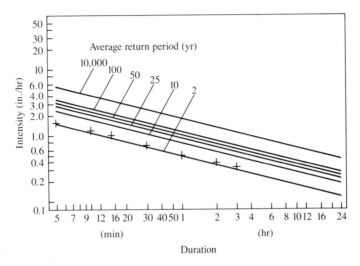

Figure 2-16 Intensity-duration-frequency curves for Grizzley Flat, California (7)

rainfall at a point must be described statistically using a mean for the period (such as mean annual or mean monthly) and its variation about that mean. Precipitation also varies with time during an individual storm, and the duration of rainfall (total time during which rain falls) varies from one storm to another. Thus, to describe rainfall quantitatively it is necessary to include both the amount of rainfall (depth) and the corresponding elapsed time (duration). This is called a depth-duration analysis if depths of rainfall are considered, or alternatively, an intensity-duration analysis of intensities are considered.

To evaluate average frequency of occurrence (or recurrence interval), it is necessary to perform further analyses of rainfall data for each gauge. To make a frequency analysis of intensity-duration data, the data must have been continuously recorded so that various durations of rainfall can be studied. Records for each year are first scanned for intervals of time. For short-duration storms, the data are first scanned using 5-min increments or smaller. The largest single 5-min depth would be selected and assigned order $m = 1$; the second largest, order $m = 2$; the third largest, $m = 3$; and so on. Eventually, a set of 5-min rainfall depths will have been selected equal in number to the number of years of record N. The apparent return frequency of each point is calculated using Eq. (2-13). A curve of frequency versus rainfall depth is then plotted on probability paper. The same process is repeated for each duration of interest until curves of frequency versus depth have been developed for all durations of interest. That family of curves can then be used to prepare a set of depth-duration-frequency curves, as shown in Fig. 2-16.

The foregoing type of analysis can cover a point, a region, or an entire nation. The National Weather Service (NWS) has prepared intensity-duration-

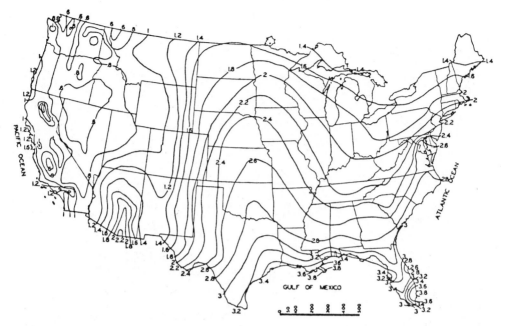

Figure 2-17   10-year 1-hour rainfall for the United
States (36)

frequency values for most recording stations in the United States (35). Those
results and more recent analyses have been used to construct depth-duration-
frequency values of rainfall for the United States (37). Figures 2-17, 2-18, and
2-19, show depth-duration values for a 10-yr return interval and three durations.
Further analysis has led to separate NWS publications for each of the states
on a scale more detailed than is shown in Figs. 2-17 through 2-19.

Synthetic Storm Design

For drainage design of a particular project, selecting a design storm is
necessary. For small drainage projects such as a parking lot or an industrial site,
a single intensity is usually used. Thus, in designing the storm drainage system
for a housing area, the engineer might select a 10-yr intensity from the U.S.
National Weather Service (35). Later we will discuss methods for choosing the
proper duration.

For a project involving a drainage area larger than several square miles,
it is necessary to develop a design storm with a duration as long as 24 to 72 hr.
Intensity is constant only for durations of a few minutes and varies significantly
over a long duration. Past design practices used the most severe recorded storms

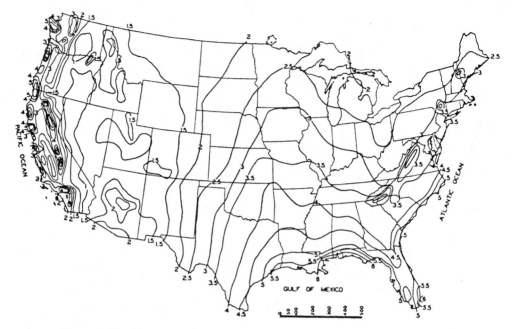

Figure 2-18   10-year 6-hour rainfall for the United
States (36)

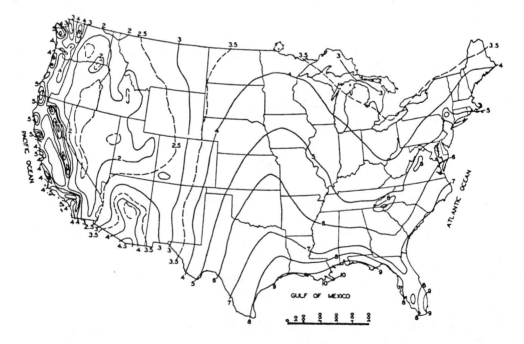

Figure 2-19   10-year 24-hour rainfall for the United
States (36)

Table 2-5 24-Hour Rainfall Depths as a Function
of 2-Year, 24-Hour Depths (10)

Return Period (yr)	Depth (% of 2-yr depth)
1	0.78
2	1.00
5	1.40
10	1.60
25	1.82
50	2.00
100	2.22

in the region. However, to develop a design storm having a given return interval, it is usually necessary to develop a hypothetical or synthetic storm. Since precipitation records are generally available in many more locations than stream gauges, it is often necessary to develop design floods using calculated runoff from these design storms.

To construct a design storm for a given return interval, select the depths of precipitation for each duration from maps like those shown in Figs. 2-17, 2-18, and 2-19. Table 2-5 provides acceptable empirical guides to proportion 24-hr rainfalls for various return intervals.

The depths can be plotted against duration for various return intervals as shown in Fig. 2-20. Smooth curves drawn through the points provide the means to interpolate for different durations or return periods. Table 2-6 shows tabulated values of rainfall for a 100-yr 24-hr storm at the common eastern corner of Arkansas and Missouri. Values in the table (column 2) were interpolated from Fig. 2-20.

The sequence of rainfall depths in a synthetic storm is more or less arbitrary. As shown in Fig. 2-12, the time distribution of an actual storm can be irregular. Nevertheless, the hydrologist must rearrange the incremental values in column 3 of Table 2-6 to represent a reasonable storm pattern. In general, the peak increment should be placed at approximately the one-third point of the storm's duration. The other increments are arranged more or less arbitrarily to provide for a continuous increase in intensity before the peak and a continuous decrease afterward. Column 4 of Table 2-6 shows a possible distribution. Specific arrangements have been adopted by certain firms and agencies. Table 2-7 shows a distribution used by the U.S. Soil Conservation Service for storms west of the Sierra Nevada and Cascade Mountains (Type 1) and storms in other parts of the United States (Type 2). Later, when we discuss runoff, further reasoning will be given for the distribution of incremental rainfall depths in a design storm.

Point rainfall during a storm is generally larger than the average rainfall over the area. Thus, values taken from Figs. 2-17, 2-18, and 2-19, which are point-rainfall values, must be corrected to represent averages over the drainage area. A study made by the NWS led to the development of Fig. 2-21, which

Table 2-6 Construction of a 100-Year Storm for a 100 Square-Mile Drainage Area at the Common Eastern Corner of Arkansas and Missouri

(1) Time (hr)	(2) Accumulated Rainfall (in.)	(3) Increment (in.)	(4) Selected Rainfall Sequence* (in.)	(5) Rainfall Sequence Average Over Area (in.)
0	0	0	0	0
2	4.10	4.10	0.10	0.09
4	4.80	0.70	0.20	0.19
6	5.40	0.60	0.30	0.28
8	5.85	0.45	0.45	0.42
10	6.15	0.30	4.10	3.85
12	6.40	0.25	0.70	0.66
14	6.65	0.25	0.60	0.56
16	6.90	0.25	0.25	0.24
18	7.10	0.20	0.25	0.24
20	7.30	0.20	0.25	0.24
22	7.40	0.10	0.20	0.19
24	7.50	0.10	0.10	0.09
Totals	7.50	7.50	7.50	7.05

* This selected distribution is a rearrangement of incremental values from column 3 to provide a realistic sequence representative of actual storms.

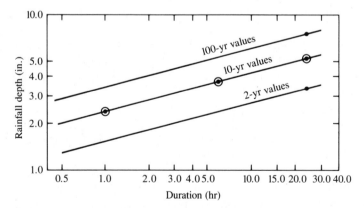

Figure 2-20   Interpolation of depth-duration-frequency for rainfall for a point at the common eastern corner of Arkansas and Missouri.*
* Values shown as ⊙ are taken from Figs. 2-17, 2-18, and 2-19. Values shown as ● are computed from 10-yr values using Table 2-5. The lines for 100-yr and 2-yr values were drawn parallel to that for the 10-yr values.

Table 2-7 Accumulation of Rainfall up to 24 hours (34)

Time (hr)	P_x/P_{24}* Type 1	P_x/P_{24}* Type 2	Time (hr)	P_x/P_{24}* Type 1	P_x/P_{24}* Type 2
0	0	0	11.0	0.624	0.235
2.0	0.035	0.022	11.5	0.654	0.283
4.0	0.076	0.048	11.75	...	0.387
6.0	0.125	0.080	12.0	0.682	0.663
7.0	0.156	...	12.5	...	0.735
8.0	0.194	0.120	13.0	0.727	0.772
8.5	0.219	...	13.5	...	0.799
9.0	0.254	0.147	14.0	0.767	0.820
9.5	0.303	0.163	16.0	0.830	0.880
9.75	0.362	...	20.0	0.926	0.952
10.0	0.515	0.181	24.0	1.000	1.000
10.5	0.583	0.204			

* P_x/P_{24} is the ratio of accumulated rainfall at time x to the accumulated rainfall in 24 hours.

has been widely used as a means to adjust point rainfall depths to averages over an area (36). Values in column 5 of Table 2-6 were obtained by multiplying values in column 4 by 0.94, as obtained from Fig. 2-21 for a 100-sq-mi drainage area.

Snowmelt

In many of the colder climates of the world, such as the Sierra Nevada mountains between Nevada and California, a major part of the annual precipitation occurs as snow. Melting snow during the spring and early summer

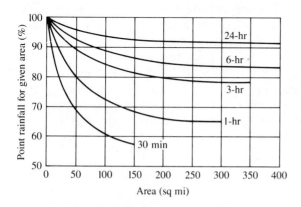

Figure 2-21 Area average versus point rainfall for use with depth-duration values (36)

months forms much of the annual runoff. In climates like this, major floods are often produced by rainfall on snow. The rain containing more heat per unit of mass tends to melt the snow. Newly fallen snow has a specific gravity of approximately 0.1, thus, 10 in. of new snow would have a water equivalent of 10% or only 1 in. As winter progresses, the snow pack compresses and reaches water equivalents of 40 to 50%.

With the arrival of spring, the air temperature warms and the snowpack melts producing streamflow. The scientific basis of snowmelt is a function of at least incoming and back radiation, air temperature, and wind (38). Only rarely are records of all these phenomena available, and snowmelt computations or predictions must usually use approximate analyses. The most common approach is the use of degree days. One *degree day* is defined as a 24-hr period during which the average temperature is 1°F above 32°F. A degree-day factor is determined by dividing the total snowmelt runoff (in inches) from a drainage basin by the number of degree days for that period of runoff. Thus, the degree-day factor is an empirical factor implying the snowmelt (in inches of water) per degree day occurring on a particular drainage basin. Degree-day factors of 0.05 to 0.15 in. per degree day are common. Since temperature varies with altitude, degree-day factor analysis can confidently be used only on basins with uniform snow coverage and moderate relief. Example 2-10 shows a snowmelt computation for a basin with considerable relief.

EXAMPLE 2-10   The accompanying table shows drainage area versus elevation for a particular snow-covered basin. Assuming that temperature decreases at a rate of 3°F/1000 ft, calculate the snowmelt depth to be expected for a given day when the reference mean-daily temperature is 38°F at an elevation of 6500 ft. Assume a degree-day factor of 0.1. The lowest elevation of snow (snowline) is at 5000 ft.

SOLUTION   The solution is done as shown in the accompanying table. The temperature is 38°F at the reference station, which is at an elevation of 6500 feet. Consider the calculation for an elevation of 7500 feet. The incremental area for this elevation is 50 mi² (area between 7000 ft and 8000 ft) and the average temperature is 38°F − 3°F = 35°F. Thus, the degree days are 1 day·(35°F − 32°F) = 3°F-days or 3°FD, and the resulting snowmelt will be equal to 3°FD multiplied by the degree-day factor. Using the given degree-day factor of 0.1 in. per degree day, the snowmelt in inches of water depth will be

$$3°\text{FD} \times \frac{0.1 \text{ in.}}{°\text{FD}} = 0.3 \text{ in.}$$

The incremental volume of snowmelt $\forall$ for this elevation is calculated by multiplying the area by the depth of melt:

Elevation (ft)	Area Above Elevation (mi²)	Incremental Area (mi²)	Average Temperature (°F)	Degree Days	Snowmelt (mi²-in.)
11,000	0	20	26	0	0
10,000	20	30	29	0	0
9,000	50	40	32	0	0
8,000	90	50	35	3	15.0
7,000	140	60	38	6	36.0
6,000	200	70	41	9*	63.0
5,000	270				Total = 114.0 mi²·in.
					= 6080 acre-ft

* This calculation assumes that the depth of snowpack is great enough to provide 0.9 in. of water equivalent.

$\forall = 50 \text{ mi}^2 \times 0.3 \text{ in.}$

$= 15.0 \text{ mi}^2\text{-in.}$

$= 800 \text{ Acre-ft or } 800 \text{ AF}$

The complete solution is given in the table above.   ■

Rain melts snow in accordance with calorimetric theory, and since 144 BTU of heat are required to melt 1 lb of ice at 32°F, Eq. (2-17) yields the number of inches M of melt produced by a rainfall of R inches having a Fahrenheit temperature Tw:

$$M = \frac{R(Tw - 32)}{144} \tag{2-17}$$

The temperature of falling rain is usually close to the wet-bulb temperature and is a useful fact for snowmelt calculations. Equation (2-17) assumes that the snow temperature is 32°F at the time rain begins.

Actually, because of intermolecular forces, a snowpack can contain approximately 5% liquid water by weight in a stable condition. Thus, for deep snowpacks considerable meltwater can be stored in the snowpack before runoff actually begins.

Probable Maximum Precipitation

When one considers the frequency analysis of rainfall or peak floods, the normally limited data plots somewhat as is shown in Fig. 2-16. For a given duration, the depth of rainfall appears to increase continuously with increasing recurrence interval. How much could fall in the largest possible rainstorm? Extending the frequency curve appears to indicate that the maximum depth would be infinite. However, physical limitations in the occurrence of rainfall limit that depth to what has become known as the *probable maximum precipitation* (PMP).

Figure 2-11 shows a plot of the world's record rainfall. The straight line drawn upward to the right in Fig. 2-11, page 43, falls just above all the plotted points. The equation of this line is

$$R = 15.3D^{0.486} \tag{2-18}$$

where R is rainfall depth in inches, and D is duration in hours. Rainfall cannot occur faster than moisture can be supplied to the air column above the point of interest. Thus, the volume of moisture (usually expressed as depth) in the air column, the rate that wind brings moisture into the column, and the physical efficiency of the condensation-rainfall process limit the maximum depth of rainfall that can occur at a point for a given duration. Looking at frequency curves like Fig. 2-11 leads to the conclusion that there is an upper limit to precipitation depth for a given frequency.

Computing PMP uses records of intense storms for which relative humidity data are available. Actual moisture available in a column of air is computed based on the relative humidity occurring at the time (37). Thus, the steps in estimating a PMP depth are as follows:

1. Obtain records of rainfall depth R and relative humidity for severe storms that have occurred at or near the site.
2. Calculate the moisture P available in a column of air at the time the data on rainfall depth R were taken (36).
3. Calculate the probable maximum moisture M that could have been available had the relative humidity been a maximum. Maximum relative humidity must be determined on the basis of records at or near the site.
4. Calculate a maximizing factor as $K = M/P$.
5. Estimate the PMP as PMP = $R \times K$.

Many specific considerations go into estimating the PMP, and the illustration given is only a simple consideration. The severe storms chosen must be exceptional storms for which the rainfall efficiency can be assumed to be as high as that which would occur under probable maximum conditions. If a storm is to be transposed to another area, you must be certain that the storm could occur in the area of interest. This judgment requires the knowledge of an experienced meteorologist.

In some areas, generalized estimates of PMP have been prepared by analyzing storms in regions. These studies have been done by the U.S. National Weather Service (28, 29). Figure 2-22 shows the results of one of these studies. To use Fig. 2-22 to construct a probable maximum storm sequence, one must

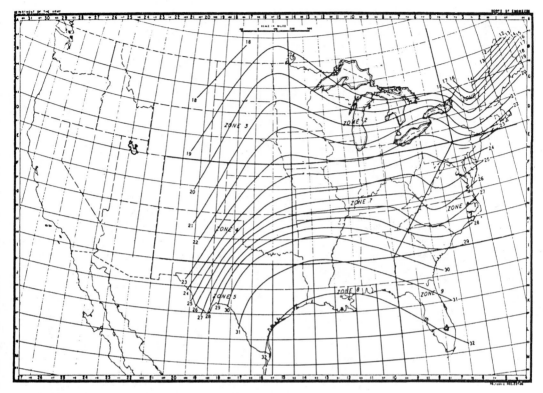

Figure 2-22   Probable maximum 24-hr precipitation
for 200 square miles for the United States east of the
105th meridian (29)

adjust the precipitation for the size of the drainage area involved and for duration. Figure 2-23 provides the means to make such adjustments (29).

The PMP is used as a design event when a large flood would result in hazards to life or great economic loss. Thus, large dams upstream from populated
centers, under current design standards, must be designed with spillways adequate to protect the dam against failure during the flood produced by the PMP
falling on the drainage area.

2-6   Surface Runoff

As we mentioned in our discussion of the hydrologic cycle, a portion
of precipitation falling on a drainage area runs off. That portion moves initially
as overland flow in very shallow depths before eventually reaching a stream
channel where streamflow is produced. In engineering, this portion of the hydrologic cycle is of the greatest interest, since it is surface runoff that fills reser-

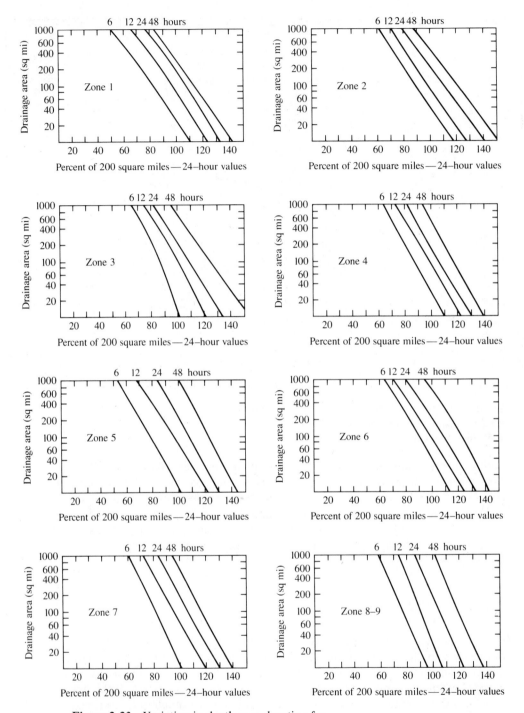

Figure 2-23   Variation in depth-area-duration for probable maximum precipitation (29)

voirs, fills rivers, and produces floods. Probably the most frequent hydrologic engineering calculation is the estimation of peak flows during flooding. Peak flows are used in designing storm drains, flood-control levees, spillways, and bridge openings.

Figure 2-4, page 25, shows that a portion of rainfall reaching the surface of the earth is held as depression storage, and that another portion infiltrates below the earth's surface. If rainfall reaches the earth at a rate larger than the infiltration rate, and if the rainfall volume exceeds the depression storage, runoff occurs. Once the rainfall stops, infiltration and evaporation continue and eventually exhaust the depression storage.

Infiltration

Infiltration is a complex process that depends at least on soil type, soil grain size, land use, and soil cover. The condition of the soil surface governs the ability or rate at which water passes into the soil, whereas the internal structure of the soil governs the rate at which water moves through the soil. During a dry period, the soil and vegetal root system shrink, opening passages for easy entry into the soil. As the moisture content of the soil increases, the passages close. Thus, the actual infiltration rate is often a maximum at the beginning of a storm and decreases with time after the storm begins.

Transmission of water through the soil depends on the size, shape, and percent of voids in the soil. A measure of this quality is the soil's *permeability*.

Vegetation covering the soil tends to increase total infiltration during a given storm, since raindrops are broken up and prevented from directly striking the ground, compacting the soil and closing its voids. Thus, the actual infiltration rate is greater on a vegetated soil than it would be for the same soil without cover, and since some rain is also intercepted and held by the vegetation, surface runoff is a greater percentage of total rainfall for bare ground.

The actual infiltration rate depends on both soil and cover conditions and rainfall intensity. The maximum rate at which infiltration could occur is the *infiltration capacity*. The actual rate of infiltration can approach the infiltration capacity only if the rainfall intensity is large enough to exceed the rate at which rainfall is intercepted and retained as depression storage.

A basic formulation for the infiltration capacity as a function of time was developed by Robert Horton (12) as

$$f = f_c + (f_0 - f_c)e^{-k_f t} \tag{2-19}$$

where f = maximum infiltration rate at time t (infiltration capacity)
f_0 = initial infiltration rate at $t = 0$
f_c = minimum infiltration rate
k_f = a constant
e = the natural logarithm base (2.718)

Table 2-8 Soils Groups and Corresponding Minimum Infiltration Rates (34)

Group	Minimum Infiltration Rate (in./hr)	(mm/hr)	Soil Description
A	0.30–0.45	7.6–11.4	Soils having a high infiltration rate. They are chiefly deep, well-drained sands or gravels, deep loess, or aggregated silts. They have *low runoff* potential.
B	0.15–0.30	3.8–7.6	Soils having a moderate infiltration rate when thoroughly wet. They are chiefly moderately deep, well-drained soils of moderately fine to moderately coarse texture such as shallow loess and sandy loam.
C	0.50–0.15	1.2–3.8	Soils having a slow infiltration rate when wet. They are soils with a layer that impedes downward movement of water and soils of moderately fine to fine texture such as clay loams, shallow sandy loam, soils low in organic content, and soils high in clay content.
D	0.00–0.05	0.00–1.2	Soils having a very slow infiltration rate. They are chiefly clay soil with a high swelling potential, soils with a permanent high water table, soils with a claypan at or near the surface, shallow soils over nearly impervious material, heavy plastic clays, and certain saline soils. They have *high runoff* potential.

Although this equation is generally accepted as an empirical but reasonable quantification of the infiltration process, it is difficult to use since values for the constants f_c, f_0, and k_f are not easily determined. Minimum infiltration rates have been evaluated by the U.S. Soil Conservation Service (34) and are shown in Table 2-8 as a function of soil type. The value of f_0 depends on the surface condition of the soil as well as on the content of soil moisture and, therefore, varies with time since the last rain.

An average infiltration capacity called the ϕ index has been widely used because of its simplicity. The ϕ index is simply an average infiltration capacity that when applied to a particular rainfall event, yields the proper volume of runoff. Figure 2-24 illustrates the ϕ index and Horton's relationship. In separating runoff from infiltration, the value of ϕ is subtracted from each incremental rainfall amount with any resulting negative values set to 0.

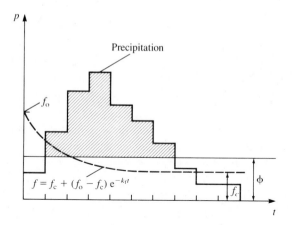

Figure 2-24 Illustration of ϕ index and Horton's equation for infiltration*
* The ordinate is precipitation in inches or millimeters per hour while the abscissa is time. Runoff is indicated by the area above the ϕ index line (shaded) or by the area above the curved line (Horton's Equation).

Another method commonly used considers infiltration to be made up of a fixed initial volume (abstraction) plus continuing infiltration at a fixed rate. This method is preferred by some over the ϕ index method because the abstraction represents an approximation of the noninfiltration quantities of interception and detention.

Modern methods of determining the parameters for use in the estimation of infiltration include analysis of the physical properties of the soil and the use of experimental infiltrometers. The experimental infiltrometers include flooding and sprinkling instruments (1).

Evaporation and Evapotranspiration

Evaporation from water and soil surfaces and transpiration through plants, as shown in Figs. 2-3 and 2-4, can account for significant volumes of water. Evaporation is the process by which water transforms into vapor. The process occurs at the water surface where molecules of water develop sufficient energy to escape bonds with the water and become vapor molecules in the air. Evaporation from a water body is a function of air and water temperatures, the moisture gradient at the water surface, and wind. Wind moves the moisture away from the lake's surface and, thus, increases the moisture gradient, increasing the rate of evaporation. Natural evaporation rates vary from as much as 90 in. per year in the hot dry climate of southeast California to 20 in. per year in the cool climate of northern Maine or the wet climate of northwest Washington. Figure 2-25 shows average annual lake evaporation in the United States.

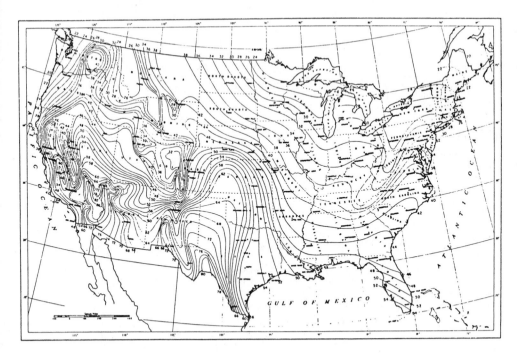

Figure 2-25   Average annual lake evaporation in the
United States in inches (6) (Enlarged map on page 650)

Several different methods are used to estimate evaporation. The so-called
Standard Class A Land Pan is used by the U.S. National Weather Service at
its official stations. It is a metal pan 4 ft in diameter and 10 in. deep. Evaporation
from the pan is measured daily by measuring the water level in the pan using
a hook gage. The pan is mounted on a wooden frame 4 in. off the ground. The
pan is drastically different aerodynamically and thermodynamically from a lake
and yields a higher rate of evaporation than is normally experienced by a lake.
An analysis of pan and lake evaporation by Kohler (15) showed that multiplica-
tion of measured pan evaporation by a "pan coefficient" equal to 0.7 yields
annual lake evaporation rates accurate to within about 15% if the pan is subject
to the same climatic conditions as the lake.

Extensive field investigations of evaporation, made at Lake Hefner, Oklahoma, from 1950 to 1953, yielded the following equation for lake evaporation, E:

$$E = 0.00241(e_s - e_8)V_8 \tag{2-20}$$

where E is the rate of evaporation in inches/day, e_s is the vapor pressure in
inches of mercury at the water surface, and e_8 and V_8 are the vapor pressure
and wind velocity (miles per day) 8 m above the lake surface. It is possible to

have large errors in predicted evaporation if data are not taken accurately and are not local to the site.

Evapotranspiration is the total moisture that leaves an area by evaporation from soil, snow, and water surfaces plus that transpired by plants. The potential evapotranspiration E_p can be estimated as being equal to the lake evaporation during the same period, since moisture is removed from leaves of plants by the same process as it is evaporated from water surfaces. C.W. Thornthwaite (24) established a procedure for estimating actual evapotranspiration E_A by using the following water-balance equation:

$$P - R - G_o - \Delta M = E_A \qquad (2\text{-}21)$$

where P is precipitation, R is surface runoff, G_o is subsurface outflow, and ΔM is the change in moisture storage within the soil. Equation (2-21) is applied to a given area, and moisture must be measured in the underlying soil. Consistent units must be used for each variable to yield E_A in either a volume or a depth of evapotranspiration in a given time.

Rainfall-Runoff Relations

The portion of rainfall that enters the stream quickly is called surface runoff. When rainfall starts, detention storage is filled first, and if the rainfall intensity is larger than the infiltration capacity, surface runoff occurs. The surface runoff flows first as a thin sheet overland and eventually reaches a rivulet or channel where the flow concentrates. Because the relative importance of viscous resistance to flow decreases as the hydraulic radius of the channel increases, the flow velocity in the channel is greater than that of the overland flow. As the flow moves downstream, flows from other channels join the stream, and the flow rate increases. The time required, after beginning of rainfall, for the most distant point in the drainage area to begin contributing runoff at the outlet of the basin is called the *time of concentration* t_c.

The continuous record of rate of flow as a function of time is called a stream *hydrograph*. Figure 2-26 shows a hydrograph for the Sacramento River at Red Bluff, California. The shape of the hydrograph reflects physical characteristics of the drainage basin as well as that of the storm producing the hydrograph. If the drainage basin is compact around the point of interest, streamflow at the outlet of the basin will peak rapidly since all points in the basin tend to be close to the outlet. By contrast, for a basin that is not compact, streamflow will peak more slowly. Figure 2-27 shows typical hydrographs for characteristic drainage basins having equal areas but different shapes.

Physical laws dictate that all other things being equal, steep slopes will produce more runoff than flat slopes; vegetated drainage areas will produce less runoff than bare areas; and areas where soils are relatively impermeable such as compacted clay will produce more runoff than sandy soils with high per-

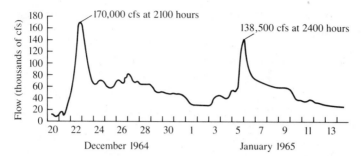

Figure 2-26   Hydrograph for the Sacramento River at
Red Bluff, California, 1964–1965 flood. (23)

meability. Analytical methods for computing runoff cannot yet consider all
characteristics of a drainage area exactly. However, many empirical methods
have been developed with which to estimate runoff and a great deal of experience
has been accumulated in using these methods, which provides confidence in their
application. The most simple of these is the so-called rational equation mentioned earlier:

$$Q_p = CIA \qquad\qquad (2\text{-}1)$$

where Q_p is the peak rate of runoff, I is rainfall intensity, and A is drainage area.
This equation has been widely used in the design of small drainage systems,
such as those for airports, city blocks, or parking lots. The runoff coefficient C
is essentially the proportion of rainfall volume that runs off the area. Thus, this
runoff coefficient has a value approaching unity for smooth, impermeable surfaces such as small areas covered by concrete surfaces. Conversely, C has a much

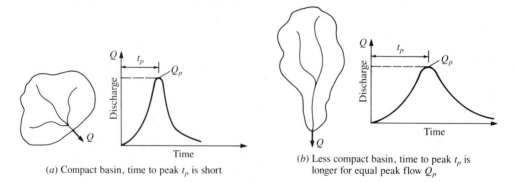

Figure 2-27   Influence of basin shape on runoff
hydrograph. (a) Area A_2 — Compact basin, short time
to peak. (b) Area A_1 — Less compact basin, longer time
to peak

Table 2-9 Runoff Coefficients Recommended
for Use in the Rational Equation (19)

Type of Area or Development	C
Type of development	
Urban business	0.70–0.95
Commercial office	0.50–0.70
Residential development	
Single-family homes	0.30–0.50
Condominiums	0.40–0.60
Apartments	0.60–0.80
Suburban residential	0.25–0.40
Industrial development	
Light industry	0.50–0.80
Heavy industry	0.60–0.90
Parks, greenbelts, cemetaries	0.10–0.30
Railroad yards, playgrounds	0.20–0.40
Unimproved grassland or pasture	0.10–0.30
Type of surface areas	
Asphalt or concrete pavement	0.70–0.95
Brick paving	0.70–0.80
Roofs of buildings	0.80–0.95
Grass-covered sandy soil	
Slopes 2% or less	0.05–0.10
Slopes 2% to 8%	0.10–0.16
Slopes over 8%	0.16–0.20
Grass-covered clay soils	
Slopes 2% or less	0.10–0.16
Slopes 2% to 8%	0.17–0.25
Slopes over 8%	0.26–0.36

smaller value for permeable surfaces and approaches 0 for sandy desert areas. Table 2-9 lists runoff coefficients for various surfaces.

Equation (2-1) is known as the rational equation primarily because it is dimensionally homogeneous in contrast to many other empirical equations for runoff that have been proposed and used in drainage design. The runoff coefficient C is a dimensionless constant. Intensity has the dimensions of depth (length) divided by time and is usually used in inches per hour or millimeters per hour. The area A is commonly used in acres or square miles. Thus, if intensity I is expressed in inches per hour and the area A is in acres, the runoff rate Q_p will be computed in acre-inches per hour, which is nearly equivalent numerically to cubic feet per second.

Since only peak rate of flow is computed by Eq. (2-1), use of the rational equation assumes that rainfall duration is at least equal to the time of concentration. If rainfall intensity were actually constant, a constant discharge would occur only after all storage and initial losses have been satisfied and flow is being contributed from all parts of the basin. The time of concentration for a drainage basin is made up of the longest combination of overland flow time plus the accumulated flow time in the stream channels to the outlet of the basin. To estimate times of concentration, several empirical expressions have been developed, including the following form of Kirpich's equation (14):

$$t_c = \left(\frac{3.35 \times 10^{-6} \, L^3}{h} \right)^{0.385} \tag{2-22}$$

where t_c = time of concentration in minutes
      L = stream length in feet
      h = difference in elevation in feet between the upper and lower
          limits of the drainage basin

A second empirical expression is the following equation modified from the original proposed by Hathaway (9):

$$t_c = \left(\frac{2Ln}{3\sqrt{S}} \right)^{0.47} \tag{2-23}$$

where t_c = time of concentration in minutes
      L = channel length in feet
      S = mean slope of the basin
      n = Manning's roughness coefficient

Table 2-10 provides values of n to be used in Eq. (2-23).

   In using the rational method, it is first necessary to compute the time of concentration. Either Eq. (2-22) or (2-23) may be used initially. However, because these equations are truly empirical, considerable care must be exercised before accepting their results. It is always advisable to make a check calculation by measuring the longest length of stream and determining the flow time in the stream using an estimated average velocity of flow. The average velocity of flow can be approximated by using the Manning equation (Eq. 4-7a). A reasonable assumption for the roughness coefficient must be made using values from

Table 2-10 Values of the Roughness Coefficient n
to be Used in Eq. (2-23)

Surface	n
Smooth pavements	0.02
Bare packed soil, free of stones	0.10
Poor grass cover or moderately rough surface	0.20
Average grass cover	0.40
Dense grass cover	0.80

Table 4-1. One-dimensional flow with a depth of 6 to 12 in. can be assumed in the velocity calculation. The velocity calculation is described in Sec. 4-2 of Chapter 4.

Once a time of concentration has been computed and accepted, a precipitation intensity is chosen from the previously determined intensity-duration values, such as the graph shown in Fig. 2-16 using the calculated time of concentration as the duration. Incorporating the drainage area, the determined rainfall intensity, and the properly chosen runoff coefficient in Eq. (2-1) yields the peak rate of runoff.

Because of its simplicity, the rational equation is widely used, particularly for calculations of urban runoff. However, there are some inherent inaccuracies in the use of the equation:

1. The runoff coefficient C cannot have a constant value throughout the storm, since as we mentioned, certain storages must be filled by rainfall before runoff can begin. Moreover, antecedent moisture conditions govern the initial infiltration rate.
2. In computing a peak flow from a precipitation intensity of a given return period, it is tacitly assumed that the return interval of the peak flow is the same as that of the chosen precipitation intensity. This is not strictly true, but for the general design range up to approximately 20 yr, it has been found to be approximately true (19).

Because the rational equation assumes that rainfall duration is at least equal to the time of concentration, it should not be used for areas greater in size than about 1 sq mi. For larger areas the entire drainage area almost certainly will not be covered by rainfall of equal intensity, thus violating the assumptions inherent in the rational equation. Other methods presented in the next section should be used for larger areas.

EXAMPLE 2-11   Use the rational equation to find the 10-yr peak rate of runoff for the two areas with the characteristics shown in the accompanying figure and for the channel at point b. The arrows show that all flow from area A_1

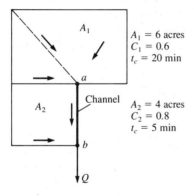

$A_1 = 6$ acres
$C_1 = 0.6$
$t_c = 20$ min

$A_2 = 4$ acres
$C_2 = 0.8$
$t_c = 5$ min

enters the channel at point a, and that all flow from area A_2 enters the channel at point b. Assume that the intensity-duration-frequency curves of Fig. 2-16 are applicable for the site, and that the time of flow from a to b is 8 min. *Note*: In using the rational equation, the units of Q are dependent on the units used for I and A.

SOLUTION Figure 2-16, page 50, shows the intensity of rainfall is 1.2 in. per hr, for a duration of 20 min. For area A_1, the peak rate of runoff occurs when the storm duration is equal to the time of concentration, t_c (20 min). Thus, the peak rate of flow from area A_1 into the channel at a is

$$Q_a = CIA = 0.6(1.2)6 = 4.3 \text{ acre-in./hr}$$
$$= 4.3 \text{ cfs}$$

Similarly, for area A_2, the peak rate of runoff entering the channel at b is

$$Q_b = 0.8(2.3)4 = 7.4 \text{ cfs}$$

since the intensity of rain is 2.3 in. per hr for a duration (time of concentration, t_c) of 5 min.

To find the peak rate of flow at b, we must consider both areas simultaneously. The critical time of concentration for area A_1 plus the time of flow from a to b is

$$t_b = 20 + 8 = 28 \text{ min}$$

We must add the rate of flow from area A_2 for a duration of 28 min to the peak rate of flow from area A_1 for a duration of 20 min. Thus, the peak flow at b is

$$Q_b = 4.3 + 0.8(1.0)4 = 4.3 + 3.2 = 7.5 \text{ cfs}$$

because the rainfall intensity for a 28-min duration is 1.0 in. per hr. We use the peak rate of flow from area A_1 for the 20-min duration because it takes that peak inflow from A_1 8 min to flow from a to b, at which point it combines with the flow coming from area A_2 after a duration of $20 + 8 = 28$ min. The peak flow at point a is actually somewhat diminished by the time it reaches point b because part of the flow volume is stored in the channel. This effect, called routing, is neglected in using the rational equation. ■

Soil-Cover Complex Method

One of the more recent and widely accepted methods for estimating the amount of runoff from a given rainstorm was developed by the U.S. Soil Conservation Service (SCS) and is generally referred to as the soil-cover complex method, or more commonly, the curve-number method (34). The method is used where runoff records are not available and assumes that runoff can be determined directly in terms of a single parameter called a curve number (CN), a quantitative parameter for the drainage area of interest. Values of the curve number, and the method in general, were developed based on field measurements of the amount of runoff from drainage areas for which the rainfall, soil characteristics, cover, and usage were studied. The curve number attempts to account for both the *initial abstraction* I_a of rainfall (interception and depression storage plus the amount of infiltration that occurs before runoff begins) and the infiltration rate after runoff begins. Values of the curve number CN, have been developed and evaluated in terms of soil type, soil cover, land use, hydrologic condition, and antecedent moisture. Table 2-8, page 62, classifies different soils into groups in terms of their minimum infiltration rate, and Table 2-11, page 72, lists CN values for each soil group in terms of land use and hydrologic condition. The CN values of Table 2-11 are for an antecedent moisture condition, called condition II, which is described by the SCS as a condition that has been found to frequently precede the occurrence of maximum annual floods.

The analytical development of the curve-number method is empirical and is based on the observed behavior of runoff as a function of precipitation. If accumulated runoff is plotted against accumulated precipitation, the initial abstraction I_a (interception plus depression storage) causes the accumulated runoff to initially be zero because I_a is equal to the accumulated precipitation before runoff begins. As time passes and precipitation continues, the interception and depression storage volumes are filled, and if the rainfall rate is greater than the infiltration capacity of the soil, surface runoff begins. If the rainfall intensity continues to be larger than the infiltration capacity, the runoff rate will also increase. The SCS curve number method assumes that, as time passes and the rainfall rate continues to exceed the infiltration capacity, the curve developed by plotting accumulated runoff versus accumulated precipitation will become parallel to a 45° line (on an arithmetic plot). That is, this model assumes that the infiltration rate ultimately becomes zero, and thus, the incremental increase in rainfall excess (surface runoff) becomes equal to the incremental rainfall. This assumption, of course, is not valid, but it does make the analysis more simple and does yield reasonable results, particularly for short-duration storms.

For a case where the initial abstraction is zero ($I_a = 0$), it is known that F, the accumulated retention, will increase with time but at an ever-decreasing rate of increase. Thus, the ratio F/S_* will start with a zero value and increase toward a value of unity as time passes. Here S_* is the potential maximum retention. Likewise, as time passes, p (precipitation excess) will increase with time as will p/P with p/P tending toward unity as t becomes very large (P is the potential maximum rainfall excess). Since both of these ratios, F/S_* and p/P, start with zero values at $t = 0$ and approach unity at large values of time, one possible simple assumption is that their values are equal at all times:

Table 2-11 Curve Numbers for Soil Groupings in Terms of Use for Antecedent Moisture Condition II (34)

Land Use	Cover Treatment or Practice	Hydrologic Condition	Hydrologic Soil Group A	B	C	D
Fallow	Straight row	—	77	86	91	94
Row crops	Straight row	Poor	72	81	88	91
	Straight row	Good	67	78	85	89
	Contoured	Poor	70	79	84	88
	Contoured	Good	65	75	82	86
	Contoured and terraced	Poor	66	74	80	82
	Contoured and terraced	Good	62	71	78	81
Small Grain	Straight row	Poor	65	76	84	88
	Straight row	Good	63	75	83	87
	Contoured	Poor	63	74	82	85
	Contoured	Good	61	73	81	84
	Contoured and terraced	Poor	61	72	79	82
	Contoured and terraced	Good	59	70	78	81
Close-seeded	Straight row	Poor	66	77	85	89
Legumes* or	Straight row	Good	58	72	81	85
Rotation	Contoured	Poor	64	75	83	85
Meadow	Contoured	Good	55	69	78	83
	Contoured and terraced	Poor	63	73	80	83
	Contoured and terraced	Good	51	67	76	80
Pasture or		Poor	68	79	86	89
Range		Fair	49	69	79	84
		Good	39	61	74	80
	Contoured	Poor	47	67	81	88
	Contoured	Fair	25	59	75	83
	Contoured	Good	6	35	70	79
Meadow		Good	30	58	71	78
Woods		Poor	45	66	77	83
		Fair	36	60	73	79
		Good	25	55	70	77
Farmsteads		—	59	74	82	86
Roads						
Dirt[†]		—	72	82	87	89
Hard surface[†]		—	74	84	90	92

* Close-drilled or broadcast-seeded
[†] Including right-of-way

$$\frac{F}{S_*} = \frac{p}{P} \tag{2-24}$$

In years of extensive use, the assumption of the equality of Eq. (2-24) has been shown to be reasonable. However, the assumption does lead to inaccuracies about infiltration rates, particularly for storms of several days duration.

For Eq. (2-24), S_* is assumed to be a constant for a given storm and is the maximum that can occur if the storm continues indefinitely. The accumulated retention F is the difference between the potential accumulated rainfall excess and the actual. Thus,

$$F = P - p \tag{2-25}$$

Equation (2-24) can therefore be written as

$$\frac{P - p}{S_*} = \frac{p}{P} \tag{2-26}$$

and can be solved for p as

$$p = \frac{P^2}{P + S_*} \tag{2-27}$$

Equation (2-27) represents the relationship between accumulated excess precipitation and accumulated precipitation for the particular case where the initial interception and depression storage volumes are zero.

If the initial abstraction I_a (initial interception plus depression storage) is considered, Eq. (2-24) can be modified as

$$\frac{F}{S} = \frac{p}{P - I_a} \tag{2-28}$$

where S is the sum of S_* and the initial abstraction, or

$$S = S_* + I_a \tag{2-29}$$

For Eq. (2-28) to hold, F/S and $p/(P - I_a)$ must be less than unity. The equivalents to Eqs. (2-25) and (2-26) are then

$$F = (P - I_a) - p \tag{2-30}$$

and $$\frac{(P - I_a) - p}{S} = \frac{p}{P - I_a} \tag{2-31}$$

Finally, Eq. (2-31) can be solved for the precipitation excess to give

$$p = \frac{(P - I_a)^2}{(P - I_a) + S} \tag{2-32}$$

Equation (2-32) provides a relationship between accumulated rainfall excess and accumulated rainfall for particular values of the initial abstraction and the potential maximum retention.

Investigations of small drainage areas have shown that a relationship between the initial abstraction and the potential maximum accumulated retention can be approximated by

$$I_a = 0.2S \tag{2-33}$$

Using Eq. (2-33), the precipitation excess can be expressed in terms of accumulated precipitation depth P as

$$p = \frac{(P - 0.2S)^2}{P + 0.8S} \tag{2-34}$$

At this point, the curve number CN is introduced to approximately relate hydrologic parameters to drainage-basin parameters. CN is defined in terms of S such that a number CN value of 100 will yield 100% surface runoff ($p = P$). The empirical relationship between CN and S is

$$CN = \frac{1000}{10 + S} \tag{2-35}$$

Equations (2-34) and (2-35) have been combined and plotted in Fig. 2-28 for ready use. In that figure, the curve number 100 corresponds to a case of zero initial abstraction and zero infiltration, and the rate of runoff is equal to the rate of rainfall. The curve numbers of less than 100 correspond to soil conditions where the initial abstraction is greater than zero. For example, consider the curve number 70. Figure 2-28 shows that for such a soil, the initial abstraction I_a is approximately 1 in. (no runoff occurs until rainfall exceeds approximately 1 in.) Then, as the rainfall continues (P increases), the rainfall excess p becomes an increasingly larger portion of the rainfall. This process indicates that the infiltration rate is steadily decreasing with additional rainfall, which is in agreement with Horton's concept of infiltration as shown in Fig. 2-24. For the curve number 70, Fig. 2-28 indicates that rainfall excess (runoff) would be approximately 6.2 in. for a total rainfall of 10 in.

The variable S, used in developing Eq. (2-34) is called the potential maximum retention and includes all the precipitation that does not run off (interception, depression storage, and infiltration). Examining Eq. (2-34) shows that

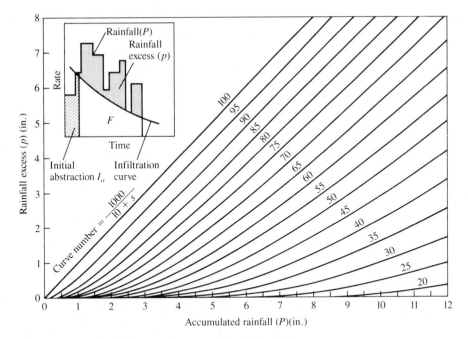

Figure 2-28 Rainfall excess in terms of rainfall and curve number for $I_a = 0.2S$ (34)

rainfall excess (runoff) will equal precipitation if $S = 0$, which is the condition for $CN = 100$. Equation (2-35) shows that $CN = 100$ for $S = 0$. Thus, the runoff model given by Eqs. (2-34) and (2-35) is seen to be very much empirical but generally in agreement with the accepted concept of infiltration and other losses of rainfall.*

The curve numbers given in Table 2-11 are for an antecedent moisture content referred to as condition II. Condition II is a state that has been found to occur frequently before many large storms. If drier conditions or wetter conditions occur before the storm, then less or more runoff will occur, respectively. An empirical method has been developed to obtain curve numbers for drier (condition I) or wetter (condition III) antecedent moisture. The conversion is given in Table 2-12, page 76.

The procedure for using the curve number method is as follows:

1. Study the soil, cover, and use characteristics of the drainage area. Then select the proper soil group from Table 2-8, page 62. For this process, it may be desirable to subdivide the drainage area into subareas having

---

* To gain further insight into the development of this useful concept, refer to the *National Engineering Handbook* of the U.S. Soil Conservation Service (34).

Table 2-12 Transformation of Curve Numbers in Terms of Antecedent Moisture Condition (34)

Curve Number for Condition II	Corresponding Curve Number for Condition I	Corresponding Curve Number for Condition III
100	100	100
95	87	99
90	78	98
85	70	97
80	63	94
75	57	91
65	45	83
60	40	79
55	35	75
50	31	70
45	27	65
40	23	60
35	19	55
30	15	50
25	12	45
20	9	39
15	7	33
10	4	26
5	2	17
0	0	0

reasonably homogeneous conditions and a separate soil-group classification may be necessary for each.

2. Select a *CN* value from Table 2-11, where the definition of "hydrologic conditions" is as follows:
   Poor: Heavily grazed, no mulch, or less than 1/2 of area with plant cover
   Fair: Moderately grazed with plant cover on 1/2 to 3/4 of the area
   Good: Lightly grazed with plant cover over more than 3/4 of the area

3. If the runoff event to be analyzed is for a drier (less conservative in terms of runoff volume) or a wetter (more conservative) estimate, select a different *CN* value using Table 2-12. If the drainage area has been subdivided, a separate *CN* value will be required for each subarea.

4. Determine an average *CN* value for the entire area by calculating an area-weighted average of subarea values.

5. Tabulate the accumulated precipitation of the storm from which runoff is to be estimated.

6. Estimate the accumulated runoff (rainfall excess) using the selected *CN* and Fig. 2-28.

EXAMPLE 2-12   Determine 2-hr increments of runoff for the 100-yr, 24-hr storm of Table 2-6 for a 100-acre drainage area in good hydrologic condition having the following characteristics: 25 acres is composed of pastureland having a shallow sandy-loam soil, and 75 acres is forest land with a clay loam soil. Assume considerable rain has preceded the storm to be analyzed.

SOLUTION

	25-acre Subarea		75-acre Subarea
(Table 2-8)	Soil group C		Soil group C
(Table 2-11)	$CN = 74$		$CN = 70$

For condition II:

$$CN \text{ (average)} = \frac{74(25) + 70(75)}{100} = 71$$

Since we have been asked to assume that very wet conditions prevailed before the storm, we must correct the condition II curve number to a condition III (wetter) state.

Interpolation is required in using Table 2-12. A CN value of 71 is 0.6 of the interval between CN values of 65 and 75. Thus, for condition III, the desired value is 0.6 of the interval between 83 and 91. For condition III (From Table 2-12):

$$CN = 83 + (91 - 83)0.6 = 88$$

The values of accumulated precipitation in the accompanying table are developed from column 4 of Table 2-6. It is assumed that rainfall on the 100-acre area is the same depth as for point rainfall. Each value of runoff for each value

Time (hr)	Accumulated Precipitation (in.)	Runoff (in.)
0	0	0
2	0.10	0
4	0.30	0
6	0.60	0.1
8	1.05	0.3
10	5.15	3.8
12	5.85	4.5
14	6.45	5.1
16	6.70	5.3
18	6.95	5.5
20	7.20	5.9
22	7.40	6.0
24	7.50	6.1

of accumulated precipitation is read from Fig. 2-28 using a curve number of 88 (interpolated between $CN = 85$ and $CN = 90$).

Thus, the total runoff from the 100-yr, 24-hr storm is 6.1 in., and 1.4 in. $(7.5 - 6.1)$ was lost to interception, depression storage, and infiltration. The runoff was $6.1/7.5 = 0.81$ or 81% of rainfall.   ■

The CN method is a reasonable and widely used representation of the runoff process and the only procedure that provides a means of assessing runoff on the basis of drainage-area characteristics without recorded data on runoff and rainfall for actual storms. In practice, the procedure has a basic fault in that it theoretically assumes that the infiltration rate eventually goes to zero. Theoretically, the actual infiltration rate should probably approach a constant minimum rate, as is indicated in Table 2-8. Thus, the curve number method may be slightly conservative when used for predicting runoff from long-duration storms. Because of this limitation, its use is probably questionable for areas greater than perhaps 5 to 10 sq mi since drainage areas that size or larger have times of concentration that may be longer than the time required for the infiltration capacity to reach a minimum. The method, however, has been widely used for much larger areas.

2-7   Streamflow

As we discussed in Sec. 2-2, hydrologic design criteria for an engineering project usually requires the determination of a peak streamflow that will be equalled or exceeded on the average only once in a specified number of years. As we mentioned in Sec. 2-4, that peak flow can be determined through statistical means, provided sufficient streamflow data are available for the watershed in question. More often, however, it is necessary to determine a peak rate of flow or a streamflow hydrograph when a peak rainfall rate or the variable rainfall rate of a design storm is known, but streamflow records are not available. In Sec. 2-6, we discussed the rational method, by which the peak rate of flow from a small drainage area can be determined based on the time of concentration and the degree of imperviousness of the surface. In this section, we discuss the development of the streamflow hydrograph, given a particular rain storm.

Hydrographs

The response of each watershed to rainfall tends to be unique and physically quite complex, but the general characteristics can be readily described. Figure 2-29 represents streamflow occurring at a particular location on a stream as a result of rainfall on the drainage area. The hydrograph is generally divided into the baseflow, the rising limb, the peak segment, and the falling limb, as shown in Fig. 2-29. Segment AB, the baseflow segment, is a recession curve,

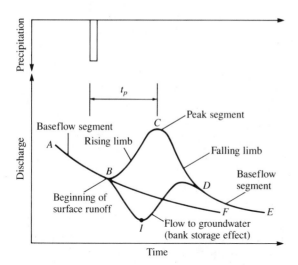

Figure 2-29 Hypothetical single-peaked hydrograph

which is generally due to flow of groundwater into the stream when no surface runoff is occurring at the time. As runoff begins, water first flows overland to the stream and then flows downstream to the point of measurement. The elapsed time between the occurrence of rainfall and the time the peak streamflow occurs is the *time to peak* t_p and is generally measured from the centroid of the rainstorm to the peak flow of the hydrograph.

The hydrograph shown in Fig. 2-29 is typical of that produced by a short rainfall of nearly uniform intensity. Line *BF* represents the streamflow that would have occurred had no rainfall occurred. Point *B* is the point in time at which surface runoff first reaches the observation point. At point *C*, the stream-flow has peaked, rainfall has stopped, and the rate of runoff begins to decrease.

Once streamflow begins to increase (point *B*), the water depth in the stream increases as well. This increase in depth may begin to counteract the flow of groundwater into the channel. If the depth increases enough, water will flow from the stream into the bank creating *bank storage*. This flow into the bank effectively reduces the baseflow as shown by curve *BID* in Fig. 2-29. Once stream-flow passes the peak, the flow into the bank begins to decrease (18). At point *D*, the surface runoff has ceased, and the streamflow follows a recession curve approximately geometrically similar to that occurring before the storm.

The streamflow hydrograph integrates all the physical properties of the drainage area (for example, soil, size, shape, slope) that govern the process of run-off. For that reason, it has been used as a single characteristic of the drainage area in what has been called the *unit hydrograph* (20). The unit hydrograph, or unit graph, is defined as the timewise distribution of 1 in. of surface runoff from a given drainage area for a particular rainfall duration. Theory of the unit graph also implies that two storms of equal duration but different intensities will pro-

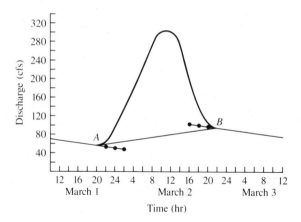

Figure 2-30   Hydrograph used in illustrating the separation of surface runoff and baseflow and the construction of a unit hydrograph (Points plotted near A and B are those computed in column 4 of Example 2-13)

duce hydrographs having similar shapes but with flow ordinates in direct proportion to the total volume of runoff.

   Two procedures are used for constructing a unit hydrograph. The first to be discussed involves developing a unit hydrograph using an actual stream hydrograph for which recorded information is available relative to the rainfall that produced the hydrograph. Since the unit graph represents surface runoff, it is necessary to separate baseflow and surface runoff for the given hydrograph. This process is subject to much interpretation, but a reasonable method assumes that a straight line can be drawn from the beginning of runoff (point A in Fig. 2-30) to the end of runoff (point B in Fig. 2-30). Locating points A and B involves analyzing the recession curve. The recession curve must be extrapolated forward from point A and backward from point B, as is shown in Fig. 2-30. One method of doing this assumes that the recession curve can be expressed as

$$Q_{t+\Delta t} = Q_t \ K^{\Delta t} \tag{2-36}$$

where Q_t = the flow rate at time t
$\quad\ Q_{t+\Delta t}$ = the flow rate at time $t + \Delta t$
$\quad\quad\ K$ = an empirical constant
$\quad\quad\ \Delta t$ = an increment in time

   Equation (2-36) is used to evaluate the coefficient K using the recorded values of Q once Δt is chosen. It is convenient to choose $\Delta t = 1$. In computing values of K, it is important to use values of Q only from those segments of the hydrograph that actually represent recession curves. The following example illustrates the use of Eq. (2-36).

EXAMPLE 2-13 Given the streamflow hydrograph of Fig. 2-30, determine the ordinates of the unit graph. Assume that the storm duration was 4 hr and that the drainage area was 460 acres.

SOLUTION To separate the runoff from the baseflow, we must examine the recession curves on the hydrograph. The hydrograph ordinates taken from Fig. 2-30, are given in the accompanying table.

(1) Time	(2) Q (cfs)	(3) K	(4) Q (cfs)	(5) Baseflow (cfs)	(6) Surface Runoff (cfs)	(7) Unit Graph Ordinate (cfs)
10:00	70			70	0	0
12:00	68	0.985		68	0	0
14:00	66	0.985		66	0	0
16:00	64	0.985		64	0	0
18:00	62	0.984		62	0	0
20:00	60	0.984		60	0	0
22:00	64	1.030	58	61	3	0.5
24:00	94	1.210	56	64	30	4.9
2:00	140		55	67	73	11.9
4:00	190			70	120	19.5
6:00	238			73	165	26.9
8:00	280			76	204	33.2
10:00	297			78	219	35.7
12:00	300			80	220	35.8
14:00	280			83	197	32.1
16:00	210	ᐧ 0.866	104	86	124	20.2
18:00	142	0.822	101	89	53	8.6
20:00	109	0.876	98	92	17	2.8
22:00	95	0.933		95	0	0
24:00	92	0.984		92	0	0
2:00	89	0.984		89	0	0
4:00	87	0.989		87	0	0
6:00	84	0.983		84	0	0
8:00	82	0.988		82	0	0
10:00	79	0.982		79	0	0
12:00	77	0.987		77	0	0
				Totals	1425	232.1

Values of K have been determined using Eq. (2-36) with $\Delta t = 2$ and are shown in column 3. Note that the value of K changes drastically between 20:00 and 24:00 on March 1. The consistent values of K average 0.985. The values at 16:00 and 18:00 of March 2, being much smaller, indicate that they are within the period when surface runoff is in recession. To determine the points where runoff begins (point A) and ends (point B), Eq. (2-36) is used to extend the recession curves both ahead and backward in time. The calculated values are shown in column 4 and are plotted in Fig. 2-30. Points A and B are thus located by the point at which the extrapolated recession curves deviate from the hydrograph. A straight line AB is then drawn to approximately separate surface runoff from baseflow. Runoff can be calculated as the ordinate between the hydrograph and the straight line and is shown in column 6. Column 5 shows the baseflow or the ordinates below the straight line AB. The total surface runoff volume is calculated by multiplying the sum of column 6 by 2, the increment in time used in calculation.

$$1425 \times 2 = 2850 \text{ cfs-hours} = 118.75 \text{ cfs-days}$$

$$= 2826.4 \text{ acre-in.}$$

The average depth of precipitation excess over the drainage area is the runoff volume divided by the area or $2826.4/460 = 6.14$ in. The unit graph ordinates are obtained by dividing the original runoff ordinates by 6.14, the depth of runoff. Thus, the total of the unit graph ordinates (as given in column 7) should indicate a total runoff of 1.0 in. Check this:

$$\text{Unit graph volume} = 232.1 \times 2 = 464.2 \text{ cfs-hours} = 19.3 \text{ cfs-days}$$

$$= 460 \text{ acre-in.} \qquad \blacksquare$$

Once completed, the unit hydrograph can be used to develop a streamflow hydrograph for increments of rainfall excess having the same duration as that for which the unit graph was developed but having depths of rainfall greater or less than unity. Unfortunately, storms of the same duration and average intensity do produce different unit graphs due to different directions of storm travel, sequence of rainfall increments, antecedent moisture conditions, and seasons. A certain amount of nonlinearity exists in the process although the assumption is made that the hydrograph ordinates are linearly proportional to rainfall excess. To offset this tendency, several unit graphs should be developed and an average unit graph constructed from them. Figure 2-31 illustrates this process. In developing the average unit graph, the times to peak and the peak flows are averaged, and the average graph is then sketched. The resulting ordinates should be checked and, if necessary, adjusted to make certain that the average unit graph does contain a runoff volume of 1 in.

Synthetic Unit Graphs

Because of the lack of either rainfall or average streamflow records for many drainage basins of interest, constructing a unit graph by the foregoing

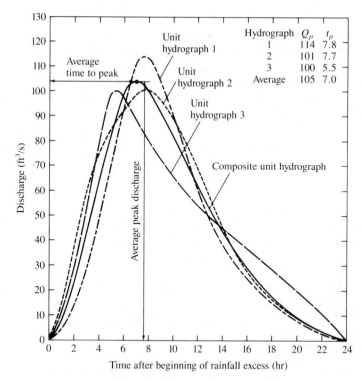

Figure 2-31   Averaging time to peak and peak rate of
flow to develop composite unit hydrograph

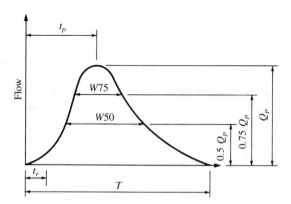

Figure 2-32   Snyder unit hydrograph

procedures frequently is not possible. Methods of constructing synthetic unit graphs have been developed for use on ungauged basins.

Probably the most frequently used synthetic unit graph is that developed by Snyder (22). Figure 2-32 illustrates the various parameters of the Snyder unit graph. The parameters are related as follows:

$$t_p = 0.95C_t(LL_{ca})^{0.3} + 0.74t_r \tag{2-37}$$

$$Q_p = \frac{640AC_p}{t_p} \tag{2-38}$$

$$T = 3 + \frac{t_p}{8} \tag{2-39}$$

where T = base time in days
$\quad t_p$ = time in hours to the hydrograph peak
$\quad t_r$ = rainfall duration in hours
$\quad Q_p$ = peak flow rate in cfs
$\quad L_{ca}$ = stream miles measured from basin outlet to a point opposite the centroid of the drainage area
$\quad L$ = stream miles measured from downstream to the upper limits of the drainage area
$\quad A$ = drainage area in square miles
$\quad C_p, C_t$ = empirical coefficients dependent on the drainage area as follows:

Area	C_p	C_t
Mostly rolling hills such as Appalachian Highlands	0.63	2.0
Steep slopes such as those in the mountains of Southern California	0.94	0.4
Very flat slopes such as those along the Eastern Gulf of Mexico	0.61	8.0

Equation (2-39) provides a reasonable estimate for very large drainage basins. However, the constant 3 days in Eq. (2-39) provides an overly large base for unit hydrographs for small basins. In general, for small watersheds, the base time can be estimated as approximately four times the time to peak.

In constructing any hydrograph, numerical accuracy requires that the storm duration for the unit graph be not greater than 20% of the time of concentration or time to peak. Using a longer duration will result in peak flows that are too high and times to peak that are too long. Additional parameters have been developed by the U.S. Army Corps of Engineers (26) to aid in shaping the Snyder unit graph. These values are shown in Fig. 2-32. As a general rule, the W_{50} and W_{75} widths, which are the widths of the hydrograph at 50% and 75% of Q_p respectively, as given in Fig. 2-33 should be positioned so that approximately one third of their width is ahead of the hydrograph peak. Once the ordinates of the Snyder unit graph have been approximated, using Eqs. (2-37), (2-38), and (2-39) and the parameters in Fig. 2-33, the unit graph must be checked to

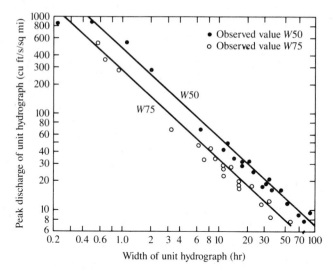

Figure 2-33 Widths to be used in shaping the Snyder unit hydrograph for use with Fig. 2-32 (26).

see that volume of runoff is indeed 1 in. times the drainage area. The ordinates will usually need to be adjusted slightly to meet the unit runoff condition.

The unit graph can be used to develop a composite streamflow hydrograph for a general storm. The process is illustrated in Example 2-14.

EXAMPLE 2-14 Develop a composite hydrograph for an 8-hr storm having two successive 4-hr periods of rainfall excess of 0.6 in. and 2.0 in., respectively. The drainage area is 460 acres, and the unit graph is that computed in Example 2-13.

SOLUTION The table, on page 86, shows the unit hydrograph ordinates (from Example 2-13) in column 2. The time in column 1 is the time measured from the beginning of the hydrograph rise (point A in Fig. 2-29, page 79).

Ordinates for the 0.6-in. and 2.0-in. rainfall excess are obtained by multiplying the unit graph ordinate by the respective rainfall depth. The composite hydrograph ordinates result from adding the two incremental ordinates for common times. The ordinates for runoff due to the 2.0-in. rainfall are lagged by 4 hr because the second period of rainfall follows the first. ∎

The construction of the unit hydrograph assumes that rainfall covers the entire basin. Thunderstorms generally can be assumed to cover areas up to approximately 40 to 50 sq mi. For general frontal storms, the coverage can be as much as 2000 to 3000 sq mi. In general, use of the unit graph should be limited to drainage areas less than these limiting sizes for the respective storm types.

Table for Example 2-14

Time (hr)	Unit Graph Ordinate (cfs)	Hydrograph Ordinates		
		0.6-in. Rainfall (cfs)	2.0-in. Rainfall (cfs)	Composite (cfs)
0	0	0	0	0
2	0.5	0.3	0	0.3
4	4.9	2.9	0	2.9
6	11.9	7.1	1.0	8.1
8	19.5	11.7	9.8	20.5
10	26.9	16.1	23.8	39.9
12	33.2	19.9	39.0	58.9
14	35.7	21.4	53.8	75.2
16	35.8	21.5	66.4	87.9
18	32.1	19.3	71.4	90.7
20	20.2	12.1	71.6	83.7
24	8.6	5.2	64.7	69.0
26	2.8	1.7	40.4	42.1
28	0	0	19.2	19.2
30			5.6	5.6
32			0	0

Probable Maximum Flood Determination

Peak flow of floods of a certain frequency can be calculated statistically according to methods presented in Sec. 2-4. However, the development of a hydrograph of particular return period is somewhat different.

As we mentioned in Sec. 2-5, major projects such as dams whose failure could cause a flood that would jeopardize many people must be designed to be safe from failure during the probable maximum flood (PMF) created by the probable maximum precipitation. Other projects such as nuclear power plants or dikes to contain radioactive material must also be designed to be safe from failure during the probable maximum flood.

In recent years, the safety of dams has become an important issue. To develop a consistent method by which a proper design flood can be chosen, the U.S. Army Corps of Engineers developed the guidelines for design floods for dams shown in Table 2-13.

Table 2-13 recommends that the design flood be chosen in accordance with the hazard that might be created by a dam and the reservoir of water stored behind the dam. In general, hazard is difficult to define quantitatively. Table 2-13 provides guidelines for quantitatively assessing the hazard in terms of reservoir volume and dam height as well as potential loss of life or economic value that might occur if the dam were to fail. Certainly these measures are

Table 2-13 Recommended Guidelines for Choosing
Design Floods for Dams and Reservoirs (25)

Category	Size Classification Reservoir Storage Volume S (acre-ft)	Dam Height H (ft)
Small	$50 < S < 1000$	$25 < H < 40$
Intermediate	$1000 < S < 50{,}000$	$40 < H < 100$
Large	$50{,}000 < S$	$100 < H$
Hazard-Potential Classification		
Category	Loss of Life	Economic Loss
Low	None expected	Minimal
Significant	Few	Appreciable
High	More than a few	Excessive
Recommended Spillway Design Flood		
Hazard	Size	Design Flood
Low	Small	50 to 100 yr
	Intermediate	100 yr to $\frac{1}{2}$ PMF
	Large	$\frac{1}{2}$ PMF to PMF
Significant	Small	100 yr to $\frac{1}{2}$ PMF
	Intermediate	$\frac{1}{2}$ PMF to PMF
	Large	PMF
Large	Small	$\frac{1}{2}$ PMF to PMF
	Intermediate	PMF
	Large	PMF

inexact, but they do provide guidance for assessing the relative need for large or small design floods. Thus, a 35-ft high dam that stores 900 acre-ft of water upstream from an unoccupied valley could have freeboard and spillway capacity designed for a 50-yr inflow flood. However, at least one half of a probable maximum flood should be used if the dam were located upstream from a community where many people might be killed by the flood resulting from a dam failure.

Return periods of design floods used for other structures and developments that have become more or less standard are indicated in Table 2-14.

Determination of a probable maximum flood follows the same guidelines used to develop a flood hydrograph produced by a given storm:

1. The probable maximum precipitation (PMP) is chosen using methods discussed in Sec. 2-5.
2. A design storm is developed using the total PMP determined in step 1 and procedures discussed in Sec. 2-5.
3. Rainfall excess is calculated using procedures discussed in Sec. 2-6.

Table 2-14 Typical Recurrence Intervals for
Design Floods for Various Projects

Type of Project or Structure	Return Period (yr)
Highway culverts	
Rural roads	5–10
Secondary highways	10–25
Interstate highway bridges	100
Airfield drainage	5
Urban storm drainage	2–10
Flood-control levees	
Urban areas	100–250
Agriculture areas	2–50
Design grade elevation	
Industrial plant	50–100
Coal-fired power plant	50–100
Nuclear power plant	PMF

4. The unit hydrograph is chosen for a storm duration not to exceed 20% of the time of concentration for the drainage area of interest.
5. A unit hydrograph is developed using procedures discussed in this section.
6. The unit hydrograph is used to generate hydrographs for each increment of rainfall excess as shown in Example 2-14 on hydrograph construction.
7. The composite hydrograph is constructed from the summation of the hydrographs generated in step 6 as was also done in Example 2-14.

Low-Flow Analysis

Section 2-4 dealt with the determination of recurrence intervals of peak flows or other hydrologic events. The low flows that a stream will experience are also of particular interest, but for an entirely different reason. If water is to be withdrawn from a stream for use in a project as a water supply, it is important for both engineering design and environmental reasons to know the flow rate at low flow and the estimated probability of the actual streamflow being equal to or less than a given magnitude.

In most areas of the world, it is now recognized that streamflow should be retained at or above a particular level, or the instream environment may suffer serious degradation. The instream environment may include resident fish, anadromous fish such as salmon or striped bass, and local resident aquatic life.

Low flows differ from peak flood flows specifically in that they may have a readily apparent value of zero. An ephemeral stream is one that has zero flow for part of most years. Nearly all streams in desert regions are ephemeral, and some are dry except for a few hours during and after a rainstorm. Because the ultimate minimum instantaneous flow of any stream may be zero, it is the duration of any particular flow rate that is important. For example, it is important in a city's planning to know for how many days flow in the stream that is their source of water may be equal to or below a certain value. Frequently, the 7-day low flow is compared between streams and is often quoted in hydrologic descriptions for environmental impact statements or other environmental or engineering documents. The *7-day low flow* represents the average low flow for a 7-day duration. To have complete meaning, it must be associated with a particular recurrence interval such as the 10-year, 7-day low flow.

To develop low-flow frequency curves, the records of average daily flows for the stream of interest are examined for various durations. The record is examined to find first the lowest flow of record for 1 day (the 1-day low flow), and then the second lowest for 1 day, and so on. The apparent probability of occurrence of the flow is calculated using Eq. (2-40) as:

$$P = \frac{M}{N + 1} \tag{2-40}$$

where M = the order number (1 for the lowest)
N = the number of years of record.

Next, the 2-day average low flows are examined and ordered in the same fashion. The same procedure is repeated for all durations of interest. Figure 2-34 shows a set of low-flow duration curves calculated for the Eel River at Scotia, California. In the low-flow selections, it is important to understand

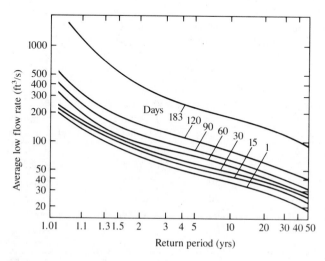

Figure 2-34 Low-flow-duration-frequency curves for the Eel River at Scotia, California 1912–1960 (21)

that each day in the record can appear in only one period for each duration. Thus, a record 700 days in length can be analyzed as not more than 100 7-day durations.

Nonsequential Drought

The set of low-flow duration curves can be used to size a reservoir that must be provided to ensure a sufficient supply of water during a drought period. The process is similar to the construction of a design storm of particular recurrence interval. If, for example, it is desired to size a reservoir to provide sufficient water during a 5-yr drought (the dry period that occurs on the average once in 5 yr) the procedure in the following example is followed.

EXAMPLE 2-15   Use the low-flow duration curves of Fig. 2-34 to determine the necessary size of a reservoir to provide an average flow of 400 cfs during a 5-yr drought.

SOLUTION   Reading the 5-yr values of Fig. 2-34 provides the low-flow values shown in column 2 of the accompanying table. Column 3 is the accumulated volume of inflow to the reservoir, which is determined by multiplying the entries in column 1 by the flow entries in column 2. Column 4 is the accumulated demand at 400 cfs for the corresponding duration. The difference between inflow (column 3) and demand (column 4) volumes is shown in column 5. Accumulated inflow to the reservoir and accumulated demand are shown in the accompanying figure. The difference between the two curves is the storage required to meet the demand without a shortage. Thus, if the reservoir were full when the drought started, it would need to have a minimum of 36,000 cfs-days (71,405 acre-ft) of storage so that the demand of 400 cfs could be met throughout the drought. *Note*, the table indicates slightly less storage is required.

(1) Duration (days)	(2) Average Flow (ft^3/s)	(3) Volume of Inflow (cfs-days)	(4) Volume of Demand (cfs-days)	(5) Inflow Less Demand (cfs-days)
7	50	350	2800	−1950
15	54	810	6000	−5190
30	60	1800	12000	−10200
60	70	4200	24000	−19800
120	100	12000	48000	−36000
183	230	42090	73200	−31110
365	700	255500	146000	109500

■

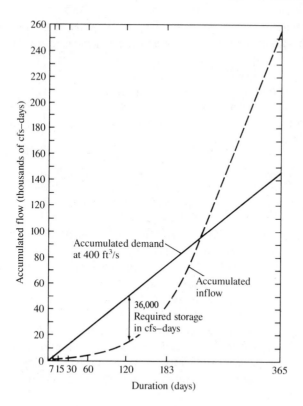

The volumes used in Example 2-15 are in cfs-days for ease of calculation. The analysis shows that inflow to the reservoir under the assumed drought conditions is less than the outflow for more than one half a year. Thus, if the reservoir is assumed to be full at the beginning of the drought, its volume must be large enough to satisfy the demand through more than 120 days. Additional volume must, of course, be added to make up estimated losses due to seepage and evaporation.

2-8   Obtaining Hydrologic Data

The planning and design of any form of water-resources project will always depend on available hydrologic data such as streamflow rates, precipitation depths, and groundwater levels. Generally, when a major project is started, the original analysis of flooding potential and/or water availability will be performed using data collected at or near the site, but data-collection facilities will be established at pertinent locations in order to collect site-specific data. In any case, locating, obtaining, and analyzing the data will nearly always be an engineering responsibility.

Data-collection activities are generally the responsibility of federal agencies with cooperative programs operated by state or local agencies or private developers such as investor-owned utilities. In the United States, the U.S. Geological Survey has the responsibility for collection, analysis, and distribution of data on both quantity and quality of streamflow and groundwater. Table 2-15 is a copy of a summary data sheet published by the U.S. Geological Survey (10). Data on water quantity and quality is published by the U.S. Geological survey every year for each of the 50 states.

The U.S. Geological Survey also publishes many special reports, such as their Open-File Reports, that summarize and analyze streamflow and groundwater data. Figure 2-34 is an example of information available on low flow. Other information on streamflow is available from state water-resource agencies, such as the California Department of Water Resources.

Data obtained by the U.S. Geological Survey is stored in the National Water Data Storage and Retrieval System (WATSTORE) (27). Streamflow data for the United States can be obtained directly from this system in either printed form or on computer-compatible magnetic tape. Moreover, several analyses of the data can be obtained directly such as annual-peak-flow-frequency analysis or low-flow-duration analysis.

Information on precipitation is generally available from the U.S. National Weather Service. The data from all precipitation gauges maintained by NWS are stored at their data processing facility in Asheville, North Carolina, and are available in monthly, daily, and annual summaries of precipitation depth at each station.

Data on snow depth are available from state water-resource agencies and from the U.S. Soil Conservation Service in Washington, D.C.

In countries outside the United States, water-resource data are generally the responsibility of the national government. For the industrialized nations, data-collection systems are generally good, and records are readily available. For the less developed nations, water data vary greatly in quantity, quality, and availability.

2-9 Computer Programs for Hydrology

Use of modern digital computers is ideally suited to hydrologic calculations where, in general, a large amount of data must be handled and extensive digital calculations are necessary. Programs referred to as *watershed runoff models* have been developed that approximately simulate the part of the hydrologic cycle involving the generation of runoff. The Hydrologic Simulation Program-Fortran (HSPF) is a numerical model that continuously simulates not only surface runoff but the entire part of the hydrologic cycle involving interception, depression storage, evaporation, and infiltration (5). In using this program (or model), recorded average rainfall for the basin is used as input, and resulting runoff and streamflow are calculated. Simultaneous precipitation and runoff

Table 2-15 Example of Daily Streamflow Data (11)

Streams Tributary to Lake Michigan
04087138 Menomonee River at Milwaukee, WI

LOCATION — Lat 43°01′28″, long 87°57′36″, in SE 1/4 NW 1/4 sec. 36, T.7 N., R.21 E., Milwaukee County, Hydrologic Unit 04040003, on left bank 10 ft downstream from pedestrian walkway over the Menomonee River, 0.1 mi upstream from bridge at 35th Street, at Milwaukee

DRAINAGE AREA—134 mi^2

PERIOD OF RECORD—December 1981 to current year

GAUGE—Water-stage recorder. Datum of gauge is 576.23 ft National Geodetic Vertical Datum of 1929

REMARKS—Records are poor except for the period November through April 18, which is fair to good. Stage-discharge relation affected by seiche from Lake Michigan Oct. 1–8, 11–18, 21–31, Dec. 12, 13, Jan. 14, Feb. 2, 3, Mar. 21, 26, 27, Apr. 19–30, May 1, 3–6, 9–18, 27, 28, June 2 to Aug. 16, Aug. 19 to Sept. 5, Sept. 7–30

EXTREMES FOR PERIOD OF RECORD — Maximum discharge, 7240 ft^3/s Aug. 17, 1983, gauge height, 14.66 ft, from rating curve extended above 1500 ft^3/s on basis of four step-backwater determinations, $Q10$, $Q50$, $Q100$, $Q500$ obtained from Oct. 3, 4, 5, 1982

EXTREMES OUTSIDE PERIOD OF RECORD — High water of July 13, 1981, reached a stage of 13.16 ft, present datum, from high-water marks; discharge, 5910 ft^3/s, from rating curve extended as explained above

EXTREMES FOR CURRENT YEAR — Maximum discharge, 7240 ft^3/s Aug. 17, gauge height, 14.66 ft, from rating curve extended as explained above; minimum daily, 14 ft^3/s (seiche affected), Oct. 3, 4, 5, determined by applying drainage area ratio to the corresponding daily discharge for Menomonee River at Wauwatosa

Discharge, in Cubic Feet per Second, Water Year October 1982 to September 1983
Mean Values

Day	Oct	Nov	Dec	Jan	Feb	Mar	Apr	May	Jun	Jul	Aug	Sep
1	15	690	72	62	33	104	508	65	155	90	16	22
2	15	335	1180	53	46	108	2390	254	112	38	17	20
3	14	147	983	43	36	120	1650	111	97	32	17	19
4	14	86	425	51	34	130	1090	90	86	212	17	17
5	14	62	693	49	28	128	907	75	74	48	19	19
6	15	49	520	48	28	252	772	82	103	36	17	537
⋮	⋮	⋮	⋮	⋮	⋮	⋮	⋮	⋮	⋮	⋮	⋮	⋮
26	21	59	119	31	121	110	83	107	26	19	34	81
27	20	50	106	29	104	201	77	83	54	19	28	52
28	20	129	217	32	101	171	84	77	62	20	26	50
29	24	92	126	38	—	159	70	399	33	20	24	41
30	21	76	134	51	—	195	65	214	36	20	26	37
31	20	—	79	35	—	318	—	189	—	17	27	—
Total	1457	4100	6718	1297	3347	5591	14110	4469	1965	1210	3391	2205
Mean	47.0	137	217	41.8	120	180	470	144	65.5	39.0	109	73.5
Max	318	690	1180	62	350	409	2390	435	155	212	1930	537
Min	14	41	56	29	28	94	65	59	26	17	15	17
Cfsm	0.35	1.02	1.62	0.31	0.90	1.34	3.51	1.08	0.49	0.29	0.81	0.55
in.	0.40	1.14	1.86	0.36	0.93	1.55	3.92	1.24	0.55	0.34	0.94	0.61

Cal yr	1982	Total	47189	Mean	129	Max	2150	Min	14	Cfsm	0.96	In.	13.10
Wtr yr	1983	Total	49860	Mean	137	Max	2390	Min	14	Cfsm	1.02	In.	13.84

records are required for initial calibration of parameters in the model. Empirical program parameters, which indirectly control the simulation of the various hydrologic components, are adjusted during the calibration process until the program output closely duplicates the measured runoff. Once this procedure is complete, it is assumed that the model can be used to generate runoff from other rainfall events as well. Watershed models, such as HSPF, can be used to generate long sequences of streamflow for a given set of recorded rainfall or can be used to generate a flood hydrograph from a given rain storm. For the latter case, however, special calibration using rainfall and runoff data from severe storms is necessary.

Another watershed model, TR-20 (Computer Program for Project Hydrology) developed by the U.S. Department of Agriculture, Agricultural Research Service, also simulates runoff but uses the curve number method described earlier (5).

Other computer programs for general use are available. The U.S. Army Corps of Engineers Hydrologic Engineering Center (HEC) in Davis, California, has developed several hydrologic programs that have received wide usage. One of these, HEC-1 (Flood Hydrograph Package), has several subroutines that are used to determine an optimal unit hydrograph, loss rates, or streamflow routing parameters by matching recorded and simulated hydrograph values. Other subroutines are used to perform computations for snowmelt, unit hydrograph usage, hydrograph routing and combining, hydrograph balancing, as well as rainfall sequencing and the generation and routing of floods produced by a hypothetical dam failure (5).

Once a basic understanding has been developed, the wide range of computer programs available for hydrologic analysis can and should provide important tools for design and analysis. As the development of microcomputers continues, these programs become easier and cheaper to use and provide an indispensable means for hydrologic analysis.

PROBLEMS

2-1   A reservoir used for both flood control and irrigation contains the following storage volumes: flood control, 210,000 acre-ft; and conservation, 323,100 acre-ft. What are equivalent volumes expressed in ft^3 and m^3?

2-2   In surface-water hydrology, it is convenient to use units that are often unfamiliar. Convert 1310 cfs to the equivalent flow in cms and acre-ft/day.

2-3   A total of 13 in. of precipitation falls on a 35-mi^2 drainage area during a 24-hr period.
   a. Calculate the total volume of rainfall in mi^2-inches, acre-ft, ft^3, and m^3.
   b. If 70% of rainfall runs off, calculate the total volume of runoff in acre-feet, and the average rate of runoff if runoff lasts for 72 hr.

2-4   A dam has been constructed with a total storage volume of 86,000 acre-ft. If the average annual flow from the drainage area upstream from the dam is 40 cfs, how many years of average runoff will the reservoir hold?

2-5   In the reservoir of Prob. 2-4, if the average annual loss due to seepage and evaporation is 40 in., and the average annual precipitation on the reservoir is 21 in., how many years (on the average) would it take to fill the reservoir? Assume there are no withdrawals, and the average surface area of the reservoir is 1000 acres.

2-6   Calculate the apparent average annual precipitation for Glenwood Springs, Colorado, for the two years given in Fig. 2-2, page 23. Calculate the average monthly precipitation. What is the maximum deviation from the average monthly precipitation?

2-7   A flood control levee is to be built in an urban area to withstand any discharge up to the 75-yr flood magnitude. The planned useful project life of the levee is 60 yr. What is the probability that the levee will be overtopped exactly twice in its 60-yr project life?

2-8   A temporary flood wall has been constructed to protect several homes in a flood plain. The wall was built to withstand any discharge up to the 20-yr flood magnitude. The wall will be removed at the end of the 3-yr period after all the homes have been relocated. Determine the probability in each of a.–d. that
a. The wall will be overtopped in any year
b. The wall will not be overtopped during the relocation operation
c. The wall will be overtopped at least once before all the homes are relocated
d. The wall will be overtopped exactly once before all the homes are relocated.
e. What return period must a highway engineer use in his design of a critical underpass drain if he is willing to accept only a 10% risk that flooding will occur in the next 5 yr?

2-9   You are to design a highway culvert for a secondary highway. Current practice dictates a 25% acceptable risk that flooding will occur at least once over its 30-yr project life. What return period must you use in its design?

2-10   The following parameters were determined for a series of annual maximum stream flows: mean of $Q = 750$ cfs, standard deviation of $Q = 110$ cfs, skew of $Q = 0.0$. Assuming that the data fit the normal distribution, what is the discharge for a flood with a recurrence interval of 100 yr?

2-11   Using Fig. 2-8, what is the apparent annual average precipitation for your hometown? What is the average annual precipitation in Phoenix, Arizona?

2-12   A record of flood flows for Touchet River at Bolles, Washington, is given in the table.
   a. Using both the Pearson Type III and the normal distributions, find the magnitude of the 10-, 50-, and 100-year floods using Eq. (2-12).
   b. What is the probability of a flood equal to or greater than the 20-yr flood during the next 3 yr?

Peak Annual Flood Flow Rates for Touchet River at Bolles, Washington. (Drainage area = 361 mi^2)

Water Year	Discharge (cfs)	Water Year	Discharge (cfs)
1925	2910	1964	1820
1926	2850	1965	9350
1927	3690	1966	1250
1928	4470	1967	2080
1929	879	1968	2520
1951	—	1969	7160
1952	3440	1970	3570
1953	3030	1971	7140
1954	1810	1972	6110
1955	925	1973	2750
1956	3410	1974	4740
1957	2390	1975	3540
1958	2420	1976	3980
1959	2790	1977	315
1960	1220	1978	2040
1961	2700	1979	2680
1962	2340	1980	2090
1963	2070	1981	3920

2-13   Records for a 110-mi^2 drainage area upstream from a stream gauge on the next page show the following monthly values of average precipitation and evapotranspiration. The average daily flow rate at the stream gauge is also shown.
   a. Calculate the total annual volume of precipitation that falls on the drainage basin.
   b. Calculate the total volume of water that flows from the basin in each month, and for the year.
   c. Calculate the volume of water that does not evaporate from the basin in each month, and for the year.
   d. Calculate the percentage of rainfall that flows from the drainage area each month, and for the year.
   e. How do you explain that during some months evaporation exceeds rainfall?
   f. How do you explain that during some months total flow out of the basin exceeds rainfall?

Table for Problem 2-13

Month	Precipitation (in.)	Evapotranspiration (in.)	Streamflow (cfs)
Oct	2.2	4.0	37
Nov	4.3	2.1	50
Dec	6.5	0.6	61
Jan	6.8	0.4	11
Feb	5.9	0.5	18
Mar	5.0	1.0	80
Apr	3.1	1.6	350
May	1.4	1.9	465
Jun	0.4	2.2	201
Jul	0.2	2.8	185
Aug	0.2	2.8	147
Sep	0.6	2.0	80

2-14   Using Fig. 2-11, page 43, calculate the total volume of precipitation that would fall on a 10-mi² drainage area during a world-record 2-hr storm.

2-15   The accompanying figure shows a drainage area and the location of several precipitation gauges. During a given rainstorm, precipitation depth was recorded at the gauges as follows:

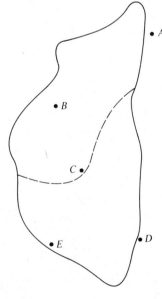

Gauge	Precipitation (in.)
A	3.2
B	2.8
C	4.1
D	1.6
E	2.4

PROBLEM 2-15

Calculate the average precipitation for the basin using the following methods:
a. Station average method
b. Thiessen polygon method

2-16   Using the data and figure for Prob. 2-15, calculate the average precipitation for the basin using the isohyetal method and
a. assume a linear variation between stations.
b. assume that the broken line is the 4.0-in. isohyet.

2-17   For the weather station nearest your hometown (or as specified by your instructor), record the average precipitation for each month of the year, and plot a bar graph (one bar for each month). What other records are taken at that station? The source of this information can be obtained from *Climatological Data*, a U.S. National Weather Service publication.

2-18   Determine from published data the mean annual precipitation at all weather stations or near a river basin close to your hometown (or as specified by your instructor). Choose a basin at least 400 mi² in area.

2-19   For the river basin of Prob. 2-18, determine the mean annual precipitation for the basin by
a. The arithmetic mean of stations in or near the basin.
b. The Thiessen polygon method. For the network, use stations in and near the basin. Show all of your work, including the Thiessen polygons and your computations.
c. The isohyetal method. Data from stations near the basin will help you in drawing the isohyets.

2-20   The following rainfall data were recorded during an intense storm:

Time	Accumulated Depth (in.)
12:15	0
12:25	0.1
12:35	0.2
12:45	0.4
12:55	0.7
13:05	1.0
13:15	1.4
13:25	2.0
13:35	2.1
13:45	2.2
13:55	2.2
14:05	2.4
14:15	2.5

Calculate the intensity of rainfall for durations varying from 10 min to 2 hr. Plot a graph using intensity as the ordinate and time as the abscissa.

2-21   Using Fig. 2-17, 2-18, and 2-19 and Table 2-5, pages 51-3, develop 100-yr, 10-yr, and 2-yr depth-duration-frequency curves for your school's location.

2-22   Using the curves developed in Prob. 2-21, tabulate accumulated 10-yr precipitation depths for durations varying from 0 to 12 hr using 1-hr increments.

2-23   Assuming a drainage area of 22 mi^2, tabulate a reasonable rainfall sequence for a 12-hr storm using the depths calculated for Prob. 2-22. Plot the incremental rainfall depth for your selected rainfall sequence versus time. Calculate two new precipitation sequences using the information in Table 2-7, page 55. Plot these sequences on the same graph.

2-24   Construct a 100-yr, 24-hr storm as it might occur in your hometown.

2-25   For the storm of Prob. 2-24, what area factor would you use to apply it over a 200-mi^2 area?

2-26   Estimate the annual water loss due to evaporation (in acre-feet) from Lake Mead (reservoir behind Hoover Dam). Lake Mead has a surface area 162,700 acres when full.

2-27   The table, below left, provides drainage area and snow-pack depth as a function of altitude for a given snow-covered basin. For an assumed rainfall of 2 in., calculate the expected snowmelt in acre-feet. Assume a degree-day factor of 0.1, the average temperature of the falling rain is 41°F, the snow temperature is 32°F when the rain begins, the lapse rate is 3°F/1000 ft, and the temperature at 5000 ft is 51°F. Also assume the original snow pack has a moisture content of 30%.

Elevation (ft)	Area Above Elevation (mi^2)	Snow Depth (in.)	Day	Avg. Temp. (°F)	Avg. Daily Flow Rate (cfs)
10600	0	62	Mon	33	39.2
10000	5.5	60	Tues	38	302.4
9000	10.6	54	Wed	49	952.0
8000	20.1	54	Thur	43	739.2
7000	24.6	51	Fri	39	470.4
6000	40.1	41	Sat	38	336.0
5000	51.2	31	Sun	33	72.8

2-28   The average temperature on a snow-covered drainage area for a 1-week period is given in the table above right, along with the average daily flow rate at a stream gauge. Assuming that all streamflow was due to snowmelt, calculate the average degree-day factor for the week. The drainage area is 21 mi^2.

2-29   The average rainfall over a 120-acre watershed for a particular storm was determined to be as follows:

Hour	Hourly Rainfall (in.)
1	0.2
2	0.4
3	1.5
4	1.0
5	0.5
6	0.2
7	0

The volume of runoff from this storm was determined to be 19 acre-ft. What is the ϕ index?

2-30   The hourly rainfall depths for a storm over a 100-acre basin is shown below. If the ϕ index for the basin is assumed to be 0.20 in./hr, what will be the *volume* of surface runoff in acre-feet from the basin for this storm?

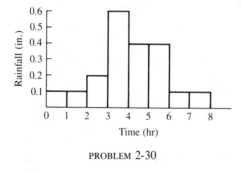

Time (hr)

PROBLEM 2-30

Hour	Hourly Rainfall (in.)
0	0
1	0.1
2	0.3
3	0.7
4	1.9
5	3.6
6	1.1
7	0.4
8	0

PROBLEM 2-31

2-31   During an actual storm that occurred on a 200-acre drainage area, the average hourly precipitation depths were as follows:
The volume of surface runoff from this storm was measured as 100 acre-ft. Calculate the apparent ϕ index for the drainage area using the table above.

2-32   If in Prob. 2-31, the soil is a clay loam, estimate a correct value for k_f assuming a reasonable value for f_c and assuming f_0 is 1.5 in./hr.

2-33   Using Fig. 2-22, page 59, calculate a 48-hr probable maximum precipitation depth for a 36-mi^2 area at the southwest corner of Kansas.

2-34   Arrange the 48-hr probable maximum precipitation depth calculated in Prob. 2-33 in a reasonable sequence of 2-hr rainfall increments.

2-35 What is the probable maximum precipitation for a 12-hr storm on a 400-mi^2 drainage basin just north of Houston, Texas?

2-36 Using the soil-cover-complex (SCS) method, determine the total volume of runoff to be expected from a 50-yr, 24-hr rainfall occurring at the southwest corner of the state of Kansas. Assume the soil is a clay loam, the land has been used as pastureland, prior conditions have been wet, the land has been very heavily grazed for some time, and the drainage area is 200 mi^2.

2-37 Repeat Prob. 2-36, but assume the hydrologic conditions are
a. fair.
b. good.

2-38 Using the soil-cover-complex method, determine 2-hr increments of rainfall excess for a 10-yr, 24-hr storm on a 27-mi^2 basin near Spokane, Washington. Assume the soil is claylike when wet. The cover is as follows: 70% wheat, 25% pasture, 5% roads. Assume the basin is in poor hydrologic condition, and considerable rain precedes the 10-yr storm. The terrain in which this basin is located is gentle sloping hills.

2-39 The following meteorologic and hydrologic information has been collected for a waste-isolation pond.

Month	Surface Inflow (cfs-days)	Average Air Temp. (°F)	Precipitation (in.)	Pan Evaporation (in.)
Mar	190	42	3.1	2.9
Apr	84	49	2.1	4.6
May	82	57	3.6	4.7
Jun	7	72	2.8	5.1
Jul	21	75	1.9	5.2
Aug	9	70	1.3	5.1
Sep	29	66	0.6	4.1
Oct	17	61	0.9	3.0

The pond has an approximately constant surface area of 620 acres and a change in storage capacity of 430 acre-ft/ft of change in depth. Calculate a net change in the volume of water stored in the pond during the 8-month period.

2-40 Calculate the peak rate of runoff from a 20-acre concrete hard-stand area at an airport if the time of concentration for the area is 15 min, and the design rainfall intensity in inches/hr is given as $i = (3.6)/(3 + t)$, where t is the duration of rainfall in hours.

2-41   Given the hypothetical flood hydrograph for a basin having an area of 10 mi^2, what is the peak discharge for a unit hydrograph obtained from this flood?

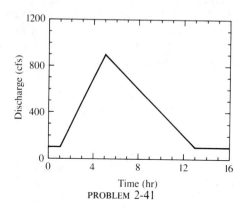

Time (hr)

PROBLEM 2-41

2-42   Calculate the runoff hydrograph for the 50-yr, 24-hr storm developed in Prob. 2-36. Assume the length of the stream in the drainage is 18 mi, and the difference in elevation between the upper and lower elevations is 180 ft. Use a Snyder unit hydrograph and assume topographic conditions are similar to those of the Appalachian Mountains.

2-43   The following streamflow hydrograph was recorded for a stream draining a drainage area of 50 mi^2 for a 24-hr storm. Develop a unit hydrograph for this basin.

Time (hr)	Q (cfs)	Time (hr)	Q (cfs)	Time (hr)	Q (cfs)
8	54	36	254	64	96
10	54	38	251	66	90
12	53	40	240	68	86
14	53	42	230	70	82
16	52	44	218	72	79
18	58	46	200	74	76
20	66	48	190	76	74
22	76	50	177	78	72
24	88	52	157	80	71
26	108	54	142	82	70
28	138	56	130	84	69
30	178	58	120	86	69
32	208	60	111	88	68
34	233	62	103	90	68

2-44   Use the unit hydrograph developed in Prob. 2-43 to develop a stream-flow hydrograph for the following 72-hr rainfall excess.

Hours	August Rainfall (in.)
0–24	1.1
24–48	1.9
48–72	0.7

2-45   The data below are the stream discharges from a 4-hr storm on the basin of Prob. 2-36. Plot the flood hydrograph. Determine the direct runoff (in inches) for this storm, and determine and plot the 4-hr unit hydrograph for the basin.

Date	Hour	Q (cfs)	Date	Hour	Q (cfs)
Feb. 16	0200	90		2200	300
	0400	80		2400	245
	0600	75	Feb 17	0200	210
	0800	450		0400	175
	1000	700		0600	145
	1200	750		0800	125
	1400	800		1000	100
	1600	600		1200	85
	1800	450		1400	75
	2000	350		1600	70

2-46   Determine a synthetic unit hydrograph for the basin of Prob. 2-38. For this basin, the stream length is 11 mi and L_{CA} is 5.5 mi.

2-47   The hydrograph shown resulted from a 2-hr rainstorm.
   a. Using reasonable assumptions, determine the volume of surface runoff and the volume of baseflow. What was the depth of runoff if the drainage area is 8000 acres?

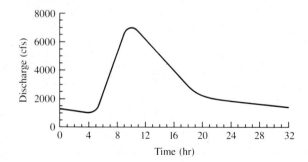

   b. Develop and plot a unit hydrograph for the drainage basin.

2-48 The parking lot shown is to have a drainage system designed to convey peak runoff from a 5-yr storm. If the inlet is at *A*, determine the peak rate of inflow to the pipe. Assume that water flows over the lot at 0.5 ft/s and that the parking lot is paved.

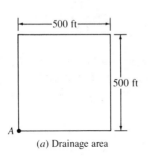

(a) Drainage area

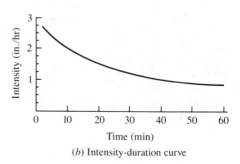

(b) Intensity-duration curve

2-49 Low-flow-duration curves for Yellow Creek near Hammondsville, Ohio, are given in the accompanying figure.

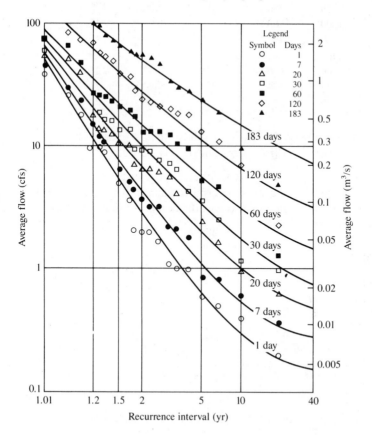

a. Determine the necessary size of a reservoir (in acre-feet) to provide a flow of not less than 25 cfs during a 6-yr drought.

b. How often on the average might one expect the flow over the driest 30-day period to be equal to or less than 80 acre-ft?

REFERENCES

1. Amerman, C.R. "Infiltration Measurement." *Proc. of the National Conference on* Advances in Infiltration, American Society of Agricultural Engineers, Chicago, Ill. (December 1983).

2. Beard, L.R. *Statistical Methods in Hydrology.* U.S. Army Engineer District, Corps of Engineers, Sacramento, Calif. (January 1962).

3. Biswas, A.K. *A History of Hydrology.* North Holland Publishing, Amsterdam, 1970.

4. Brooks, N.H. "Storms, Floods, and Debris Flows in Southern California and Arizona 1978 and 1980." *Proc. of a Symposium, September 17–18, 1980,* Environmental Quality Laboratory, California Institute of Technology, 1982.

5. Cassidy, J.J. "Hydraulic and Hydrologic Computer Applications." *ASCE Journal of the Technical Councils.* TCS, paper 17471 (November 1982).

6. Farnsworth, R.K., E.S. Thompson, and E.L. Peck, NOAA Technical Report, NWS 33, Evaporation Atlas for the Contiguous (48) United States, U.S. Department of Commerce, National Weather Service, Washington, D.C., (June 1982).

7. Goodridge, J.D. *Rainfall Depth-Duration-Frequency for California.* State of California Department of Water Resources, Sacramento, Calif. (November 1982).

8. Haan, C.T. *Statistical Methods in Hydrology.* Iowa State University Press, Ames, Iowa, 1979.

9. Hathaway, G.A. "Design of Drainage Facilities." *Trans. ASCE,* 110 (1945).

10. Hershfield, David. M. Personal communication. Kensington, Md. (October 1986).

11. Holmstrom, B.K., C.A. Harr, and R.M. Erickson. U.S. Geological Survey Water-Data Report WI-83-1. Water Resources Data Wisconsin Water Year 1983, Madison, Wis., 1984.

12. Horton, R.E. "The Role of Infiltration in the Hydrologic Cycle." *Trans. AGU* 14 (1933), pp. 446–60.

13. Jennings, A.H. "World's Greatest Observed Point Rainfall." *Monthly Weather Review,* 78, (1950).

14. Kirpich, P.A. "Time of Concentration of Small Agricultural Watersheds." *Civil Engineering,* 10, no. 6 (June 1940).

15. Kohler, M.A., T.J. Nordenson, and D.R. Baker. Evaporation Maps for the United States. U.S. Weather Bureau Tech. Paper No. 37, Washington, D.C., 1959.

16. Myers, V.A. "The Estimation of Extreme Precipitation as the Basis for Design Floods, Resume of Practice in the United States." *Proc. of Symposium, Int. Assoc. of Sci. Hydrol.,* Leningrad, 1967.

17. Perrault, Pierre. *Origin of Fountains.* Trans. A. La Rogue. Hafner Press, Paris, France, 1967.

18. Pogge, E. "Analysis of Bank Storage." Ph.D. dissertation. Department of Mech. and Hydraul., University of Iowa, Iowa City, Iowa (1967).

19. Schaake, J.C., Jr., and J.W. Knapp. "Experimental Examination of the Rational Method." *Proc. ASCE, J. Hyd. Div.,* 93, no. HY6, (November 1967).

20. Sherman, L.K. "Streamflow from Rainfall by the Unit Graph Method." *Engineering News Record*, 108 (April 7, 1932).
21. Smith, W., and C.F. Hains. *Flow Duration and High and Low-Flow Tables for California Streams*. U.S. Geological Survey, Menlo Park, Calif. (October 1961).
22. Snyder, F.M. "Synthetic Unit Graphs." *Trans. AGU* 19 (1938).
23. State of California Department of Water Resources. *Flood*, bulletin no. 161, Sacramento, Calif. (January 1965).
24. Thornthwaite, C.W. "The Moisture Factor in Climate." *Trans. AGU* 27 (1946).
25. U.S. Army Corps of Engineers. *Recommended Guidelines for Safety Inspection of Dams*. Office of the Chief Engineer, Washington, D.C., 1976.
26. U.S. Army, Office of the Chief of Engineers. *Engineering Manual for Civil Works*, part 2, chapter 5 (April 1946).
27. U.S. Geological Survey. *WATSTORE Users Guide*. Open-File Report 75-426 (August 1975).
28. U.S. National Weather Service. Hydrometeorological Report No. 43, Probable Maximum Precipitation, Northwest States. Washington, D.C. (November 1966).
29. U.S. National Weather Service. Hydrometeorological Report No. 33, Probable Maximum Precipitation Estimates — United States East of the 105th Meridian. Washington, D.C. (1978).
30. U.S. National Weather Service. *Observing Handbook No. 2, Substation Observations*. Department of Commerce, Washington, D.C., 1972.
31. U.S. National Weather Service. *Precipitation Atlas of the Western United States*. Department of Commerce, Washington, D.C., 1973.
32. U.S. National Weather Service. *Rainfall-Intensity-Frequency Regime*, Technical Paper No. 29. Washington, D.C., 1957.
33. U.S. National Weather Service. Tech. Memo. NWS HYDRO 33, *Greatest Known Rainfall Depths for the Contiguous United States*. Department of Commerce, Washington, D.C. (December 1976).
34. U.S. Soil Conservation Service. *National Engineering Handbook*. Department of Agriculture, Washington, D.C., 1964.
35. U.S. Weather Bureau. "Rainfall-Intensity-Duration-Frequency," Technical Paper No. 25. Washington, D.C., 1955.
36. U.S. Weather Bureau. *Rainfall Frequency Atlas of the U.S.*, Technical Paper No. 40. Department of Commerce, Washington, D.C., 1961.
37. Wiesner, C.J. *Hydrometeorology*. Chapman and Hall, London, 1970.
38. Wilson, W.T. "An Outline of the Thermodynamics of Snow Melt." *Trans. AGU* 22 (1941).

Big Spring in Carter County Missouri. The spring
rises from the Eminence Dolomite and has an annual
average flow of 438 cubic feet per second. (Photo
by Mr. James E. Vandike of the State of Missouri,
Department of Natural Resources, Division of Geology
and Land Survey, Rolla, Missouri)

Groundwater

3-1   General Considerations

As was shown in Fig. 2-3, page 24 groundwater, in general, is that portion of the earth's water occuring below the surface of the earth. Groundwater (that is, underground water) is one of the earth's most important resources. Its use accounts for approximately 40% of the water used on the earth with the exception of that used for hydropower generation and cooling thermal power plants. Despite its wide use, groundwater is also the least understood resource. There are many common misconceptions about groundwater, including the belief that it runs in rivers beneath the surface of the earth. The only groundwater occurrence resembling the concept of a surface stream are the large solution caverns often found in massive limestone formations. The Carlsbad Caverns in New Mexico and the Meramac Caverns in Missouri are well-known examples of solution caverns in regions underlain by limestone.

Groundwater, which makes up about 14% of the earth's fresh water and amounts to approximately 4 million cu km, moves through the openings that exist within the natural materials forming the earth's surface (7). *Groundwater hydrology* is the science of the occurrence, movement, and quality of water beneath the surface of the earth. It has become increasingly important to understand the science of groundwater.

Groundwater has always been one of the most important sources of water supply. Virtually all parts of the earth are underlain by water, and wells have been constructed throughout recorded history to provide a water supply when surface water was not readily available. In early times, the water wells were excavated by hand and were rarely more than a few meters deep. Today, modern water wells are drilled and in some localities are more than 300 m deep.

Water wells have played an increasingly important role in irrigation as long as water has been available, energy has been relatively cheap, and prices for agricultural products have been high enough to make irrigation systems feasible. Favorable economic conditions between 1950 and 1970 led to extensive irrigation development using well water in the high plains of Texas, the central valley of California, central Arizona, eastern Washington, eastern Colorado, western Kansas, and central Oklahoma. In all these areas, pumping rates have substantially exceeded natural recharge rates, resulting in continuously dropping groundwater elevations. As the groundwater levels have dropped, the energy required to maintain the pumping rates has increased correspondingly. Moreover, the larger initial cost of constructing and maintaining deeper wells, and the sharp increases in the cost of energy since 1973, have resulted in significant increases in the cost of the pumped water. Because of this, lands formerly irrigated have been abandoned in many areas. In other areas, such as the central valley of California, the falling groundwater levels have produced substantial subsidence of the land surface.

Other activities that have influenced attention given to groundwater development include the injection of liquid wastes into very deep aquifers for permanent disposal and the use of groundwater for heat storage. Disposal of waste products in locations where products leached from them enter the groundwater and are transported toward wells or springs often contaminates water supplies and causes serious concern.

All these activities make it important for engineers to be aware of the hydraulics of groundwater.

3-2 Occurrence of Groundwater

Figure 3-1 illustrates the occurrence of water beneath the earth's surface. In the *unsaturated zone*, which adjoins the earth's surface, the spaces between soil and rock particles are filled with air, water, and water vapor. In the *saturated zone*, the interconnected openings of the supporting material are almost invariably filled with water. In a strict technical sense groundwater is that within the saturated zone. The soil or rock containing the saturated zone is called an *aquifer*.

The *soil zone*, the upper part of the unsaturated zone, is generally less than 1 or 2 m in thickness and encompasses that part of the earth's surface used by plant roots, rodents, eartn burrowing insects, and worms. It often is altered by the activities of both man and nature. Within this zone, water may be present in the form of *hygroscopic water* (water absorbed from the air), *capillary water* (water held in the soil by capillary action), or *gravitational water* (water draining downward through the soil). The lowest part of the unsaturated zone contains the *capillary zone*. In this zone, water is pulled from the saturated zone by capillary action within the voids of the porous material, which may be rock,

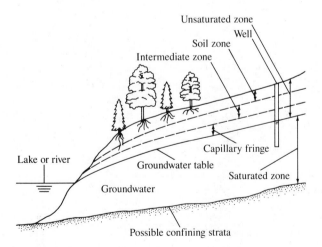

Figure 3-1 Schematic of the occurrence of groundwater

gravel, or soil depending on the local geology. The smaller the voids in the capillary zone, the higher the water is pulled and the thicker the capillary zone becomes. The water in this capillary zone is under negative gauge pressure, since the upper surface of the saturated zone is normally at atmospheric pressure.

The *intermediate zone* is essentially a connecting zone between the soil zone and the capillary zone. Water passes through this zone under the action of gravity and moves downward to the saturated zone.

As we discussed in Sec. 2-3, and as shown in Fig. 2-3, groundwater is replenished from precipitation that infiltrates through the earth's surface. If sufficient infiltration occurs to fill the "field capacity," deep infiltration occurs, and groundwater replenishment takes place. *Deep infiltration* is sometimes referred to as *percolation*, or the motion of water downward through the soil zone.

Within the unsaturated zone, the movement of water is complicated by the presence of capillary and gravitational forces. If layers of low vertical permeability exist, the influence of these gravity and capillary forces can produce lateral movement through the soil called *interflow*. This interflow ends where the layer terminates and can result in flow directly from the unsaturated soil to channels or drains.

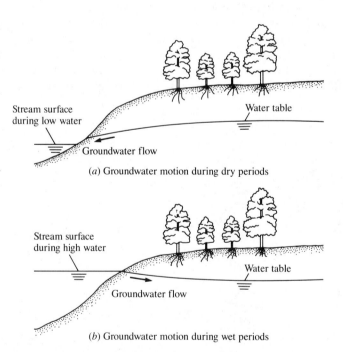

Stream surface
during low water

Water table

Groundwater flow

(*a*) Groundwater motion during dry periods

Stream surface
during high water

Water table

Groundwater flow

(*b*) Groundwater motion during wet periods

Figure 3-2   Behavior of groundwater during periods
of high and low streamflow (a) Groundwater flow
during dry periods (b) Groundwater flow during
wet periods

The thickness of the unsaturated zone varies depending on the amount of water in storage. After significant rainfall and infiltration, the top of the saturated level will be high. During dry periods, as stored groundwater drains to streams or other water bodies, the saturated level will generally fall.

Below the groundwater table, water generally moves very slowly. As shown in Fig. 3-2, water can flow either into or out of streams, springs, or lakes depending on the gradient of the groundwater surface. Springs occur wherever the groundwater table intersects the earth's surface. Springs can be intermittent if the water table rises and falls above and below the spring's elevation, since the amount of rainfall and infiltration vary throughout the year. In some locations, streams gain water from groundwater inflow during the rainy season and lose water to the groundwater zone during the dry season.

Water within the saturated zone fills the natural voids that occur within the solid material, which may be soil, gravel, or rock depending on the local geology. If voids make up a large percent of the bulk volume of the solid material, a greater volume of water can be contained. Thus, an aquifer of saturated gravel contains a much greater volume of water in a unit volume of the solid material than does an equal volume of hard rock. All natural materials are porous to some degree. Table 3-1 lists typical values of *porosity* (volume of voids divided by total volume) for several natural materials. The equation used to calculate porosity (n) is

$$n = \frac{\forall_v}{\forall_t} \tag{3-1}$$

where $\forall_v$ is the volume of the voids, and $\forall_t$ is the total volume.

Using the definition of porosity, the volume of water $\forall_w$ contained in a saturated total volume of material $\forall_t$ is

$$\forall_w = n\forall_t \tag{3-2}$$

Table 3-1 Typical Values of Mechanical Properties for Various Natural Materials (6)

Material	Porosity	Specific Yield	Specific Retention	Hydraulic Conductivity (m/day)
Soil	0.55	0.40	0.15	$10^{-3}-5$
Clay	0.50	0.02	0.48	$10^{-7}-10^{-4}$
Sand	0.25	0.22	0.03	$0.06-120$
Gravel	0.20	0.19	0.01	$100-7000$
Limestone	0.20	0.18	0.02	$10^{-4}-5000$
Sandstone	0.11	0.06	0.05	$10^{-5}-0.5$
Basalt	0.11	0.08	0.03	$10^{-8}-1000$
Granite	0.001	0.0009	0.0001	$10^{-8}-5$

However, during pumping or draining of a given aquifer, not all the water originally held during saturated conditions will be drained. Part of the water will be held within the solids by molecular attraction. The physical forces (surface tension) creating this attraction are the same as those that pull water from the water table forming the capillary fringe. In groundwater terminology, the volume of water contained in the ground is divided into two parts. The fraction that drains from the ground under the action of gravity is called *specific yield*, whereas the fraction retained as a film around particle surfaces or in very small openings is called *specific retention*. Porosity is related to specific yield (S_y) and specific retention (S_r):

$$n = S_y + S_r \tag{3-3}$$

$$S_y = \frac{\forall_d}{\forall_t} \tag{3-4}$$

$$S_r = \frac{\forall_r}{\forall_t} \tag{3-5}$$

where $\forall_r$ is the volume retained, and $\forall_d$ is the volume drained from the total volume $\forall_t$. Table 3-1 lists typical values of specific yield and specific retention.

3-3   Principles of Groundwater Flow

Water can flow through all natural materials. However, the velocity at which it can move is inversely proportional to the size of the openings through which it moves. Figure 3-3 illustrates a situation in which water from one reservoir is flowing to another through a conduit filled with a permeable material. If the energy equation* is written between point 1 and a point inside the pipe, the following equation arises:

$$\frac{V_1^2}{2g} + \frac{p_1}{\gamma} + z_1 = \frac{V^2}{2g} + \frac{p}{\gamma} + z + h_L \tag{3-6}$$

where V = average velocity within the conduit at the point of interest (V is further defined as Q/A, where Q is the flow rate through the conduit, and A is the cross-sectional area of the conduit.)

V_1 = the average velocity across a section at point 1
p = pressure at the point of interest in the conduit
p_1 = the pressure at point 1 (atmospheric)
z = the elevation of the centerline of the conduit at x

---

* This form of the one-dimensional energy equation should be familiar from the student's first course in fluid mechanics. It is presented again in a different content in Sec. 5-2, page 241.

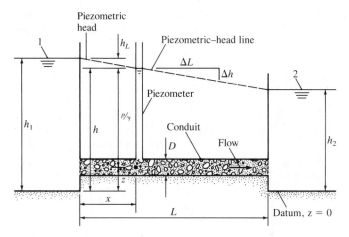

Figure 3-3   Flow through a conduit filled with permeable material

z_1 = the elevation of point 1
g = acceleration due to gravity
γ = specific weight of the flowing fluid
h_L = energy loss which occurs as a result of the flow between the two points

For most flow through porous media, the average water velocity is very small (in the range of inches or feet per day). Thus, the terms involving the square of the velocity are also very small and, because the other terms in the equation are much larger by comparison, can usually be neglected.

The term $(p/\gamma + z)$ is referred to as the piezometric head, and we shall use h to represent it. The piezometric head is the distance above the datum to which the water surface would rise if free to do so (See Fig. 3-3). Thus, if a vertical tube (a piezometer) is connected to the horizontal pipe as shown in Fig. 3-3, the water in the tube will rise until its free surface is p/γ above the centerline of the pipe. The broken line in Fig. 3-3 represents the piezometric head for the water in the pipe between reservoirs 1 and 2. We will consider piezometric head further as we examine cases of groundwater flow.

If the velocity terms are neglected in Eq. 3-6, we have (between points 1 and 2)

$$z_1 - z_2 = h_1 - h_2 = h_L \tag{3-7}$$

since $p_1 = p_2$ = atmospheric pressure.

For very small velocities such as those normally encountered in ground-water flow, the flow is usually laminar, and the energy loss is linearly pro-portional to the velocity V:

$$h_L = \frac{LV}{K} \tag{3-8}$$

where L is the conduit length and the constant of proportionality K, known as the *hydraulic conductivity*, is a function of the size and shape of the voids between the particles making up the porous media as well as the viscosity of the water (13). Substituting Eq. (3-8) into Eq. (3-7) produces

$$V = -K\frac{h_2 - h_1}{L} \tag{3-9}$$

Equation (3-9) can be written as

$$V = -K\frac{dh}{dL} \tag{3-10}$$

because the derivative dh/dL (the gradient of the piezometric head) is a constant for the situation shown in Fig. 3-3, where the average velocity V is constant throughout the length of the tube.

Using the principle of continuity, Eq. (3-10) can be written as

$$Q = -KA\frac{dh}{dL} \tag{3-11}$$

where Q = the total rate of flow through the cross-sectional area A

Equations (3-10) and (3-11) are two forms of Darcy's law. In 1856, Henri Darcy first concluded that velocity of flow through a porous media was directly proportional to the gradient of piezometric head (hydraulic gradient) (8). The negative sign arises because we have assumed the positive direction of velocity is in the direction of decreasing piezometric head.

As we noted, Eq. (3-8), and thus Darcy's law, Eq. (3-11), are valid only for laminar flow. In developing Eq. (3-11), we have defined the velocity V as the average velocity calculated as if the fluid were flowing throughout the entire cross-sectional area occupied by both solid particles and the openings be-tween particles. Defined this way, average velocity is sometimes referred to as the Darcy velocity. The actual velocity of flow in the tortuous path formed by the interconnected openings would be larger than the Darcy velocity because the cross-sectional area formed by the openings between particles is much smaller than the total area of the conduit. The cross-sectional area of voids can be calculated by multiplying the porosity by the conduit cross-sectional area.

Laminar flow is limited to conditions for which the Reynolds number

$$\text{Re} = \frac{VD_{50}\rho}{\mu} \tag{3-12}$$

does not exceed approximately 60 to 700. In Eq. (3-12), V is the Darcy velocity, D_{50} is the diameter of the average particle making up the porous media, ρ is the fluid density, and μ is the dynamic viscosity of the fluid.

For most cases of groundwater flow, Re does not exceed unity and Eq. (3-11) is valid. A Reynolds number of unity implies a very small average velocity. For example, for water with a temperature of 60°F, the ratio of $\rho/\mu = 1.2 \times 10^5$. If the average particle size of the porous media is 1/64 in., the velocity must not exceed 0.009 ft/s (778 ft/day).

The hydraulic conductivity (also called the coefficient of permeability or simply permeability) has the dimensions of a velocity and is generally thought of as a measure of the permeability or impermeability of a porous material. Table 3-1, page 111, lists typical values of hydraulic conductivity for various natural materials and shows that a wide variation can exist for all materials depending on their state. Basalt, for example, can vary in hydraulic conductivity from 10^{-8} to 1000 m per day. The lower value is applicable to unfractured basalt in its natural state. Because rock such as basalt is frequently fractured as a result of weathering, cooling, or other geologic processes, openings and passageways develop that provide paths with much less resistance to water movement than is provided by the voids in the unfractured material. A large value of hydraulic conductivity is typical of a highly fractured rock.

Expansive soils, such as some clays, also exhibit wide ranges in hydraulic conductivity depending on their state. When the clay is very dry, cracks may form that provide paths along which water readily flows, and the resulting hydraulic conductivity can be large. Once the clay becomes wet again, it may swell significantly, reducing the size of the interstitial openings and greatly reducing the hydraulic conductivity.

The viscosity of groundwater has a definite effect on the hydraulic conductivity, as might be expected for laminar flow. Generally, values of hydraulic conductivity are given for a temperature of 20°C (68°F). The hydraulic conductivity is approximately proportional to viscosity, and values of K at temperatures other than 20°C can be computed by

$$K_t = K_{20}\left(\frac{\mu_{20}}{\mu_t}\right) \tag{3-13}$$

where t and 20 refer to the temperature of interest and 20°C, respectively.

Darcy's law is valid for most conditions of groundwater flow. However, for some cases where the openings between particles are very small, such as in dense clays, the effects of relative electrical charges between clay and water

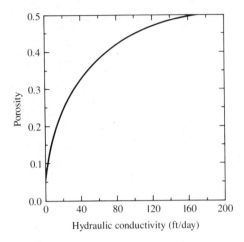

Figure 3-4   Variation of hydraulic conductivity with porosity (17)

molecules can be important, and a nonlinearity between flow rate and hydraulic gradient may exist.

For a given material, the hydraulic conductivity tends to decrease as porosity decreases. This is illustrated in Fig. 3-4, which shows the variation of hydraulic conductivity of mixed-grain sand for different porosities. The data for Fig. 3-4 were developed by measuring the permeability for given sands with varying degrees of vibratory compaction (17). In the compaction process, the volume of solids remains constant, but the volume of the voids decreases with additional compaction.

Figure 3-3 is schematically similar to a device commonly used to measure hydraulic conductivity in the laboratory. However, it is difficult, if not impossible, to obtain "undisturbed" samples of natural material that can be tested in the laboratory. Hydraulic conductivities for use in analyzing the motion of groundwater in its natural condition are usually determined by field pumping tests, which we will describe later. The so-called undisturbed samples are obtained by driving a cylindrical tube into the material to retrieve a core or by core drilling in the case of rock.

3-4   One-Dimensional Steady Groundwater Flow

Equations (3-9), (3-10), and (3-11) can be readily applied to steady one-dimensional groundwater flows. Figure 3-5 illustrates a situation in which an aquifer has been fully penetrated by a trench. The river is assumed to maintain a constant head (or water-surface elevation) h_1. Water is to be pumped at a constant rate Q to maintain a constant head h_2. The thickness of the aquifer

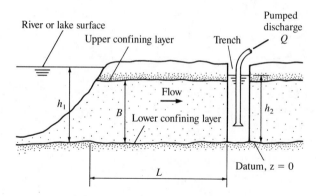

Figure 3-5   Flow to a trench through a confined
aquifer

is B, and we assume that the layers above and below the aquifer are level and
have a much smaller hydraulic conductivity than the aquifer. The datum $z = 0$
is the elevation of the lower confining layer. Thus, the aquifer is known as a
confined aquifer.

Application of Eq. (3-11) gives

$$q = -KB \frac{h_2 - h_1}{L} \tag{3-14}$$

where q is the pumping rate per unit (foot or meter) of trench length. The total
pumped discharge Q is equal to q multiplied by the length of the trench. *Note*:
The area through which flow occurs is simply B for a unit width of the aquifer.

This solution assumes the aquifer thickness B is constant, water flows only
from the aquifer on the river side of the trench, the hydraulic conductivity of
the porous material within the aquifer is constant throughout, and nonuniform
flow conditions near the trench and the river occur over a distance much smaller
than L. Thus, the flow is assumed to be essentially one-dimensional through
the aquifer.

If the aquifer is not fully confined (the groundwater surface does not touch
the upper confining layer throughout), as in Fig. 3-6 on the next page, Eq. (3-11)
can still be used to analyze the flow. Partially confined aquifers are typically
encountered at construction sites where groundwater is being pumped to maintain dry conditions within an excavation, a process called dewatering.

The flow through the aquifer between the river and the point where the
aquifer no longer flows full (the portion that is confined) is analyzed in a manner
similar to that used for Fig. 3-5. The following equation arises:

$$q = KB \frac{h_1 - B}{L_1} \tag{3-15}$$

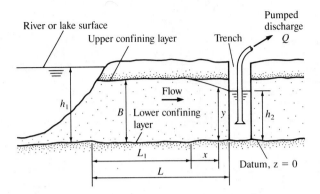

Figure 3-6 Flow to a trench through a partially confined aquifer

The form of the gradient term in Eq. (3-15) arises since the piezometric head at L_1 is equal to the aquifer depth B. To analyze flow along the remaining length of the aquifer where the velocity is not constant, the differential form of Darcy's law, Eq. (3-10) must be used:

$$q = VA = -KA\frac{dh}{dx} \tag{3-16}$$

In this case, the area A (with unit width) is a variable equal to y. The piezometric head h at a distance x from the section at L_1 is likewise equal to y if the bottom of the aquifer is assumed to be horizontal.

Thus, Equation (3-16) becomes

$$q = -Ky\frac{dy}{dx} \tag{3-17}$$

which can be integrated over the distance from L_1 to L as

$$q\int_{L_1}^{L} dx = -K\int_{B}^{h_2} y\,dy \tag{3-18}$$

which upon integration becomes

$$q = K\frac{B^2 - h_2^2}{2(L - L_1)} \tag{3-19}$$

showing that the surface of the saturated portion of the aquifer between L_1 and L is parabolic.

Because the flow rate q is the same in both the confined and unconfined lengths of the aquifer, Eqs. (3-15) and (3-19) can be equated, giving

$$KB\frac{h_1 - B}{L_1} = \frac{K}{2} \cdot \frac{(B^2 - h_2{}^2)}{L - L_1} \tag{3-20}$$

which when solved for L_1, gives

$$L_1 = 2LB \cdot \frac{h_1 - B}{2B(h_1 - B) + (B^2 - h_2{}^2)} \tag{3-21}$$

Substituting Eq. (3-21) into Eq. (3-15) gives

$$q = \frac{K}{2L}\left[2B(h_1 - B) + (B^2 - h_2{}^2)\right] \tag{3-22}$$

In the analysis of flow in Fig. 3-6, we have made the same assumptions as were used in developing Eq. (3-14). Moreover, we have assumed that the flow is still essentially one-dimensional in the unconfined zone, a valid assumption as long as velocities are small and the slope dy/dx is not large.

The foregoing analysis of one-dimensional groundwater flows is applied to the following two examples, in which the bottom of the aquifer is at constant elevation. A sloping aquifer can be analyzed similarly if it is recognized that h in Eqs. (3-10) and (3-11) is in reality the piezometric head $(p/\gamma + z)$. Thus, Eq. (3-14) can be applied to the situation shown in Fig. 3-7 to yield

$$q = -KB\frac{h_2 + z_2 - h_1 - z_1}{L} \tag{3-23}$$

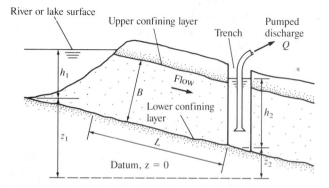

Figure 3-7   Flow to a trench through a confined, saturated, sloping aquifer

EXAMPLE 3-1 A trench 1000 ft long is to be excavated parallel to and 800 ft from a river. An aquifer, similar to that shown in Fig. 3-5, exists and is known to have a hydraulic conductivity of 15 ft per day and a depth of 15 ft. If the water level in the trench must be kept 10 ft below the water level in the river (but still above the top of the aquifer), determine the rate at which water must be pumped from the trench.

SOLUTION Since the aquifer is confined throughout its length, Eq. (3-14) is applicable. Thus, the flow through each unit width of the aquifer is

$$q = -15(15)\frac{-10}{800}$$

$$= 15(15)\frac{10}{800} = 2.81 \text{ ft}^3/\text{day}$$

The rate at which water must be pumped from the trench is

$$Q = 2.81(1000) = 2810 \text{ ft}^3/\text{day}$$

$$= 0.0325 \text{ ft}^3/\text{s}$$

$$= 14.6 \text{ gal/min}$$

Note: This solution assumes no flow from that portion of the aquifer on the side of the trench away from the river. Such a situation would occur only after a long period of pumping when the water table on the land side had subsided to a near equilibrium position at height h_2, or if a barrier, such as a sheet-piling wall, were constructed to prevent flow from that side of the trench. ■

EXAMPLE 3-2 For the trench and aquifer of Example 3-1, assume a change in the construction requires that the water surface in the trench be maintained at a depth of only 3 ft. The water surface in the river is 25 ft above the bottom of the aquifer. At what rate must water now be pumped out of the trench?

SOLUTION Since a portion of the aquifer will be unwatered in this case, Eq. (3-22) must be used. Thus, the flow through each unit width of the aquifer is

$$q = \frac{15}{2(800)}[2(15)(25 - 15) + (15^2 - 3^2)]$$

$$= 4.84 \text{ ft}^3/\text{day/ft}$$

The rate at which water must be pumped from the trench is

$$Q = 4.84(1000) = 4840 \text{ ft}^3/\text{day}$$
$$= 0.056 \text{ ft}^3/\text{s}$$
$$= 25 \text{ gal/min}$$

■

3-5   Well Hydraulics

Wells are the principle means by which groundwater is extracted. Wells such as Fig. 3-8 are designed to penetrate an aquifer and to deliver a desired rate of water flow. Wells have been used throughout recorded history to obtain a water supply. In arid lands, villages were developed around wells that yielded a dependable water supply. For the most part, older wells were dug by hand and penetrated only shallow aquifers. Because shallow aquifers are easily polluted by infiltration of contaminated runoff, discharge from inefficient or faulty sewage-treating septic tanks, or leakage from underground storage tanks, many shallow wells have had to be abandoned and replaced by much deeper modern wells that penetrate aquifers carrying water of good quality. A modern well includes a hole that is normally drilled, a casing that prevents caving in of the

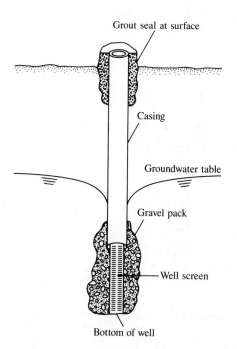

Figure 3-8   Typical well installation in an unconfined aquifer

sides· or inflow of water from an undesirable aquifer, a pump and its appurtenances, and a section of perforated casing (well screen) that is located so that
water is drawn from the desired level. Figure 3-8 shows a typical modern well
installation. The grout seal at the surface prevents possibly contaminated surface
runoff from following a low-resistance path along the casing and eventually
reaching the source aquifer. The gravel pack around the screen provides a local
zone of large hydraulic conductivity, which enhances flow to the well screen.

Unconfined Steady Flow

Equation (3-11) can be used to analyze axially symmetric flow into a
well in a manner similar to that used to analyze the previous cases of one-
dimensional flow. Figure 3-9 illustrates the flow conditions for a well that fully
penetrates the aquifer. The velocity of the flow (Darcy velocity) is steady and
equal to V at a distance r from the center of the well, and the depth of saturated
flow at that point is h. The original depth of groundwater is h_o. The radius of the
well is r_w, and the steady-state height of the water in the well is h_w. We will assume

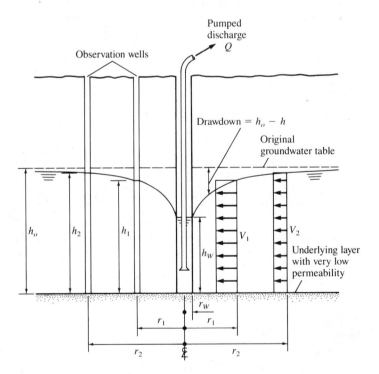

Figure 3-9 Axially-symmetric steady flow to a well
in an unconfined aquifer

that the aquifer is homogeneous and underlain by an impervious layer, but unconfined on top, and that the flow is steady.

These wells are normally relatively shallow. Equation (3-11) becomes

$$Q = 2\pi Krh \frac{dh}{dr} \tag{3-24}$$

where the area A has been replaced by $2\pi rh$, and since the positive direction of r is opposite to the positive direction of velocity, the gradient dh/dr is equal to $-dh/dL$. Equation (3-24) can be integrated as

$$Q \int_{r_1}^{r_2} \left(\frac{1}{r}\right) dr = 2\pi K \int_{h_1}^{h_2} h \, dh \tag{3-25}$$

to yield

$$Q = \frac{\pi K(h_2{}^2 - h_1{}^2)}{\ln\left(\dfrac{r_2}{r_1}\right)} \tag{3-26}$$

The *drawdown* of the water table is $h_o - h$ and is the vertical distance at a point r distant from the well centerline that pumping has lowered the water table from its original undisturbed level. The water surface formed by the drawdown has the form of a funnel or an inverted cone as shown in Fig. 3-9. This surface is called the *cone of depression*. The development of Eq. (3-26) has assumed that velocity is steady and that water continually moves toward the well from some undefined surrounding source so that the water level remains at a constant level h at a distance r from the well. In actuality, this seldom occurs. As we pointed out earlier, recharge of water in the aquifer comes from the infiltration of precipitation or from streams or lakes. Thus, the infiltration is a very unsteady (nearly stochastic) event and can vary significantly from year to year. If the well is pumped at a steady rate equal to the average recharge rate, the average water level at any point will drop during dry periods when the pumping rate is more than the recharge rate and rise during wet periods when the recharge rate is greater than the pumping rate. Since recharge is a slow process for most aquifers, the resulting fluctuation in groundwater levels is also a slow process.

If the average annual pumping rate is greater than the average annual recharge rate for the groundwater basin, the level of the groundwater and the volume in storage in the aquifer will continuously fall. This situation, common in many areas today, is referred to as *groundwater mining*. Because the absolute extent of an aquifer and its total recharge rate are difficult to determine with certainty, the rate of mining is also difficult to determine. In general, Eq. (3-26) gives reasonably accurate results if the maximum drawdown is not more than about one half the aquifer depth.

If Eq. (3-26) were used to estimate the water depth in the well, substituting r_w for r_1, then headloss due to flow through the well screen would need to be considered. A headloss occurs as water flows from the aquifer through the well screen and into the well. If r_w is substituted for r in Eq. (3-26) an estimate is obtained for the groundwater depth immediately outside the well casing. To obtain an estimate of the water depth in the well for a given Q, it is necessary to subtract the headloss from h_w the water depth outside the well. Figure 3-9 does not show this headloss.

To use Eq. (3-26) to estimate the hydraulic conductivity of an aquifer, information obtained during a pumping test is required. A well is constructed as shown in Fig. 3-8, and one or two smaller observation wells are drilled at different distances from the pumped well, such as r_1 and r_2 as shown in Fig. 3-9. The water levels are observed and recorded before pumping at a known rate begins, and the approximately steady-state values achieved after continuous pumping are substituted into Eq. (3-26) to obtain estimates of K. However, because the shape of the drawdown surface usually varies with time, values of K determined through using Eq. (3-26) must be considered approximations. If the rate of decline of the groundwater levels becomes slow, the indicated value of hydraulic conductivity can be reasonable. However, if the groundwater levels continue to fall rapidly, the flow is obviously very unsteady, and errors in the value of hydraulic conductivity, as determined from the steady-state equations, can be significant. Experience has shown that a concept called the radius of influence can often be invoked where the surface of the groundwater table at an appreciably large distance from the well (usually 500 r_w or approximately 1000 ft) is assumed to be at a constant elevation. Values of Q, h_1, and h_2 (measured during pumping) are substituted into Eq. (3-26) to calculate the hydraulic conductivity. This assumption of a radius of influence provides reasonably accurate values of the hydraulic conductivity if the observation well located at r_2 is sufficiently far from the pumped well and if conditions are such that the observed water table at r_2 is declining at a very slow rate.

The development of Eq. (3-26) tacitly assumed that the velocity is always horizontal, which is definitely not the case near the well. Moreover, velocities become large near the well and may actually reach turbulent rather than laminar-flow conditions, with the result that resistance to flow is greater, and the gradient dh/dr of the water table can be much steeper than is given by Darcy's equation. Thus, drawdown outside the well casing is always greater than that predicted by Eq. (3-26) although using the equation gives a reasonable approximation for an aquifer having large hydraulic conductivity such as would occur for coarse gravels. Using Eq. (3-26) to calculate ordinates of the groundwater surface will not yield accurate values for points near the well. However, for values of r greater than approximately $1.5h_o$, using Eq. (3-26) will usually provide satisfactory values for groundwater elevations.

EXAMPLE 3-3   A 48-in. diameter well ($r_w = 2$ ft) is drilled as shown in Fig. 3-9. The water table is initially at a depth of 300 ft above the bottom of the aquifer, which has a hydraulic conductivity of 0.80 ft/day. If 100 gal/min (19,251 ft³/day) is pumped from the well, what will be the depth of water in the well if the drawdown is essentially zero in the observation well 1000 ft from the well? What is the drawdown at a distance of 300 ft from the well?

SOLUTION   The aquifer is unconfined, so Eq. (3-26) can be used. Equation (3-26) is first solved for h_1. Because we have only one observation well, h_w must then be substituted for h_1. Thus, the depth in the well will be approximately

$$h_w = \left[(300)^2 - \left(\frac{19{,}251}{0.80\,\pi} \right) \ln\left(\frac{1000}{2} \right) \right]^{1/2} = 206 \text{ ft}$$

Note: The flow rate has been converted to ft³/day for use in Eq. (3-26) with the hydraulic conductivity in ft/day. The actual value of h_w will be somewhat less than 206 ft because of headloss through the screen.

Equation (3-26) can be rewritten to calculate the drawdown at 300 ft from the well.

$$h_o^2 - h^2 = \frac{Q}{\pi K} \ln\left(\frac{r_o}{r} \right)$$

Thus, $$h_o^2 - h^2 = \frac{19{,}251}{0.80\,\pi} \ln\left(\frac{1000}{300} \right) = 9222$$

$$h = [(300)^2 - 9222]^{1/2} = 284 \text{ ft}$$

The drawdown at 300 ft from the well is

$$h_o - h = 300 - 284 = 16 \text{ ft} \qquad \blacksquare$$

Confined Steady Flow

Equation (3-11) can also be used to analyze the behavior of the hydrostatic pressures (piezometric head) in a confined aquifer under pressure (artesian conditions), such as that illustrated in Fig. 3-10 on the next page. In this case, the aquifer is confined both above and below by essentially impermeable layers. Equation (3-11) becomes

$$Q = 2\pi K B r \frac{dh}{dr} \tag{3-27}$$

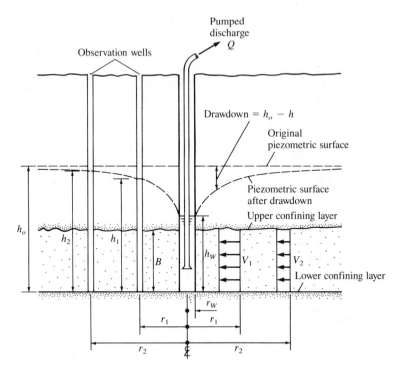

Figure 3-10 Axially symmetric steady flow to a well in a confined aquifer

which can be integrated as

$$Q \int_{r_1}^{r_2} \left(\frac{1}{r} \right) dr = 2\pi K B \int_{h_1}^{h_2} dh \tag{3-28}$$

or

$$Q = 2\pi K B \frac{h_2 - h_1}{\ln \dfrac{r_2}{r_1}} \tag{3-29}$$

where h_2 and h_1 refer to the piezometric heads at r_2 and r_1, respectively.

In this case, we have assumed that the depth of saturation does not change after pumping. Only the piezometric head declines in the area around the well, and we assume that it does not fall below the top of the aquifer. For the confined aquifer the drawdown $h_o - h$ forms a geometric surface which is the top of the piezometric head. The inverted cone formed by the drawdown is *analogous to the cone of depression* formed by the groundwater surface in Fig. 3-9, page 122.

In actuality, if pumping takes place at a rate greater than that at which recharge of the aquifer occurs, portions of the aquifer could be dewatered, and the conditions assumed for development of Eq. (3-29) would no longer exist.

However, a case like this can be at least approximately analyzed using an approach similar to that used to develop Eq. (3-22) for the flow conditions shown in Fig. 3-6, page 118.

Frequently, large reductions in piezometric head within the aquifer are produced by intensive groundwater pumping. The reduction in pressure results in compression of the aquifer by the weight of overburden, which originally was partially supported by the groundwater under pressure. For very thick aquifers, the amount of compression may be large and can result in significant subsidence of the land surface. Subsidence of more than 30 ft has been measured in areas such as the San Joaquin valley in California where extensive pumping has taken place (10). The withdrawal of oil from oil-bearing strata has produced the same effect in some locations, such as near Long Beach, California, where subsidence of the coastal area has been enough to result in flooding by Pacific Ocean tides (10).

EXAMPLE 3-4 Determine the hydraulic conductivity of a confined aquifer from which water is being pumped by a fully penetrating well. The aquifer is 100 ft thick, and the well is being pumped at a rate of 1500 gal/min. Water surfaces in two observation wells 700 ft and 70 ft from the pumped well are respectively 1 ft and 10 ft below their levels prior to the beginning of pumping.

SOLUTION Since the aquifer is fully confined, Eq. (3-29) may be rearranged to obtain

$$K = \frac{Q \ln (r_2/r_1)}{2\pi B(h_2 - h_1)}$$

In order to maintain consistent units, the 1500 gal/min must be divided by 7.48 gal/ft^3.

Thus, $K = \dfrac{(1500/7.48) \ln (700/70)}{2\pi(100)(9)} = 0.0816 \text{ ft/min} = 118 \text{ ft/day}$ ■

Unsteady Well Hydraulics

As we noted in the previous section, drawdown of an aquifer is not a steady-state occurrence. Even in very infrequent situations where all the conditions assumed for the development of Eq. (3-26) exist, a long period of pumping would be required before the cone of depression around the well would reach a steady-state shape. To obtain accurate estimates of hydraulic conductivity for conditions where the flow is appreciably unsteady, analyzing the flow to the well under unsteady conditions is necessary. Theis analyzed this situation in 1935 (18).

Unsteady flow in a confined aquifer can be formulated by inserting Darcy's Law into the continuity equation. The resulting equation which is a partial differential equation with time as an independent variable can be solved exactly for the case of a single fully penetrating well having a constant pumping rate. The solution is obtained in the form of an exponential integral and can be expressed as follows (18):

$$h_o - h = \left(\frac{Q}{4\pi KB}\right) W(u) \tag{3-30}$$

where $h_o - h$ = drawdown at a point r distance from the well

$\quad\quad Q$ = pumping rate

$\quad\quad B$ = aquifer thickness

$\quad\quad K$ = hydraulic conductivity

$\quad\quad W(u)$ = a dimensionless mathematical function called the well function

The well function $W(u)$ which is dimensionless can be expressed in a series as

$$W(u) = -0.577216 - \ln u + u - \frac{u^2}{2 \cdot 2!} + \frac{u^3}{3 \cdot 3!} - \frac{u^4}{4 \cdot 4!} + \cdots \tag{3-31}$$

where the argument u is

$$u = \left(\frac{S}{4KB}\right)\frac{r^2}{t} \tag{3-32}$$

and r = distance between the pumped and observation wells

$\quad\quad t$ = time after pumping starts

$\quad\quad S$ = a dimensionless constant called the *storage coefficient*

In Eqs. (3-30) and (3-32), the units for K, Q, S, r, and t must be chosen consistently. Thus, if Q is measured in ft^3/sec, K should be measured in ft/sec, r measured in ft, t measured in sec, and B in ft. Various forms of Eq. (3-30) and (3-32) have been published where the units are not consistent, and constants of proportionality are required.

The dimensionless constant S, called the storage coefficient, is defined as the volume of water yielded by a prism of the aquifer having a projected unit area in the horizontal plane if withdrawal is sufficient to cause a unit drop in the hydrostatic head. For an unconfined aquifer, the drained water comes primarily from gravity drainage of the aquifer, and for this case, the storage coefficient is essentially equal to the specific yield Sy.

For the case of a confined aquifer, the volume of water draining out to produce a unit drop in head comes entirely from expansion of the water and compression of the aquifer. The volume expansion of water for a unit change in head is very small. For a confined aquifer having a porosity of 0.2 and containing water at 15°C, approximately 3×10^{-7} m^3 of water are forced out of 1 m^3 of aquifer for a 1 m decline in head. If the aquifer were 100 m thick, the storage coefficient would be

$$S = 100(3 \times 10^{-7}) = 3 \times 10^{-5}$$

The storage coefficient of most aquifers ranges from 0.00001 to 0.001. The more compressible the aquifer, the larger the value of the storage coefficient.

In general, it is necessary to solve Eqs. (3-30), (3-31), and (3-32) for the hydraulic conductivity K and the storage coefficient S for a set of measurements of h at a given distance r from the well for periods of time t after pumping begins at a constant rate Q. Theis devised a convenient graphical method by which the solution can be achieved. The graphical solution utilizes the fact that the terms in parentheses in Eqs. (3-30) and (3-32) are constants for a given aquifer and an observation well. As a result, graphs of $W(u)$ versus u and $(h_o - h)$ versus (r^2/t) will have geometrically similar shapes. In the graphical solution, graphs of $W(u)$ versus u are prepared using Eq. (3-31) or Table 3-2, which presents calculated values of $W(u)$ as functions of u. The measured values of $h_o - h$, r, and t are used to separately plot the graph of $(h_o - h)$ versus (r^2/t). The two graphs are then superimposed and shifted as necessary, holding the coordinate axes parallel, until the curves are coincident. A point on the coincident curves is selected, and numerical values of $W(u)$, u, $h_o - h$, and r^2/t are

Table 3-2 Values of $W(u)$ (6)

u	1.0	2.0	3.0	4.0	5.0	6.0	7.0	8.0	9.0
$\times 1$	0.219	0.049	0.013	0.0038	0.0011	0.00036	0.00012	0.000038	0.00
$\times 10^{-1}$	1.82	1.22	0.91	0.70	0.56	0.45	0.37	0.31	0.26
$\times 10^{-2}$	4.04	3.35	2.96	2.68	2.47	2.30	2.15	2.03	1.92
$\times 10^{-3}$	6.33	5.64	5.23	4.95	4.73	4.54	4.39	4.26	4.14
$\times 10^{-4}$	8.63	7.94	7.53	7.25	7.02	6.84	6.69	6.55	6.44
$\times 10^{-5}$	10.94	10.24	9.84	9.55	9.33	9.14	8.99	8.86	8.74
$\times 10^{-6}$	13.24	12.55	12.14	11.85	11.63	11.45	11.29	11.16	11.04
$\times 10^{-7}$	15.54	14.85	14.44	14.15	13.93	13.75	13.60	13.46	13.34
$\times 10^{-8}$	17.84	17.15	16.74	16.46	16.23	16.05	15.90	15.76	15.65
$\times 10^{-9}$	20.15	19.45	19.05	18.76	18.54	18.35	18.20	18.07	17.95
$\times 10^{-10}$	22.45	21.76	21.35	21.06	20.84	20.66	20.50	20.37	20.25
$\times 10^{-11}$	24.75	24.06	23.65	23.36	23.14	22.96	22.81	22.67	22.55
$\times 10^{-12}$	27.05	26.36	25.96	25.67	25.44	25.26	25.11	24.97	24.86
$\times 10^{-13}$	29.36	28.66	28.26	27.97	27.75	27.56	27.41	27.28	27.16
$\times 10^{-14}$	31.66	30.97	30.56	30.27	30.05	29.87	29.71	29.58	29.46
$\times 10^{-15}$	33.96	33.27	32.86	32.58	32.35	32.17	32.02	31.88	31.76

recorded for that point. The aquifer parameters are then determined from the following equations derived from Eqs. (3-30) and (3-32):

$$\frac{Q}{4\pi KB} = \frac{h_o - h}{W(u)} \tag{3-33}$$

$$\frac{S}{4KB} = \frac{u}{r^2/t} \tag{3-34}$$

Rearranging Eqs. (3-33) and (3-34) yields

$$KB = \frac{Q}{4\pi}\left(\frac{W(u)}{h_o - h}\right) \tag{3-35}$$

and $$S = 4KB\left(\frac{u}{r^2/t}\right) \tag{3-36}$$

In these equations, the product KB is frequently used as a single variable called *transmissivity*. Special forms of the Theis method have been developed for wells that only partially penetrate the aquifer, for wells in unconfined aquifers, and for wells in leaky confined aquifers (3, 5, 8, 11, and 20).

A solution for unsteady flow to a well in an unconfined aquifer is difficult because the depth of groundwater flow is not constant but decreases with both time and nearness to the well. Moreover, the streamlines become steeper near the well, making the velocity deviate from the horizontal. If, however, the drawdown $h_o - h$ is small compared to h_o, the Theis method gives approximate but adequate results for the unconfined aquifer.

A more simple version of Eq. (3-30) can be written for the particular case where u has a value less than 0.01. For this condition only, the first two terms of Eq. (3-31) are important. Substituting those two terms into Eq. (3-30) and substituting Eq. (3-32) for u gives

$$h_o - h = \frac{Q}{4\pi KB}\ln\frac{2.25KBt}{r^2 S} \tag{3-37}$$

Equation (3-37) can be satisfactorily used to directly evaluate the transmissivity KB from pumping-test data as long as values of u do not exceed 0.01, a condition that will be more common for very thick aquifers or aquifers having large values of hydraulic conductivity.

EXAMPLE 3-5   Determine the transmissivity (KB) and the storage coefficient S for a confined aquifer from which 500 gal/min is being pumped. The water level measurements shown in the accompanying table were made at an observation well 100 ft from the pumped well.

(1) Time (hr)	(1) Time (min)	(2) Distance to Water Level (ft)	(3) Drawdown (ft)	(4) (r^2/t)
0	0	10.2	0	
1	60	11.7	1.5	166.67
2	120	13.5	3.3	83.33
4	240	24.0	13.8	41.67
6	360	31.5	21.3	27.78
8	480	37.8	27.6	20.83
12	720	46.7	36.5	13.89

SOLUTION Drawdown is computed as the distance from the initial water level, as column 3 of the accompanying table shows. Using the given data on time and drawdown and the given 100 ft radius, values of r^2/t are computed, as shown in column 4, for each measured water level. As shown in the second table values of the argument u are chosen arbitrarily. Values of $W(u)$ are calculated using Eq. (3-31) or, for the chosen values of the argument u, Table 3-2.

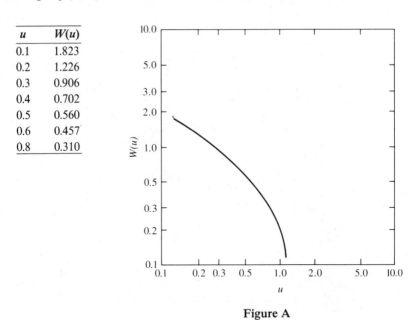

u	$W(u)$
0.1	1.823
0.2	1.226
0.3	0.906
0.4	0.702
0.5	0.560
0.6	0.457
0.8	0.310

Figure A

The values of $W(u)$ and u are then plotted on log-log paper, as shown in Fig. A, whereas values of drawdown $(h_o - h)$ are plotted against r^2/t, as in Fig. B. The two figures are then superimposed and positioned until the curves closely coincide, as in Fig. C.

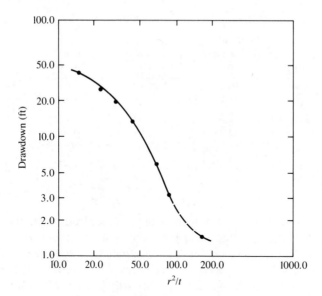

Figure B

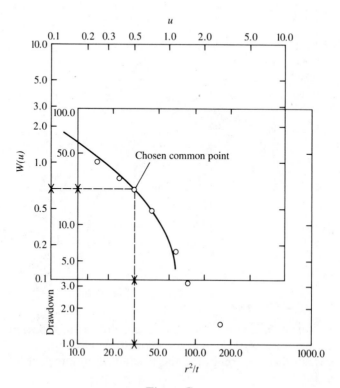

Figure C

For the common point shown in Fig. C, the following values of the variables are read

$$u = 0.500, \qquad W(u) = 0.59, \qquad \frac{r^2}{t} = 28.0, \qquad h_o - h = 22$$

Equation (3-35) then yields

$$KB = \frac{(500/7.48)(0.59/22)}{4\pi} = 0.143 \text{ ft}^2/\text{min}$$

and Eq. (3-36) gives

$$S = \frac{4(0.143)(0.500)}{28} = 0.010 \qquad\qquad \blacksquare$$

3-6 Pumping Tests

The most important aspect of quantitatively predicting the movement of groundwater or in determining the yield of a well is accurate knowledge of the aquifer characteristics, including the hydraulic conductivity K, the aquifer thickness B, and the storage coefficient S. In the previous section, Examples 3-4 and 3-5 presented simple cases where drawdown and discharge data were used to estimate aquifer characteristics. The basic tool for collecting this data is the pumping test, which consists of pumping directly from one well while water levels are recorded in one or more observation wells. Figure 3-11, page 134, shows a pumping test in operation.

Observation wells are drilled solely for observing the drawdown as pumping of the test well proceeds. They are often referred to as piezometers, since their sole function is to monitor piezometric levels of the groundwater. Having at least three observation wells each at different distances from the pumped well is desirable. For unconfined aquifers, the observation wells should be at least 1.5 times the aquifer thickness away from the pumped well to avoid error due to the effects of nonhorizontal velocity. In general, it is desirable to space the wells approximately at 1.5, 2, and 4 times the thickness of the aquifer from the test well. Financial considerations will often limit the number of observation wells that can be constructed. In many cases, there may be existing wells nearby that can be used as observation wells.

Determining K, B, and S from pumping-test data is based on the following assumptions, most of which are also inherent in the development of the steady and unsteady well-hydraulic equations:

1. The aquifer is homogeneous, isotropic, and of infinite horizontal extent.
2. The flow in the aquifer is horizontal only.

Figure 3-11 Pumping test in progress.
Pump drive is in left background. Tank in foreground
is used as volume measurement tank for use in
measuring pumping rate. (Photo courtesy of U.S.
Geological Survey, Boise, Idaho)

3. For unsteady conditions, water is released from storage ın the aquifer in immediate response to a drop in piezometric head.
4. The only flow in the aquifer is that produced by the well.
5. Pumping is at a constant rate.
6. The volume of water initially in the well that is removed by pumping during unsteady conditions is much smaller than the volume coming from the aquifer.
7. The well completely penetrates the aquifer and is open to inflow throughout the entire depth of the aquifer.

For steady-state conditions Eqs. (3-26) and (3-29) are used to compute the hydraulic conductivity of the aquifer for unconfined and confined conditions, respectively. In theory, of course, the water levels in the observation wells will never reach equilibrium. However, they often approach steady-state conditions sufficiently close to yield accurate estimates of K. If at least two observation wells are used, an inherent advantage arises: Even though the water table is falling, the difference in drawdown between the two observation wells will be nearly constant. Under these conditions, Eqs. (3-26) or (3-29) will give reasonably constant values of KB for each of the simultaneous drawdown measurements.

Once KB has been determined using the steady-state equations, Eqs. (3-30), (3-31), and (3-32) can be used to calculate the storage coefficient S. With a measured value of drawdown for an observation well at a particular time, Eq. (3-30) can be solved for $W(u)$. Table 3-2 can then be entered with the value of $W(u)$ to obtain an associated value of u. Using the resulting value of u, the computed KB, and the associated values of r and t, Eq. (3-32) can then be solved for S.

The procedures developed by Theis, and demonstrated in Example 3-5, should be used whenever unsteady effects are important and the various drawdown observations do not produce consistent values of KB using the pertinent steady-state equations.

Several techniques have been developed for approximately evaluating local aquifer characteristics using so-called rate of use or slug tests. In these techniques, a volume of water is quickly removed from the well, and the rate of rise of the water surface in the well is carefully observed after the water removal. The rate of rise can then be related to the local value of the hydraulic conductivity. These techniques are much cheaper and quicker to perform than pumping tests and are valuable for use in preliminary investigations. Bouwer (2) describes these tests and their use in detail.

3-7 Recharging or Injection Wells

In areas where recharge of aquifers is desirable, it may sometimes be accomplished by excavation of surface material to expose the aquifer and directing surface runoff to it. Alternatively, a well may be used to inject water into the aquifer. Wells are frequently used for injecting water (and sometimes steam) into oil-bearing strata to aid in recovering oil and to prevent subsidence of the aquifer and the overlying ground. Figure 3-12, on the next page, shows a well in an unconfined aquifer into which water is being injected at a rate Q. Assume that the well fully penetrates the aquifer and that the screen is long enough to allow flow into the full depth of the unbounded aquifer. Darcy's law is used to analyze the steady-state flow as in Sec. 3-5, for the well pumping from the unconfined aquifer. The following equation arises from Eq. 3-26:

$$h = \left[h_w{}^2 - \frac{Q}{\pi K} \ln \left(\frac{r}{r_w} \right) \right]^{1/2} \tag{3-38}$$

Equation (3-38) can be used to calculate the coordinates of the mounding by solving the equation for h for chosen values of r. As was the case in using Eq. (3-26), the calculated ordinates will be approximate near the well, where the actual velocities are not horizontal. The rise of the groundwater surface at any point is $h - h_o$ and the resulting surface can be called the *cone of mounding*.

If pumping is stopped, the flow becomes unsteady and the mound gradually flattens out, approaching the level h_o after a theoretically infinite time. Obviously, for the real case of a bounded aquifer, the groundwater table will rise above h_o reflecting the volume of recharge.

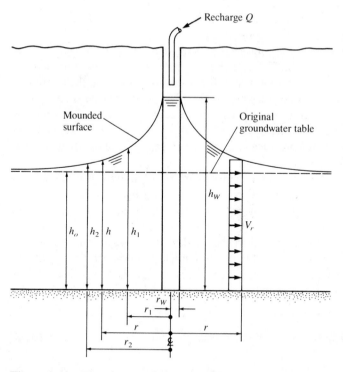

Figure 3-12 Flow from an injection or recharge well to an unconfined aquifer

Recharging a confined aquifer can be readily analyzed by the method used to analyze a pumping well. The following equation is formed:

$$h = h_w - \frac{Q}{2\pi KB} \ln\left(\frac{r}{r_w}\right) \tag{3-39}$$

where h again refers to piezometric head, as in Eq. (3-29).

3-8 Boundaries of Aquifers

In the analysis of steady and unsteady well hydraulics, the assumption was tacitly made that the aquifer extended an infinite distance radially in all directions. In actuality, although some aquifers are very large, none totally satisfy the assumption. All aquifers are bounded both in the horizontal and vertical directions. Layers of low-hydraulic conductivity can exist both above and below the aquifer and may effectively cut off vertical flow. Horizontal flow is affected

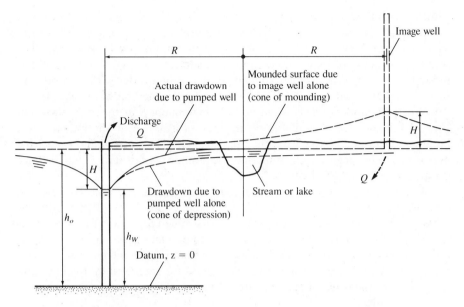

Figure 3-13   Analysis of drawdown for a pumped well near a recharging body of water

by at least two types of boundaries. A recharge boundary is one where water enters the aquifer providing a source of supply. Figure 3-13 illustrates a case where the aquifer is recharged by a perennial stream or lake. If a well is located as shown in Fig. 3-13, pumping will cause drawdown that will ultimately reach the river. The continuous pumping induces recharge from the stream, and the cone of depression cannot fall below the water surface at the stream as long as the streamflow is sufficient to satisfy flow to the aquifer. Such a stream is commonly referred to as a "losing" stream. In some cases, the stream can be dried up by flow to the aquifer.

Drawdown around a well near a recharging body of water can be analyzed using a technique known as the method of images (13). In the analysis, the river is treated as a line source, and a recharge well (negative image of a pumped well) is placed across the river and at the same distance from the river as the pumped well, as in Fig. 3-13. The final steady-state drawdown at any point is obtained by adding the drawdown calculated for the image well (negative) and the drawdown calculated for the pumped well. Since both wells are equidistant from the plane of symmetry (a vertical plane along the river), the cone of depression due to the pumped well is as far below the river surface as the cone of mounding due to the image well is above the river surface. Thus, when image-well and pumped-well drawdowns (calculated with Eqs. (3-38) and (3-26), respectively) are added, the calculated cone of depression due to the pumped well will coincide with the river surface along the centerline of the river. Final drawdown values at any point between the well and the stream are computed by algebraically adding values of image-well mounding and pumped-well drawdown. The resulting ordinates of the actual groundwater surface are calculated by subtracting the final drawdown values from the initial groundwater surface ordinates. More complex situations can be analyzed using the method of images (4).

When an aquifer ends at an impermeable boundary, such as that shown in Fig. 3-14, less water is available on the side of a well near the impermeable boundary than is available from the opposite side. Thus, one would expect the drawdown to be greater at the impermeable boundary than at an equal distance on the opposite side of the well.

Analysis of this situation is again made using an image well. However, in this case, a pumped well is used for the image well. Using Eq. (3-26), the ordinates of the cones of depression of the two wells are added algebraically, which produces twice the depression at the impermeable boundary as would be calculated for the pumped well alone. Coordinates of the resulting groundwater surface are computed by adding the drawdowns computed for each well and subtracting the sum from the original groundwater elevation.

3-9   Well Fields

If wells are constructed near one another, the drawdown of each well will be affected by the other wells in a manner that produces increased drawdown. The effect is illustrated in Fig. 3-15, where the drawdown curves of the wells overlap.

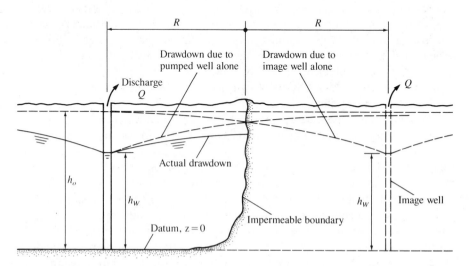

Figure 3-14   Analysis of drawdown for a pumped well near an impermeable barrier

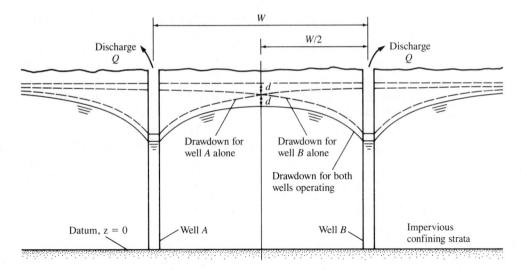

Figure 3-15   Determination of resultant drawdown
for two pumped wells with mutual interference

To calculate the drawdown due to both wells at any point, an independent
drawdown is calculated (using Eq. (3-26) for an unconfined aquifer or Eq. (3-29)
for a confined aquifer) for each well neglecting the effect of any other well. The
actual drawdown at any point is then calculated as the algebraic sum of the
calculated independent drawdown values for that point. Fig. 3-15 graphically
illustrates this process for two wells, but it can readily be applied to any number
of wells. In general, the total drawdown predicted by this method gives accurate
results provided the total drawdown does not exceed approximately one half
the height of the original piezometric head. This restriction is usually stated
because the combination of limitations assumed in using Darcy's law, Eq. (3-11),
may be violated for greater drawdown.

EXAMPLE 3-6   Calculate the drawdown half way between the two wells
shown in Fig. 3-15 if the wells are 300 ft apart, the original groundwater table is
210 ft above the bottom of the aquifer, and the hydraulic conductivity of the
aquifer is 1.0 ft/day. Each well is pumped at a rate of 200 gal/min.

SOLUTION   Equation (3-26) can be used to calculate the drawdown for
each well. Thus, for one well, we can write

$$h_o^2 - h^2 = \frac{Q}{\pi K} \ln\left(\frac{r_o}{r}\right)$$

where r_o is the distance outward to where the groundwater table is at the original elevation h_o. In theory, r_o is at infinity, and we have already noted that h_o will decrease slowly with time if pumping exceeds recharge. However, in actual practice r_o will usually not be more than approximately 1000 ft. Equation (3-26) is not very sensitive to the value used for r_o, so we will assume 1000 ft.

Thus, for one well alone, the depth of the groundwater at a distance of 150 ft from the well will be

$$h^2 = h_o{}^2 - \frac{(200 \text{ gal/min})(60 \text{ min/hr})(24 \text{ hr/day})}{(7.48 \text{ gal/ft}^3)(1 \text{ ft/day})\pi} \ln\left(\frac{1000}{150}\right)$$

$$h^2 = (210)^2 - 23{,}251 = 20{,}849$$

$$h = 144 \text{ ft}$$

The drawdown due to one well is thus $d = 200 - 144 = 56$ ft. The drawdown due to the two wells will be $d + d = 56 + 56 = 112$ ft. ∎

3-10 Groundwater Recharge and Safe Yield

As we mentioned, the extent of a groundwater aquifer is difficult to determine with certainty. A detailed study of the geologic structure must be made to determine if and where the aquifer, if it is confined, strikes the surface. Since some aquifers may extend for several hundred miles, such as the Ogallala aquifer in Kansas and Oklahoma, the outcrops may be at a great distance from the point of interest. Even confined aquifers receive some recharge from infiltration at the earth's surface, part of which will eventually leak through the upper confining strata and into the aquifer.

A water balance equation for all water in a drainage area can be written as

$$Q_P \cdot \Delta t = (Q_G - Q_s - E_T \cdot A_s) \cdot \Delta t + P \cdot A_s - \Delta S_G - \Delta S_S \qquad (3\text{-}40)$$

where Q_P = average rate of groundwater withdrawal from the aquifer during the time interval Δt

P = average precipitation depth over the drainage area during the time interval Δt

A_s = surface drainage area contributing to the aquifer

Q_s = average rate at which surface flow leaves the area during Δt

E_T = average evapotranspiration rate over the drainage area during Δt

Q_G = rate at which groundwater flows into the area

ΔS_G = change in volume of stored groundwater during Δt

ΔS_S = change in surface water storage during Δt

All terms in Eq. (3-40) are difficult to evaluate with accuracy. The average precipitation P and the streamflow Q_s must be determined by direct measurement using several precipitation and streamflow recording stations to provide the necessary accuracy. The change in surface-water storage volume must be determined by developing surface-area elevation information for all ponds, lakes, and impoundments in the area and installing stage recorders in each. The change in groundwater volume is still more difficult to assess. The extent of the aquifer must be defined through geologic mapping. Water-surface elevations must be monitored in all wells, and the change in stored volume calculated from the change in groundwater elevation integrated over the area of the aquifer. Proper consideration must of course be given to the specific yield or the storage coefficient for the aquifer.

Knowing the rate at which groundwater flows into the basin is necessary because surface drainage areas do not always coincide with the groundwater drainage area. Inflow of groundwater can be assessed only if the inflowing aquifers are defined and wells are drilled to determine the groundwater gradient, the hydraulic conductivity of the material in the aquifer, and the cross-sectional area of the aquifer.

The area-average evapotranspiration rate E_T can only be estimated. An average rate must usually be estimated based on assumed parameters for the different types of vegetation. Frequently, evapotranspiration rates are assumed to equal evaporation rates, and evaporation pans are located and monitored within the drainage area.

The term *safe yield* is defined as the rate at which groundwater can be pumped from the basin without endangering the groundwater supply. Since, as we have pointed out, precipitation is erratic, recharge and groundwater levels are also erratic. Figure 3-16 shows the fluctuation in groundwater level at an observation well and precipitation at a nearby station. Recharge of the aquifer

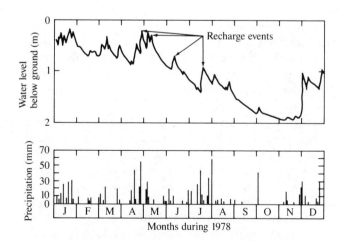

Figure 3-16   Fluctuations of observed water table at an observation well on the coastal plain of North Carolina (6)

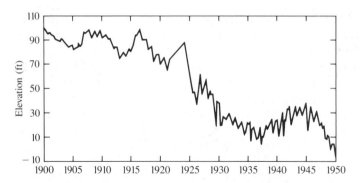

Figure 3-17   Long-term declines in groundwater
levels in the Santa Ana River Basin in California* (14).
* The short-term fluctuations are the result of changes in
   pumping rate.

was erratic, but discharge occurred continuously, as is shown in Fig. 3-16 by
the declining water surface from August through November when relatively
little rain fell. Observations made over several years will show that average
levels fluctuate significantly from year to year depending on the amount of
average annual rainfall. Moreover, the long-period observations, like those
shown in Fig. 3-17, will exhibit continuous declines in average water-table levels
when average pumping rates exceed long-term average recharge.

   The safe yield of an aquifer is generally considered a matter of economics.
If declines in groundwater levels, which result in deeper and more expensive
wells and greater energy costs for pumping, do not make the use of the water
uneconomical, pumping will likely continue despite the declines. However,
groundwater levels will continue to decline where pumping rates exceed recharge rates; eventually, the stored groundwater can be exhausted or levels can
drop so far that the pumping of groundwater becomes so expensive that only
municipal and industrial users may be able to economically justify continued
use. Thus, safe yield is a matter of definition. Groundwater cannot be withdrawn
indefinitely at greater than the recharge rate. Temporary withdrawal rates
greater than recharge rates are certainly permissible if fluctuation of the groundwater table is acceptable.

   As we noted, many areas of the United States have experienced ground-
water-table drawdowns, producing important concerns about the future availability of groundwater supply. As a result, recharge wells (see Sec. 3-7) or
spreading basins have been constructed to help recharge the depleted aquifers.
Figure 3-17 shows the historical declines in the groundwater levels in the Santa
Ana River basin in Southern California. To help recharge the groundwater,
large spreading basins were constructed along the Santa Ana River. Surface
runoff is impounded in these spreading basins and allowed to infiltrate the
underlying aquifer. Some of these basins are shown in Fig. 3-18 during a period
when recharge was occurring.

Figure 3-18 Spreading basins on the Santa Ana
River for recharging the underlying groundwater
basin. (Photo courtesy of the Orange County
Water District, Fountain Valley, California)

3-11 Other Analytical Tools

The analytical solutions considered in Secs. 3-4, 3-5, and 3-6 resulted
in ordinary differential equations that could be solved in closed form. However,
the general analysis of groundwater flow must consider significantly more complicated flow problems that may be two-dimensional, three-dimensional, and
unsteady. In these cases, numerical solutions are often the only ones possible. A
wide variety of programs for digital computers (so-called groundwater models)
have been developed for special purposes (1, 9, 12).

The development of these programs is beyond the level of this text. However, to illustrate the type of mathematical analysis required and the general
concept of groundwater modeling, we will consider briefly the numerical
modeling.

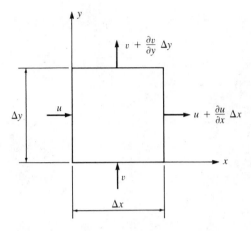

Figure 3-19   Control volume for two-dimensional flow in a horizontal plane.

To develop the general partial differential equation for flow of groundwater, we will consider two-dimensional flow in a horizontal plane. Figure 3-19 illustrates a control volume in the x–y plane. For steady flow, the rate of flow into and out of the control volume must be equal. Thus,

$$u\,\Delta y + v\,\Delta x - \left(u + \frac{\partial u}{\partial x}\,\Delta x\right)\Delta y - \left(v + \frac{\partial v}{\partial y}\,\Delta y\right)\Delta x = 0 \qquad (3\text{-}41)$$

where u and v are the velocity components and $\partial u/\partial x$, and $\partial v/\partial y$ are their rates of change in the x and y directions, respectively. Equation (3-41) reduces to

$$-\frac{\partial u}{\partial x}\,\Delta x\,\Delta y - \frac{\partial v}{\partial y}\,\Delta x\,\Delta y = 0$$

or
$$\frac{\partial u}{\partial x} + \frac{\partial v}{\partial y} = 0 \qquad (3\text{-}42)$$

which is the familiar differential form of the equation of continuity. Equation (3-10), Darcy's equation, tells us that the velocity components u and v can be expressed as

$$u = -K\frac{\partial h}{\partial x} \qquad (3\text{-}43)$$

and
$$v = -K\frac{\partial h}{\partial y} \qquad (3\text{-}44)$$

Substituting Eqs. (3-43) and (3-44) into Eq. (3-42) then yields

$$\frac{\partial^2 h}{\partial x^2} + \frac{\partial^2 h}{\partial y^2} = 0 \tag{3-45}$$

as long as the hydraulic conductivity K is a constant throughout the aquifer. Equation (3-45) is a classic partial differential equation known as the Laplace equation. It is linear in the piezometric head h, and its solution depends solely on the values of h on the boundaries of the flow field in the x–y plane. Thus, the value of h at any point within a flow field can be determined uniquely in terms of the values of h on the boundaries.

Many classic solutions have been developed for Eq. (3-45) for the flow of groundwater (2, 3, 20). Moreover, the same equation arises in many other areas of interest such as hydrodynamics, elasticity, electricity, and the flow of heat (19). You may also find classical solutions to this equation in several texts on applied mathematics (15, 16).

Equations (3-43) and (3-44) show that the velocity of flow is normal to the lines of constant piezometric head. In the derivations of Eqs. (3-26) and (3-29) for flow to wells from unconfined and confined aquifers, respectively, we have used this fact but without calling attention to it. For example, in Figs. 3-5, 3-6, 3-7, 3-9, and 3-10, the velocity is in the plane of the page, and we assumed it to be horizontal in our analyses. The velocity is normal to lines of constant h, which in Figs. 3-5, 3-6, and 3-7, would be normal to the page. Figure 3-20 is an example of a hypothetical flow in which the velocity vector V has the x and y components u and v, respectively, and is normal to the lines of constant h. V is directed in the direction of decreasing h, which agrees with our formulation of Darcy's equation.

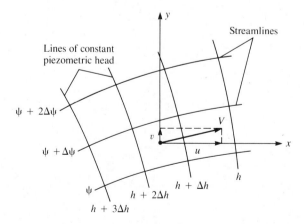

Figure 3-20 A hypothetical flow net for groundwater flow in the x–y plane.

If lines are drawn tangent to the velocity vector at every point in a steady flow field, a set of lines called *streamlines* is formed that are normal to the lines of constant piezometric head. This family of lines (streamlines) is called the streamfunction ψ, and in steady flow, the streamlines mark the paths of the flowing particles of fluid.

Because the streamlines are everywhere normal to the lines of constant piezometric head, the velocity components can be expressed in terms of the streamfunction ψ as

$$u = \frac{\partial \psi}{\partial y} \tag{3-46}$$

$$v = -\frac{\partial \psi}{\partial x} \tag{3-47}$$

Substituting Eqs. (3-46) and (3-47) into Eq. (3-42) yields

$$\frac{\partial^2 \psi}{\partial x^2} + \frac{\partial^2 \psi}{\partial y^2} = 0 \tag{3-48}$$

which again is the Laplace equation, this time expressed in terms of the streamfunction rather than the piezometric head.

Since streamlines are lines that are everywhere tangent to the velocity vectors, there can be no flow across the streamlines in steady flow, and the rate of flow is constant between any two streamlines. Figure 3-21 illustrates the relationship between the streamfunction and the velocity. We can determine

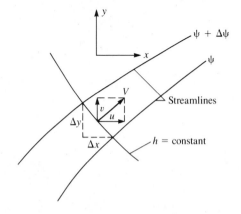

Figure 3-21   Definition sketch for relating velocity and discharge to the stream function

the rate of flow between the streamlines as follows where the flow field is assumed to have a depth of 1:

$$v \, \Delta n = u \, \Delta x - v \, \Delta y \tag{3-49}$$

or $\qquad u \, \Delta x - v \, \Delta y = \Delta q$ (3-50)

where q is the unit discharge, Δn is the spacing of the streamlines along a line normal to the velocity vector, and Δx and Δy are the projections of the normal increment Δn on the x–y plane. In differential form, Eq. (3-49) becomes

$$u \, dx - v \, dy = dq \tag{3-51}$$

Substituting Eqs. (3-46) and (3-47) for u and v, respectively, we have

$$\frac{\partial \psi}{\partial x} \, dx + \frac{\partial \psi}{\partial y} \, dy = dq \tag{3-52}$$

Equation (3-52) implies that

$$dq = d\psi \tag{3-53}$$

or that in two-dimensional flow, the value of the streamfunction is numerically equal to the unit discharge, and that the increment in unit discharge between two streamlines is equal to the change in the value of the streamfunction between two streamlines.

The use of streamlines, lines of constant piezometric head, and the definition of the streamfunction provide a graphic means of illustrating complex two-dimensional groundwater flows. For example, Fig. 3-22 on the next page shows the plan view of flow toward a well in an infinite aquifer.

In this case, the radial lines are the streamlines and show the direction of motion, whereas the circular lines are the lines of constant piezometric head. Figure 3-22 is actually a plan view of the flow pattern shown in Fig. 3-10.

A much more complex flow would be created by a well in an aquifer where groundwater is moving at a constant velocity in the x direction. Figure 3-23 on page 149 shows the resulting streamline and constant piezometric head lines for such a flow. Without the well, the streamlines would be straight lines parallel to the x-axis, and the lines of constant piezometric head would be straight and parallel to the y-axis. Actually, the flow pattern shown in Fig. 3-23 was formulated by superimposing the streamline pattern of Fig. 3-22 onto the streamline pattern for a uniform flow field. Known as the method of superposition, this technique can be readily used to analyze well fields (13), and in Secs. 3-8 and 3-9, it was used to analyze well fields, wells near a river, and wells near an impermeable barrier.

Equation (3-48) provides the basis for numerical solution of complex two-dimensional groundwater flows. It can be expressed as a finite-difference or

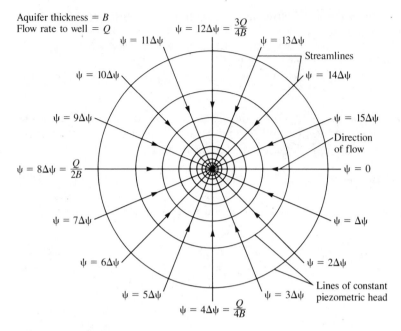

Figure 3-22   Streamlines and lines of constant piezometric head for flow to a well in an infinite aquifer

finite-element approximation (3). Figure 3-24 shows a finite-difference network in the x–y plane. In the finite-difference approximation, values of ψ or h will be known at points all around the outside boundary of the flow field. To develop a solution that will give velocities and directions of flow, the value of ψ or h at each interior node in the field must be determined. For convenience, we will consider only the streamfunction ψ in our development, but the same analysis could be performed using only h. To formulate the finite-difference equation, the partial derivatives $\partial^2\psi/\partial x^2$ and $\partial^2\psi/\partial y^2$ must be approximated numerically. This is done by assuming that the variation in ψ is linear between nodes (intersections of the x- and y-coordinate lines) in the field, an assumption whose accuracy depends on the spacing of the finite-difference grid. Thus, the first-order differential $\partial\psi/\partial x$ is approximated at a point halfway between point i, j and point $i + 1, j$ (Fig. 3-24) as

$$\frac{\partial\psi}{\partial x} = \frac{\psi_{i+1,j} - \psi_{i,j}}{\Delta x} \tag{3-54}$$

Aquifer thickness $= B$
Pumped discharge from well $= Q$

Direction of flow

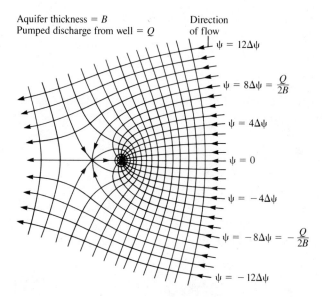

$\psi = 12\Delta\psi$

$\psi = 8\Delta\psi = \dfrac{Q}{2B}$

$\psi = 4\Delta\psi$

$\psi = 0$

$\psi = -4\Delta\psi$

$\psi = -8\Delta\psi = -\dfrac{Q}{2B}$

$\psi = -12\Delta\psi$

Figure 3-23   Streamlines and lines of constant piezometric head for a pumped well in a uniform groundwater flow parallel to the x axis

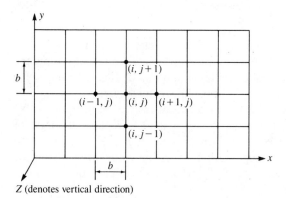

Z (denotes vertical direction)

Figure 3-24   Square finite difference mesh for development of Eq. (3-59)

A similar expression formulated at a point halfway between $i - 1, j$ and i, j is

$$\frac{\partial \psi}{\partial x} = \frac{\psi_{i,j} - \psi_{i-1,j}}{\Delta x} \tag{3-55}$$

The second-order differential at point i, j formulated in the x direction as the rate of change in $\partial \psi / \partial x$ is as follows

$$\frac{\partial^2 \psi}{\partial x^2} = \left[\frac{\psi_{i+1,j} - \psi_{i,j}}{\Delta x} - \frac{\psi_{i,j} - \psi_{i-1,j}}{\Delta x} \right] \frac{1}{\Delta x}$$

or $\qquad \dfrac{\partial^2 \psi}{\partial x^2} = \dfrac{\psi_{i+1,j} + \psi_{i-1,j} - 2\psi_{i,j}}{(\Delta x)^2} \tag{3-56}$

Similarly, the second-order differential at point i, j in the y direction is

$$\frac{\partial^2 \psi}{\partial y^2} = \frac{\psi_{i,j+1} + \psi_{i,j-1} - 2\psi_{i,j}}{(\Delta y)^2} \tag{3-57}$$

Substituting Eqs. (3-56) and (3-57) into Eq. (3-48) yields

$$\frac{\psi_{i+1,j} + \psi_{i-1,j} - 2\psi_{i,j}}{b^2} + \frac{\psi_{i,j+1} + \psi_{i,j-1} - 2\psi_{i,j}}{b^2} = 0 \tag{3-58}$$

where b has been substituted for Δx and Δy, since we are using a square grid with a mesh spacing of b.

After rearranging and combining terms, Eq. (3-58) becomes

$$\psi_{i,j} = \frac{\psi_{i+1,j} + \psi_{i-1,j} + \psi_{i,j+1} + \psi_{i,j-1}}{4} \tag{3-59}$$

To solve a two-dimensional groundwater flow problem using Eq. (3-59), we first establish values of the streamfunction ψ at all points around the boundary. These values are boundary values or initial conditions. Estimated values of ψ are then calculated at each interior point by linear interpolation or simply by assumption. Equation (3-59) is then applied successively at each point within the flow field, and the calculated value for $\psi_{i,j}$ is used to replace the previous value of ψ at point i, j. This is continued until values of ψ at all interior points remain constant from iteration to iteration to within some small numerical tolerance. This numerical process, generally referred to as relaxation, we demonstrate in Example 3-7.

EXAMPLE 3-7   A line of wells is to be drilled in an alluvial aquifer as shown in Fig. A. The wells are equally spaced along a line 600 ft from the river.

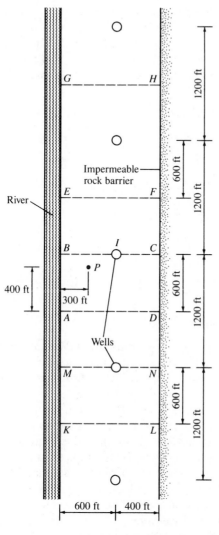

Figure A

An impermeable barrier is located parallel to the river and 1000 ft from the river. Each well is pumped at a rate of 12 ft^3/s. Using numerical methods, determine the streamline pattern, the piezometric head pattern, and the velocity at point P. Assume that the aquifer is confined with a depth of 100 ft.

SOLUTION   Considering the geometry of the flow field will show us that the flow pattern in $AEFD$ will have a flow pattern exactly like that within $EGHF$. Thus, we need to solve for the flow pattern within $AEFD$ only, since that solution will apply to *EGHE* and other similar areas such as *ADKL*. Further consideration of the geometry of area *AEFD* also shows that we can expect the flow pattern within *ABCD* to be the mirror image of that within *BEFC* or *ADMN*. Therefore, we need to solve numerically for the flow pattern within *ABCD* only, since that solution will apply to all other similar areas in the flow field.

I (Well)

B	6	6	6	6	6	6	0	0	0	0	0 C
	5	5	5	5	5	5	4	3	2	1	0
	4	4	4	4	4	4	4	3	2	1	0
	3	3	3	3	3	3	3	3	2	1	0
	2	2	2	2	2	1	1	1	1	1	0
	1	1	1	1	1	1	1	1	1	1	0
A	0	0	0	0	0	0	0	0	0	0	0 D

Figure B

Figure B shows the boundary of the flow field *ABCD*. We must now establish values of the streamfunction completely around the boundary. This is done by first considering flow at the boundaries themselves. Groundwater will flow from the river toward the well and cannot cross the solid barrier *CD*. Thus, *BI* must be a bounding streamline and so must *ADCI*. Since the total flow rate passing through the *ABCD* field must equal half the pumping rate of well *I*, we can set the maximum value of ψ as 6 cfs. Thus, the value of the streamfunction at all points on line *BI* is equal to 6. The value of the streamfunction is 0 along the streamline *ADCI* which we determine from the fact that the total difference in the value of ψ between points *A* and *B* must equal 6 cfs.

To establish the boundary values of ψ along *AB*, we need to recognize that the streamlines will be normal to the river along *AB*, and that the value of the streamfunction will vary linearly along *AB* from 0 at *A* to 6 at *B*. We arrive at this conclusion by realizing that the flow out of the river into the aquifer is uniform along the river. We can arbitrarily choose the size of the finite-difference mesh we wish to use in our solution. A very fine grid will produce the greatest accuracy in determining the flow pattern but also will require greater effort to solve Eq. (3-59). Conversely, a coarse finite-difference grid gives less accuracy but provides for a more speedy solution because fewer numbers are involved. In this case, we will arbitrarily use a spacing created by dividing the line *AD* into ten increments each 100 ft long. Using the same spacing in the vertical direction, line *AB* is divided into six equal increments. As shown, the value of the difference in streamfunction between successive meshes along *AB* must equal 1. The boundary values of the streamfunction are now established entirely around the field, and approximate values must be assumed for all interior points. In Fig. B, these values have been assumed arbitrarily, considering only that all values must be between 0 and 6.

Once all boundary values have been established and initial interior values approximated, Eq. (3-59) must be applied to each interior point. In this case, we will apply the equation row by row beginning at the bottom and moving to the right and up. Consider the first point at the left of the first row.

Equation (3-59) gives

$$\psi = \frac{1 + 2 + 1 + 0}{4} = 1.00$$

which is the value we have already assumed for that point; therefore, our assumption appears to be good. At other points, however, our guess is not as good. For example, consider the last point to the right on the first row. Applying Eq. (3-59) at that point gives

$$\psi = \frac{1 + 1 + 0 + 0}{4} = 0.50$$

I (Well)

B	6	6	6	6	6	6	0	0	0	0	0
	5	5	5	5	5	4.74	2.80	1.88	1.19	0.53	0
	4	4	4	4	3.99	3.98	3.47	2.72	1.86	0.93	0
	3	3	3	3	2.94	2.91	2.90	2.39	1.70	0.86	0
	2	2	2	2	1.75	1.69	1.67	1.67	1.42	0.73	0
	1	1	1	1	1	1	1	1	1	0.50	0
A	0	0	0	0	0	0	0	0	0	0	0

The top-right corner is labeled C and the bottom-right corner is labeled D.

Figure C

Thus, the assumed value of 1 at that point must be replaced by 0.50. We make this replacement before applying Eq. (3-59) to the next point. Figure C shows the results of applying Eq. (3-59) to all interior points in succession for one complete iteration. Thus, Fig. C provides a first iteration to the solution. To complete the solution, Eq. (3-59) is applied successively to each point until the interior values of ψ remain constant from iteration to iteration to within a chosen numerical tolerance. Figure D shows the interior values of ψ after 29 iterations. The interior values at the end of the 29th iteration changed less than 0.01 from the 28th iteration. The values of ψ shown in Fig. D are, thus, a solution of our problem to within a numerical precision of ± 0.01 cfs.

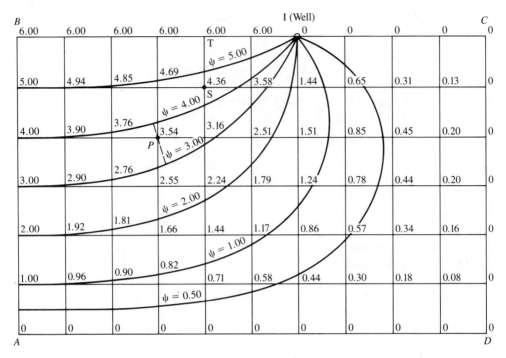

Figure D

The velocity at point P can be calculated by using Eq. (3-46) and (3-47) to calculate, respectively, the x and y components of velocity. The partial derivatives are calculated from the numerical values in Fig. D as

$$\left(\frac{\partial \psi}{\partial x}\right)_P = \frac{(3.16 - 3.54) + (3.54 - 3.76)}{200} = -\frac{0.6}{200} = -0.003$$

$$v = -\frac{\partial \psi}{\partial x}\frac{1}{B} = \frac{0.003}{100} = 0.00003 \text{ ft/s}$$

$$\left(\frac{\partial \psi}{\partial y}\right)_P = \frac{(4.69 - 3.54) + (3.54 - 2.55)}{200} = \frac{2.14}{200} = 0.0107$$

$$u = \frac{\partial \psi}{\partial y}\frac{1}{B} = \frac{0.0107}{100} = 0.000107 \text{ ft/s}$$

The thickness of the aquifer $B = 100$ had to be included because flow takes place throughout the entire depth of the aquifer. Thus, the y component of velocity is 0.00003 ft/s in the positive y direction, whereas the x component of velocity is 0.000107 ft/s in the positive x direction. The total velocity is

$$V = \sqrt{u^2 + v^2} = \sqrt{(0.00003)^2 + (0.000107)^2} = 0.00011 \text{ ft/s}$$

The velocity at point P can be determined graphically as well. Streamlines can be drawn as shown in Fig. D. To locate the streamlines, interpolation was performed along each vertical and horizontal grid line in Fig. D. For example, to locate the streamline for $\psi = 5$, between points S and T in Fig. D the streamline must be above point S, a distance of

$$\Delta y = 100 \frac{5 - 4.36}{6 - 4.36} = 39.0 \text{ ft}$$

Similar interpolation was done at each remaining vertical grid line that would be crossed by the $\psi = 5$ streamline. The remaining streamlines were located using the same method but for each of the other streamlines, interpolation had to be performed along the horizontal grid lines as well. The resulting streamline pattern is sketched in Fig. D. An additional streamline $\psi = 0.5$ has been located as well to provide more detail of the flow pattern.

The velocity at point P can also be determined using the principle of continuity. Since the streamfunction is numerically and dimensionally equal to unit discharge, we can determine the average velocity between the streamlines by dividing the total discharge between two streamlines by the normal distance between them. The normal spacing between the lines $\psi = 4$ and $\psi = 5$ at point P on Fig. D scales as 88 ft. Thus, the velocity is

$$V = \frac{1}{88(100)} = 0.00011 \text{ ft/s}$$

which agrees with the value calculated using Eq. (3-46) and (3-47).   ∎

The streamline pattern sketched in Fig. D of Example 3-7 is a good representation of the two-dimensional flow pattern, particularly near the left side of the field. However, near the well (point I), the streamlines develop a strong curvature, and the spacing of the streamlines appears somewhat irregular. This is because the grid spacing is too coarse in that region to give accurate values of the streamfunction at each mesh point (the numerical approximation to the partial derivatives in Eq. (3-48) is not accurate because of the rapid curvature). The mesh could be subdivided near the well, and Eq. (3-59) could be applied to that finer grid just as it was applied to the total grid in producing the numbers shown on Fig. D for Example 3-7. The result would be a better definition of the streamfunction where rapid curvature occurs.

The streamline pattern of Example 3-7 shows that the velocity is very small in the region to the right near the impermeable barrier because the streamlines are widely spaced there. Theoretically the velocity must be zero at points C and D and infinite at the well (point I).

The numerical procedures discussed in Example 3-7 are used to model steady-state flow patterns for complex groundwater flow patterns where more simple closed-form solutions cannot be obtained. Contaminant transport can also be modeled to the degree that chemical and physical actions between the aquifer material, the groundwater, and the contaminant are understood and amenable to analytical modeling.

Problems

3-1   How many cubic meters of water are contained in a 100-m^3 volume of saturated clay having a porosity of 0.48?

3-2   In a laboratory test, a one cubic foot sample of an aquifer was found to weigh 85 lb. After being allowed to drain thoroughly, the sample weighed 73 lb. After being crushed and thoroughly dried, the sample weighed 51 lb. If the sample was saturated initially, calculate the indicated specific yield, porosity, specific retention, and specific gravity of the solids.

3-3   Estimate the number of acre-feet of water that could theoretically be withdrawn from a saturated sandstone aquifer with an average surface area of 1500 sq mi and an average depth of 4500 ft.

3-4   If a spill of radioactive material occurred over an aquifer, estimate the length of time required for the material to reach a river 2 mi away if the aquifer is
a. Gravel
b. Clay
c. Sandstone
d. Granite
Groundwater elevation at the spill site is 100 ft above the water surface in the river.

3-5   Make the same estimate as requested in Prob. 3-4, but assume the aquifer is badly fractured basalt.

3-6   A permeameter (similar to the one shown in Fig. 3-3, page 113) was used to test three different materials. The horizontal tube of the permeameter is 5 ft long with an inside diameter of 4 in. The head measurements were 185 in., 77 in., and 39 in. for h_1, and 34 in., 35 in., and 36 in. for h_2 for the three materials, respectively. What was the indicated hydraulic conductivity for each sample if the flow rate was 0.227 gal/hr for each test?

3-7   The indicated hydraulic conductivity for a material tested in .he permeameter of Prob. 3-6 is 10 ft/day for a gradient of 0.005. If the p)rosity of the material is 0.30, what is the average velocity of flow within the voids of the sample?

3-8   In using a permeameter like that shown in Fig. 3-3, page 113, it is sometimes convenient to make a "falling-head" test. The horizontal tube is 1 m long and 10 cm in diameter and is filled with a sample material. During a test, the upstream head h_1 falls from 200 to 180 cm in 4 hr, while h_2 is held constant at 15 cm. The diameter of the vertical tube in which h_1 is measured is 40 cm. What is the indicated average hydraulic conductivity? Estimate the range between maximum and minimum values of hydraulic conductivity.

3-9   Laboratory testing indicates that a sample of an aquifer material has a porosity of 0.5. The sample is recompacted and tested in the permeameter whose physical dimensions are as given in Prob. 3-8. The differential head during the permeameter test is 100 cm, and 2 liters of water were discharged in 10 min. Was the porosity of the recompacted sample different from its original value?

3-10   In Fig. 3-5, page 117, the aquifer thickness is 10 ft, the hydraulic conductivity of the aquifer is 0.06 m/day, and the river water surface is 7 m above the bottom of the aquifer. If it is necessary to maintain a water level in the trench not more than 10 m deep, at what rate must water be pumped out of a 100-m long trench? The distance between the edge of the trench and the river is 300 m. Assume all water comes from the river.

3-11   A trench 100 ft long is excavated parallel to a river, as shown in Fig. 3-5, page 117. The aquifer thickness B is 19 ft, and the trench is 400 ft from the river. During a pumping test, a 10-ft differential head was maintained while pumping was steady at a rate of 10 gal/min. What is the indicated hydraulic conductivity of the aquifer?

3-12   A trench 100 ft long is excavated 300 ft away and is parallel to a river bank. A sand aquifer is 20 ft thick and similar to that shown in Fig. 3-6, page 118. The river water surface is 30 ft above the horizontal bottom of the aquifer. If water depth in the trench must be maintained at 6 ft, at what rate must water be pumped from the trench? How far from the trench will the aquifer cease to be saturated?

3-13   An aquifer slopes away from a river, as shown in Fig. 3-7, page 119. The hydraulic conductivity of the aquifer which is 20 ft thick is known to be 0.1 ft/day. The water is 20 ft deep in the river, and the bottom of the trench is 30 ft below the bottom of the river and 600 ft away from the edge of the river. The trench is 40 ft deep. Will water flow out of the trench without pumping? If so, what will be the rate of flow?

3-14   A 10-in. diameter well is drilled to the bottom of an unconfined aquifer. The water table is originally 500 ft above the bottom of the aquifer whose hydraulic conductivity is known to be 0.25 ft/day. If 100 gal/min is pumped from the well, what will be the equilibrium depth of water in the well if the radius of influence is 1000 ft? How much drawdown should be expected at a distance of 250 ft from the well?

3-15 In a pumping test, 100 gal/min is pumped from a 12-in. diameter well. The groundwater table is originally 300 ft above the bottom of the unconfined aquifer. Drawdown in the well is 110 ft and 5 ft at a distance of 400 ft from the well when approximately steady-state conditions are achieved. What is the indicated hydraulic conductivity for the aquifer?

3-16 In a confined aquifer (see Fig. 3-10, page 126), a 50-cm diameter well fully penetrates the aquifer. The hydraulic conductivity of the aquifer was approximately 0.8 m/day. Originally, the piezometric head was 300 m above the top of the aquifer, which is 30 m thick. If 0.8 m^3/min is pumped from the well, what should be the depth of water in the well? What will the height of the piezometric head be (above the top of the aquifer) at a distance of 50 m from the well? Assume the radius of influence to be 1000 ft.

3-17 A confined aquifer has a thickness of 85 ft. An 86-in. diameter well is drilled to the bottom of the aquifer, as shown in Fig. 3-10, page 126. If the hydraulic conductivity of the aquifer is 10 ft/day, and if water is originally 170 ft deep in the well, calculate approximately the maximum rate of flow that can be pumped from the well without dewatering any part of the aquifer. Assume a reasonable value for the radius of influence. What will be the average velocity (Darcy velocity) of flow in the aquifer at a distance of 20 ft from the center of the well?

3-18 The following measurements were made at an observation well 70 ft from a well being pumped at 150 gal/min. Determine the storage coefficient and transmissivity of the aquifer being pumped.

Time (hr)	Distance From Top of Well to Water (ft)
0	15.5
0.02	15.6
0.03	15.7
0.04	15.8
0.06	16.0
0.11	16.3
0.25	16.8
0.43	17.1
1.00	17.7
3.90	18.5

3-19 The following data were taken during the pumping test of a well that was pumped at a rate of 0.41 ft^3/s. The two observation wells were located

320 ft and 640 ft from the well. Calculate the storage coefficient and the transmissivity of the confined aquifer.

Time	Drawdown (ft)	
(hr)	Well at 320 ft	Well at 640 ft
0.024	0.27	0.06
0.120	0.54	0.32
0.240	0.81	0.48
1.200	1.23	0.88
2.400	1.41	1.06
12.000	1.83	1.47
24.000	2.01	1.65
120.000	2.43	2.07
240.000	2.61	2.25

3-20   For Prob. 3-19, assume that the argument of the well function u is less than 0.01, and solve for the transmissivity and the storage coefficient for each of the drawdown values. Compare this with the values obtained in the Theis method used in Prob. 3-19.

3-21   A groundwater recharge well, as shown in Fig. 3-11, page 134, is injected with a flow of 100 gal/min. If the water level in the well is originally 100 ft deep and rises to an elevation 15 ft above the original water table, estimate the hydraulic conductivity of the unconfined aquifer. Calculate the distance that water will be mounded above the original water table at a distance of 80 ft from the well. The well diameter is 24 in.

3-22   What is the volume of water contained above the original water surface if injection of 30 gal/min into an injection well results in a quasi-steady water surface in the 24-in. diameter well 30 ft above the original ground water table? The original water depth is 200 ft. Assume that the mounded surface essentially coincides with the original groundwater table at 600 ft from the well.

3-23   A 60-in. diameter well is located 300 ft from a river and fully penetrates a gravel aquifer having a hydraulic conductivity of 40 ft/day. It is pumped at a rate of 400 gal/min. Calculate coordinates and plot the equilibrium groundwater surface along a line perpendicular to the river bank and passing through the well. Assume a 1000-ft radius of influence on the side away from the river. The river water surface is 15 ft above the bottom of the aquifer.

3-24   For the well and pumping conditions described in Prob. 3-23, calculate the drawdown to be expected at a point halfway between the well and the river and for a point 150 ft from the well on the side away from the river.

3-25 For the well and pumping conditions given in Prob. 3-23, calculate the drawdown to be expected at a point 100 ft from the well and 300 ft from the river.

3-26 A well is pumped to unwater an excavation site in an unconfined aquifer. The water level is originally 10 ft below the top of the well. The well extends to the bottom of the aquifer and is 100 ft deep. A sheet-pile wall is driven completely through the aquifer at a distance of 80 ft from the well. If the hydraulic conductivity of the aquifer is 0.8 ft/day, calculate the drawdown at the sheet-pile wall and at a point 80 ft away from the well but on the side away from the wall. The pumping rate is 7.5 gal/min. Assume a radius of influence of 1000 ft on the side away from the wall, and that the wall is long enough to fully cut off flow in the aquifer.

3-27 For the well conditions given in Prob. 3-26, assume that the pumping rate is increased to 150 gal/min. What will the drawdown become at the wall under the new pumping rate?

3-28 Two wells are drilled 1300 ft apart. Both fully penetrate a confined aquifer 800 ft thick. They are both pumped at a rate of 20 gal/min. If the aquifer has a hydraulic conductivity of 3×10^{-2} ft/day, and the piezometric head is originally 400 ft above the top of the aquifer, calculate the drawdown to be expected halfway between the wells. Assume a radius of influence of 1000 ft.

3-29 Three wells are drilled into an unconfined aquifer. The level, impervious bottom of the aquifer is 500 ft below the earth's surface. The groundwater table is initially 100 ft below the top of the wells, all of which are at the same elevation. Hydraulic conductivity of the aquifer is 20 ft/day. The three wells are arranged as shown in the accompanying figure and are all pumped at a rate of 5000 gal/min. Assume a radius of influence of 1000 ft.

(continued p. 161)

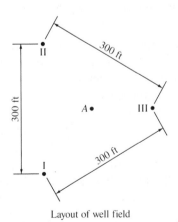

Layout of well field

Calculate the drawdown at point *A* midway between the three wells. Calculate the ordinates and plot the equilibrium groundwater surface along a line between wells I and II.

3-30   A drainage basin 400 sq mi in area completely covers a limestone aquifer, which averages 300 ft in thickness. Estimate the total volume of water stored in the aquifer if it were saturated. If average annual precipitation over the area is 22 in./yr, evapotranspiration averages 8 in./yr, and runoff is 20% of precipitation, estimate the annual yield that could be pumped from groundwater without a long-term overdraft.

3-31   Water flows through a confined aquifer 50-ft thick from a river at *AC* to a pond at *BD*, as shown in the accompanying figure. Hydraulic conductivity is 4 ft/day. Piezometric heads are as shown on the boundaries of the flow field. Using Eq. (3-59) (and using piezometric head instead of streamfunction), determine the values of the piezometric head for the interior of the field. What is the velocity of flow at point *M*? Determine the flow rate between the river and the pond.

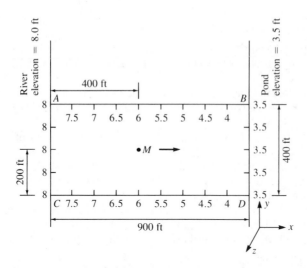

PROBLEM 3-31

3-32   Flow infiltrates from a pond to a confined aquifer as shown in the figure on the next page. The infiltration rate is a constant 12 in./day. Assuming that the flow is two-dimensional and that the upper and lower boundaries are level and parallel, establish values of the streamfunction around the boundary. Dividing the vertical dimension of the aquifer into five finite-difference increments and using a square finite-difference grid, calculate the value of the streamfunction at each interior grid point. What is the velocity (continued p. 162)

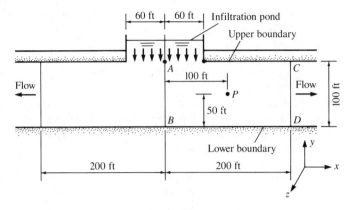

at point *P*? What is the velocity at any point along *CD*? Hydraulic conductivity is 0.1 ft/day.

REFERENCES

1. Bennett, G.D. "Introduction to Groundwater Hydraulics." Chapter B2 of Book 3, *Applications of Hydraulics*. U.S. Dept. of the Interior, Geological Survey, Washington, D.C., 1976.
2. Bouwer, H. *Groundwater Hydrology*. McGraw-Hill, New York, 1978.
3. Davis, S.N., and R.J.M. DeWiest. *Hydrogeology*. John Wiley & Sons, New York, 1966.
4. Ferris, J.G., D.B. Knowles, R.H. Brown, and R.W. Stallman. "Theory of Aquifer Tests." Water Supply Paper 1536, U.S. Geological Survey, Alexandria, Vir., 1962.
5. Harr, M.E. *Groundwater and Seepage*. McGraw-Hill, New York, 1962.
6. Heath, R.C. *Basic Ground-Water Hydrology*. Water Supply Paper 2220, U.S. Geological Survey, Alexandria, Vir., 1984.
7. L'vovich, M.I. *World Water Resources and Their Future*. Trans. edited by R.L. Nace. American Geophysical Union, Washington, D.C., 1979.
8. Muskat, M. "The Flow of Homogeneous Fluids Through Porous Media." J.W. Edwards, 1946.
9. Pinder, G.F., and W.G. Gray. *Finite Element Simulation in Surface and Subsurface Hydrology*. Academic Press, New York, 1977.
10. Poland, J.F., and G.H. Davis. *Land Subsidence Due to Withdrawal of Fluids*. Geological Society of America, Boulder, Colo., 1969.
11. Polubarinova-Kochina, P.Y. *Theory of Groundwater Movement*. Trans. R.J.M. DeWiest. Princeton University Press, Princeton, N.J., 1962.
12. Prickett, T.A., and C.G. Lonnquist. "Selected Digital Computer Techniques for Groundwater Resources Evaluation." Illinois State Water Survey, Bulletin 55, Urbana, Ill., 1971.
13. Rouse, H. (ed.). *Engineering Hydraulics*. John Wiley & Sons, New York, 1950.
14. "Santa Ana River Investigations." Bulletin 15, State of California Department of Water Resources, Sacramento, Calif. (February 1959).

15. Sokolnikoff, I.S., and E.S. Sokolnikoff. *Higher Mathematics for Engineers and Physicists.* McGraw-Hill, New York, 1941.

16. Sokolnikoff, I.S., and R.M. Redheffer. *Mathematics of Physics and Modern Engineering.* McGraw-Hill, New York, 1958.

17. Terzagi, K., and R.B. Peck. *Soil Mechanics In Engineering Practice.* John Wiley & Sons, New York, 1948.

18. Theis, C.V. "The Relationship Between Lowering of the Piezometric Surface and the Rate and Duration of Discharge of a Well Using Ground-Water Storage." *Trans. AGU,* 16 (1935).

19. Vallentine, H.R. *Applied Hydrodynamics.* Butterworth Scientific Publications, London, 1959.

20. Walton, W.C. *Groundwater Resource Evaluation.* McGraw-Hill, New York, 1970.

4

This trapezoidal canal is part of the Colorado River
Aqueduct that conveys water to Southern California.
This canal is 55 feet across at the top, 25 feet wide
at the bottom, 11 feet deep and delivers water at a
maximum rate of 1,850 cfs. (Courtesy of Metropolitan
Water District of Southern California.)

Open Channel
Flow

4-1   General Considerations

Open channel flow is flow of a liquid in a conduit in which the upper surface of the liquid (that is, the free surface) is in contact with the atmosphere. Water flow in rivers and streams are obvious examples of open channel flow in natural channels. Other occurrences of open channel flow are flow in irrigation canals, sewer lines that flow partially full, storm drains, and street gutters.

The engineer may be required to solve problems having to do with either natural or manmade channels. In natural channels, the problem may be predicting the water surface profile along an extended reach of the river or stream given a certain discharge. Or one may be asked to estimate velocity and depth of flow in a local region of the channel. A more complex type of problem would be one in which the discharge, velocity, and depth along the channel are simultaneously a function of time and distance.

The same types of problems exist for already constructed manmade channels. However, one of the most challenging problems is the design and construction of the channel itself. In this case, as in all designs, the task is to produce a structure that will accomplish the desired result with minimum cost. For example, one might be asked to design a canal to convey water with a given maximum discharge from a reservoir to a power plant at some distance from the reservoir. Although it may be easy to design the canal for the stated conditions, the designer must consider all aspects of the problem, and many may not be envisioned at the outset. The design engineer will often have to consider complex situations that result from unusual natural hazards or abnormal operating procedures—for example, landslides into the reservoir or canal, accidental gate operation, or a breach of a canal embankment. Thus, the designer must envision the whole problem as well as find a satisfactory solution to the primary design objective.

4-2   Steady-Uniform Flow in Open Channels

Definition and Description of Uniform Flow

Uniform flow in a channel exists when there is no change of velocity along the channel. Under this condition, the convective acceleration is zero, and the streamlines are straight and parallel. Because the velocity does not change, the velocity head will be constant; therefore, the energy grade line and water surface will have the same slope as the channel bottom. For the flow to be uniform, the channel must be straight and without change in slope or cross section along the length of the channel. Such a channel is called a *prismatic* channel. When flow is uniform, the depth in the channel is called *normal depth*. Figure 4-1 depicts this condition.

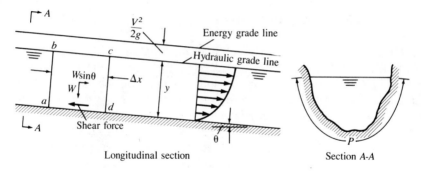

Figure 4-1   Uniform flow in a channel

Because the acceleration of fluid is zero for uniform flow, the net force acting on a mass of fluid, *abcd*, as shown in Fig. 4-1, will be zero. That is, the motive force (gravitational force component) will be equal and opposite to the resisting force of the channel bottom and wall. The motive force is equal to the component of weight of the fluid mass in the direction of flow ($W \sin \theta$). The resistance is equal to the product of the shear stress, τ_0, and the surface area of the channel ($P \Delta x$) that is in contact with the liquid. Thus, the force balance equation is

$$W \sin \theta - \tau_0 P \Delta x = 0$$

or $$\gamma A \Delta x \sin \theta - \tau_0 P \Delta x = 0$$

$$\tau_0 = \gamma \frac{A}{P} \sin \theta \tag{4-1}$$

In Eq. (4-1), A/P is the hydraulic radius, R, and $\sin \theta$ is the slope of the channel S_0. The shear stress, τ_0, can be expressed as a function of the mass density, ρ, the mean velocity, V, and a resistance coefficient, c_f: $\tau_0 = c_f \rho(V^2/2)$. Therefore, Eq. (4-1) can be written as

$$c_f \rho \frac{V^2}{2} = \gamma R S_0$$

or $$V = \sqrt{\frac{2g}{c_f}} \sqrt{R S_0} \tag{4-2}$$

Equation (4-2) can also be given in the form

$$V = C\sqrt{R S_0} \tag{4-3}$$

This equation was first developed by Chezy, a French engineer of the eighteenth century. The coefficient in Eq. (4-3) is called the Chezy *C*.

Resistance Effects Using the Friction Factor

The resistance coefficient, c_f, and the Chezy coefficient C are both functions of the roughness of the channel bottom and wall and the depth of flow whose values are determined experimentally. In pipe flow, the resistance coefficient, c_f, is one fourth the value of the Darcy-Weisbach friction factor $f(c_f = f/4)$; therefore, Eq. (4-2) can be expressed as

$$V = \sqrt{\frac{8g}{f}} \sqrt{RS} \qquad (4\text{-}4)$$

Equation (4-4) is a form of the Darcy-Weisbach equation applied to flow in open channels. In Eq. (4-4), the friction factor f is a function of the Reynolds number Re and the relative roughness, $k_s/4R$, and is usually given in graphical form, such as in the Moody diagram (see Fig. 5-5 on page 248). To obtain the discharge equation for open channels, simply multiply both sides of Eq. (4-4) by the cross-sectional area A, yielding

$$VA = \sqrt{\frac{8g}{f}} A\sqrt{RS_0}$$

or $\qquad Q = \sqrt{\frac{8g}{f}} A\sqrt{RS_0} \qquad (4\text{-}5)$

For fairly straight rock-bedded streams, the larger rocks produce most of the resistance to flow. Limerinos (13) has shown that the resistance coefficient f can be given in terms of the size of rock in the stream bed as

$$f = \frac{1}{\left(1.2 + 2.03 \log\left(\dfrac{R}{d_{84}}\right)\right)^2} \qquad (4\text{-}6)$$

where d_{84} is a measure of the rock size.*

E X A M P L E   4 - 1 Determine the discharge in a long, rectangular concrete channel that is 5 ft wide, that has a slope of 0.002, and in which the water depth is 2 ft.

---

* Most river-worn rocks are somewhat elliptical in shape. Limerinos (13) showed that the intermediate dimension correlates best with f. The d_{84} refers to the size of rock (intermediate dimension) for which 84% of the rocks in the random sample are smaller than the d_{84} size. Details for choosing the sample are given by Wolhman (25). The basic procedure entails sampling at least 100 rocks on the channel bottom. For example, a grid with 100 points could be laid out on the channel bottom, and the rock under each grid point would be a rock of the sample from which the d_{84} is determined.

SOLUTION   Assume $k_s = 5 \times 10^{-3}$ ft. This is an intermediate value of roughness as given in Chapter 5 (page 248). The hydraulic radius is

$$R = \frac{A}{P} = \frac{(2 \times 5)}{[(2 \times 2) + 5]} = 1.11 \text{ ft}$$

Then $\dfrac{k_s}{4R} = \dfrac{(5 \times 10^{-3})}{(4 \times 1.11)} = 1.13 \times 10^{-3}.$

Using the Moody diagram (see Fig. 5-5), the f value is found to be about 0.020 for a Reynolds number of about 10^6. Use this f for the first computation of Q:

$$Q = \sqrt{\frac{8 \times 32.2}{0.020}} \times (2 \times 5) \times 1.11 \times 0.002$$

$$Q = 53.5 \text{ cfs} \qquad V = Q/A = 5.35 \text{ ft/s}$$

Assume the water temperature is 60°F, then, the kinematic viscosity will be 1.22×10^{-5} ft²/s, and the Reynolds number $V \times 4R/v$ will be found to be 1.95×10^6. On checking the Moody diagram with the Reynolds number of 1.95×10^6, we find that f is indeed 0.020; therefore, with the given assumptions, the discharge will be 53.5 cfs.   ■

The Manning Equation

The discharge equation most often used by hydraulics engineers is the Manning equation, named after an Irish engineer of the nineteenth century. In the Manning equation, using the English system of units, the Chezy coefficient of Eq. (4-3) is given as $C = (1.49/n) \times R^{1/6}$, where n is a resistance factor. Thus, the Manning discharge equation is

$$Q = \frac{1.49}{n} AR^{2/3} S_0^{1/2} \tag{4-7a}$$

In the SI system of units,

$$Q = \frac{1}{n} AR^{2/3} S_0^{1/2} \tag{4-7b}$$

The resistance factor, n, is a function of a number of variables, the primary one being the roughness of the channel.* To assist the engineer in choosing n, Table 4-1 gives n values for various types and conditions of channels. Figures 4-2, 4-3, and 4-4 are photos of actual channels along with measured n values.

---

* For a more complete discussion of n values, see Chow (4).

Table 4-1 Typical Values of the
Roughness Coefficient n

Lined Canals	n
Cement plaster	0.011
Untreated gunite	0.016
Wood, planed	0.012
Wood, unplaned	0.013
Concrete, troweled	0.012
Concrete, wood forms, unfinished	0.015
Rubble in cement	0.020
Asphalt, smooth	0.013
Asphalt, rough	0.016
Natural Channels	
Gravel beds, straight	0.025
Gravel beds plus large boulders	0.040
Earth, straight, with some grass	0.026
Earth, winding, no vegetation	0.030
Earth, winding	0.050

Figure 4-2   Clark Fork River near St. Regis,
Montana — $n = 0.028$ for $R = 16$ ft, $d_{84} = 205$ mm

Figure 4-3   Moyie River above left near Eastport,
Idaho—$n = 0.038$ for $R = 7.0$ ft

Figure 4-4   Boundary Creek above right near Porthill,
Idaho—$n = 0.073$ for $R = 4.0$ ft, $d_{84} = 375$ mm

Flow in Conduits of Circular Cross Section

Highway culverts and city sewers are common examples of open channel conduits of circular cross section. For a given slope and resistance coefficient
(n = constant), the discharge will be proportional to $AR^{2/3}$, as can be seen by
inspecting Eqs. (4-7a and 4-7b). Since $V = Q/A$, it can also be deduced that the
velocity will be proportional to $R^{2/3}$. Therefore, one can easily determine $Q/Q_0 =
(AR^{2/3})/(A_0R_0^{2/3})$, where Q is the discharge for a given depth of flow, and Q_0
is the discharge for the completely full conduit. Figure 4-5 is a plot of the relative
discharge (Q/Q_0) versus relative depth (y/d_0) for a circular conduit. The relative
velocity is obtained by dividing the relative discharge by the relative area (A/A_0).
The relative velocity is also shown in Fig. 4-5. *Note*: The maximum discharge
and velocity occur at a depth less than that for full flow condition. This occurs

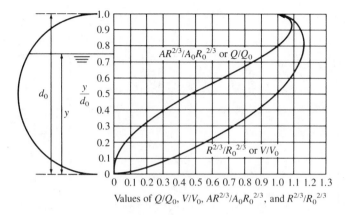

Figure 4-5 Flow characteristics of a circular section assuming constant n value

because as the conduit becomes nearly full, the perimeter increases much faster than the increase in cross-sectional area, thus decreasing the hydraulic radius and discharge.

Figure 4-5 along with Fig. 4-6, page 172, (a nomograph for solving Manning's equation) can be used for easily solving uniform flow problems in circular conduits.

EXAMPLE 4-2 Determine the discharge in a 3-ft sewer pipe if the depth of flow is 1.00 ft, and the slope of the pipe is 0.0019. Assume $n = 0.012$.

SOLUTION For this example we use Fig. 4-6 by drawing a straight line through the points for $n = 0.012$ and $S = 0.0019$. Note where the straight line intersects the match line. Then draw a line through the point of intersection on the match line and the 3 ft diameter point. It is noted that this line intersects the discharge scale at $Q_0 = 30$ cfs which is the discharge for the full flow condition. The relative depth y/d_0 is $1.00/3 = 0.333$; therefore, $Q/Q_0 = 0.20$ (from Fig. 4-5), so $Q = 0.20 \times 30 = 6.0$ cfs. ■

Flow in Conduits of Trapezoidal Cross Section

To assist in solving problems involving the flow in trapezoidal channels, the factor $AR^{2/3}$ of Eq. (4-7) is plotted in Fig. 4-7, page 173, in dimensionless form as a function of the relative depth (y/b), where b is the bottom width of the channel.

EXAMPLE 4-3 Determine the normal depth for a trapezoidal channel with side slopes of 1 vertical to 2 horizontal, a bottom width of 8 ft, discharge of 200 cfs, channel bottom slope of 1.0 ft in 1000 ft, and $n = 0.012$.

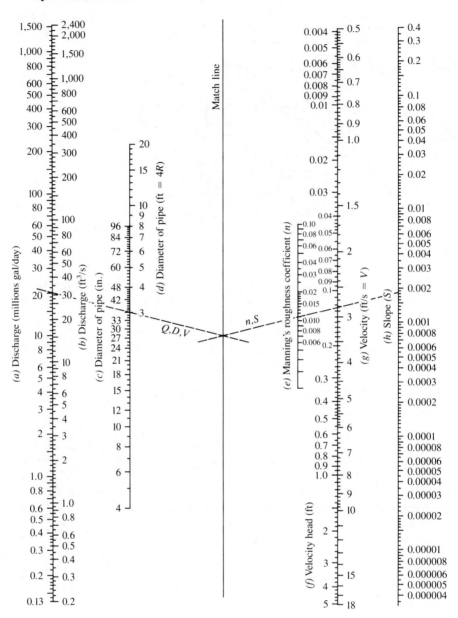

Figure 4-6   Alignment chart for flow of water in pipes
flowing full (1)

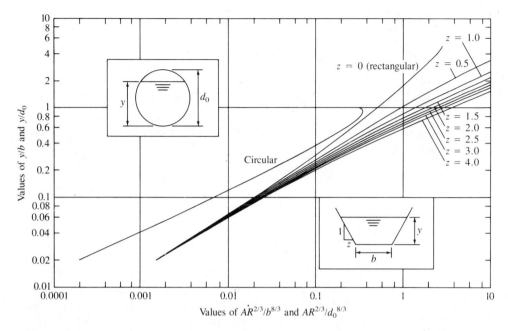

Figure 4-7 Curves for determining the normal depth
[adapted from *Open Channel Hydraulics* by Chow (4)
Copyright © 1959, McGraw-Hill Book Company,
New York; used with permission of McGraw-Hill
Book Company.]

SOLUTION With a little algebraic manipulation so that we can use Fig. 4-7,
Eq. (4-7a) can be written as

$$AR^{2/3} = \frac{Qn}{(1.49\, S_0^{1/2})}$$

or $\quad \dfrac{AR^{2/3}}{b^{8/3}} = \dfrac{Qn}{1.49\, S_0^{1/2} b^{8/3}}$ (4-8)

Since $Q = 200$ cfs, $n = 0.012$, $S_0 = 0.001$, and $b = 8$ ft, we can evaluate the right-hand side of Eq. (4-8):

$$\frac{AR^{2/3}}{b^{8/3}} = \frac{200 \times 0.012}{1.49 \times (0.001)^{1/2} \times 8^{8/3}} = 0.199$$

Then for $AR^{2/3}/b^{8/3} = 0.199$, one determines that $y/b = 0.33$ (from Fig. 4-7).
Thus,

$$y_n = 0.33b \quad \text{or} \quad y_n = 2.64 \text{ ft} \qquad \blacksquare$$

Table 4-2 Channel Wall Slopes for Different Materials (4)

Material	Side Slopes
Rock	Nearly vertical
Stiff clay or earth with concrete lining	$\frac{1}{2}$:1 to 1:1
Firm soil	1:1
Loose sandy soil	2:1
Sandy loam	3:1

Table 4-3 Maximum Permissible Velocities and n Values for Different Materials (5)

Material	V (ft/s)	n
Fine sand	1.50	0.020
Sandy loam	1.75	0.020
Silt loam	2.00	0.020
Firm loam	2.50	0.020
Stiff clay	3.75	0.025
Fine gravel	2.50	0.020
Coarse gravel	4.00	0.025

Design of Erodible Channels

If a channel is constructed in erodible material, such as an irrigation canal through soil or fine gravel, then the possibility exists that the channel will erode if the water velocity is too large. Two methods can be used for designing erodible channels: the permissible velocity method and the tractive force method. In this text, we will discuss only the *permissible velocity method*.*

Assuming that the channel will be of trapezoidal cross section, the first decision the designer must make is to choose the appropriate side slope for the channel. A slope should be chosen that will be stable under all conditions. Given the type of material, one can apply basic soil mechanics to determine a suitable slope. Table 4-2 gives approximate permissible side slopes for different materials, and Table 4-3 gives approximate permissible velocities for different materials.

Once Q, V, n, S_0, and the basic channel shape have been determined, we can solve for the depth and width of the channel. Example 4-4 illustrates that procedure.

EXAMPLE 4-4   An unlined irrigation canal is to be constructed in a firm loam soil. The slope is to be 0.0006, and it is to carry a water flow of 100 cfs. Determine an appropriate cross section for this canal.

SOLUTION   Using Table 4-2 as a guide, choose a side slope of $1\frac{1}{2}$ horizontal to 1 vertical (this is a conservative choice). From Table 4-3, choose a maximum permissible $V = 2.50$ ft/s, and $n = 0.020$. Now we get the hydraulic radius R from the Manning velocity equation:

$$V = \frac{1.49}{n} R^{2/3} S_0^{1/2}$$

---

* For information on the more sophisticated *tractive force method*, see Chow (4).

$$R = \left(\frac{nV}{1.49S_0^{1/2}}\right)^{3/2} = \left(\frac{(0.02 \times 2.50)}{1.49 \times (0.0006)^{1/2}}\right)^{3/2} = 1.60$$

From the continuity equation, $A = Q/V$:

$$A = \frac{100}{2.5}$$

$$= 40 \text{ ft}^2$$

Because $R = A/P$, we then have $P = A/R$ or

$$P = \frac{40}{1.60} = 25 \text{ ft}$$

The area $A = by + 1.5y^2$, and $P = b + 3.61y$; therefore, we can solve for b and y, yielding

$$b = 18.1 \quad \text{and} \quad y = 1.91 \text{ ft}$$

For ease of construction, use $b = 18$ ft and $y = 2.0$ ft.   ■

Best Hydraulic Section

The quantity $AR^{2/3}$ in Manning's equation (Eq. 4-7) is called the section factor in which $R = A/P$; therefore, the section factor relating to uniform flow is given by $A(A/P)^{2/3}$. Thus, for a channel of given resistance and slope, the discharge will increase with increasing cross-sectional area but decrease with increasing wetted perimeter P. For a given area, A, and a given shape of channel, for example, rectangular cross section, there will be a certain ratio of depth to width (y/b) for which the section factor will be maximum. Such a ratio establishes the *best hydraulic section*. That is, the best hydraulic section is the channel proportion that yields a minimum wetted perimeter for a given cross-sectional area.

EXAMPLE 4-5   Determine the best hydraulic section for a rectangular channel.

SOLUTION   For the rectangular channel, $A = by$, and $P = b + 2y$. Let A be constant then let us minimize P. But

$$P = b + 2y$$

or $$P = \frac{A}{y} + 2y$$

Thus, we see that the perimeter varies only with y for the given conditions. If we differentiate P with respect to y and set the differential equal to zero, we will have the condition for minimizing P for the given area A:

$$\frac{dP}{dy} = \frac{-A}{y^2} + 2 = 0$$

or $$\frac{A}{y^2} = 2$$

But $A = by$, so

$$\frac{by}{y^2} = 2 \quad \text{or} \quad y = \frac{1}{2}b$$

Thus, the best hydraulic section for a rectangular channel occurs when the depth is one half the width of the channel.   ■

It can be shown that the best hydraulic section for a trapezoidal channel is half a hexagon; for the circular section, it is the half circle, and for the triangular section, it is half of a square. Of all the various shapes, the half circle has the best hydraulic section.

The best hydraulic section can be relevant to the cost of the channel. For example, if a trapezoidal channel were to be excavated and if the water surface were to be at ground level, the minimum amount of excavation would result if the channel of best hydraulic section were used, and the minimum cost would result if only excavation were involved. However, many other factors are involved in designing the most economical channel. For example, if the water surface lies below ground level, the best hydraulic section may not result in minimum volume of excavation. Thus, the best hydraulic section should be used only as a guide or starting point in designing a channel.

Project Scope

In the preceding sections, we discussed methods for considering uniform flow in channels of various shapes and roughness. However, water-resources projects may include other structures in addition to the channels. For example, consider an irrigation project for which the irrigation water is drawn from a river and distributed to the irrigable land. For a project like this, *intake works, flumes, checks, drops,* and *transitions* may all be included as part of the water distribution system. The next paragraphs scope the need for such structures.

Figure 4-8 shows the layout of a typical irrigation project. Across the river, a *diversion dam* is constructed so as to maintain a water level high enough in the river to be able to always divert the required flow of water into the main irri-

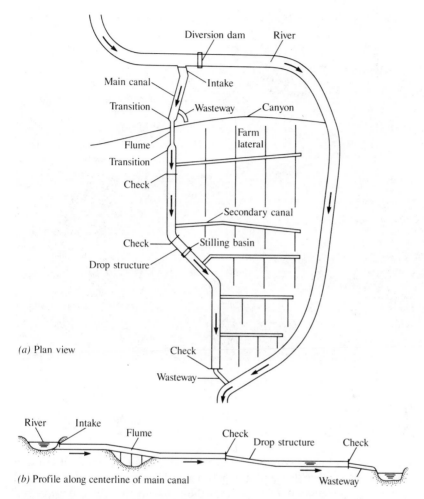

(a) Plan view

(b) Profile along centerline of main canal

Figure 4-8   Layout of irrigation system (a) Plan view
(b) Profile along centerline of main canal

gation canal. The intake to the main canal consists of a canal entrance structure, including gates for controlling the discharge into the canal. Figure 4-9 on the next page is a photograph of a diversion dam and intake structure for the Riverton Irrigation Project in Wyoming.

Wasteways are often provided at intervals along the main canal to prevent overtopping of the canal if an emergency develops where downstream water use is stopped. Then the unused water is wasted into a natural channel such as in the canyon shown in Fig. 4-8.

Figure 4-9   Wind River diversion dam and intake structure for the Riverton Irrigation project in west central Wyoming (20)

When the water must be conveyed across a depression or canyon either a flume or an *inverted siphon* (large pipe) can be used. Then *transitions* are required to provide a smooth passage of flow from the canal to the flume or inverted siphon and back again.

A *check structure* is a concrete lined part of the canal in which a gate or stop logs are installed to maintain a high enough water surface level in the canal so that water can be diverted into a secondary canal. Some checks are equipped with automatic water level devices.

If there is a significant drop in the land surface, increased slope of a canal in soil may lead to undesirable erosion. In these cases, a concrete *drop structure* may be needed. Figure 4-10 shows the basic features of a baffled drop structure.

The secondary canals draw water from the main canal through their own intake structures, and farm laterals take water from the secondary canals.

The preceeding paragraphs are to help you understand how various structures can be incorporated to achieve a complete workable project. The example chosen (irrigation system) includes the basic structures that would be encountered in almost any open channel project even though it may not be agricul-

Figure 4-10   Drop Structures (21) (a) Above is a
baffled drop structure before backfilling (b) Below is a
baffled drop structure in operation

turally oriented. For example, the cooling water system of a major thermal power plant would use many of the basic structures described for the irrigation system. Therefore, if you can develop an appreciation for the basic function of each structure rather than view it as an isolated part of a specific project then your awareness of design considerations will be greatly expanded.

We discuss further details about open channel structures in Chapter 7.

4-3   Steady-Nonuniform Flow in Open Channels

Energy Relations

THE ENERGY EQUATION   The one-dimensional energy equation for open channels (see Fig. 4-11) is

$$\frac{p_1}{\gamma} + \alpha_1 \frac{V_1{}^2}{2g} + z_1 = \frac{p_2}{\gamma} + \alpha_2 \frac{V_2{}^2}{2g} + z_2 + h_L \tag{4-9}$$

We see from Fig. 4-11 that the following equalities hold:

$$\frac{p_1}{\gamma} + z_1 = y_1 + S_0 \, \Delta x \quad \text{and} \quad \frac{p_2}{\gamma} + z_2 = y_2$$

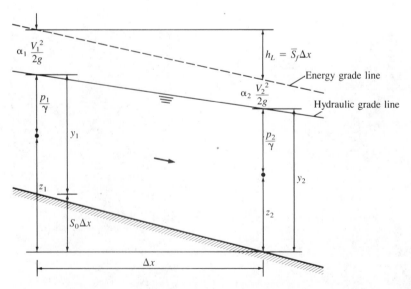

Figure 4-11   Definition sketch for flow in open channels

Here S_0 is the slope of the channel bottom, and y is the depth of flow. Then if we assume $\alpha_1 = \alpha_2 = 1.0$, we can write Eq. (4-9) as

$$y_1 + \frac{V_1^2}{2g} + S_0 \Delta x = y_2 + \frac{V_2^2}{2g} + h_L \tag{4-10}$$

Now, if we consider the special case where the channel bottom is horizontal ($S_0 = 0$), and the head loss is zero ($h_L = 0$), Eq. (4-10) becomes

$$y_1 + \frac{V_1^2}{2g} = y_2 + \frac{V_2^2}{2g} \tag{4-11}$$

SPECIFIC ENERGY   The sum of the depth of flow and the velocity head is the specific energy:

$$E = y + \frac{V^2}{2g} \tag{4-12}$$

Thus Eq. (4-11) states that the specific energy at section 1 is equal to the specific energy at section 2, or $E_1 = E_2$. The continuity equation between sections 1 and 2 will be

$$A_1 V_1 = A_2 V_2 = Q \tag{4-13}$$

Therefore, Eq. (4-11) can be expressed as

$$y_1 + \frac{Q^2}{2gA_1^2} = y_2 + \frac{Q_2}{2gA_2^2} \tag{4-14}$$

Because A_1 and A_2 are both functions of the depth y, the magnitude of the specific energy at section 1 or 2 is solely a function of the depth at each section. If, for a given channel and given discharge, one plots depth versus specific energy, a relationship such as shown in Fig. 4-12, page 182, is obtained. By studying Fig. 4-12 for a given value of specific energy, we can see that the depth may be either large or small. In a physical sense, this means that for the low depth, the bulk of the energy of flow is in the form of kinetic energy ($Q^2/2gA^2$); whereas for a larger depth, most of the energy is in the form of potential energy. Flow under a *sluice gate* (Fig. 4-13, page 182) is an example of flow in which two depths occur for a given value of specific energy. The large depth and low kinetic energy occurs upstream of the gate; the low depth and large kinetic energy occurs downstream. The depths as used here are called *alternate depths*. That is, for a given value of E, the large depth is alternate to the low depth, or vice versa. Returning to the flow under the sluice gate, we find that if we main-

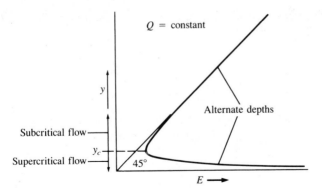

Figure 4-12   Relation of depth versus specific energy

tain the same rate of flow but set the gate with a larger opening, as in Fig. 4-13b, the upstream depth will drop, and the downstream depth will rise. Thus we have different alternate depths and a smaller value of specific energy than before. This is consistent with the diagram in Fig. 4-12.

Finally, it can be seen in Fig. 4-12 that a point will be reached where the specific energy is minimum and only a single depth occurs. At this point, the flow is termed *critical*. Thus one definition of critical flow is the flow that occurs when the specific energy is minimum for a given discharge. The flow for which the depth is less than critical (velocity is greater than critical) is termed *super-critical flow*, and the flow for which the depth is greater than critical (velocity is less than critical) is termed *subcritical flow*. Using this terminology, we can see

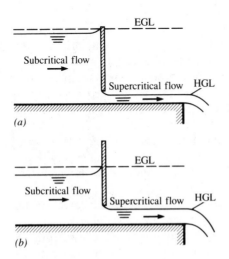

Figure 4-13   Flow under a sluice gate

that subcritical flow occurs upstream and supercritical flow occurs downstream of the sluice gate in Fig. 4-13. We will consider other aspects of critical flow in the next section.

Characteristics of Critical Flow

We have already seen that critical flow occurs when the specific energy is minimum for a given discharge. The depth for this condition may be determined if we solve for dE/dy from $E = y + Q^2/2gA^2$ and set dE/dy equal to zero:

$$\frac{dE}{dy} = 1 - \frac{Q^2}{gA^3} \cdot \frac{dA}{dy} \qquad (4\text{-}15)$$

However, $dA = T\,dy$, where T is the width of the channel at the water surface as shown in Fig. 4-14. Then Eq. (4-15), with $dE/dy = 0$, will reduce to

$$\frac{Q^2 T_c}{gA_c^{\,3}} = 1 \qquad (4\text{-}16)$$

or $\qquad\dfrac{A_c}{T_c} = \dfrac{Q^2}{gA_c^{\,2}} \qquad (4\text{-}17)$

$$\frac{A_c}{T_c} = \frac{V_c^2}{g} \qquad (4\text{-}18)$$

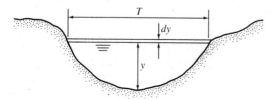

Figure 4-14   Open channel relations

E X A M P L E   4 - 6   Determine the critical depth in a trapezoidal channel for a discharge of 500 cfs. The width of the channel bottom is 20 ft, and the sides slope upward at an angle of 45°.

S O L U T I O N   Starting with Eq. (4-16),

$$\frac{Q^2 T_c}{gA_c^{\,3}} = 1$$

or $\dfrac{A_c{}^3}{T_c} = \dfrac{Q^2}{g}$

Then, for $Q = 500$ cfs

$$\frac{A_c{}^3}{T_c} = \frac{500^2}{32.2} = 7764 \text{ ft}^2$$

For this channel, $A = y(b + y)$ and $T = b + 2y$. Then by iteration (choose y and compute A^3/T), we can find y that will yield an A^3/T equal to 7764 ft². Such a solution yields $y_c = 2.57$ ft. ■

If the channel is of rectangular cross section, then A_c/T_c is the critical depth, and $Q^2/A_c{}^2 = q^2/y_c{}^2$, so the formula for critical depth (Eq. 4-17) becomes

$$y_c = \left(\frac{q^2}{g}\right)^{1/3} \tag{4-19}$$

where q is the discharge per unit width of channel.

If we apply Eq. (4-18) to a rectangular channel, divide it by $A_c/T_c = y_c$, and then take the square root of both sides, we obtain

$$\frac{V_c}{\sqrt{gy_c}} = 1 \tag{4-20}$$

The left side of Eq. (4-20) is the Froude number; therefore, the Froude number is equal to unity when the flow is critical.

Originally, the term *critical flow* probably related to the unstable character of the flow for the condition. If we refer to Fig. 4-12, we see that only a slight change in specific energy will cause the depth to increase or decrease a significant amount; this is a very unstable condition. In fact, observations of critical flow in open channels show that the water surface consists of a series of standing waves. Because of the unstable nature of the depth in critical flow, designing canals so that normal depth is either well above or well below critical depth is usually best. The flow in canals and rivers is usually subcritical; however, the flow in steep chutes or over spillways is supercritical.

Occurrence of Critical Depth

Critical flow occurs when a liquid passes over a broad-crested weir (Fig. 4-15a). The principle of the broad-crested weir is illustrated by first considering a closed sluice gate that prevents water from being discharged from

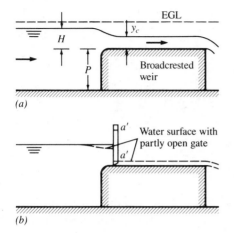

Figure 4-15 Flow over a broad-crested weir

the reservoir (Fig. 4-15b). If the gate is opened a small amount (gate position $a' - a'$), the flow upstream of the gate will be subcritical, and the flow downstream will be supercritical (like the condition first introduced in Fig. 4-13). As the gate is opened further, a point is finally reached where the depths upstream and downstream are the same. This is the critical condition. At this gate opening and beyond, the gate has no influence on the flow; this is the condition shown in Fig. 4-15a, the broad-crested weir. If the depth of flow over the weir is measured, the rate of flow can easily be computed from Eq. (4-19):

$$q = \sqrt{gy_c^3}$$

or $\qquad Q = L\sqrt{gy_c^3} \qquad\qquad\qquad\qquad$ (4-21)

where L is the length of the weir crest.

Because $y_c/2 = V_c^2/2g$ [from Eq. (4-20)], it is easily shown that $y_c = 2/3E$, where E is the total head above the crest ($H + V_{approach}^2/2g$); hence Eq. (4-21) can be rewritten as

$$Q = L\sqrt{g}\left(\frac{2}{3}\right)^{3/2} E^{3/2}$$

or $\qquad Q = 0.385L\sqrt{2g}E_c^{3/2} \qquad\qquad\qquad$ (4-22)

For high weirs, the upstream velocity of approach is almost zero. Hence Eq. (4-22) can be expressed as

$$Q_{theor} = 0.385L\sqrt{2g}H^{3/2} \qquad\qquad\qquad$$ (4-23)

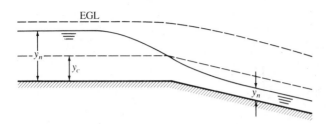

Figure 4-16 Critical depth at break in grade

Equations (4-22) and (4-23) are the basic theoretical equations for a broad-crested weir, and they may be used for roughly estimating the discharge. However, because the discharge will also be influenced by head loss and the shape of the weir, a coefficient of discharge should be used to reflect these effects.*

The depth also passes through a critical stage in channel flow where the slope changes from a mild one to a steep one. Here, a *mild slope* is a slope for which the normal depth y_n is greater than y_c. Likewise, a *steep slope* is one for which $y_n < y_c$. This condition is shown in Fig. 4-16. Note that y_c is the same for both slopes in the figure because y_c is a function of the discharge only. However, normal depth (uniform flow depth) for the mild upstream channel is greater than critical, whereas the normal depth for the steep downstream channel is less than critical; hence it is obvious that the depth must pass through a critical stage. Experiments show that critical depth occurs a very short distance upstream of the intersection of the two channels.

Another place where critical depth occurs is upstream of a free overfall at the end of the channel with a mild slope (Fig. 4-17). Critical depth will occur at a distance of 3 to 4 y_c upstream of the brink. Such occurrences of critical depth (at a break in grade or at a brink) are useful in computing surface profiles because they provide a point for starting surface-profile calculations.[†]

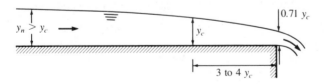

Figure 4-17 Critical depth at a free overfall

---

* We discuss these effects in more detail in Sec. 4–5.

[†] The procedure for making these computations starts on page 201.

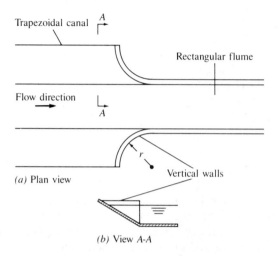

(a) Plan view

(b) View A-A

Figure 4-18   Cylinder quadrant inlet transition from trapezoidal canal to rectangular channel

Channel Transitions

PURPOSE AND TYPES OF TRANSITIONS   A structure designed to convey water smoothly from a conduit of one shape to one of a different shape is called a *transition*. A common transition for open channel flow is used between a canal of trapezoidal cross section and a flume of rectangular section, as shown in Fig. 4-18. Transitions are also used between open channels and inverted siphons (pipe used to convey water across depressions or under highways). If the transition is from a conduit of large cross section to one of smaller cross section, it is an *inlet transition* or a *contraction*. If the transition is from a smaller one to a larger one, it is an *expansion*. In this text, we consider only subcritical flow transitions.*

The simplest type of transition is a straight wall constructed normal to the flow direction, as shown in Fig. 4-19, page 188. This type of transition can work, but it will produce excessive head loss because of the abrupt change in cross section and ensuing separation that would occur. To prevent excessive head losses and to reduce the possibility of erosion in the case of an expansion to an erodible channel, a more gradual type of transition is usually used. Three common types of gradual transitions are the *cylinder-quadrant*, the *wedge* (often called *broken-back*), and the *warped-wall transition* (Fig. 4-20, page 188). All three of these can be successfully used for inlet transitions, but because they are

---

* For more information on transitions in general and supercritical flow in particular, see Chow (4) and Rouse (18).

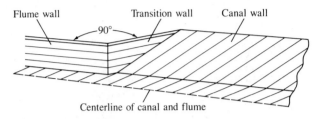

Flume wall   Transition wall   Canal wall

90°

Centerline of canal and flume

Figure 4-19   Simplest type of transition between a
canal and a flume

more gradual expansions the wedge and warped wall are best suited for expansions if head loss or erosion is significant. The warped wall is generally more expensive to build than the others because of the complicated form work required.

If either the wedge or warped-wall transition is used, experience has shown that the expansion should have a more gradual change in section than the inlet transition (9, 21). Quantitatively, the recommendations are manifested in the value of the expansion or contraction angle θ as shown in Fig. 4-21. For the wedge transition, the recommended angle is 27.5° and 22.5°, respectively, for inlets and expansions. For the warped-wall transition, the recommended angle, is 12.5° for both the inlet and expansion.

DESIGN OF TRANSITION TO JOIN CANAL AND FLUME   Before the transition itself is designed, one must be given the depth and velocity in both the flume and canal and the water surface elevation in the upstream channel for the case of an inlet (downstream water surface elevation for expansion). Then

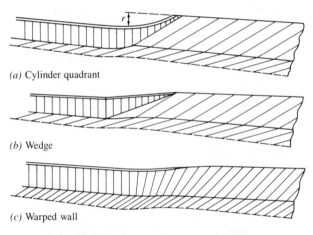

(a) Cylinder quadrant

(b) Wedge

(c) Warped wall

Figure 4-20   Half sections of commonly used
transitions

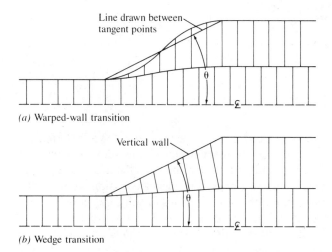

(a) Warped-wall transition

(b) Wedge transition

Figure 4-21 Expansion (or contraction) angles

the transition can be designed step by step as follows:

1. Choose the type of transition to be used (cylinder quadrant, wedge, or warped wall).

2. For an inlet transition, calculate the water surface elevation at the downstream end of the transition. For an expansion, calculate the upstream water surface elevation. This is done by applying the energy equation between the upstream and downstream ends of the transition. The energy equation will include a head loss term for the transition. For an inlet transition, the head loss is given as $K_I V^2/2g$, where K_I is the head loss coefficient for the transition, and V is the velocity in the downstream conduit (the highest mean velocity). For an expansion, the head loss is given as $K_E(V_1^2 - V_2^2)/2g$, where K_E is the expansion loss coefficient, V_1 is the mean velocity at the upstream end of the expansion, and V_2 is the mean velocity at the downstream end. Table 4-4 shown on the next page lists typical loss coefficients for various types of transitions.

3. For an inlet transition, calculate the downstream invert elevation. For an expansion, calculate the upstream invert elevation.* The invert elevation is simply the water surface elevation at that section minus the depth of water in the flume.

4. Establish invert elevations along the transition by making a straight-line elevation change between the upstream and downstream ends of the transition.

---

* This assumes that the invert elevation (elevation of the bottom of the conduit) of the canal is essentially fixed. Therefore, the invert elevation of the flume must be solved for.

Table 4-4 Transition Loss
Coefficients

Type of Transition	K_I	K_E
Cylinder quadrant	0.10	0.50
Wedge	0.20	0.50
Warped wall	0.10	0.30

5. Establish water surface elevations through the transition. As a first approximation, assume a straight-line change in water surface elevation between the upstream and downstream ends. Then, using velocities based on the assumed water surface elevations, apply the energy equation from the upstream end of the transition to the other sections to solve for more accurate water surface elevations at prescribed sections. For these calculations, a head loss in proportion to distance along the transition is usually used (for example, the head loss halfway down the transition would be assumed to be one half of that for the entire transition).

EXAMPLE 4-7 A transition is needed between a trapezoidal canal carrying water (depth = 3.00 ft; velocity = 2.30 ft/s) and a flume of rectangular cross section. The flume will convey the water around a steep hill. The canal has a bottom width of 10 ft and side slopes of 1 vertical to 2 horizontal. The invert elevation of the canal at its downstream end (upstream end of transition) is 1000 ft. The flume velocity is to be 5.90 ft/s. Determine the proportions for the flume (width and wall height) to keep the Froude number below 0.50, and design a transition to join the canal and flume.

SOLUTION

1. Choose the type of transition.

For this case, let us use a wedge transition. With this type, we use a head loss coefficient of 0.20 ($K_I = 0.20$). Figure A shows the plan view of the transition. The cross-sectional flow area in the channel is $A = 10 \times 3 + 6 \times 3 = 48 \text{ ft}^2$.

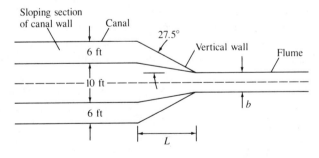

Figure A

2.   Determine the flume dimensions and depth of flow in the flume.

$$Q = V_{flume} A_{flume}$$

Let $b/d = 1.0$, then $A = d^2$, or

$$d = \sqrt{\frac{Q}{V}} = \sqrt{\frac{2.3 \times 48}{V}} = \sqrt{\frac{110.40}{V}}$$

$$\sqrt{V}d = \sqrt{110.40} \tag{4-24}$$

The Froude number is to be limited to 0.50, so

$$\mathrm{Fr} = \frac{V}{\sqrt{gd}} = 0.5$$

or $$V = 0.5\sqrt{gd} \tag{4-25}$$

Solving Eqs. (4-24) and (4-25) for d yields $d = 4.326$ ft.
     For design, let $b_{flume} = 4.40$ ft. Then

$$d_{flume} = \frac{Q}{b \times 5.9} = 4.25 \text{ ft}$$

3.   Determine the length L of transition.

$$\tan 27.5° = \frac{11 - 0.50 \times 4.40}{L}$$

or $$L = 16.90 \text{ ft}$$

For design, let $L = 17.0$ ft.

4.   Determine the water surface elevation in the flume.
     Use the energy equation written from canal to flume and assume that
$\alpha_1 = \alpha_2 = 1.1$.

$$z_1 + \alpha_1 \frac{V_1^2}{2g} = z_2 + \alpha_2 \frac{V_2^2}{2g} + \sum h_L$$

$$1003.00 + 1.1 \times \frac{2.30^2}{64.4} = z_2 + 1.1 \times \frac{5.9^2}{64.4} + 0.2 \times \frac{5.9^2}{64.4}$$

$$z_2 = 1002.39 \text{ ft}$$

5.   Determine the invert elevation of the flume.

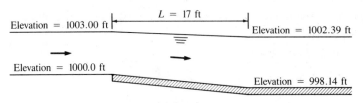

Figure B

$$\text{Invert elevation} = \text{water surface elevation} - d_{\text{flume}}$$

$$= 1002.39 - 4.25$$

$$= 998.14 \text{ ft}$$

Figure B shows the water surface and invert profiles.

6.   Now check the velocities at sections *A* and *B* where these two sections are 5 ft and 10 ft, respectively, downstream of the upstream end of the transition. Figure C shows the plan view of the transition along with dimensions at sections *A* and *B*.*

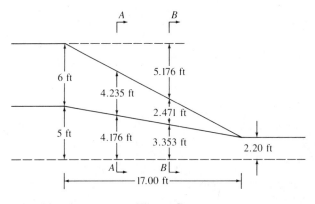

Figure C

First assume a plane water surface between the upstream end of the transition and the downstream end. With this assumption, we can calculate the depths at sections *A* and *B*.

---

* In Fig. C, the transition wall height should include freeboard (extra height added to wall as a safety factor) to account for uncertainties in the design (for example, the assumed loss coefficient may not be exactly the same as the "as built" coefficient) and to provide for waves that were not considered in the design calculations. The height of freeboard is usually from 0.5 to 1.0 ft and will usually match the freeboard used for the canal and flume.

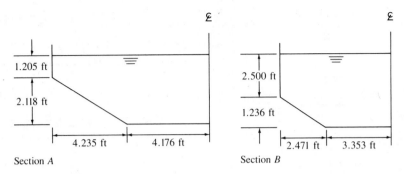

Figure D

$$d_A = 3.00 + \left(\frac{5}{17}\right)[(1002.39 - 998.14) - (1003.00 - 1000.00)]$$

$$= 3.368 \text{ ft}$$

In a similar manner, d_B is calculated.

$$d_B = 3.735 \text{ ft}$$

Figure D shows the cross-sectional flow area (half section) at section A.

$$A_A = 2\left[(4.176 \times 3.368) + (1.205 \times 4.235) + \frac{1}{2} \times (2.118 \times 4.235)\right]$$

$$= 47.306 \text{ ft}^2$$

Then $$V_A = \frac{Q}{A_A} = \frac{2.3 \times 48}{47.306} = 2.33 \text{ ft/s}$$

Similar calculations for section B yield:

$$A_B = 40.456 \text{ ft}^2$$
$$V_b = 2.729 \text{ ft/s}$$

7.   With the velocities obtained above, we can more accurately determine the water surface elevation at sections A and B by using the energy equation and assuming that the head loss is linearly distributed along the transition.

$$z_1 + \alpha_1 \frac{V_1^2}{2g} = z_A + \alpha_A \frac{V_A^2}{2g} + h_{L_1 \rightarrow A}$$

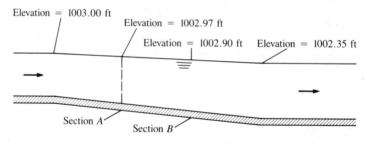

Figure E

$$1003.0 + 1.1 \times \frac{2.3^2}{64.4} = z_A + 1.1 \times \frac{2.33^2}{64.4} + \left(\frac{5}{17}\right) \times 0.20 \times \frac{5.9^2}{64.4}$$

$$1003.0 + 0.090 = z_A + 0.093 + 0.0318$$

$$z_A = 1002.97 \text{ ft}$$

A similar calculation for z_B yields $z_B = 1002.90$ ft.

Figure E shows the corrected water surface profile.   ■

The Hydraulic Jump

OCCURRENCE OF THE HYDRAULIC JUMP   When the flow is supercritical in an upstream section of a channel and is then forced to become subcritical in a downstream section (the change in depth can be forced by a sill in the channel or by just the prevailing depth in the stream further downstream), a rather abrupt change in depth usually occurs and considerable energy loss accompanies the process. This flow phenomenon, called the *hydraulic jump* (Fig. 4-22),

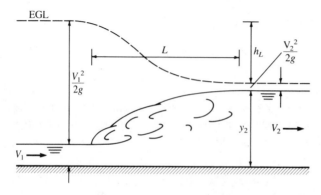

Figure 4-22   Definition sketch for the hydraulic jump

is often considered in the design of open channels and spillways. For example, many spillways are designed so that a jump will occur on an apron of the spillway, thereby reducing the downstream velocity so that objectionable erosion of the river channel is prevented. If a channel is designed to carry water at supercritical velocities, the designer must be certain that the flow will not become subcritical prematurely. If it did, overtopping of the channel walls would undoubtedly occur, with consequent failure of the structure. Because the energy loss in the hydraulic jump is initially not known, the energy equation is not a suitable tool for analysis of the velocity-depth relationships. Therefore, the momentum equation is applied to the problem.

DERIVATION OF DEPTH RELATIONSHIPS   Consider flow as shown in Fig. 4-22. Here it is assumed that uniform flow occurs both upstream and downstream of the jump and that the resistance of the channel bottom is negligible. The derivation is for a horizontal channel, but experiments show that the results of the derivation will apply to all channels of moderate slope ($S_0 < 0.02$). We start the derivation by applying the momentum equation in the x direction to the control volume shown in Fig. 4-23.

$$\sum F_x = \sum_{cs} V_x \rho \mathbf{V} \cdot \mathbf{A}$$

The forces are the hydrostatic forces on each end of the system, thus the following is obtained:

$$\bar{p}_1 A_1 - \bar{p}_2 A_2 = \rho V_1(-V_1 A_1) + \rho V_2(V_2 A_2)$$

or $\bar{p}_1 A_1 + \rho Q V_1 = \bar{p}_2 A_2 + \rho Q V_2$ (4-26)

In Eq. (4-26), $\bar{p}_1$ and $\bar{p}_2$ are the pressures at the centroids of the respective areas A_1 and A_2.

A representative problem might be to determine the downstream depth y_2 given the discharge and upstream depth. The left-hand side of Eq. (4-26) would be known because V, A, and p, are all functions of y and Q, and the right-hand side is a function of y_2; therefore, y_2 can be solved for.

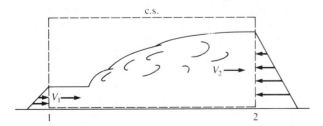

Figure 4-23   Control-volume analysis for the hydraulic jump

EXAMPLE 4-8   Water flows in a trapezoidal channel at a rate of 300 cfs. The channel has a bottom width of 10 ft and side slopes of 1 to 1. If a hydraulic jump is forced to occur where the upstream depth is 1.00 ft, what will be the downstream depth and velocity?

SOLUTION   For the upstream section, the area A_1 is equal to 11 ft². The depth of the centroid of A_1 is found to be 0.470 ft; therefore, the pressure at the centroid is 62.4 lb/ft³ × 0.470 ft = 29.3 lb/ft². Further $V = Q/A_1 = 27.3$ ft/s. Substituting these values into Eq. (4-26), we get (to 4 significant digits):

$$29.3 \times 11 + 1.94 \times 300 \times 27.3 = \bar{p}_2 A_2 + \rho Q V_2$$

or

$$\bar{p}_2 A_2 + \rho Q V_2 = 16{,}210$$

$$\gamma \bar{y}_2 A_2 + \frac{\rho Q^2}{A_2} = 16{,}210$$

where $A_2 = b y_2 + y_2{}^2$

$$\bar{y}_2 = \frac{\sum A_i \bar{y}_i}{A_2} = \left[b\left(\frac{y_2}{2}\right) + \left(\frac{1}{3}\right) y_2{}^2 \right] \bigg/ (b + y_2)$$

By trial and error, we can solve for A_2 and y_2. The solution gives

$$y_2 = 5.75 \text{ ft}$$

and $$V_2 = \frac{Q}{A_2} = 3.31 \text{ ft/s}$$ ■

HEAD LOSS DUE TO A HYDRAULIC JUMP   Because of the intense turbulent mixing that occurs in a hydraulic jump, the head loss due to the jump is relatively large. This is shown in Fig. 4-22. One can determine the head loss due to the hydraulic jump by writing the energy equation across the jump and solving for h_L.*

HYDRAULIC JUMP IN A RECTANGULAR CHANNEL   If a jump occurs in a rectangular channel, the solution of Eq. (4-26) yields a formula for y_2 as a function of y_1 and the Froude number of the upstream flow:

$$y_2 = \frac{y_1}{2} \left(\sqrt{1 + 8\text{Fr}_1{}^2} - 1 \right) \tag{4-27}$$

where $\text{Fr}_1 = \dfrac{V_1}{\sqrt{g y_1}}$.

---

* For a discussion of other head loss characteristics associated with the hydraulic jump, see Sec. 7-2 of Chapter 7.

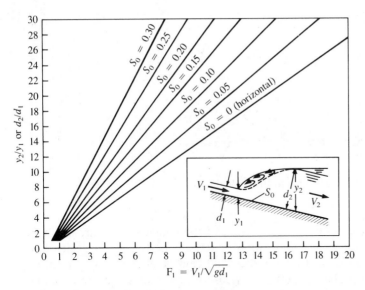

Figure 4-24 Experimental relations between F_1 and y_2/y_1 or d_2/d_1 for jumps in sloping channels [Adapted from *Open Channel Hydraulics* by Chow (4) Copyright 1959, McGraw-Hill Book Company, New York; used with permission of McGraw-Hill Book Company.]

A graph of Eq. (4-27) is shown in Fig. 4-24 (the case for $S_0 = 0$). Figure 4-24 also shows the relations between depth and Froude numbers for hydraulic jumps in channels that are not horizontal.

LENGTH OF THE HYDRAULIC JUMP   The length of the hydraulic jump is the distance measured from the front face of the jump to a point on the surface immediately downstream of the roller, as shown in Fig. 4-22. No general theoretical solution exists for this length. However, experiments in rectangular channels (4) show that $L = 6y_2$ for $4 < \mathrm{Fr}_1 < 20$. For Froude numbers outside this range, the length is somewhat less than $6y_2$.

STILLING BASINS   We have already shown that the transition from supercritical to subcritical flow produces a hydraulic jump, and that the relative height of the jump (y_2/y_1) is a function of Fr_1. Because flow over the spillway of a dam invariably results in supercritical flow at the lower end of the spillway, and because flow in the channel downstream of a spillway is usually subcritical, it is obvious that a hydraulic jump must form near the base of the spillway (see Fig. 4-25). The downstream portion of the spillway must be designed so that the hydraulic jump always forms on the concrete structure itself. If the hydraulic

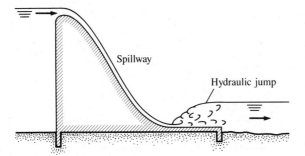

Figure 4-25   Spillway of dam and hydraulic jump

jump were allowed to form downstream of the concrete structure, as in Fig. 4-26, severe erosion of the foundation material from the high velocity supercritical flow could undermine the dam and cause complete failure of it. One way to solve this problem might be to incorporate a long sloping apron into the design of the spillway, as shown in Fig. 4-27. A design like this would work very satisfactorily from the hydraulics point of view. For all combinations of Fr_1 and water surface elevation in the downstream channel, the jump would always form on the sloping apron. However, its main drawback is cost of construction. Construction costs will be reduced as the length (L) of the stilling basin is reduced. Much research has been devoted to the design of stilling basins that will operate properly for all upstream and downstream conditions and yet be relatively short to reduce the cost of the project. Research by the U.S. Bureau of Reclamation and the U.S. Corps of Engineers has resulted in sets of standard designs that can be used.*

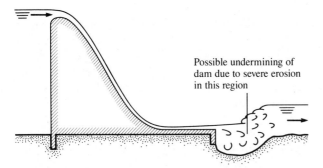

Figure 4-26   Hydraulic jump occurring downstream of spillway apron

---

* For details of some of these designs, see Sec. 7-2, Chapter 7.

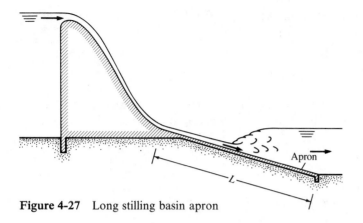

Figure 4-27   Long stilling basin apron

Gradually Varied Flow in Open Channels

BASIC DIFFERENTIAL EQUATION FOR GRADUALLY VARIED FLOW   There
are a number of cases of open channel flow in which the change in water surface
profile is so gradual that it is possible to integrate the relevant differential equation from one section to another to obtain the desired change in depth. This
may be either an analytical integration or, more commonly, a numerical
integration. Considering the energy equation (Eq. 4-10) written for a reach of
Δx in length, and defining S_f as the friction slope, h_f/L, Eq. (4-10) becomes

$$y_1 + \frac{V_1^2}{2g} + S_0 \Delta x = y_2 + \frac{V_2^2}{2g} + S_f \Delta x \qquad (4\text{-}28)$$

Now, let $\Delta y = y_2 - y_1$ and

$$\frac{V_2^2}{2g} - \frac{V_1^2}{2g} = \frac{d}{dx}\left(\frac{V^2}{2g}\right)\Delta x$$

Eq. (4-28) then becomes

$$\Delta y = S_0 \Delta x - S_f \Delta x - \frac{d}{dx}\left(\frac{V^2}{2g}\right)\Delta x$$

Dividing through by Δx and taking the limit as Δx approaches zero gives us

$$\frac{dy}{dx} + \frac{d}{dx}\left(\frac{V^2}{2g}\right) = S_0 - S_f \qquad (4\text{-}29)$$

The second term is rewritten as $[d(V^2/2g)/dy]\,dy/dx$ so that Eq. (4-29) simplifies to

$$\frac{dy}{dx} = \frac{S_0 - S_f}{1 + d(V^2/2g)/dy} \tag{4-30}$$

To put Eq. (4-30) in a more usable form, we express the denominator in terms of the Froude number. This is done by observing that

$$\frac{d}{dy}\left(\frac{V^2}{2g}\right) = \frac{d}{dy}\left(\frac{Q^2}{2gA^2}\right) \tag{4-31}$$

After differentiating the right side of Eq. (4-31), the equation becomes

$$\frac{d}{dy}\left(\frac{V^2}{2g}\right) = \frac{-2Q^2}{2gA^3}\cdot\frac{dA}{dy}$$

But $dA/dy = T$(top width), and $A/T = D$ (hydraulic depth); therefore,

$$\frac{d}{dy}\left(\frac{V^2}{2g}\right) = \frac{-Q^2}{gA^2D}$$

or

$$\frac{d}{dy}\left(\frac{V^2}{2g}\right) = -\mathrm{Fr}^2$$

Hence, when $-\mathrm{Fr}^2$ is substituted for $d(V^2/2g)/dy$ in Eq. (4-30), we obtain

$$\frac{dy}{dx} = \frac{S_0 - S_f}{1 - \mathrm{Fr}^2} \tag{4-32}$$

This, the general differential equation for gradually varied flow, is used to describe the various types of water-surface profiles that occur in open channels. In the derivation of the equation, S_0 and S_f were taken as positive when sloping downward in the direction of flow. Since y is measured from the bottom of the channel, $dy/dx = 0$ if the slope of the water surface is equal to the slope of the channel bottom, and dy/dx is positive if the water surface slope is less than the channel slope for positive S_0. With these definitions in mind, we will now consider the different forms of water surface profiles.

CLASSIFICATION OF SURFACE PROFILES   Water surface profiles are classified two different ways: according to the slope of the channel (*mild, steep, critical, horizontal,* or *adverse*) and according to the actual depth of flow in relation to the critical and normal depths (zone 1, 2, or 3). The first letter of the type of slope (M, S, C, H, or A) in combination with 1, 2, or 3 defines the type of surface profile.

If the slope is so small that the normal depth (uniform flow depth) is greater than critical depth for the given discharge, then the slope of the channel is *mild*, and the water surface profile is given an M classification. Similarly, if the channel slope is so *steep* that a normal depth less than critical is produced, then the channel is steep, and the water surface profile is given an S designation. If the slope's normal depth equals its critical depth, then we have a *critical slope*, denoted by C. *Horizontal* and *adverse* slopes, denoted by H and A, respectively, are special categories because normal depth does not exist for them. An adverse slope is characterized by a slope upward in the flow direction. The 1, 2, and 3 designations of water surface profiles indicate if the actual flow depth is greater than both normal and critical depths (zone 1), between the normal and critical depths (zone 2), or less than both normal and critical depths (zone 3). The basic shapes of the various possible profiles are shown in Fig. 4-28, page 202. Figure 4-29 on page 203 shows typical examples of physical situations that produce the various profiles.

With the foregoing introduction to the classification of surface profiles, we can now refer to Eq. (4-32) to describe the shape of the profiles. For example, if we consider the M3 profile, it is known that $Fr > 1$ because the flow is supercritical $(y < y_c)$ and $S_f > S_0$ because the velocity is greater than normal velocity. Inserting these relative values into Eq. (4-32), we see that both the numerator and denominator are negative; thus, dy/dx must be positive (the depth increases in the direction of flow). As critical depth is approached, the Froude number approaches unity; hence, the denominator of Eq. (4-32) approaches zero. Therefore, as the depth approaches critical depth, $dy/dx \rightarrow \infty$. What actually occurs when the critical depth is approached in supercritical flow is that a hydraulic jump forms, and a discontinuity in profile is thereby produced.

Certain general features of profiles, as shown in Fig. 4-28, are evident. First, as the depth becomes very great the velocity of flow approaches zero, and $Fr \rightarrow 0$ and $S_f \rightarrow 0$. Hence it follows from Eq. (4-32) that dy/dx approaches S_0. That is, the depth increases at the same rate that the channel bottom drops away from the horizontal. Thus, the water surface approaches the horizontal. The curves that tend this way are M1, S1, and C1. A physical example of the M1 curve is that of the water surface profile behind the dam, as shown in Fig. 4-29. This is often called a *backwater curve*. The second general feature of several of the profiles is that the depth approaches normal depth asymptotically. This is shown in the S2, S3, M1, and M2 profiles. For these cases, it is seen that in supercritical flow $y \rightarrow y_n$ going downstream, and in subcritical flow $y \rightarrow y_n$ going upstream. In Fig. 4-28, profiles approaching critical depth are shown by broken lines. This is done because near critical depth discontinuities develop (hydraulic jump), or the streamlines are very curved (such as near a brink); therefore, the surface profiles cannot be accurately predicted because Eq. (4-32) is based on one-dimensional flow, which in these regions is invalid.

QUANTITATIVE EVALUATION OF SURFACE PROFILES   In practice, most surface profiles are generated by numerical integration, that is, by dividing the

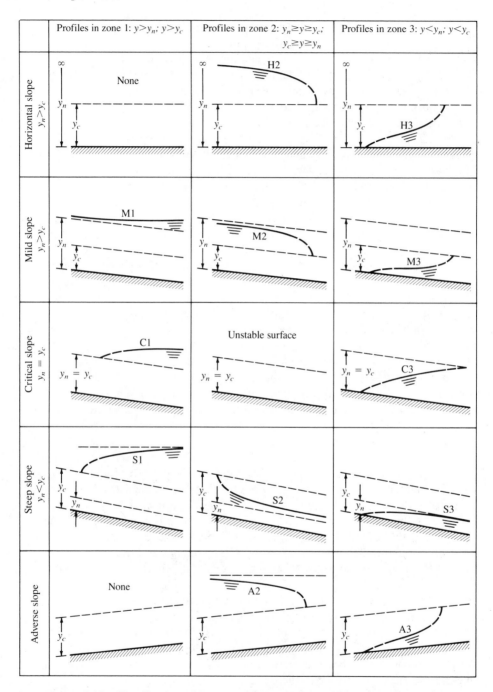

Figure 4-28 Classification of water-surface profiles of gradually varied flow [Adapted from *Open Channel Hydraulics* by Chow (4) Copyright 1959, McGraw-Hill Book Company, New York; used with permission of McGraw-Hill Book Company.]

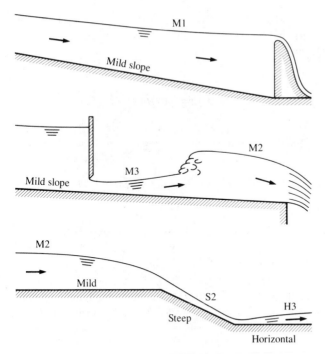

Figure 4-29   Water surface profiles associated with flow behind a dam, flow under a sluice gate, and flow in a channel with a change in grade

channel into short reaches and carrying the computation for water surface elevation from one end of the reach to the other. The two most common approaches are the direct step method and the standard step method.

   In the *direct step method*, the depth and velocity are known at a given section of the channel (one end of the reach), and one arbitrarily chooses the depth at the other end of the reach. Then the length of the reach is solved for. The applicable equation is Eq. (4-28). In that equation, if we let $y_1 + \alpha_1 \dfrac{V_1{}^2}{2g} = E_1$ and $y_2 + \alpha_2 \dfrac{V_2{}^2}{2g} = E_2$ and solve for Δx, we get

$$\Delta x = \frac{E_1 - E_2}{S_f - S_0} \tag{4-33}$$

The procedure for evaluating a profile is to first ascertain the type of profile that applies to the given reach of channel (we use the methods of the preceding section). Then, starting from a known depth, a finite value of Δx is computed for an arbitrarily chosen change in depth. The process of computing Δx, step by step in the upstream direction (negative Δx) or in the downstream direction (positive Δx) along the channel is repeated until the full reach of channel has been covered. Usually small changes of y are taken so that the friction slope is approximated by the following:

$$S_f = \frac{n^2 V^2}{2.22 R^{4/3}}$$ (Manning equation-English units)

or $$S_f = \frac{f V^2}{8 g R}$$ (Darcy-Weisbach equation)

In the above equations, V and R are the mean values in the reach; $V = (V_1 + V_2)/2$ and $R = (R_1 + R_2)/2$.

EXAMPLE 4-9   Water discharges from under a sluice gate into a horizontal channel at a rate of 1 m³/s per meter of width as shown in Fig. A. What is the classification of the water surface profile? Quantitatively evaluate the profile downstream of the gate and determine whether or not it will extend all the way to the abrupt drop 80 m downstream. Make the simplifying assumption that the resistance factor f is equal to 0.02 and that the hydraulic radius R is equal to the depth y.

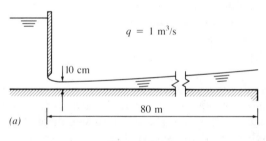

Figure A

SOLUTION   First, determine the critical depth y_c:

$$y_c = \left(\frac{q^2}{g}\right)^{1/3} = \left[\left(\frac{1^2 \text{ m}^4/\text{s}^2)}{(9.81 \text{ m/s}^2)}\right)\right]^{1/3} = 0.467 \text{ m}$$

The depth of flow from the sluice gate is less than the critical depth. Hence the water surface profile is classified as type H3.

To solve for the depth versus distance along the channel, we apply Eq. (4-33) using a numerical approach. In this example we chose to make the change in depth 0.04 m. The results of the computations are shown on the next page. From the numerical results, we plot the profile shown in Fig. B on page 206, and we see that the *profile extends to the abrupt drop.*

Section Number Downstream of Gate	Depth y (m)	Velocity at Section V (m/s)	Mean Velocity in Reach $(V_1+V_2)/2$	V^2	Mean Hydraulic Radius $R_m = (y_1+y_2)/2$	$S_f = \dfrac{fV^2_{mean}}{8gR_m}$	$\Delta x = \dfrac{(y_1-y_2)+\dfrac{(V_1^2-V_2^2)}{2g}}{(S_f-S_0)}$	Distance From Gate x (m)
1 (at gate)	0.1	10	—	100	—	—	—	0
	—	—	8.57	73.4	0.12	0.156	15.7	—
2	0.14	7.14	—	51.0	—	—	—	15.7
	—	—	6.35	40.3	0.16	0.064	15.3	—
3	0.18	5.56	—	30.9	—	—	—	31.0
	—	—	5.05	25.5	0.20	0.032	15.1	—
4	0.22	4.54	—	20.6	—	—	—	46.1
	—	—	4.19	17.6	0.24	0.019	13.4	—
5	0.26	3.85	—	14.8	—	—	—	59.5
	—	—	3.59	12.9	0.28	0.012	12.4	—
6	0.30	3.33	—	11.1	—	—	—	71.9
	—	—	3.13	9.8	0.32	0.008	10.9	—
7	0.34	2.94	—	8.6	—	—	—	82.8

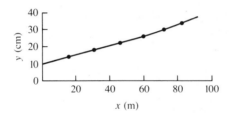

Figure B   ■

The direct step method is ideally suited for prismatic channels because the channel cross section is independent of position in the reach. However, in non-prismatic channels such as in natural channels, the channel cross sections are not independent of distance along the channel. In these channels, a field survey is usually made to provide the data required at the sections considered in the computation. The computations are made between the sections for which the data are available. Therefore, the computation procedure now involves the determination of depth for a given Δx. This computational procedure is called the *standard step method*.

To develop the procedure for the standard step method, we use Eq. (4-28) with the kinetic energy correction factors included plus another head loss term, h_ℓ, where h_ℓ are head losses in the reach in addition to losses resulting from surface resistance alone. These additional losses may be caused by such things as abrupt expansions, contractions, or bends. Thus the relevant equation is

$$y_1 + \frac{\alpha_1 V_1{}^2}{2g} + S_0 \Delta x = y_2 + \frac{\alpha_2 V_2{}^2}{2g} + S_f \Delta x + h_\ell \qquad (4\text{-}34)$$

The method of solution is an iterative one. All information is known at one section (section 1), and one assumes a depth for the other section (section 2). Then with this assumed depth, V_2 can be calculated along with S_f and h_ℓ. After the quantities on the right-hand side of Eq. (4-34) have been calculated, a check is made to see if the equation is satisfied. If not, a new value of y_2 is assumed, and so on. The process is continued until a value for y_2 is found that satisfies Eq. (4-34). Because of the repetitive nature of water surface profile computations,

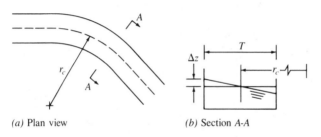

(a) Plan view   *(b)* Section A-A

Figure 4-30   Flow in a channel bend

solution by computer is often done. In recent years, a number of sophisticated computer programs have been written to solve water surface profiles. One popular program developed by the Hydrologic Engineering Center of the U.S. Army Corps of Engineers is HEC-2. It uses the standard step method of solution.

For either the direct step method or the standard step method, the direction in which the computation proceeds is upstream for subcritical flows and downstream for supercritical flows.

Flow in Bends With Subcritical Flow

WATER SURFACE ELEVATION CHANGE   Whenever there is a bend in a channel, the water is accelerated toward the center of curvature of the bend; therefore, a force must be applied to the flowing water toward the center of curvature. This force is produced by a rise in the water surface toward the outside of the bend and a drop in the water surface toward the inside (Fig. 4-30). Therefore, when designing a bend, the outside wall of the channel must be made high enough to accommodate the increase in water surface elevation due to the bend. The amount of elevation change is determined by using Euler's equation applied along a radial line:

$$-\frac{\partial h}{\partial r} = \frac{a_r}{g} \tag{4-35}$$

In Eq. (4-35), the acceleration, a_r, is equal to V^2/r toward the center of curvature (in the negative r direction), and h is the piezometric head ($p/\gamma + z$); therefore, if we consider the change in elevation of the water surface ($p = 0$), we obtain

$$\frac{\partial z}{\partial r} = \frac{V^2}{gr}$$

Thus it can be shown that the increase in water surface elevation from the channel centerline to the outer wall will be approximately

$$\Delta z = \frac{V^2}{gr_c} \cdot \frac{T}{2} \tag{4-36}$$

where T = water surface width in channel
r_c = radius of curvature of centerline of channel.

HEAD LOSS in BENDS   The head loss due to the bend (in addition to the normal head loss due to friction) can be expressed as

$$h_{L(\text{bend})} = C_{\text{bend}} \frac{V^2}{2g} \tag{4-37}$$

For most canals, the depth is considerably less than the width of the canal, and the radius of curvature is quite large. Therefore, the head loss in addition to the normal head loss due to friction is quite small. For example, the state of California (6) used values of 0.01 and 0.02 for the bend coefficient, C, for 90° bends with radii of curvature of $5T$ and $2T$, respectively. For 45° bends, half these values of C were used.

SUPERCRITICAL FLOW IN BENDS   In supercritical flow in bends, cross waves can develop thereby producing a greater change in water surface elevation than one would predict for subcritical flow in bends. Banking of the channel bottom (used for rectangular channels) or carefully designed spiral curve transitions can be used to prevent or suppress cross waves.*

4-4   Measurement of Discharge in Open Channels

Velocity-Area-Integration Method

The discharge past a flow section is given as $Q = VA$. Here the mean velocity, V, is the component of mean velocity normal to the section area, A. In a river, the velocity varies with depth and position across the river. Therefore, to measure the discharge in a river, standard practice is to make velocity measurements at various stations across the river and to apportion to each station the flow-section area that is closest to that particular station. Then the total discharge in the river is given by

$$Q = \sum V_i A_i$$

where V_i = mean velocity at a particular station
   A_i = the section area assigned to the station.

In rivers, theory and experimental evidence both show that the mean velocity in a vertical section is closely approximated by the average of the velocity taken at 0.2 depth and 0.8 depth below the surface. Moreover, if the stream is quite shallow so that it may be difficult to measure the velocity at 0.8 depth, a single measurement of velocity at 0.6 depth is a good approximation to the mean velocity in the vertical section. Figure 4-31 depicts a section in a river where velocity measurements at 0.2 and 0.8 depths have been made in a number of vertical sections across the river. At section 4, the velocities at 0.2 depth and 0.8 depth were measured to be 8.5 ft/s and 6.6 ft/s, respectively; therefore, the mean velocity for that vertical section would be 7.55 ft/s, and the area apportioned to the that section would be 8 ft × $[(20/2) + (15/2)] = 140$ ft². Then the

---

* For details on the design of bends for supercritical flow see Chow (4) and Rouse (18).

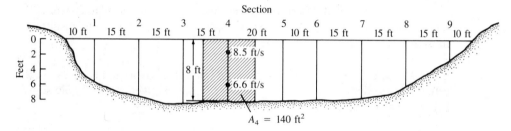

Figure 4-31   River section

ΔQ for section 4 would be given as $V_4 A_4$ or $7.55 \times 140 = 1057$ cfs. The total river discharge is determined by summing all the $V_i A_i$'s across the section.

The most common velocity meter used in rivers in the United States is the Price current meter (Fig. 4-32). Cups on a wheel mounted on a vertical axis cause the wheel to rotate when water flows past it. The meter is calibrated so that the frequency of rotation can be converted to velocity in ft/s or m/s. The meter can be hand held on the end of a rod, or for deep rivers it can be suspended by a cable from a bridge, cable car, or boat. When used in this latter manner, a lead weight is attached to the bottom of the meter (see Fig. 4-32) to stabilize it in the flowing stream.

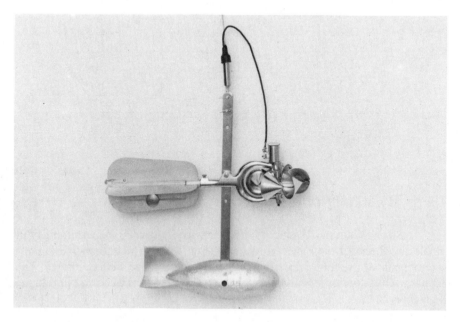

Figure 4-32   Current meter with weight (Courtesy of
Teledyne Gurley, Troy, New York)

The procedure just described for determining the discharge in a river is the same procedure used by the U.S. Geological Survey and leads to results like those given in Table 2-15, page 93.

EXAMPLE 4-10   Consider the river section shown in Fig. 4-31. If the velocities and depths at sections 1 through 9 are as given in the Table shown below, what is the discharge in the river?

	Section								
	1	2	3	4	5	6	7	8	9
Depth (ft)	5.5	7.3	8.5	8.0	7.9	7.9	7.8	6.3	3.2
Velocity at 0.2 depth (ft/s)	7.7	8.2	8.4	8.5	8.4	8.3	8.1	7.7	
Velocity at 0.8 depth (ft/s)	6.0	6.4	6.5	6.6	6.5	6.6	6.3	5.8	
Velocity at 0.6 depth (ft/s)									4.9

SOLUTION   The solution is done in tabular form as shown below.

Section	Depth (ft)	Width (ft)	ΔA (ft^2)	Avg. V (ft/s)	ΔQ (ft/s)
1	5.5	12.5	68.8	6.85	471
2	7.3	15.0	109.5	7.30	799
3	8.5	15.0	127.5	7.45	950
4	8.0	17.5	140.0	7.55	1057
5	7.9	15.0	118.5	7.45	883
6	7.9	12.5	98.8	7.45	736
7	7.8	15.0	117.0	7.20	842
8	6.3	15.0	94.5	6.75	638
9	3.2	12.5	40.0	4.90	196
				$Q =$	6572 cfs

■

Weirs, Flumes, Spillways, and Gates

SHARP-CRESTED WEIRS   A simple device for discharge measurement in canals and flumes is the *sharp-crested weir* (Fig. 4-33). When atmospheric pressure prevails above and below the nappe, it is said to be well ventilated. The discharge equation for a *rectangular weir* (a weir that spans a rectangular flume) is given as

$$Q = K\sqrt{2g}LH^{3/2} \tag{4-38}$$

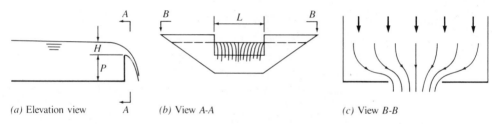

(a) Elevation view   (b) View A-A   (c) View B-B

Figure 4-33   Rectangular weir

In Eq. (4-38), K is the flow coefficient of the weir and is given as

$$K = 0.40 + 0.05 \frac{H}{P} \qquad (4\text{-}39)$$

Based on experimental work by Kindsvater (11), this is valid up to an H/P value of about 10.

Often the weir section does not span the entire width of the channel (Fig. 4-33). Therefore, there will be a contraction of the flow section just downstream of the weir so that the effective length of the weir will be somewhat less than L. Experiments show that this effective reduction in length is approximately equal to $0.20H$ when $L/H > 3$. Thus the formula for a *contracted weir* (one with flow contraction due to end walls) is given as

$$Q = K \sqrt{2g}(L - 0.20H)H^{3/2} \qquad (4\text{-}40)$$

The formula for the *suppressed weir* (a weir without end walls; suppressed contractions) may also be used for the contracted weir if the sides of the weir are inclined as shown in Fig. 4-34. The degree of inclination is made so that the reduction of flow caused by the end contractions is counterbalanced by the increased flow in the "notches" (regions ABC of Fig. 4-34) at either end of the weir. The Cipolletti weir (often used in irrigation canals in the western United States) has an angle of wall inclination of $28°$ to effect the desired result.

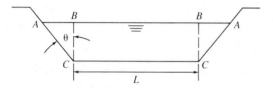

Figure 4-34   Trapezoidal weir

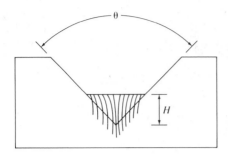

Figure 4-35 Triangular weir

With low flow rates, it is common to use a triangular weir (Fig. 4-35). The basic weir equation is given as

$$Q = \frac{8}{15} K \sqrt{2g} \tan\left(\frac{\theta}{2}\right) H^{5/2} \tag{4-41}$$

where K is the flow coefficient that is primarily a function of the head H. For weirs with θ values of 60° and 90°, Lenz (12) showed that the flow coefficient values varied from 0.60 to 0.57 as the head varied from 0.20 to 2.0 ft.

BROAD-CRESTED WEIRS   If the weir is long in the direction of flow so that the flow leaves the weir in essentially a horizontal direction, the weir is a *broad-crested weir* (see Fig. 4-15, page 185). The basic theoretical equation for the broad-crested weir is shown in (Eq. 4-23, page 185). However, to account for head loss and shape of weir, a discharge coefficient should be applied to Eq. (4-23). If the ratio of the actual discharge, Q, to the theoretical discharge, Q_{theor}, is given by C, then

$$Q = 0.385CL\sqrt{2g}H^{3/2} \tag{4-42}$$

For low weirs, the velocity of approach can be significant, and this effect will tend to make C greater than unity. However, the frictional resistance over the length of the weir will tend to make C less than unity. When these two effects are combined, the resulting C values are as shown in Fig. 4-36. These are for a weir with a vertical upstream face and a sharp corner at the intersection of the upstream face and the weir crest. If the upstream face is sloping at a 45° angle, the discharge coefficient should be increased 10% over that given in Fig. 4-36. Rounding of the upstream corner will also produce a coefficient of discharge as much as 3% greater.

VENTURI FLUME   Disadvantages of the broad-crested weir are that it produces considerable head loss, and sediment can accumulate in front of it. To reduce both of these detrimental effects, the Venturi flume was developed and calibrated by Parshall (15). Critical flow is produced by reducing the width of

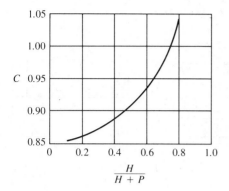

Figure 4-36 Form and resistance coefficient for a broad-crested weir for $0.1 < H/L < 0.8$ (16)

the channel (the Venturi effect) and by increasing the slope of the bottom in the contracted section (Fig. 4-37, pages 214–15). Thus the contracted section serves as a *control* and a predictable head-discharge relationship exists if the depth downstream of the contracted section is low enough to allow "free flow" through the contracted section. The criterion for free flow is that the ratio of downstream head to upstream head, H_d/H_u, shall not exceed 0.70. Then with this condition the discharge through the flume is given as

$$Q = K\sqrt{2g}WH_u^{3/2} \tag{4-43}$$

where H_u is the head above the floor level of the flume measured at the location shown in Fig. 4-37, and K is a function of H_u/W as given in Fig. 4-38, page 215.* Equation (4-43) is valid for flumes with throat widths (W) from 1 ft (0.31 m) to 8 ft (2.4 m)[†].

SPILLWAYS ON DAMS   The spillway crest serves as a control section; therefore, it can be used for discharge measurements. The general form of the discharge equation is the same as for weirs, $Q = KL\sqrt{2g}H^{3/2}$. The head, H, is measured from the crest of the spillway (Fig. 4-39, page 215), and L is the length of the spillway. The value of the flow coefficient depends on the shape of the spillway and the ratio of the head to height of dam.

---

* The discharge equation for the Parshall flume is normally expressed as $Q = 4BH_u^{1.522W^{0.026}}$, where Q is in cfs, and W and H are in feet. However, this equation is not dimensionally homogenous; therefore, Fig. 4-38 was derived from the Parshall equation, which is dimensionally homogenous and can be applied with any system of units.

[†] For information on the use of Parshall flumes outside the range of size noted above or for use under submerged conditions ($H_d/H_u > 0.70$), see Parshall (15) or Chow (4).

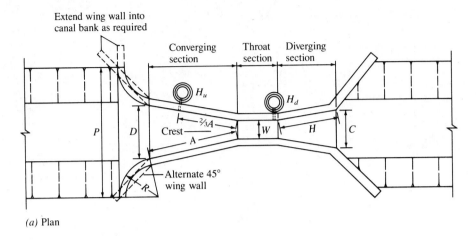

(a) Plan

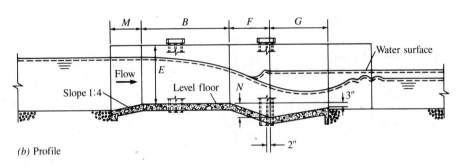

(b) Profile

W		A		$\frac{2}{3}A$		B		C		D		E	
(ft)	(in.)	(ft)	(in.)	(ft)	(in.)	(ft)	(in.)	(ft)	(in.)	(ft)	(in.)	(ft)	(in.)
0	6	2	$\frac{7}{16}$	1	$4\frac{5}{16}$	2	0	1	$3\frac{1}{2}$	1	$3\frac{3}{8}$	2	0
	9	2	$10\frac{5}{8}$	1	$11\frac{1}{8}$	2	10	1	3	1	$10\frac{5}{8}$	2	6
1	0	4	6	3	0	4	$4\frac{7}{8}$	2	0	2	$9\frac{1}{4}$	3	0
1	6	4	9	3	2	4	$7\frac{7}{8}$	2	6	3	$4\frac{3}{8}$	3	0
2	0	5	0	3	4	4	$10\frac{7}{8}$	3	0	3	$11\frac{1}{2}$	3	0
3	0	5	6	3	8	5	$4\frac{3}{4}$	4	0	5	$1\frac{7}{8}$	3	0
4	0	6	0	4	0	5	$10\frac{5}{8}$	5	0	6	$4\frac{1}{4}$	3	0
5	0	6	6	4	4	6	$4\frac{1}{2}$	6	0	7	$6\frac{5}{8}$	3	0
6	0	7	0	4	8	6	$10\frac{3}{8}$	7	0	8	9	3	0
7	0	7	6	5	0	7	$4\frac{1}{4}$	8	0	9	$11\frac{3}{8}$	3	0
8	0	8	0	5	4	7	$10\frac{1}{8}$	9	0	11	$1\frac{3}{4}$	3	0

Figure 4-37 Standard Parshall Flume dimensions (21)

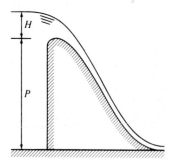

Figure 4-38 Flow coefficient for the Parshall flume

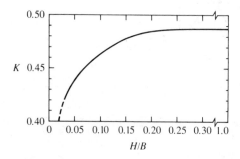

Figure 4-39 Flow over spillway of a dam

F		G		M		N		P		R		Free-Flow Capacity	
												Minimum	Maximum
(ft)	(in.)	(ft)	(in.)	(ft)	(in.)	(ft)	(in.)	(ft)	(in.)	(ft)	(in.)	(cfs)	(cfs)
1	0	2	0	1	0	0	$4\frac{1}{2}$	2	$11\frac{1}{2}$	1	4	0.05	3.9
1	0	1	6	1	0		$4\frac{1}{2}$	3	$6\frac{1}{2}$	1	4	0.09	8.9
2	0	3	0	1	3		9	4	$10\frac{3}{4}$	1	8	0.11	16.1
2	0	3	0	1	3		9	5	6	1	8	0.15	24.6
2	0	3	0	1	3		9	6	1	1	8	0.42	33.1
2	0	3	0	1	3		9	7	$3\frac{1}{2}$	1	8	0.61	50.4
2	0	3	0	1	6		9	8	$10\frac{3}{4}$	2	0	1.3	67.9
2	0	3	0	1	6		9	10	$1\frac{1}{4}$	2	0	1.6	85.6
2	0	3	0	1	6		9	11	$3\frac{1}{2}$	2	0	2.6	103.5
2	0	3	0	1	6		9	12	6	2	0	3.0	121.4
2	0	3	0	1	6		9	13	$8\frac{1}{4}$	2	0	3.5	139.5

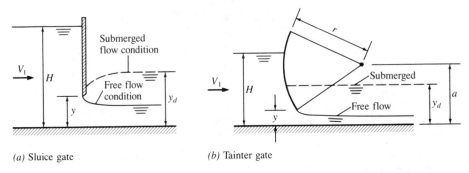

(a) Sluice gate   (b) Tainter gate

Figure 4-40   Flow below underflow gates

The discharge is also influenced by piers that are often used in the dam to support gates and for supporting a roadway across the dam. Their effect is usually to cause some reduction of flow (it is a contraction effect), and the degree of reduction is a function of the shape and thickness of pier.*

SLUICE AND TAINTER GATES   Both the sluice and Tainter gates are used extensively for controlling water in canals, in flumes, and on spillways. They fall in the category of *underflow gates*, as shown in Fig. 4-40. The discharge through underflow gates can be given as

$$Q = C_c C_v L y \sqrt{2g} \sqrt{\left(H + \frac{V_1^2}{2g} - C_c y\right)} \tag{4-44}$$

where C_c, the contraction coefficient, is a function of the relative gate opening and the shape of the gate. The velocity coefficient C_v has a value slightly less than unity. For convenience Eq. (4-44) is simplified to

$$Q = KLy\sqrt{2gH} \tag{4-45}$$

where K is the flow coefficient that is a function of the same parameters that C_c is a function of. If the downstream jet is submerged, as shown in Fig. 4-40, then the discharge rate is a function of the downstream depth, y_d, as well. The functional relationships between K and the other relevant parameters are shown in Fig. 4-41 for both the sluice gate and the Tainter gate. In Fig. 4-41a, Fr_0 is the Froude number based on the velocity through the gate opening $(\text{Fr}_0 = V/\sqrt{gy})$.[†]

---

* We discuss these effects and other characteristics of flow over spillways such as design details in Chapter 7.

† For more details, about underflow gates, see Henry (8) and Toch (19). Chapter 7 also includes information about the use of Tainter gates on spillways.

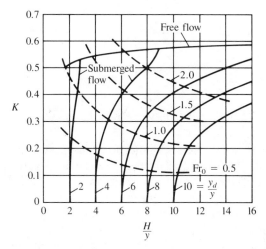

(a) Discharge coefficient for vertical sluice gate (8)

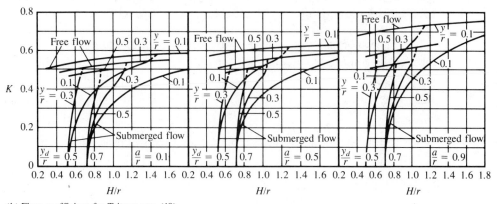

(b) Flow coefficient for Tainter gate (19)

Figure 4-41   Flow coefficient for underflow gates

4-5   Unsteady-Nonuniform Flow in Open Channels

Flood Routing Through a Reservoir

QUALITATIVE DESCRIPTION   Consider the reservoir shown in Fig. 4-42. Flow is coming into the reservoir from a stream and leaving it over a spillway. If a flood were to occur on the incoming stream, then as that flood flow comes into the reservoir, some of the flood volume is stored in the reservoir while the outflow discharge increases as the reservoir rises. However, the outflow rate is never as great as the peak inflow rate because of the attenuation of the inflow due to the fact that much of the flow is temporarily stored in the reservoir. The

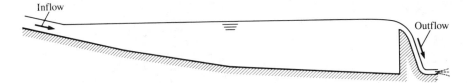

Figure 4-42   Flow into and out of a reservoir

stored amount will be released after the peak passes. The quantitative evaluation of outflow hydrograph from the reservoir as a function of the inflow is
termed *flood routing* through the reservoir.

In a completely general problem, we would have to consider local acceleration of the water in the reservoir because of velocity changes with time.*
However, because the depths are usually large in reservoirs and because the
velocities are small, accelerations in the reservoir are negligible. Therefore, in
reservoir flood routing, we simply use the general form of the continuity equation
to effect a solution. We also assume that the reservoir water surface is horizontal.

DEVELOPMENT OF THE FLOOD ROUTING PROCEDURE   The general
form of the continuity equation is given as

$$\sum_{cs} \rho \mathbf{V} \cdot \mathbf{A} = -\frac{d}{dt} \int_{cv} \rho \, d\forall \tag{4-46}$$

When Eq. (4-46) is applied to the control volume shown in Fig. (4-43), we have

$$I - O = \frac{d}{dt} (\forall_{\text{reservoir}}) \tag{4-47}$$

Equation (4-47) states that the volume rate of inflow of water to the reservoir
minus the outflow rate is equal to the rate of change of the volume of water in
the reservoir. If we call the volume of water in the reservoir storage, S, then Eq.

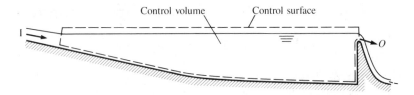

Control volume   Control surface

I   O

Figure 4-43   Control volume for a reservoir

---

* We discuss the more general problem in Chapter 12.

(4-47) becomes

$$I - O = \frac{dS}{dt} \qquad (4\text{-}48)$$

For a finite time period Δt, Eq. (4-48) can be approximated by

$$I - O = \frac{\Delta S}{\Delta t} \qquad (4\text{-}49)$$

In general, I and O will vary with time; thus, for a given time increment, we can approximate I as $(I_i + I_{i+1})/2$ and O as $(O_i + O_{i+1})/2$. Likewise, ΔS can be expressed as $S_{i+1} - S_i$. Then when these relations are substituted into Eq. (4-49), we get

$$\frac{(I_i + I_{i+1})}{2} - \frac{(O_i + O_{i+1})}{2} = \frac{S_{i+1} - S_i}{\Delta t} \qquad (4\text{-}50)$$

or $\qquad I_i + I_{i+1} + \dfrac{2S_i}{\Delta t} - O_i = O_{i+1} + \dfrac{2S_{i+1}}{\Delta t} \qquad (4\text{-}51)$

For uncontrolled reservoirs such as the one being considered here (gates are not operated to control the outflow), both the storage, S, and outflow, O, will be a function of the water surface elevation in the reservoir. Here the S is obtained by simply calculating the volume in the reservoir below the given elevation from a topographic map of the reservoir, and the O is obtained from a discharge equation in the form $O = KL(z - z_0)^n$. In this latter equation, z is the elevation of the water surface, z_0 is the elevation of the spillway crest, L is the length of the spillway, and K is a flow coefficient for the spillway. Since O and S are specific functions of the water surface elevation in the reservoir (reservoir stage), it should be obvious that O will be functionally related to S. Moreover, if O is functionally related to S, then O will be functionally related to $2S/\Delta t$. That is, for any uncontrolled reservoir, one can determine a specific relationship between O and $2S/\Delta t$. For example, Fig. 4-44 on page 220 shows one such relationship for about a 100-acre reservoir.

We start routing a flood through the reservoir when both inflow and reservoir stage are known (therefore, we also know outflow at the start of the routing process). All values of inflow are also known for the period of routing. Therefore, referring to Eq. (4-51), we see that at the start of the routing process (and at the beginning of the first time increment) all the values for the left-hand side of Eq. (4-51) will be known. Thus, we will also know the value of the sum on the right-hand side of Eq. (4-51). Then from a relationship between O and $(2S/\Delta t) + O$, such as given in Fig. 4-44, we can evaluate the outflow at the end of the time

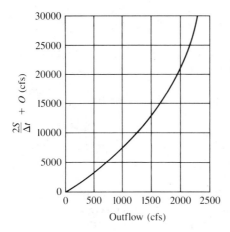

Figure 4-44   Typical relationship between O and $2S/\Delta t + O$

increment. We then use this outflow and the $(2S_i/\Delta t) - O_i$ for the next time increment. Example 4-11 explains the procedure in detail.

EXAMPLE 4-11   Using the storage-outflow relations for a reservoir (Fig. 4-44), route the hypothetical inflow (shown by the triangular hydrograph in Fig. A) through the reservoir. Assume the initial outflow rate is 0.

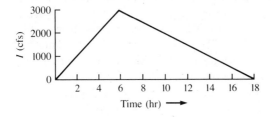

Figure A

SOLUTION   The solution is carried out in the accompanying table. The numbers in column 2 are indices for time increments. For example, $i = 3$ is the time at the start of the third hourly increment. Thus, I_i, O_i, S_i for $i = 3$ are the inflow, outflow, and storage, respectively, at the beginning of time increment number 3 or, in this case, at time = 2 hr (120 min). The arrows in rows $i = 1$ and 2 show how the calculation proceeds. The procedure is explained as follows:

1. Starting with $i = 1$, we calculate the value in column 4 by summing I_1 and I_2.

(1) Time (hr)	(2) i	(3) I_i (cfs)	(4) $I_i + I_{i+1}$	(5) $\dfrac{2S_i}{\Delta t} - O_i$	(6) $\dfrac{2S_{i+1}}{\Delta t} + O_{i+1}$	(7) O_i (cfs)
0	1	0	→500	→0←	→500—	~0
1	2	500	1,500	280←	1,780	110
2	3	1,000	2,500	1,200	3,700	290
3	4	1,500	3,500	2,640	6,140	530
4	5	2,000	4,500	4,500	9,000	820
5	6	2,500	5,500	6,680	12,180	1,160
6	7	3,000	5,750	9,300	15,050	1,440
7	8	2,750	5,250	11,690	16,940	1,680
8	9	2,500	4,750	13,320	18,070	1,810
9	10	2,250	4,250	14,290	18,540	1,890
10	11	2,000	3,750	14,700	18,450	1,920
11	12	1,750	3,250	14,650	17,900	1,900
12	13	1,500	2,750	14,140	16,890	1,880
13	14	1,250	2,250	13,250	15,500	1,820
14	15	1,000	1,750	12,100	13,850	1,700
15	16	750	1,250	10,670	11,920	1,590
16	17	500	750	9,100	9,850	1,410
17	18	250	250	7,410	7,660	1,220
18	19	0	0	5,660	5,660	1,000
19	20	0	0	4,100	4,100	780
20	21	0	0	2,900	2,900	600
21	22	0	0	2,000	2,000	450
22	23	0	0	1,320	1,320	340
23	24	0	0	820	820	250
24	25	0	0	500	500	160

2. The value of O_1 is known ($O_1 = O$); therefore, $2S_1/\Delta t$ can be found from Fig. 4-44, which for $i = 1$, yields $(2S_1/\Delta t) = 0$. Thus $(2S_1/\Delta t) - O_1 = 0$. Thus we record a zero in the first row of column 5.

3. The sum of values in columns 4 and 5 yields the value for the left-hand side of Eq. (4-51), and this is equal to $2S_2/\Delta t + O_2$, the value recorded in the first row of column 6 or $(2S_2/\Delta t) + O_2 = 500$. Once this quantity in column 6 is calculated, we go to Fig. 4-44.

4. With a value of $2S_2/\Delta t = 500$, we read off a value of approximately 110 cfs from Fig. 4-44.* The 110 cfs is the outflow rate at the end of time period 1, and it is also the outflow rate at the beginning of time period 2 ($i = 2$). Therefore, we record 110 cfs in row 2 of column 7.

5. For time step 2 ($i = 2$), we get $(2S_{i+2}/\Delta t) - O_{i+2}$ by subtracting $2O_2$ in column 7 from the value of $(2S_{i+1}/\Delta t) + O_{i+1}$ in column 6 of row 1. That

---

* This value of 110 cfs was obtained from a much larger graph than was possible to reproduce here.

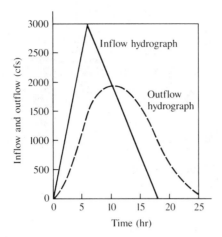

Figure B

is: $(2S_2/\Delta t) - O_2 = (2S_2/\Delta t) + O_2 - 2O_2$. In this case, we get $(2S_2/\Delta t) - O_2 = 500 - 2(110) = 280$ cfs.

6.   We continue this process, row by row, until the routing is completed.

The inflow flood hydrograph and outflow hydrograph are shown in Fig. B. As a result of routing the flood through the reservoir, the peak discharge has been reduced about 35%. ∎

Flood Routing Through Channels

DESCRIPTION OF THE PROBLEM   When routing floods through reservoirs, both the storage, S, and outflow, O, are a function of reservoir stage. Thus, the storage could also be given as a function of outflow $(S = f(O))$; therefore, the numerical form of the continuity equation could be solved. The basic continuity equation (Eq. 4-46) is also valid for channel flow. However, because the water surface elevation is not so directly linked to outflow, the problem becomes more complicated. The physical aspect of the problem is revealed if we look at a channel that has steady-uniform flow in it and compare it to the same channel when a flood wave is passing through it (Fig. 4-45). For the case of steady-uniform flow, the discharge will be consistent with the channel slope and depth that prevails. Moreover, because the water surface slope is the same as the channel bed slope, obtaining the storage in the channel is relatively easy. Thus, for any uniform flow, we could obtain the storage for the channel at the given discharge $(S = f(O))$. However, in routing a flood through the channel, the flow is not uniform; therefore, we have the situation shown in Fig. 4-45b, where the water surface slope is different from the channel slope. The case shown in Fig. 4-45b is for the rising phase of the flood wave. In general, for the flood

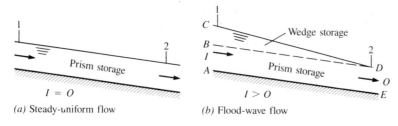

Figure 4-45   Flow in channel

routing situation, $I \neq O$, and $A_1 \neq A_2$. In relation to storage, we may visualize the storage $ABDE$ as being closely related to the downstream depth and thus to the outflow, O. This storage is sometimes called *prism* storage. Besides the prism storage, there is that part (volume BCD in Fig. 4-45b) associated with the rising part of the flood wave (or the falling part in a receding flood). This part is called *wedge* storage. We may visualize that the wedge storage is associated with the inflow, I, as well as the outflow.

MUSKINGUM METHOD OF FLOOD ROUTING   As we noted, when a flood wave is passing through a reach of a channel, the storage is a function of both inflow and outflow. In 1938, a method for channel routing was presented by McCarthy (14), which was later used by U.S. Corps of Engineers (23). The method was developed in connection with flood control work on the Muskingum river and is therefore called the *Muskingum method* of flood routing. The critical point of this method is that the storage is expressed as a function of both inflow and outflow as follows:

$$S = KO + KX(I - O) \tag{4-52}$$

In Eq. (4-52), K is a constant having units of time, and X is a dimensionless weighting factor that relates to the amount of wedge storage. In fact, we can think of the first term on the right-hand side of Eq. (4-52) as prism storage and the second term as wedge storage. This concept is reinforced because K is equivalent to the time required for an elemental discharge wave to travel the reach. The time increment in Eq. (4-51) should also be approximately this same length of time; therefore, $K \approx \Delta t$. Thus, $KO \approx \Delta tO = \Delta tVA_o = LA_o = $ prism storage.
   Equation (4-52) can be written as

$$S = K[XI + (1 - X)O] \tag{4-53}$$

Then, when the storage for a given time (let the index $= i$) is determined, we have

$$S_i = K[XI_i + (1 - X)O_i] \tag{4-54}$$

Then for an increment of time later, we have

$$S_{i+1} = K[XI_{i+1} + (1 - X)O_{i+1}] \tag{4-55}$$

Substituting Eqs. (4-54) and (4-55) in Eq. (4-51) yields

$$I_i + I_{i+1} + \frac{2K}{\Delta t}[XI_i + (1 - X)O_i] - O_i$$
$$= \frac{2K}{\Delta t}[XI_{i+1} + (1 - X)O_{i+1}] + O_{i+1} \tag{4-56}$$

On solving Eq. (4-56) for O_{i+1}, we obtain

$$O_{i+1} = C_0 I_{i+1} + C_1 I_i + C_2 O_i \tag{4-57}$$

where $$C_0 = \frac{0.5\,\Delta t - KX}{K(1 - X) + 0.5\,\Delta t} \tag{4-58}$$

$$C_1 = \frac{0.5\,\Delta t + KX}{K(1 - X) + 0.5\,\Delta t} \tag{4-59}$$

$$C_2 = \frac{K(1 - X) - 0.5\,\Delta t}{K(1 - X) + 0.5\,\Delta t} \tag{4-60}$$

From Eq. (4-57), we can see that if K, X, I_i, I_{i+1}, and O_i are known, we can easily solve for O_{i+1}. This outflow can be used for the starting outflow, O_i, for the next time increment. Then by taking succession time increments, the outflow can be obtained for the entire flood period.

DETERMINATION OF K AND X We have shown that a flood could be routed through a given reach of channel if K, X, inflow, and initial outflow are known. In practice, one often wants to route a flood through a channel that already has gauging stations along its length. For such a channel, records of historical floods would be available. Therefore, the flood routing method is reversed so as to obtain K and X; that is, given the inflow and outflow hydrographs, solve for K and X. We can do this by solving Eq. (4-56) for K:

$$K = \frac{0.5\,\Delta t[(I_i + I_{i+1}) - (O_i + O_{i+1})]}{X(I_{i+1} - I_i) + (1 - X)(O_{i+1} - O_i)} \tag{4-61}$$

The numerator in Eq. (4-61) is the increment of storage that has accumulated in time Δt, and the denominator is termed the weighted inflow and outflow. Let N be the numerator of Eq. (4–61) and D be the denominator. Then for each time increment, the historical I and O hydrographs may be analyzed to obtain N and D (then, in turn, $K = N/D$) for a given value of X. Actually, we want to find one value of K and one value of X that will be valid for the entire

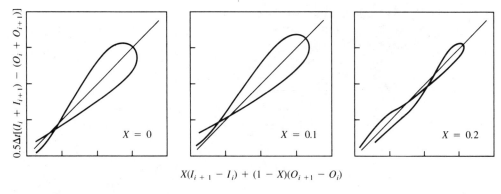

Figure 4-46 Plots for evaluating X and K

hydrograph. To ascertain the "best" values of K and X, the accumulated value of D is plotted against the accumulated value of N, as shown in Fig. 4-46. Obviously, the plot will be different depending on the value of X used in Eq. (4–61). We choose different values of X (say 0.10, 0.20, 0.30, 0.40) and make a plot for each X value. The plot giving a curve that is most nearly a straight line is the best. Thus, that value of X is the one to use, and the slope of the curve will yield the K value that should be used.*

PROBLEMS

4-1   A rectangular concrete channel is 12 ft wide, and water flows in it uniformly at a depth of 4 ft. If the channel drops 10 ft in a length of 800 ft, what is the discharge? Assume $T = 60°F$.

4-2   A concrete sewer pipe 4 ft in diameter is laid so it has a drop in elevation of 0.90 ft per 1000 ft of length. If sewage (assume the properties are the same as those of water) flows at a depth of 2 ft in the pipe, what will be the discharge?

4-3   A rectangular concrete channel 4 m wide on a slope of 0.004 is designed to carry water ($T = 10°C$) at a discharge of 25 m³/s. Estimate the uniform flow depth for these conditions.

4-4   A rectangular troweled concrete channel 12 ft wide with a slope of 10 ft in 8000 ft is designed for a discharge of 600 cfs. For a water temperature of 40°F, estimate the depth of flow.

---

* For more detailed information on this subject, see Hjelmfelt (10). The preceding sections on unsteady flow do not explicitly consider the momentum effects in the solutions for flow through reservoirs or channels. A more accurate method of solution considers these effects; however, much more detailed sets of data of the channel geometry are required to bring about a solution. We discuss the more general approach in Chapter 12.

4-5   Water flows at a depth of 6 ft in the trapezoidal, concrete-lined channel shown. If the channel slope is 1 ft in 2000 ft, what is the average velocity, and what is the discharge?

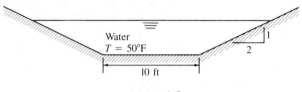

Water
$T = 50°F$

1

2

10 ft

PROBLEM 4-5

4-6   A concrete-lined trapezoidal channel having a bottom width of 10 ft and side slopes of 1 vertical to 2 horizontal is designed to carry a flow of 3000 cfs. If the slope of the channel is 0.001, what would be the depth of flow in the channel?

4-7   Estimate the discharge in a rock-bedded stream ($d_{84} = 30$ cm) that has an average depth of 2.21 m, a slope of 0.0037, and a width of 46 m. Assume $k_s = d_{84}$.

4-8   Determine the discharge in a 5-ft diameter concrete sewer pipe on a slope of 0.01 that is carrying water at a depth of 4 ft.

4-9   What will be the depth of flow in a trapezoidal concrete-lined channel that has a water discharge of 1000 cfs in it? The channel has a slope of 1 ft in 500 ft. The bottom width of the channel is 10 ft, and the side slopes are 1 V to 1 H.

4-10   Estimate the discharge in the Moyie River near Eastport, Idaho, when the depth is 4 ft, as shown in the figure below. Assume $S_0 = 0.0032$.

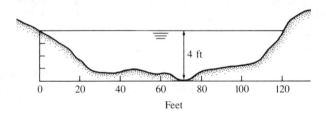

4 ft

0   20   40   60   80   100   120

Feet

PROBLEM 4-10

4-11   What discharge of water will occur in a trapezoidal channel that has a bottom width of 10 ft and sides slope 1 to 1 if the slope of the channel is 5 ft/mi and the depth is to be 5 ft? The channel will be lined with concrete.

4-12   Consider channels of rectangular cross section carrying 100 cfs of water flow. The channels have a slope of 0.001. Determine the cross-sectional areas required for widths of 2 ft, 4 ft, 6 ft, 8 ft, 10 ft, and 15 ft. Plot A versus y/b, and see how the results compare with the accepted result for the best hydraulic section.

4-13 The canal shown below has a slope of 0.004. The sides of the canal are sealed with a plastic membrane, which is protected by a 1 ft thick layer of 1-in. diameter gravel. The bottom is a 6-in. slab of rough concrete. If the water is flowing at a uniform depth of 6 ft, do you think the gravel will be scoured by the flowing water? Justify your answer by appropriate calculations.

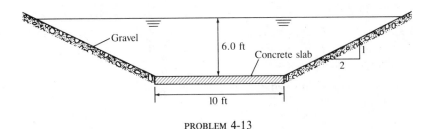

PROBLEM 4-13

4-14 A trapezoidal irrigation canal is to be excavated in soil and lined with coarse gravel. The canal is to be designed for a discharge of 200 cfs, and it will have slope of 0.0016. What should be the magnitude of the cross-sectional area and hydraulic radius for the canal if it is to be designed so that erosion of the canal will not occur? Choose a canal cross section that will satisfy the limitations.

4-15 Determine the cross section of an unlined canal excavated from sandy loam soil that is to carry 30 cfs. Assume the canal slope is to be 0.0005.

4-16 The water discharge in a rectangular channel 5 m wide is 15 m³/s. If the depth of water is 1 m, is the flow subcritical or supercritical?

4-17 The discharge in a rectangular channel 5 m wide is 10 m³/s. If the water velocity is 1.0 m/s, is the flow subcritical or supercritical?

4-18 Water flows at a rate of 10 m³/s in a rectangular channel 3 m wide. Determine the Froude number and the type of flow (subcritical, critical, or supercritical) for depths of 30 cm, 1.0 m and 2.0 m. What is the critical depth?

4-19 For the discharge and channel of Prob. 4-18, what is the alternate depth to the 30-cm depth? What is the specific energy for these conditions?

4-20 Water flows at a depth of 10 cm with a velocity of 6 m/s in a rectangular channel. Is the flow subcritical or supercritical? What is the alternate depth?

4-21 Water flows at the critical depth in a rectangular channel with a velocity of 2 m/s. What is the depth of flow?

4-22   Water flows uniformly at a rate of 9.0 m³/s in a rectangular channel that is 4.0 m wide and has a bottom slope of 0.005. If n is 0.014, is the flow subcritical or supercritical?

4-23   A rectangular channel is 6 m wide, and the discharge of water in it is 18 m³/s. Plot depth versus specific energy for these conditions. Let specific energy range from E_{min} to $E = 7$ m. What are the alternate and sequent depths to the 30-cm depth?

4-24   A long rectangular channel that is 3 m wide and has a mild slope ends in a free outfall. If the water depth at the brink is 0.250 m, what is the discharge in the channel?

4-25   A rectangular channel that is 15 ft wide and has a mild slope ends in a free outfall. If the water depth at the brink is 1.20 ft, what is the discharge in the channel?

4-26   Determine the critical depth for the canal of Prob. 4-13 for a discharge of 700 cfs.

4-27   A 48-in. concrete pipe culvert carries a discharge of water of 25 cfs. Determine the critical depth.

4-28   Derive a formula for critical depth, d_c, in the V-shaped channel shown below.

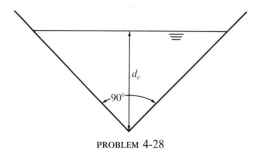

PROBLEM 4-28

4-29   A 10 ft wide rectangular channel is very smooth except for a small reach that is roughened with angle irons attached to the bottom of the channel (see figure below). Water flows in the channel at a rate of 200 cfs and at a depth of 1.00 ft. Assume frictionless flow except over the roughened part where the total drag of all the roughness (all the angle irons) is assumed to be 2000 lb. Determine the depth at the end of the roughness elements for the assumed conditions.

PROBLEM 4-29

4-30   Water flows with a velocity of 3 m/s and at a depth of 3 m in a rectangular channel. What is the change in depth and in water surface elevation produced by a gradual upward change in bottom elevation (upstep) of 30 cm? What would be the depth and elevation changes if there were a gradual downstep of 30 cm? What is the maximum size of upstep that could exist before upstream depth changes would result? Neglect head losses.

4-31   Water flows with a velocity of 2 m/s and at a depth of 3 m in a rectangular channel. What is the change in depth and in water surface elevation produced by a gradual upward change in bottom elevation (upstep) of 60 cm? What would be the depth and elevation changes if there were a gradual downstep of 15 cm? What is the maximum size of upstep that could exist before upstream changes would result? Neglect head losses.

4-32   Water flows with a velocity of 3 m/s in a rectangular channel 3 m wide at a depth of 3 m. What is the change in depth and in water surface elevation produced when a gradual contraction in the channel to a width of 2.6 m takes place? Determine the greatest contraction allowable without altering the specified upstream conditions. Neglect head losses.

4-33   Design a cylinder quadrant inlet transition to join a trapezoidal concrete-lined channel and a concrete flume of rectangular cross section. Assume the channel is on a slope of 0.0005, has side slopes 1 vertical to 2 horizontal, and the depth of flow therein is 5 ft. The bottom width of the channel is 10 ft. The depth in the flume is to be equal to the width of flume. Assume that the Froude number in the flume is to be equal to or less than 0.60.

4-34   The spillway shown has a discharge of 2.0 m³/s per meter of width occurring over it. What depth y_2 will exist downstream of the hydraulic jump? Assume negligible energy loss over the spillway.

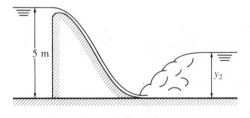

PROBLEM 4-34

4-35   The flow of water downstream from a sluice gate in a horizontal channel has a depth of 30 cm and a flow rate of 1.8 m³/s per meter of width. Could a hydraulic jump be caused to form downstream of this section? If so, what would be the depth downstream of the jump?

4-36   Consider the dam and spillway shown in Fig. 4-25, page 198. Water is discharging over the spillway of the dam as shown. The elevation difference between upstream pool level and the floor of the apron of the dam is 100 ft. If the head on the spillway is 5 ft and if a hydraulic jump forms on the horizontal apron, what is the depth of flow on the apron just downstream of the jump? Assume that the velocity just upstream of the jump is 95% of the maximum theoretical velocity. *Note*: The discharge over the spillway is given as $Q = KL\sqrt{2g}H^{3/2}$, where L is the length of the spillway, K is a coefficient (assume it has a value of 0.5), and H is the head on the spillway.

4-37   A hydraulic jump occurs in a wide rectangular channel. If the depths upstream and downstream are 15 cm and 4.0 m, respectively, what is the discharge per foot of width of channel?

4-38   Water is flowing as shown under the sluice gate in a horizontal rectangular channel that is 5 ft wide. The depths y_0 and y_1 are 65 ft and 1 ft, respectively. What will be the horsepower lost in the hydraulic jump?

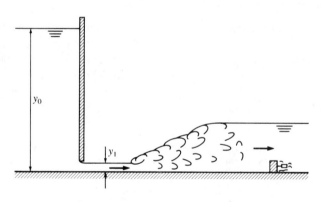

PROBLEM 4-38

4-39   Water flows uniformly at a depth $y_1 = 40$ cm in the concrete channel, which is 10 m wide. Estimate the height of the hydraulic jump that will form when a sill is installed to force it to form.

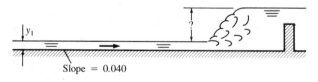

PROBLEM 4-39

4-40   For the derivation of Eq. (4-27), it is assumed that the bottom shearing force is negligible. For the conditions of Prob. 4-39, estimate the magnitude of the shearing force F_s associated with the hydraulic jump, and then determine F_s/F_H, where F_H is the net hydrostatic force on the hydraulic jump.

4-41   The normal depth in the channel downstream of the sluice gate shown is 1 m. What type of water surface profile occurs downstream of the sluice gate?

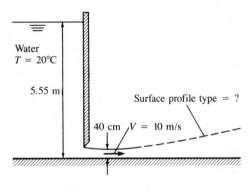

Water
$T = 20°C$

5.55 m

Surface profile type = ?

40 cm $V = 10$ m/s

PROBLEM 4-41

4-42   The partial water surface profile shown is for a rectangular channel that is 3 m wide and has water flowing in it at a rate of 5 m³/s. Sketch in the missing part of water surface profile and identify the type(s).

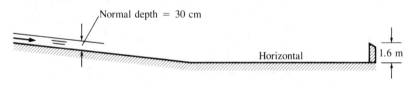

Normal depth = 30 cm

Horizontal   1.6 m

PROBLEM 4-42

4-43   Water flows from under a sluice gate into a horizontal rectangular channel at a rate of 3 m³/s per meter of width. The channel is concrete, and the initial depth is 20 cm. Apply Eq. (4-33), page 203, to construct the water surface profile up to a depth of 60 cm. In your solution, compute reaches for adjacent pairs of depths given in the following sequence: $d = 20$ cm, 30 cm, 40 cm, 50 cm, and 60 cm. Assume that f is constant with a value of 0.02. Plot your results.

4-44   A horizontal rectangular concrete channel terminates in a free outfall. The channel is 4 m wide and carries a discharge of water of 12 m³/s. What is the water depth 300 m upstream from the outfall?

4-45   Given the hydraulic jump shown on the next page for the long horizontal rectangular channel, what kind of water surface profile (classification) is

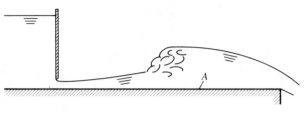

PROBLEM 4-45

upstream of the jump? What kind of water surface profile is downstream of the jump? If baffle blocks are put on the bottom of the channel in the vicinity of A to increase the bottom resistance, what changes are apt to occur given the same gate opening? Explain or sketch the changes.

4-46   A very long 10 ft wide concrete rectangular channel with a slope of 0.0001 ends with a free overfall. The discharge in the channel is 120 cfs. One mile upstream the flow is uniform. What kind (classification) of water surface profile occurs upstream of the brink?

4-47   The discharge in a 2 m wide very rough rectangular channel in which the water flows 1 m deep is 10.5 m³/s. The channel has a slope of 0.10, and the flow is uniform at this depth. A certain structure causes the flow at another part of the same channel to be locally only 0.40 m deep. At this different section of the channel, the channel slope, roughness, and cross section are the same as before. What do you conclude about the water surface profile over a reach of channel that includes this section where the depth is 0.4 m?

4-48   The horizontal rectangular channel downstream of the sluice gate is 10 ft wide, and the water discharge therein is 108 cfs. The water surface profile was computed by the direct step method. If a 2 ft high sharp-crested weir is installed at the end of the channel, do you think a hydraulic jump would develop in the channel? If so, approximately where would it be located? Justify your answers by appropriate calculations. Label any water surface profiles that can be classified.

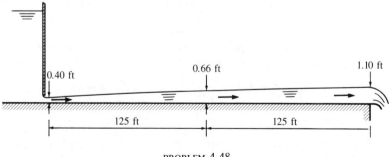

PROBLEM 4-48

4-49   The discharge per foot of width in this rectangular channel is 20 cfs. The normal depths for parts 1 and 3 are 0.5 ft and 1.00 ft, respectively. The slope for part 2 is 0.001 (sloping upward in the direction of flow). Sketch all possible water surface profiles for flow in this channel, and label each part with its classification.

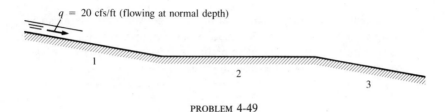

$q = 20$ cfs/ft (flowing at normal depth)

PROBLEM 4-49

4-50   Consider a rectangular channel that ends with a dam, as shown in Fig. 4-29, page 203. Assume that the slope (S_0), discharge (Q), channel roughness (n) or (k_s), height of dam (Z), and length of spillway (L), are all given. Write a computer program to solve for the backwater curve (M1 curve) for this flow situation. Solve for the backwater curve (water surface elevation versus X) for $S_0 = 0.004$, $Q = 20,000$ cfs, $Z = 50$ ft, and $L = 200$ ft, and plot the results. Submit your program with your solution. Make your own assumption about roughness. Assume the channel has the same width as spillway length.

4-51   A dam 50 m high backs up water in a river valley as shown. During flood flow, the discharge per meter of width, q, is equal to 10 m^3/s. Making the simplifying assumptions that $R = y$ and $f = 0.030$, determine the water surface profile upstream from the dam to a depth of 6 m. In your numerical calculations, let the first increment of depth change be y_c; use increments of depth change of 10 m until a depth of 10 m is reached; and then use 2-m increments until the desired limit is reached.

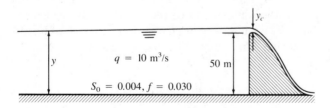

y_c

$q = 10$ m^3/s

50 m

$S_0 = 0.004, f = 0.030$

y

PROBLEM 4-51

4-52   Water flows at a steady rate of 12 cfs per foot of width ($q = 12$ cfs) in the wide rectangular concrete channel shown on the next page. Determine the water surface profile from section 1 to section 2.

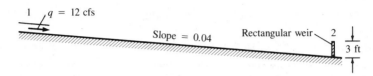

PROBLEM 4-52

4-53   Theory and experimental verification indicate that the mean velocity along a vertical line in a wide stream is closely approximated by the velocity at 0.6 depth. If the indicated velocities at 0.6 depth in a river cross section are measured, what is the discharge in the river?

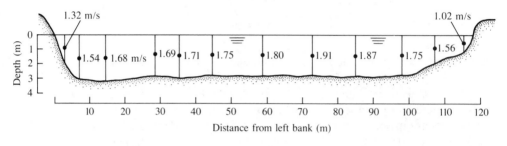

PROBLEM 4-53

4-54   Water flows over a rectangular weir that is 2 m wide and 30 cm high. If the head on the weir is 13 cm, what is the discharge in cubic meters per second?

4-55   What is the discharge over a rectangular weir 1 m high in a channel 2 m wide if the head on the weir is 25 cm?

4-56   What is the discharge over a rectangular weir 2 ft high in a channel 6 ft wide if the head on the weir is 1 ft?

4-57   A flood caused water to flow over a highway as shown below. The water surface elevation upstream of the highway (at *A*) was measured to be 101.00 ft. The elevation at the top of the crown of the pavement of the highway is 100.10 ft. Estimate the discharge over a stretch of highway with this elevation, which is 100 ft long. What was the depth of flow at the crown of the highway?

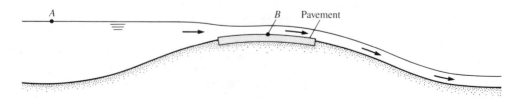

PROBLEM 4-57

4-58   The head on a 60° triangular weir is 1.5 ft. What is the discharge of water over the weir?

4-59   A rectangular irrigation canal 3 m wide carries water with a discharge of 6 m³/s. What height of rectangular weir installed across the canal will raise the water surface to a level 2 m above the canal floor?

4-60   What discharge of water will occur over a high, broad-crested weir that is 10 m long if the head on the weir is 60 cm?

4-61   The crest of a high, broad-crested weir has an elevation of 100.00 m. If the weir is 20 m long and the discharge of water over the weir is 50 m³/s, what is the water surface elevation in the reservoir upstream?

4-62   The crest of a high, broad-crested weir has an elevation of 300.00 ft. If the weir is 50 ft long and the discharge of water over the weir is 1500 cfs, what is the water surface elevation in the reservoir upstream?

4-63   Several discharge measuring devices are being considered for measurement of flow to a part of an irrigation project. For a discharge of 100 cfs, determine the head H and approximate minimum head loss across the device for the following:
   a. Rectangular weir, as shown in Fig. 4-33, page 211, with $L = 15$ ft and $P = 2$ ft
   b. Broad-crested weir, as shown in Fig. 4-15, page 185, with $L = 20$ ft and $P = 2$ ft
   c. Triangular weir with $\theta = 90°$
   d. Parshall flume with $W = 6$ ft (free flow condition)
   e. Ten foot wide sluice gate with $y = 1/2$ ft (free flow condition)
   f. Ten foot wide Tainter gate with $y = 1/2$ ft (free flow condition); also for this gate, $a = 5$ ft and $r = 10$ ft

4-64   The steep rectangular concrete channel shown is 4 m wide and 500 m long. It conveys water from a reservoir and delivers it to a free outfall. The channel entrance is rounded and smooth (negligible head loss at the entrance). If the water surface elevation in the reservoir is 2 m above the channel bottom, what will the discharge in the channel be?

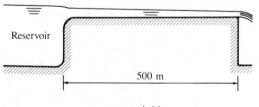

Reservoir

500 m

PROBLEM 4-64

4-65   The concrete rectangular channel shown on the next page is 3.5 m wide and has a bottom slope of 0.001. The channel entrance is rounded and

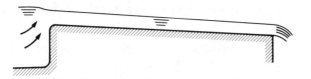

PROBLEM 4-65

smooth (negligible head loss at the entrance), and the reservoir water surface is 2.5 m above the bed of the channel at the entrance.

a. If the channel is 3000 m long, estimate the discharge in it.
b. If the channel is only 100 m long, tell how you would solve for the discharge in it.

4-66   A straight rectangular concrete ($n = 0.013$) flood control channel flows through a city in a valley. Upstream of a control section, the slope is 0.0004, the width is 40 ft, and sidewall height is 18 ft. In the upstream reach, the uniform approach flow is subcritical, whereas in the reach downstream from the control, the flow is supercritical. The slope change isolates the upstream channel from any backwater effects of the downstream channel.

An industrial developer proposes to build a railroad bridge across the channel, upstream from the control section. To match track grades, his design calls for a bridge with 8 ft deep plate girders having a 10-ft clearance above the channel bottom. (Special bulkhead gates are to be provided to close the holes breached in the channel walls when water depth in the channel approaches the bottom of the plate girders.)

a. What is the maximum flow (cfs) in the channel before the water surface rises to the bottom of the plate girders?

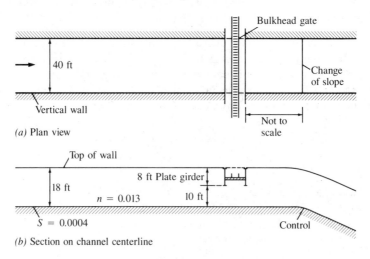

(a) Plan view

(b) Section on channel centerline

PROBLEM 4-66

b. Estimate the flow at which the girder will first be overtopped.

c. What is the horizontal force on the bridge, due to the water, at this time?

d. What is the reduction caused by the bridge in the maximum discharge that will just overtop the sidewall at the bridge location?

4-67   The sketch for this problem is a plan view of an earth fill dam and proposed centerline of a spillway channel. Because of geology and other restrictions, assume that this location is fixed. The profile of the ground surface along the channel route is also shown. The spillway should be designed to handle a maximum flood of 10,000 cfs with a water surface elevation of 200 ft in the reservoir. Sketch the spillway profile and plan showing elevations, and make necessary calculations to determine channel dimensions. What other special considerations must be addressed for the design of the spillway channel?

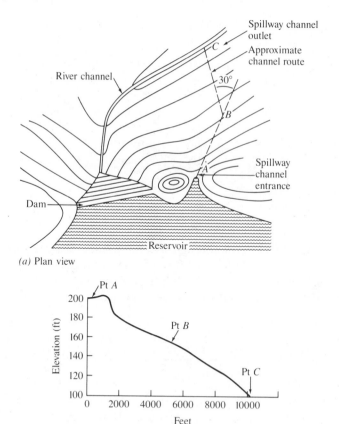

(a) Plan view

(b) Profile along centerline of channel

PROBLEM 4-67

4-68   Consider the stream, reservoir, and dam shown below. Assume that the reservoir is enclosed by vertical rock walls except at the lower end where the dam is located. The reservoir pool area is 1000 acres. The length of the spillway is 200 ft. Before the flood hit the basin, assume that the steady flow rate was 200 cfs. A flood hit the basin, and the flood discharges for the incoming stream are given in the table. Route the flood through the reservoir. Plot the inflow and outflow hydrograph on the same scale. Assume the value of the spillway flow coefficient, K, is 0.50.

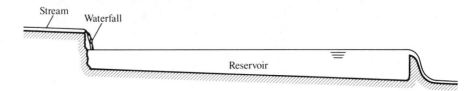

PROBLEM 4-68

Time (hr)	Q (cfs)	Time (hr)	Q (cfs)	Time (hr)	Q (cfs)
0	200	9	22,000	18	2,000
1	1,000	10	20,000	19	1,500
2	3,000	11	19,000	20	1,000
3	5,000	12	14,000	21	500
4	8,000	13	10,000	22	400
5	12,000	14	7,000	23	300
6	18,000	15	5,000	24	250
7	23,000	16	4,000	25	200
8	24,000	17	3,000		

4-69   The analysis of the flow through a certain reach of stream channel yielded the following results for the Muskingum method: $K = 25.2$ hr, $\Delta t = 12$ hr, $X = 0.3$. With these data and with the following inflow to that reach of channel, determine the outflow versus time. The initial outflow rate is 2 cfs.

Date	Hour	I (cfs)	Date	Hour	I (cfs)
1	0600	2	5	0600	13
	1800	15		1800	10
2	0600	28	6	0600	7
	1800	32		1800	5
3	0600	30	7	0600	4
	1800	25		1800	3
4	0600	20	8	0600	3
	1800	16		1800	2

REFERENCES

1. ASCE Manual of Practice No. 37. *Design and Construction of Sanitary and Storm Sewers.* Am. Soc. of Civil Engineers, New York, 1976.
2. Barnes, H.H. *Roughness Characteristics of Natural Channels.* U.S. Geological Survey, Water Supply Paper 1849, U.S. Govt. Printing Office, 1967.
3. Camp, T.R. "Design of Sewers to Facilitate Flow." *Sewage Works Journal,* 18 (January 1946).
4. Chow, V.T. *Open Channel Hydraulics.* McGraw-Hill, New York, 1959.
5. Fortier, S., and F.C. Scobey. "Permissible Canal Velocities." *Trans. ASCE,* 89, (1926), pp. 940–54.
6. Frederiksen, H.D., and J.J. DeVries. "Selection of Methods Used for Computing the Head Loss in the Open Channels of the California Aqueduct." Tech. Memo no. 18 of the Water Resources Dept. of the state of California (October 1965).
7. Henderson, F.M. *Open Channel Flow.* Macmillan, New York, 1966.
8. Henry, H.R. "Diffusion of Submerged Jets." Discussion by M.L. Albertson, Y.B. Dai, R.A. Jensen, and Hunter Rouse. *Trans. ASCE,* 115 (1950).
9. Hinds, J. "The Hydraulic Design of Flume and Siphon Transitions." *Trans. ASCE,* 92, (1928), pp. 1423–59.
10. Hjelmfelt, A.T., and J.J. Cassidy. *Hydrology for Engineers and Planners.* Iowa State University Press, Ames, Iowa, 1975.
11. Kindsvater, Carl E., and R.W. Carter. "Discharge Characteristics of Rectangular Thin-Plate Weirs." *Trans. ASCE,* 124 (1959).
12. Lenz, A.T. "Viscosity and Surface Tension Effects on V-Notch Weir Coefficients." *Trans. ASCE,* (1943), p. 759.
13. Limerinos, J.T. "Determination of the Manning Coefficient From Measured Bed Roughness in Natural Channels." Water Supply Paper 1898-B, U.S. Geological Survey, Washington, D.C., 1970.
14. McCarthy, G.T. "The Unit Hydrograph and Flood Routing." Presented at a conference of the North Atlantic Division, U.S. Army, Corps of Engineers, June 24, 1938.
15. Parshall, R.L. "The Improved Venturi Flume." *Trans. ASCE,* 89 (1926), pp. 841–51.
16. Raju, K.G.R. *Flow Through Open Channels.* Tata McGraw-Hill, New Delhi, 1981.
17. Roberson, J.A., and C.T. Crowe. *Engineering Fluid Mechanics,* 3d ed. Houghton Mifflin, Boston, 1985.
18. Rouse, H. (ed.). *Engineering Hydraulics.* John Wiley & Sons, New York, 1950.
19. Toch, Arthur. "Discharge Characteristics of Tainter Gates." *Trans. of Am. Soc. of Civil Engrs.,* 120 (1955).
20. U.S. Bureau of Reclamation. "Dams and Control Works." U.S. Govt. Printing Office, 1938.
21. U.S. Bureau of Reclamation. *Design of Small Canal Structures.* U.S. Dept. of Interior, U.S. Govt. Printing Office, 1978.
22. U.S. Bureau of Reclamation. *Hydraulic Design of Stilling Basin and Bucket Energy Dissipators.* Engr. Monograph, no. 25, U.S. Supt. of Doc., 1958.
23. U.S. Corps of Engineers. "Flood Control." The Engineer School, Fort Belvoir, Vir. 1940.
24. U.S. Corps of Engineers. *Manual of Design Practice.*
25. Wolman, M.G. "The Natural Channel of Brandywine Creek, Pennsylvania." Prof. Paper 271, U.S. Geological Survey, Washington D.C., 1954.

5

These five 15-ft diameter penstocks deliver water from
the reservoir behind Shasta Dam to the powerhouse.
At design flow the discharge in each penstock is
3,800 cfs and the head on each turbine is 380 feet
yielding a power output of 140,000 horsepower for
each turbine. The steel plate thickness of the penstocks
is 2 inches, and the anchor blocks just upstream of
the powerhouse each weigh over 2 million pounds.
(Courtesy of the U.S. Bureau of Reclamation).

Closed Conduit Flow

5-1 General Considerations

In the design and operation of a pipeline, the main considerations are head losses, forces and stresses acting on the pipe material, and discharge. Head loss for a given discharge relates to flow efficiency; that is, an optimum size of pipe will yield the least overall cost of installation and operation for the desired discharge. Choosing a small pipe results in low initial costs; however, subsequent costs may be excessively large because of high energy cost from large head losses. Forces and stresses mainly result from fluid pressure in a pipe. Forces are also created by momentum change associated with flow around bends or through other types of pipe fittings.

The basic continuity, energy, and momentum equations of fluid mechanics are used in the solution of pipe-flow problems. For example, to design a pipe, you would use the continuity and energy equations to obtain the required pipe diameter. Then applying the momentum equation will yield the forces acting on bends for a given discharge. Applications of the aforementioned equations in the design and analysis of conduits are treated in this chapter. The energy equation is considered first.

5-2 Energy Equation

The initial design of a conduit involves determining the size of the conduit with the least cost for a required discharge. This cost includes first cost plus operating and maintenance costs. We only briefly discuss the economic aspect of conduit design. The hydraulic aspects of the problem require applying the one-dimensional steady flow form of the energy equation:

$$\frac{p_1}{\gamma} + \alpha_1 \frac{V_1^2}{2g} + z_1 + h_p = \frac{p_2}{\gamma} + \alpha_2 \frac{V_2^2}{2g} + z_2 + h_t + h_L \tag{5-1}$$

where p/γ = pressure head
$\alpha V^2/2g$ = velocity head
z = elevation
h_p = head supplied by a pump
h_t = head supplied to a turbine
h_L = head loss between sections 1 and 2

A typical graphical representation of the terms of Eq. (5-1) is shown in Fig. 5-1. We give further explanation of the terms of Eq. (5-1) in the following paragraphs.

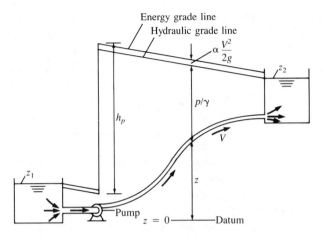

Figure 5-1   Definition sketch for terms in the energy equation

Velocity Head

In $\alpha V^2/2g$, the velocity V is the mean velocity in the conduit at a given section and is obtained by $V = Q/A$, where Q is the discharge, and A is the cross-sectional area of the conduit. The kinetic energy correction factor is given by α, and its definition is

$$\alpha = \frac{\int_A u^3\, dA}{V^3 A} \tag{5-2}$$

where u = velocity at any point in the section.

In Eq. (5-2), the integration is carried out over the cross section of the pipe. It can be shown that α has a minimum value of unity when the velocity is uniform across the section, and that it has values greater than unity depending on the degree of velocity variation across a section. It can also be shown that if laminar flow occurs in a pipe, the velocity distribution across the section will be parabolic, and α will have a value of 2.0 (36). However, if the flow is turbulent, as is the usual case for water flow in large conduits, the velocity is fairly uniform over most of the conduit section, and α has a value near unity (typically: $1.04 < \alpha < 1.06$). Therefore, in hydraulic engineering, for ease of application in pipe flow, the value of α is usually assumed to be unity, and the velocity head is then simply $V^2/2g$.

Pump or Turbine Head

The head supplied by a pump is directly related to the power supplied to the flow, as given in Eq. (5-3).

$$P = Q\gamma h_p \tag{5-3}$$

Likewise, if head is supplied to a turbine, the power supplied to the turbine will be $P = Q\gamma h_t$. In the preceding two equations, P refers to the power supplied directly to or taken directly from the flow. If you want to relate that to electrical or mechanical energy of the pump or turbine, you must include an efficiency factor. For example, the power that could be obtained from a turbine would be $P = eQ\gamma h_t$, where e is the efficiency of the turbine generator.

Head-Loss Term

The head-loss term h_L accounts for the conversion of mechanical energy to internal energy (heat). When this conversion occurs, the internal energy is not readily converted back to useful mechanical energy; therefore, it is called *head loss*. Head loss results from viscous resistance to flow (friction) at the conduit wall or from the viscous dissipation of turbulence usually occurring with separated flow, such as in bends, fittings, or outlet works.

5-3   Head Loss

Variables Affecting Head Loss

Head loss in a straight length of pipe is due to dissipation of energy caused by the resistance of the pipe wall. In the case of laminar flow, which generally occurs with the Reynolds number less than 2000, the head loss is all due to viscous resistance. The head loss is a function of the first power of the velocity. If the flow is turbulent, the head loss is related to the dissipation of the kinetic energy of turbulence, which produces a more complicated relationship between head loss and velocity. If the conduit is rough, still more variables involving the characteristics of roughness are needed to define the head loss.

Laminar-Flow Head Loss and Velocity Distribution

It can be shown that for steady-uniform flow in a pipe, the shear-stress distribution will vary linearly from a maximum at the pipe wall to zero at the

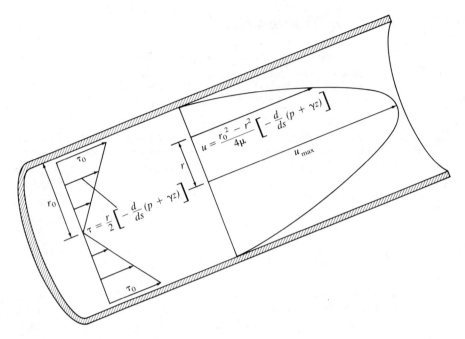

Figure 5-2   Distribution of shear stress and velocity
for laminar flow in a pipe

center (36). If the flow is laminar, the velocity distribution will be parabolically
distributed, as given by Eq. (5-4) and shown in Fig. 5-2.

$$u = \frac{r_0{}^2 - r^2}{4\mu}\left[-\frac{d}{ds}(p + \gamma z)\right] \tag{5-4}$$

where r_0 is the pipe radius, and s is the coordinate axis parallel to the pipe axis
and in the direction of flow.

By integrating the velocity across the section and using Eq. (5-1), it can be
shown that the head loss for laminar flow is given by

$$h_L = \frac{32\mu L V}{\gamma D^2} \tag{5-5}$$

Turbulent-Flow Head Loss and Velocity Distribution

SMOOTH PIPES   When the pipe flow-Reynolds number, $VD\rho/\mu$, is
greater than about 3000, one can expect the flow to be turbulent, and in this

case, the shear stress is primarily in the form of Reynolds stress, which varies linearly across the pipe section (increasing from zero at the center of the pipe) except near the pipe wall in the viscous sublayer, where the Reynolds stress decreases, and a true viscous shear stress takes over.

The velocity distribution takes different forms depending on the relative distance from the pipe wall. In the viscous sublayer, the velocity distribution is given by

$$\frac{u}{u_*} = \frac{u_* y}{v} \qquad \text{for } 0 < \frac{u_* y}{v} < 5 \tag{5-6}$$

where u = velocity
$\quad y$ = distance from pipe wall
$\quad v$ = kinematic viscosity
$\quad u_*$ = shear velocity = $\sqrt{\tau_0/\rho}$
$\quad \tau_0$ = shear stress at pipe wall

Immediately outside the viscous sublayer, the velocity distribution is of the logarithmic form

$$\frac{u}{u_*} = 5.75 \log_{10} \frac{u_* y}{v} + 5.5 \qquad \text{for } 20 < \frac{u_* y}{v} \leqslant 10^5 \tag{5-7}$$

Figure 5-3 is a plot of Eqs. (5-6) and (5-7) as well as an indication of the spread of experimental data from various sources. It has also been found that

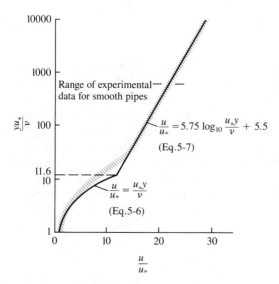

Figure 5-3 Velocity distribution for smooth pipes (37)

Table 5-1 Exponents for Power-Law Equation (37)

Re →	4×10^3	2.3×10^4	1.1×10^5	1.1×10^6	3.2×10^6
$m \rightarrow$	$\dfrac{1}{6.0}$	$\dfrac{1}{6.6}$	$\dfrac{1}{7.0}$	$\dfrac{1}{8.8}$	$\dfrac{1}{10.0}$

the velocity distribution for turbulent flow can be approximated quite well with a power law formula. This is given as

$$\frac{u}{u_{max}} = \left(\frac{y}{r_0}\right)^m \tag{5-8}$$

where u_{max} = velocity at the pipe center
r_0 = radius of pipe
m = exponent that varies with the Reynolds number, Re

The variation of m with Re is given in Table 5-1.

The head loss for turbulent flow in pipes is given by the Darcy-Weisbach formula as

$$h_f = f \frac{L}{D} \frac{V^2}{2g} \tag{5-9}$$

ROUGH PIPES   Numerous tests on flow in rough pipes show that a semilogarithmic velocity distribution is valid over most of the pipe section (36, 37). This relationship is given as

$$\frac{u}{u_*} = 5.75 \log_{10} \frac{y}{k} + B \tag{5-10}$$

where y is the distance from the rough wall, k is a measure of the height of roughness elements, and B is a function of the character of roughness, that is, B is a function of the type, concentration, and size variation of the roughness. Research by Roberson and Chen (35) shows that B can be analytically determined for artificially roughened boundaries. More recent work by Wright (49), Calhoun (15), Kumar (25), and Eldridge (19) indicate that the same theory using a numerical approach will yield solutions for B and the coefficient f for natural roughness as found in rock-bedded streams and commercial pipes.

In 1933, Nikuradse (32) carried out numerous tests on the flow in pipes roughened with uniform-sized sand grains. From these tests, he found that the value for B with this kind of roughness was 8.5. Thus, for his tests, Eq. (5-10) becomes

$$\frac{u}{u_*} = 5.75 \log_{10} \frac{y}{k_s} + 8.5 \tag{5-11}$$

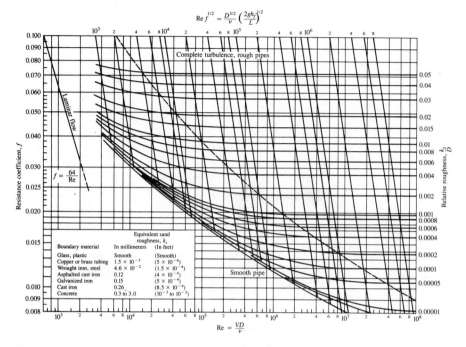

Figure 5-4   Resistance coefficient f versus Re (31)

where the distance y was measured from the geometric mean of the wall surface, and k_s was the size of the sand grains.

The uniform character of the sand grains used in Nikuradse's tests produces a dip in the f-versus-Re curve before reaching a constant value of f. However, tests on commercial pipes where the roughness is somewhat random reveal that no such dip occurs. By plotting data for commercial pipe from a number of sources, Moody (31) developed a design chart similar to that shown in Fig. 5-4.

In Fig. 5-4, the variable k_s is the symbol used to denote the *equivalent sand roughness*. That is, a pipe that has the same resistance characteristics at high Re values as a sand-roughened pipe of the same size is said to have a size of roughness equivalent to that of the sand-roughened pipe. Figure 5-4 gives approximate values of k_s and k_s/D for various kinds of pipe. This figure along with Fig. 5-5 is used to solve certain kinds of pipe-flow problems.*

---

* Besides the k_s values given in Fig. 5.5, see Sec. 5-9, where we give further information on k_s values for very large conduits.

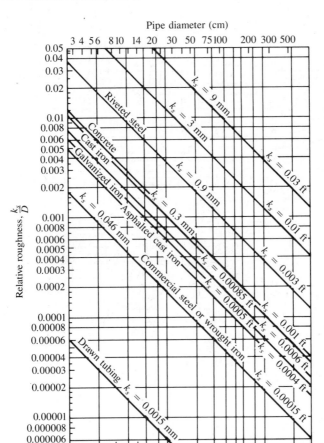

Figure 5-5 Relative roughness for various kinds of pipe (31)

In Fig. 5-4, the abscissa (labeled at the bottom) is the Reynolds number, Re, and the ordinate (labeled at the left) is the resistance coefficient f. Each solid curve is for a constant relative roughness, k_s/D, and the values of k_s/D are given on the right at the end of each curve. To find f, given Re and k_s/D, go to the right to find the correct relative-roughness curve; then look at the bottom of the chart to find the given value of Re and, with this value of Re, move vertically upward until you reach the given k_s/D curve. Finally, from this point, move horizontally to the left scale to read the value of f. If the curve for the given value of k_s/D is not plotted in Fig. 5-4, simply find the proper position on the graph by interpolation between curves of k_s/D, which bracket the given k_s/D.

For some problems, it is convenient to enter Fig. 5-4 using a value of the parameter $Re f^{1/2}$. This parameter is useful when h_f and k_s/D are known but the velocity, V, is not.

Basically three types of problems are involved with uniform flow in a single pipe.

1. Determine the head loss, given the kind and size of pipe along with the flow rate.
2. Determine the flow rate, given the head, kind, and size of pipe.
3. Determine the size of pipe needed to carry the flow, given the kind of pipe, head, and flow rate.

In the first type of problem, the Reynolds number and k_s/D are first computed and then f is read from Fig. 5-4, after which the head loss is obtained by the use of Eq. (5-9).

EXAMPLE 5-1   Water, 20°C, flows at a rate of 0.05 m³/s in a 20-cm asphalted cast-iron pipe. What is the head loss per kilometer of pipe?

SOLUTION   First compute the Reynolds number, VD/v, where $V = Q/A$. Thus

$$V = \frac{0.05 \text{ m}^3/\text{s}}{(\pi/4)(0.20^2 \text{ m}^2)} = 1.59 \text{ m/s}$$

$$v = 1.0 \times 10^{-6} \text{ m}^2/\text{s}$$

Then $$Re = \frac{VD}{v} = \frac{(1.59 \text{ m/s})(0.20 \text{ m})}{10^{-6} \text{ m}^2/\text{s}}$$

$$= 3.18 \times 10^5$$

From Fig. 5-5, $k_s/D = 0.0007$. Then from Fig. 5-4 on page 247, using the values obtained for k_s/D and Re, we find $f = 0.019$. Finally the head loss is computed from the Darcy-Weisbach equation:

$$h_f = f \cdot \frac{L}{D} \cdot \frac{V^2}{2g} = 0.019 \left(\frac{1,000 \text{ m}}{0.20 \text{ m}} \right) \left(\frac{1.59^2 \text{ m}^2/\text{s}^2}{2(9.81 \text{ m/s}^2)} \right)$$

$$= 12.2 \text{ m}$$

The head loss per kilometer is 12.2 m.   ■

In the second type of problem, if you know the value of h_f, then k_s/D and the value of $(D^{3/2}/v)\sqrt{2gh_f/L}$ can be computed so that the top scale can be used to enter the design chart of Fig. 5-4. The lines that slope down and to the right are lines of equal value of the parameter $(D^{3/2}/v)\sqrt{2gh_f/L}$. Then, once f is read

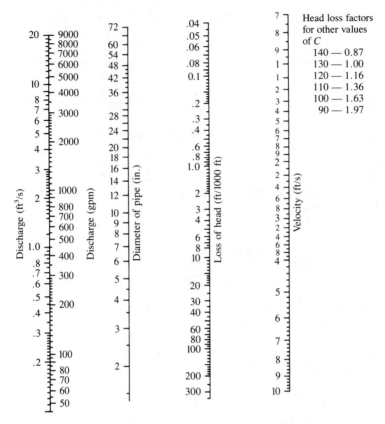

Figure 5-6 Nomograph for solving the Hazen-Williams formula with $C_h = 130$ (5) ("Pipeline Design for Water and Wastewater," a report of the TASK COMMITTEE ON ENGINEERING PRACTICE IN THE DESIGN OF PIPELINES, 1975.)

from the chart, the velocity from Eq. (5-9) is solved and the discharge is computed from $Q = VA$. This procedure yields a direct solution for Q.

However, many problems for which the discharge Q is desired cannot be solved directly. For example, a problem in which water flows from a reservoir through a pipe and into the atmosphere is of this type. Here, part of the available head is lost to friction in the pipe, and part of the head remains in kinetic energy in the jet as it leaves the pipe. Therefore, at the outset, one does not know how much head loss occurs in the pipe itself. To effect a solution, you must iterate on f. The energy equation is written and an initial value for f is guessed; then the velocity, V, is solved. With this value of V, a Reynolds number is computed that allows a better value of f to be determined through the use of Fig. 5-4 and so on. This type of solution usually converges quite rapidly because f changes more slowly than Re.

In the third type of problem, it is usually best to first assume a value of f and then solve for D, after which a better value of f is computed based on the first estimate of D. This iterative procedure is continued until a valid solution is obtained. A trial-and-error procedure is necessary because without D you cannot compute k_s/D or Re to enter the Moody diagram.

EXAMPLE 5-2 What size asphalted cast-iron pipe is needed to carry, water at a discharge of 12 cfs and with a head loss of 4 ft per 1000 ft of pipe?

SOLUTION First assume $f = 0.015$. Then

$$h_f = \frac{fL}{D} \cdot \frac{V^2}{2g} = \frac{fL}{D} \cdot \frac{Q^2/A^2}{2g} = \frac{fLQ^2}{2g(\pi/4)^2 D^5}$$

or $$D^5 = \frac{fLQ^2}{0.785^2(2gh_f)}$$

or, for this example,

$$D^5 = \frac{0.015(1000 \text{ ft})(12 \text{ ft}^3/\text{s})^2}{0.615(64.4 \text{ ft/s}^2)(4 \text{ ft})} = 13.63 \text{ ft}^5$$

$$D = 1.69 \text{ ft} = 20.3 \text{ in.}$$

Now compute a more accurate value of f:

$$\frac{k_s}{D} = 0.00025 \qquad V = \frac{Q}{A} = \frac{12 \text{ ft}^3/\text{s}}{0.785(2.86 \text{ ft}^2)} = 5.34 \text{ ft/s}$$

Then $$\text{Re} = \frac{VD}{v} = \frac{5.34 \text{ ft/s}(1.69 \text{ ft})}{1.21(10^{-5} \text{ ft}^2/\text{s})} = 7.47 \times 10^5$$

From Fig. 5-5, $f = 0.0155$. Now recompute D by applying the ratio of f's to previous calculations for D^5:

$$D^5 = \frac{0.0155}{0.015}(13.63 \text{ ft}^5) = 14.08 \text{ ft}^5$$

$$D = 1.70 \text{ ft} = 20.4 \text{ in.}$$

Use a 22-in. diameter pipe.

Note: In actual design practice, if a nonstandard size of pipe is called for as a result of the design calculation, it is customary to choose the next standard size larger that is available commercially. By doing so, the cost is less than that for an odd-sized pipe, and the pipe will be more than large enough to carry the flow. In this case, the 22-in. pipe is the next standard size larger. ∎

Head Loss Using the Hazen-Williams Formula

The head loss formulas we have presented up to now are general because they are applicable for any fluid and any system of units. Other more restrictive empirical equations are also useful for their limited range of application. The most notable one, used for decades by waterworks engineers in the United States, is the Hazen-Williams formula. In English units, the formula is given in Eq. (5-12):

$$V = 1.318 C_h R^{0.63} S^{0.54} \qquad\qquad (5\text{-}12)$$

where V = mean velocity in ft/s
C_h = Hazen-Williams friction coefficient (depends on pipe roughness)
R = hydraulic radius in ft
$S = h_f/L$ (slope of energy grade line)

To solve for head loss using the Hazen-Williams equation, a little algebraic manipulation of Eq. (5-12) yields

$$h_f = 3.02 L D^{-1.167} \left(\frac{V}{C_h}\right)^{1.85} \qquad\qquad (5\text{-}13)$$

The resistance coefficient C_h depends on the surface characteristics of the pipe

Table 5-2 Hazen-Williams C_h Values for Different Kinds of Pipe (5)

Character of Pipe	C_h
New or in excellent condition cast-iron and steel pipe with cement or bituminous linings centrifugally applied, concrete pipe centrifugally spun, cement-asbestos pipe, copper tubing, brass pipe, plastic pipe, and glass pipe	140
Older pipe listed above in good condition, and cement mortar-lined pipes in place with good workmanship, larger than 24 in. in diameter	130
Cement mortar-lined pipe in place, small diameter with good workmanship or large diameter with ordinary workmanship; wood stave; tar dipped cast-iron pipe new or old in inactive water	120
Old unlined or tar-dipped cast-iron pipe in good condition	100
Old cast-iron pipe severely tuberculated, or any pipe with heavy deposits	10–80

wall. Representative values of C_h for various kinds and conditions of pipe are given in Table 5-2.

Because of the widespread use of the Hazen-Williams formula, charts and tables have been developed for easy solution of the formula. One of these charts, in nomograph form, is shown in Fig. 5-6 on page 250.

The Hazen-Williams formula and several other empirical head loss formulas are applicable only for water flow and for the usual range of pipe sizes and discharges found in water distribution systems. Therefore, the Hazen-Williams formula may yield erroneous results for fluids other than water and for pipe diameters smaller than 2 in. and larger than 6 ft.

Head Loss in Noncircular Conduits

TYPES OF NONCIRCULAR CONDUITS One type of noncircular closed conduit commonly used in water resources projects is the tunnel. The tunnel cross section is typically rounded at the top and flat on the bottom, a horseshoe shape, as shown in Fig. 5-7a. Another noncircular cross section often used in hydraulic engineering is the *rectangular section*. The rectangular conduit may be used as a closed conduit; however, it is most often used as an open channel, as shown in Fig. 5-7b. In either case, the method for calculating head loss in these noncircular conduits is the same.

HYDRAULIC RADIUS CONCEPT If it is assumed that the wall shear stress, τ_0, is uniformly distributed around the part of the perimeter of the conduit in contact with the flowing liquid (called the *wetted perimeter*), it can be shown that the Darcy-Weisbach equation has the following form (36):

$$h_f = \frac{f}{4} \cdot \frac{L}{A/P} \cdot \frac{V^2}{2g} \tag{5-14}$$

where A = cross-sectional area of flow section
P = wetted perimeter

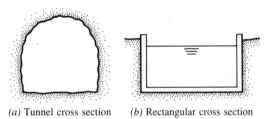

(a) Tunnel cross section   *(b)* Rectangular cross section

Figure 5-7 Noncircular conduits

The ratio A/P is called the *hydraulic radius R*; therefore, the Darcy-Weisbach equation reduces to

$$h_f = \frac{fL}{4R} \cdot \frac{V^2}{2g} \tag{5-15}$$

Equation (5-15) is the same as the original form of the equation, Eq. (5-9), except that the diameter, D, is replaced by $4R$. Experimental evidence shows that we can solve for the head loss in noncircular conduits if we apply the same methods and equations that we used for pipes but use $4R$ in place of D. Thus the relative roughness is $k_s/4R$, and the Reynolds number is defined as $V(4R)/\nu$.

EXAMPLE 5-3   A concrete-lined tunnel has a cross section described as follows. The top part of the tunnel is a 20-ft diameter semicircle, and the bottom part is a rectangular section 20 ft wide by 10 ft high. Estimate the head loss in 1-mi length of tunnel when water is flowing in it with a mean velocity of 12 ft/s.

SOLUTION   The head loss is given as $h_f = f(L/4R)V^2/2g$, where f is a function of Re and $k_s/4R$. First solve for the hydraulic radius R:

$$R = \frac{A}{P}$$

$$= \frac{(\pi 10^2/2) + (20 \cdot 10)}{20 + 2 \cdot 10 + \pi \cdot 10}$$

$$= \frac{357}{71.4} = 5.00 \text{ ft}$$

Next solve for the Reynolds number:

$$\text{Re} = V \cdot \frac{4R}{\nu}$$

Assume the water temperature is 60°F, so $\nu = 1.22 \cdot 10^{-5}$ ft^2/s (from Table A-4, page 648).

Then $$\text{Re} = 12 \cdot 4 \cdot \frac{5.00}{1.22 \cdot 10^{-5}} = 1.96 \cdot 10^7$$

Assume $k_s = 0.01$ ft. Then $k_s/4R = 0.0005$. With these values of Re and $k_s/4R$, we obtain $f = 0.017$.

The head loss is then computed:

$$h_f = 0.017 \cdot \frac{5280}{(4 \times 5.0)} \cdot \frac{12^2}{(2 \times 32.2)} = 10.0 \text{ ft/mi} \qquad \blacksquare$$

Head Loss Due to Transitions and Fittings

Besides the head loss due to the conduit itself, other losses are caused by the inlet, outlet, bends, and other appurtenances that alter the uniform flow regime in the conduit. Physically, all these head losses occur because additional turbulence is created by the particular conduit fitting, and the energy associated with the turbulence is finally dissipated into heat that produces the head loss. The head loss produced by transitions and fittings is expressed as

$$h_L = K \frac{V^2}{2g} \qquad (5\text{-}16)$$

where V is the mean velocity in the conduit, and K is the loss coefficient for the particular fitting involved. Table 5-3 on the next page gives the loss coefficients, determined by experimentation, for various transitions and fittings.

EXAMPLE 5-4 The conduit of Example 5-3 is used to convey water from a reservoir (water surface elevation 5000 ft) through hydroturbines and from there to another reservoir (water surface elevation 3000 ft). The tunnel is 5 mi long and has two long-radius 45° bends in it plus two wide-open gate valves and well-designed inlets and outlets. What power can be delivered to the turbines if we assume the flow passages associated with the turbines themselves have a loss coefficient of 0.20? Further assume the water velocity in the tunnel is the same as in Example 5-3 and that the head loss for the gate valve is negligible.

SOLUTION The energy equation is first written with $\alpha = 1$:

$$\frac{p_1}{\gamma} + \frac{V_1^2}{2g} + z_1 = \frac{p_2}{\gamma} + \frac{V_2^2}{2g} + z_2 + h_t + \sum h_L$$

In this example, let point 1 be at the upper reservoir water surface, and let point 2 be at the lower reservoir water surface. The head loss will be given as

$$\sum h_L = \frac{V^2}{2g}\left(\frac{fL}{4R} + 2K_b + K_e + K_o + 0.2\right)$$

where $\dfrac{V^2}{2g}\left(\dfrac{fL}{4R}\right) = 5 \times 10.0 = 50$ ft (from Example 5-3)

$$K_b \approx 0.1 \text{ (estimated from Table 5-3)}$$
$$K_e = 0.12 \text{ (estimated from Table 5-3)}$$
$$K_{\text{outlet}} = K_E = 0.15 \text{ (estimated from Table 5-3, assuming } \theta = 10°)$$

Then $\quad \sum h_L = \dfrac{12^2}{2 \times 32.2}(2 \times 0.1 + 0.12 + 0.15 + 0.2) + 50.0 = 51.5$ ft

Table 5-3 Loss Coefficients for Various Transitions and Fittings

Description	Sketch	Additional Data		K		Sourc
		r/d		K_e		(7)
Pipe entrance		0.0		0.50		
		0.1		0.12		
$h_L = K_e V^2/2g$		>0.2		0.03		
			K_C		K_C	
Contraction		D_2/D_1	$\theta = 60°$		$\theta = 180°$	(7)
		0.0	0.08		0.50	
		0.20	0.08		0.49	
		0.40	0.07		0.42	
		0.60	0.06		0.32	
		0.80	0.05		0.18	
$h_L = K_C V_2^2/2g$		0.90	0.04		0.10	
			K_E		K_E	
Expansion		D_1/D_2	$\theta = 10°$		$\theta = 180°$	(7)
		0.0			1.00	
		0.20	0.13		0.92	
		0.40	0.11		0.72	
		0.60	0.06		0.42	
$h_L = K_E V_1^2/2g$		0.80	0.03		0.16	
90° miter bend	Vanes	Without vanes		$K_b = 1.1$		(42)
		With vanes		$K_b = 0.2$		(42)
		r/d	K_b		K_b	(14)
			$\theta = 45°$		$\theta = 90°$	
Smooth bend		1	0.10		0.35	(22)
		2	0.09		0.19	and
		4	0.10		0.16	(30)
		6	0.12		0.21	
	Globe valve — wide open			$K_v = 10.0$		
	Angle valve — wide open			$K_v = 5.0$		
	Gate valve — wide open			$K_v = 0.2$		
Threaded	Gate valve — half open			$K_v = 5.6$		
pipe fittings	Return bend			$K_b = 2.2$		
	Tee			$K_t = 1.8$		
	90° elbow			$K_b = 0.9$		
	45° elbow			$K_b = 0.4$		

The head given up to the turbines h_t is then

$$h_t = 5000 - 3000 - 51.5 = 1948.5 \text{ ft}$$

Finally
$$P = \frac{Q\gamma h_t}{550}$$

$$= 12 \cdot \left[\left(\pi \cdot \frac{10^2}{2} \right) + (20 \cdot 10) \right] \cdot 62.4 \cdot \frac{1948.5}{550}$$

$$= 947{,}000 \text{ hp} \qquad \blacksquare$$

Explicit Equations for h_f, Q, and D

On pages 249 to 251, we presented methods by which h_f, Q, and D can be calculated. All these methods involve using the Moody diagram (Fig. 5-4). With the advent of computers and programmable calculators, it is most desirable to be able to solve similar problems without having to resort to the Moody diagram. By using the Colebrook-White formula, from which the Moody diagram was developed, Swamee and Jain (44) developed explicit formulas relating f, h_f, Q, and D. It is reported that their formulas yield results that deviate no more than 3% from those obtained from the Moody diagram for the following ranges of k_s/D and Re: $10^{-5} < k_s/D < 2 \times 10^{-2}$ and $4 \times 10^3 < \text{Re} < 10^8$. The formulas for f and Q are

$$f = \frac{0.25}{\left[\log \left(\dfrac{k_s}{3.7D} + \dfrac{5.74}{\text{Re}^{0.9}} \right) \right]^2} \tag{5-17}$$

$$Q = -2.22 D^{5/2} \sqrt{gh_f/L} \, \log \left(\frac{k_s}{3.7D} + \frac{1.78v}{D^{3/2}\sqrt{gh_f/L}} \right) \tag{5-18}$$

They also developed a formula for the explicit determination of D. A modified version of that formula, given by Streeter and Wylie (41), is

$$D = 0.66 \left[k_s^{1.25} \left(\frac{LQ^2}{gh_f} \right)^{4.75} + vQ^{9.4} \left(\frac{L}{gh_f} \right)^{5.2} \right]^{0.04} \tag{5-19}$$

If you want to solve for head loss given Q, L, D, k_s, and v, simply solve for f by Eq. (5-17) and compute h_f with the Darcy-Weisbach equation, Eq. (5-9). Straightforward calculations for Q and D can also be made if h_f is known. However, for problems involving head losses in addition to h_f, an iterative solution is required. For computing Q, you can assume an f and solve for Q from the energy equation after substituting Q/A in that equation. Then compute

Re and use that in Eq. (5-17) to get a better estimate of f and so on, until Q converges analogous to the procedure for determining Q using the Moody diagram. In this case, however, Eq. (5-17) is substituted for the Moody diagram. Similarly, D can be determined given Q, v, the change in pressure or head, and the geometric configuration.

EXAMPLE 5-5   Determine the diameter of steel pipe needed to deliver water (20°C) at a rate of 2 m³/s from a reservoir with water surface at an elevation of 60 m to a reservoir 200 m away with a water surface elevation of 30 m. Assume a square-edged inlet and outlet and no bends in the pipe. Further assume there are two open gate valves in the pipe.

SOLUTION   Writing the energy equation from the upper to lower reservoir, we have

$$\frac{p_1}{\gamma} + \frac{\alpha_1 V_1{}^2}{2g} + z_1 = \frac{p_2}{\gamma} + \frac{\alpha_2 V_2{}^2}{2g} + z_2 + \sum h_L$$

$$0 + 0 + z_1 = 0 + 0 + z_2 + \left(k_e + 2k_v + k_E + \frac{fL}{D}\right)\frac{V^2}{2g}$$

$$0 = z_2 - z_1 + \left(k_e + 2k_v + k_E + \frac{fL}{D}\right)\frac{Q^2}{2gA^2}$$

Assume $k_s = 0.046$ mm, $k_e = 0.5$, $k_v = 0.2$, $k_E = 1.0$, $f = 0.02$, and let $A = \pi D^2/4$. Then

$$0 = 30 - 60 + \frac{[1.9 + (0.02 \times 200/D)]}{2^2/[2g(\pi^2/16)D^4]}$$

Solving this for D yields $D = 0.56$ m, $V = 8.12$ m/s, and Re $\approx 4.5 \times 10^6$. Then from Eq. (5-17), we have

$$f = 0.25/[\log (0.000046/3.7 \times 0.56) + (5.74/9.82 \times 10^5)]^2$$

$$f = 0.0092 \qquad \text{Re} = 4.5 \times 10^6$$

Substituting this value of f back into the energy equation and solving for a better value of D yields $D = 0.52$ m. Another iteration still yields

$$D = 0.52 \text{ m} \qquad\qquad\qquad\qquad\qquad \blacksquare$$

In Example 5-5, we used the same iterative procedure introduced in Example 5-3 except that we replaced the Moody diagram by Eq. (5-17) and included the energy equation because head losses other than pipe friction losses were significant. Another more rapid iterative solution, presented by Streeter and Wylie (41), for D uses Eq. (5-18) when other than pipe friction losses are present. These are expressed as an equivalent length of pipe, and h_f is the total difference in head between the sections under consideration. Thus in Example 5-5, we would obtain the equivalent length of pipe for the minor losses as

$$f\left(\frac{L_e}{D}\right)\left(\frac{V^2}{2g}\right) = 1.9\frac{V^2}{2g}$$

where L_e = equivalent pipe length = $1.9D/f$. Then

$$L = L_{\text{pipe}} + L_e = L + 1.9D/f$$

You could then solve Eq. (5-19) by iteration. That is, first assume f, then solve for D, after which a better value of f is obtained from Eq. (5-19), and so on.

EXAMPLE 5-6   Solve Example 5-5 using Eq. (5-19).

SOLUTION   First assume $f = 0.02$.

Then $L = L + L_e = 200 \text{ m} + \dfrac{1.9\,D}{0.02}$

$$= 200 + 95D$$

Letting $h_f = 30$ m and $L = 200 + 95\,D$ in Eq. (5-19) and solving for D yields

$$D = 0.66\left[0.000046\left(\frac{(200 + 95D) \times 2^2}{9.81 \times 30}\right)^{4.75}\right.$$
$$\left. + 10^{-6} \times 2^{9.4}\left(\frac{200 + 95D}{9.81 \times 30}\right)^{5.2}\right]^{0.04} = 0.509 \text{ m}$$

First iteration:

With $D = 0.51$ m, Re = 5.00×10^6, $f = 0.0122$ (from Eq. 5-17)

$L_e = 1.9D/0.0122 = 155.7D$ $D = 0.509$ m

$L = 200 + 156D$ $D = 0.506$ m

Second iteration:

With $D = 0.506$ m, Re = 5×10^6, $f = 0.0122$

$L_e = 156D$

Since there is no significant change in either f or L_e, the diameter will be the same: $D = 0.51$ m.

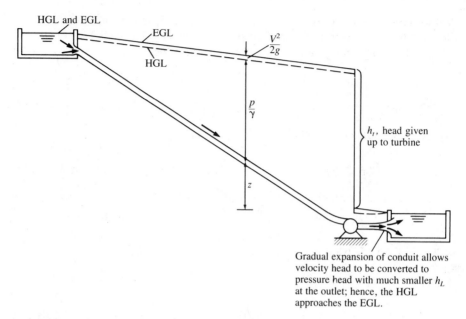

Figure 5-8   Drop in EGL and HGL due to turbine

Hydraulic and Energy Grade Lines

As we noted, the terms of Eq. (5-1) have linear dimension (feet or meters); thus we can attach a useful physical relationship to them, as shown in Fig. 5-1, page 242. If you were to tap a piezometer into the pipe of Fig. 5-1, the liquid would rise to a height p/γ above the pipe; hence the reason for the name *hydraulic grade line* (HGL). The total head $(p/\gamma + V^2/2g + z)$ in the system is greater than $p/\gamma + z$ by an amount $\alpha V^2/2g$; thus the *energy grade line* (EGL) is above the HGL a distance $\alpha V^2/2g$. The engineer who develops a visual concept of the energy equation as we explained earlier will find it much easier to sense trouble spots (usually points of low pressure) in the system.

Some hints for drawing hydraulic grade lines and energy grade lines are as follows:

1. By definition, the EGL is positioned above the HGL an amount equal to the velocity head. Thus if the velocity is zero, as in a lake or reservoir, the HGL and EGL will coincide with the liquid surface (see Fig. 5-8).
2. Head loss for flow in a pipe or channel always means the EGL will slope downward in the direction of flow. The only exception to this rule occurs when a pump supplies energy (and pressure) to the flow. Then an abrupt rise in the EGL occurs from the upstream side to the downstream side of the pump.

3. In point 2, we noted that a pump can cause an abrupt rise in the EGL be-
    cause energy is introduced into the flow by the pump. Similarly, if energy
    is abruptly taken out of the flow by, for example, a turbine, the EGL and
    HGL will drop abruptly as in Fig. 5-8. Figure 5-8 also shows that much
    of the velocity head can be converted to pressure head if there is a gradual
    expansion such as at the outlet. Thus the head loss at the outlet is reduced,
    making the turbine installation more efficient. If the outlet to a reservoir
    is an abrupt expansion, all the kinetic energy is lost; thus the EGL will
    drop an amount $\alpha V^2/2g$ at the outlet.

4. In a pipe or channel where the pressure is zero, the HGL is coincident with
    the water in the system because $p/\gamma = 0$ at these points. This fact can be
    used to locate the HGL at certain points in the physical system, such as at
    the outlet end of a pipe, where the liquid discharges into the atmosphere, or
    at the upstream end, where the pressure is zero in the reservoir (see Fig. 5-8).

5. For steady flow in a pipe that has uniform physical characteristics (diameter,
    roughness, shape, and so on) along its length, the head loss per unit of
    length will be constant; thus the slope $(\Delta h_L/\Delta L)$ of the EGL and HGL will
    be constant along the length of pipe.

6. If a flow passage changes diameter, such as in a nozzle or a change in pipe
    size, the velocity therein will also change; hence the distance between the
    EGL and HGL will change. Moreover, the slope on the EGL will change
    because the head loss per unit length will be larger in the conduit with the
    larger velocity.

7. If the HGL falls below the pipe, p/γ is negative, thereby indicating subatmo-
    spheric pressure (see Fig. 5-9). If the pressure head of water is less than the
    vapor pressure head of the water (approximately -33 ft at standard atmo-
    spheric pressure and $T = 60°F$), cavitation will occur. Generally, cavitation
    in conduits is undesirable. It increases the head loss and can cause structural
    damage to the conduit from excessive vibration and pitting of the conduit
    walls. If the pressure at a section in the pipe decreases to the vapor pressure
    and stays that low, a large vapor cavity can form leaving a gap of water
    vapor with columns of water on either side of the cavity. As the cavity

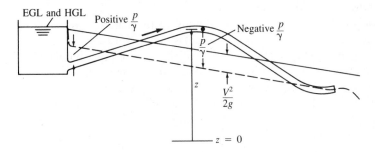

Figure 5-9   Subatmospheric pressure when pipe is
above the HGL

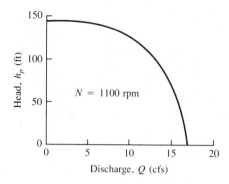

Figure 5-10   Typical performance curve for a
centrifugal pump

grows in size, the columns of water move away from each other. Often
these columns of water will rejoin later, and when they do, a very high
dynamic pressure (water hammer) can be generated, possibly rupturing the
pipe. Furthermore, if the pipe is relatively thin walled, such as thin-walled
steel pipe, subatmospheric pressure can cause the pipe wall to collapse.
Therefore, the design engineer should be extremely cautious about negative
pressure heads in the pipe.*

5-4   Head-Discharge Relations for Pump or Turbine

In the energy equation, Eq. (5-1), the terms h_p and h_t are the heads supplied by a pump or given up to a turbine, respectively, and these heads are a
function of the discharge for a machine (pump or turbine) that is operating at
a given speed. Figure 5-10 is a typical plot of h_p versus Q for a centrifugal pump.
Such a plot is one of the *performance* or *characteristic* curves of the machine.
Other performance curves, such as efficiency and power versus discharge, are
often included with the head discharge curve.

Solutions of problems involving a given pump or turbine are direct if Q is
given; the head for the machine (pump or turbine) is taken directly from the performance curve of the machine, and then one solves for the pipe diameter or
head, as the case may be. On the other hand, if one is solving for Q, a simultaneous solution of the energy equation and the h versus Q relation will yield the
desired result. For the latter type of problem, it is often convenient to approximate the hQ relation of the machine by an equation.[†]

---

* For a more detailed description of water hammer with methods of analyses, see Chapter 11.
[†] For more detail about performance curves, pumps, and turbines, see Chapter 8.

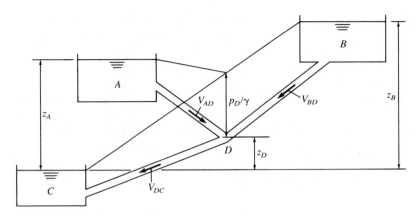

Figure 5-11   Flow in a branched-pipe system

5-5   Conduit Systems

So far, we have considered only problems involving pipes in series. Other applications include branching pipes, parallel pipes, manifolds, and pipe networks.

Branching Pipes

Consider the case shown in Fig. 5-11, where three reservoirs are connected by a branched-pipe system. The problem is to determine the discharge in each pipe and the head at the junction point D. There are four unknowns (V_{AD}, V_{BD}, V_{DC}, and p_D/γ), and a solution is obtained by solving the energy equations for the pipes (neglecting velocity heads and including only pipe losses) and the continuity equation. These equations are given as

$$z_A = z_D + \frac{p_D}{\gamma} + f_{AD}\frac{L_{AD}}{D_{AD}}\frac{V_{AD}^2}{2g} \tag{5-20}$$

$$z_B = z_D + \frac{p_D}{\gamma} + f_{BD}\frac{L_{BD}}{D_{BD}}\frac{V_{BD}^2}{2g} \tag{5-21}$$

$$z_D + \frac{p_D}{\gamma} = f_{DC}\frac{L_{DC}}{D_{DC}}\frac{V_{DC}^2}{2g} \tag{5-22}$$

$$V_{AD}A_{AD} + V_{BD}A_{BD} = V_{DC}A_{DC} \tag{5-23}$$

The preceding equations are written for the assumed flow directions. It should be obvious that the directions shown by the velocity vectors V_{BD} and V_{DC} will be valid for the final solution. However, at the outset, one cannot always discern whether the flow in pipe AD will be into or out of the reservoir. One can determine the proper flow direction in pipe AD by first assuming $z_D + p_D/\gamma = z_A$ (piezometric head at point D is the same as in reservoir A). With this assumed head at junction D, one determines Q_{BD} and Q_{DC} from the given equations. If $Q_{BD} > Q_{DC}$, then p_D/γ will have to be increased, which will then mean flow must be directed into reservoir A (piezometric head at point D will be greater than in reservoir A). If $Q_{BD} < Q_{DC}$ for $z_D + p_D/\gamma = z_A$, then flow would be out of reservoir A. The next process is to iterate on p_D/γ until all equations are satisfied.

Parallel Pipes

Consider a pipe that branches into two parallel pipes and then rejoins, as in Fig. 5-12. A problem involving this configuration might be to determine the division of flow in each pipe given the total flow rate. It can be seen that the head loss must be the same in each pipe because the pressure difference is the same. Thus we can write

$$h_{L1} = h_{L2}$$

$$f_1 \frac{L_1}{D_1} \frac{V_1^2}{2g} = f_2 \frac{L_2}{D_2} \frac{V_2^2}{2g}$$

Then
$$\left(\frac{V_1}{V_2}\right)^2 = \frac{f_2}{f_1} \frac{L_2}{L_1} \frac{D_1}{D_2}$$

or
$$\frac{V_1}{V_2} = \left(\frac{f_2}{f_1} \frac{L_2}{L_1} \frac{D_1}{D_2}\right)^{1/2}$$

If f_1 and f_2 are known, the division of flow can be easily determined; however, some trial-and-error analysis may be required if f_1 and f_2 are in the range where they are functions of the Reynolds number.

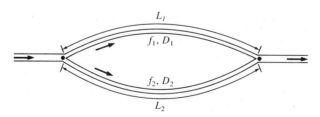

Figure 5-12   Flow in parallel pipes

Manifolds

Manifolds (pipes that branch into other pipes) can be the combining flow type (Fig. 5-13a) or the dividing flow type (Figs. 5-13b and c).

In hydraulic engineering, the dividing flow type is used more; therefore, we focus on this type of manifold. However, the basic approach is similar for both types. Examples of the dividing type include manifolds used in navigation locks to effect uniform filling and diffusers for disposal of sewage or heated effluents into large bodies of water.

Diffusers may include separate branch pipes (Fig. 5-13b) for distributing the flow, or the distribution may simply be accomplished by ports (holes) cut in the manifold (Fig. 5-13c). In general, the head will change along the length of the manifold because of friction and momentum change. The flow from each port or branch pipe also is affected by the magnitude of the velocity in the diffuser. Thus the discharge from each branch will in general be different depending on its location along the manifold. This assumes all branches are the same size.

In most manifold design problems, the objective is to distribute a given discharge fairly uniformly along the length of the manifold. The method of analysis and design of manifolds follows essentially the procedure given by Rawn (34). The basic assumptions and fluid mechanics principles involved in a manifold design problem are as follows:

1. The discharge from each port or branch pipe can be expressed as

$$q = Ka\sqrt{2gE} \qquad\qquad (5\text{-}24)$$

where K = flow coefficient
$\quad\ a$ = cross-sectional area of branch pipe or port
$\quad E = V^2/2g + \Delta h$
$\quad V$ = mean velocity in manifold
$\quad \Delta h = [(p_m/\gamma) + z_m] - [(p_o/\gamma) + z_o]$, and m and o are subscripts
$\qquad\qquad$ that refer to conditions inside and outside the manifold,
$\qquad\qquad$ respectively, at the section where the branch or port is
$\qquad\qquad$ located.

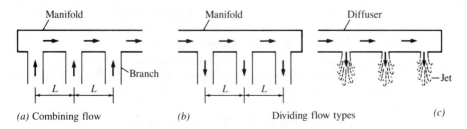

(a) Combining flow $\qquad$ *(b)* $\qquad$ Dividing flow types $\qquad\qquad$ *(c)*

Figure 5-13   Flow manifolds

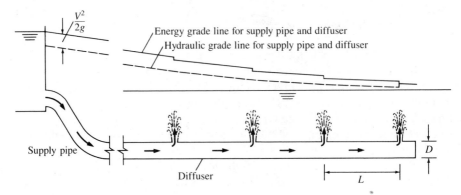

Figure 5-14   Schematic view of a diffuser

Thus Eq. (5-24) is a discharge equation for the branch pipe or port, and E is the difference in head between the manifold and the ambient fluid at the outlet of the branch pipe or port.

2. The branches or ports are spaced at intervals along the manifold, and between the branches or ports in the manifold a change in head will occur due to the head loss in the manifold. This is given as

$$h_f = f\left(\frac{L}{D}\right)\frac{V^2}{2g}$$

where L = port or branch spacing (see Fig. 5-13)
      D = diameter of manifold
      V = mean velocity in the manifold.

There may also be a small change in head across the section where a port or branch is located (from upstream of port or branch to downstream of it); however, accepted design practice assumes this change in head is negligible (34).

The design of a manifold is described, by example, for a diffuser, such as is used in the discharge of wastewater into the sea. The physical setup is shown in Fig. 5-14. The objective in the manifold design is to distribute a given total discharge of effluent fairly uniformly along the length of the manifold. The design process is iterative. The design assumes the geometric configuration (diameter of manifold, size of ports, spacing between ports) and the discharge from the end port. Then by computation, the discharge from all other ports and the total required head can be determined. If, as a result of these computations, the conditions are not to the designer's satisfaction, a new set of conditions are assumed, and the process is repeated until the desired design is achieved. Example 5-7 illustrates the procedure.

EXAMPLE 5-7 Wastewater, after primary treatment, is to be discharged into a large body of water by means of a diffuser. Design a manifold type of diffuser to discharge effluent at a rate of 4 cfs. Assume the design criteria limit the total head (above that of lake level) at the upstream end of the supply pipe that feeds the diffuser to 25 ft, and that the difference in discharge between the upstream and downstream ports is to be no greater than 10% of the discharge from the downstream port. Further assume the supply pipe will be 400 ft long.

SOLUTION Assume the manifold will be made of PVC pipe, the average discharge from each port $\bar{q}$ will be 0.20 cfs, and the spacing between ports will be 4 ft. Thus there will have to be approximately 20 ports [4.0 cfs/(0.20 cfs/port)], and the total length of diffuser itself will be 80 ft. Assume the velocity from the downstream end port will be 19.0 ft/s. Therefore, the head, E_{end}, at the dead end of the diffuser will be given by

$$ E_{end} = \left(\frac{V^2}{2g}\right) + \Delta h $$

However, $V^2/2g = 0$ at the dead end. Therefore,

$$ E_{end} = \Delta h = \left(\frac{q^2}{K^2 a^2}\right)\frac{1}{2g} $$

where $q^2/a^2 = (19.0)^2$ ft^2/s^2. K is the flow coefficient for the orifice, and as noted by Subramanya (43), it can be given as

$$ K = 0.675 \sqrt{1 - \frac{V^2}{2gE}} \tag{5-25} $$

Then for the end section where $V \approx 0$ ($K = 0.675$), the total energy head is $E_{end} = (19.0^2/0.675^2)/(2 \times 32.2) = 12.3$ ft.

Next, as a first approximation, assume the diameter of the manifold pipe and feeder pipe are the same size and are based on the head available as given by the design criteria. That is, the total head available at the inlet to the supply pipe is given as 25 ft, and the head at the dead end of the diffuser (just calculated above) is 12.3 ft. Therefore, the head available for flow in the supply pipe and the manifold is equal to $25 - 12.3$, or 12.7 ft. Because the velocity in the diffuser averaged over its length is only about one half the supply pipe velocity, the average head loss per unit of length in diffuser would be only a small fraction of that in the supply pipe. Assume this loss per unit length to be 1/4 of that in the supply pipe. Then for the total loss in both pipes, we have

$$ 12.7 = f\left(\frac{L_1}{D}\right)\frac{V^2}{2g} + \frac{1}{4}f\left(\frac{L_2}{D}\right)\frac{V^2}{2g} $$

where $L_1 = 400$ ft
$\qquad L_2 = 80$ ft

Assume $f = 0.015$ (first approximation). Then

$$12.7 = 0.015(420/D)\frac{Q^2}{(2gA^2)}$$

$$= 0.015(420/D)\frac{4^2}{\left(64.4 \cdot \frac{\pi^2}{4^2} \cdot D^4\right)}$$

$$D^5 = 0.200 \text{ ft}^5 \qquad D = 0.725 \text{ ft} = 8.7 \text{ in.}$$

The 8.7-in. size is not a standard size; therefore, choose the next standard size larger, which is a 10.0-in. size. To complete the initial geometric characteristics of the diffuser, we determine the port size. This is found from Eq. (5-25) for $q = 0.20$ cfs and assuming $E \approx 12.3$ ft and $K = 0.675$.

Now we solve for a and d_{port}.

$$a = \frac{q}{(K\sqrt{2gE})} = \frac{0.2}{(0.675\sqrt{64.4 \times 12.3})}$$

$$= 0.0102 \text{ ft}^2$$

$$d = 0.1142 \text{ ft} = 1.37 \text{ in.}$$

Thus $d = 0.1142$ ft is used for the initial port diameter.

Now that the basic geometric configurations have been assumed along with q at the end port, we can analyze the flow in the diffuser to determine the discharge distribution from the ports and the head required at the inlet to the supply pipe.

The analysis starts at the downstream end of the diffuser and works upstream (step by step) until the head at the upstream end of the diffuser is obtained. Then the head at the inlet to the supply pipe is obtained. This procedure follows that given by Rawn (34) and by Vigander (48). Fig. 5-15 shows the

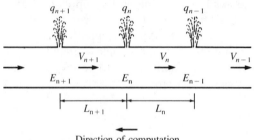

Direction of computation

Figure 5-15 Definition sketch for flow in a manifold

diffuser pipe and the notation used in the computations. For each step of the computation, the following equations are solved in the order shown.

$$V_{n-1} = \frac{Q_{n-1}}{A_{\text{diffuser}}}$$ (*Note*: This is zero at the dead end.)

$$K_n = 0.675 \sqrt{1 - \frac{V_{n-1}^2}{2gE_n}}$$ (*Note*: This has a value of 0.675 for the first port.

$$q_n = K_n a_n \sqrt{2gE_n}$$

$$\Delta V_n = \frac{q_n}{(\pi D^2/4)}$$

$$V_n = V_{n-1} + \Delta V_n$$

$$\text{Re}_n = V_n \frac{D}{\nu}$$

$$f_n = \frac{0.25}{\left[\log\left(\frac{k_s}{3.7\,D} + \frac{5.74}{\text{Re}^{0.9}}\right)\right]^2}$$

$$h_{f_n} = f\left(\frac{L}{D}\right)\left(\frac{V_n^2}{2g}\right)$$

$$E_{n+1} = E_n + h_{f_n}$$

where V_n is the mean velocity in the manifold at the nth computational step, E_n is the total energy head at the nth step, a_n is the cross-sectional area of the nth port, and ΔV_n is the change in velocity in the manifold due to the discharge from the nth port. It was assumed that $k_s = 0$ for the PVC pipe.

The table on page 270 is a summary of the computations for this diffuser.

The table shows that 20 ports provide a total discharge of 7.069 ft/s $\times$ $(\pi/4) \times (10/12)^2 = 3.86$ cfs. The percent difference in discharge between the upstream and downstream ports is given as

$$\left(\frac{0.1945 - 0.1893}{0.1945}\right) \times 100 = 2.67\%$$

The total head required at the inlet to the supply pipe is the total head at the upstream end of the diffuser plus the head loss in the supply pipe, or

$$E = E_{20} + h_{f_{\text{supply pipe}}}$$

$$= 12.4 + f\left(\frac{L}{D}\right)\left(\frac{V^2}{2g}\right)$$

Port	V_{n-1} (ft/s)	q_n (ft³/s)	ΔV_n (ft/s)	V_n (ft/s)	Re	f_n	h_{f_n} (ft)	E_{n+1} (ft)	K_n
1	0.000	0.1945	0.357	0.357	2.0×10^4	0.0250	0.0003	12.300	0.675
2	0.357	0.1945	0.357	0.714	4.0	0.0215	0.0006	12.300	0.675
3	0.714	0.1945	0.357	1.071	6.4	0.0196	0.00168	12.300	0.675
4	1.071	0.1944	0.357	1.428	8.6	0.0185	0.00281	12.302	0.675
5	1.428	0.1943	0.356	1.784	1.1×10^5	0.0176	0.0042	12.305	0.674
6	1.784	0.1942	0.356	2.140	1.3	0.0170	0.0058	12.306	0.674
7	2.140	0.1940	0.356	2.496	1.5	0.0169	0.0076	12.308	0.673
8	2.496	0.1939	0.355	2.851	1.7	0.0160	0.0097	12.310	0.672
9	2.851	0.1936	0.355	3.206	1.9	0.0157	0.01199	12.312	0.672
10	3.206	0.1934	0.355	3.561	2.1	0.0153	0.0145	12.314	0.671
11	3.561	0.1931	0.354	3.915	2.3	0.0151	0.0172	12.317	0.670
12	3.915	0.1928	0.354	4.268	2.6	0.0148	0.0201	12.320	0.668
13	4.268	0.1925	0.353	4.621	2.8	0.0146	0.0232	12.322	0.667
14	4.621	0.1921	0.352	4.973	3.0	0.0144	0.0265	12.326	0.666
15	4.973	0.1917	0.351	5.324	3.2	0.0142	0.0300	12.329	0.664
16	5.324	0.1913	0.351	5.675	3.4	0.0140	0.0337	12.332	0.663
17	5.675	0.1908	0.350	6.025	3.6	0.0139	0.0376	12.336	0.661
18	6.025	0.1904	0.349	6.374	3.8	0.0137	0.0416	12.340	0.659
19	6.374	0.1898	0.348	6.722	4.0	0.0136	0.0458	12.344	0.658
20	6.722	0.1893	0.347	7.069	4.2	0.0135	0.0502	12.348	0.656

$$E = 12.4 + 0.0134 \left(\frac{400}{(10/12)} \right) \left(\frac{7.759^2}{64.4} \right)$$

$$= 18.4 \text{ ft}$$

The total head at the upstream end of the inlet pipe is well within the original design criteria, as is the limit on the distribution of discharge from the ports. However, the total discharge from the 20 ports (3.86 cfs) is just short of the design discharge of 4.0 cfs. To determine the total head required at the inlet to yield a discharge of 4.00 cfs, multiply the 18.4 ft of head by $(4.00/3.86)^2$:

$$E = 18.4 \times \left(\frac{4.00}{3.86} \right)^2 = 19.76 \text{ ft}$$

This calculation assumes all head losses are a function of V^2 or Q^2/A^2, which is essentially the case here. Similarly, for a total head of 25 ft at the inlet, one can estimate the discharge through the system to be

$$Q = 3.86 \left(\frac{25}{18.4} \right)^{1/2} = 4.50 \text{ cfs}$$

∎

The designer may also wish to see if an 8-in. pipe size would produce results within the established criteria. If so, the same process would be repeated with that size.

All the preceding computations are easily programmed for computer solution; therefore, additional runs for different conditions are easily made. And as in all engineering design problems, the total cost of construction for each condition can be estimated. Thus the designer, by iteration, can find a solution that will satisfy the technical requirements for a minimum overall cost, which is the essence of engineering design.

Pipe Networks

The most common pipe networks are the water-distribution systems for municipalities. These systems have one or more sources (discharge of water into the system) and numerous loads: one for each household and commercial establishment. For purposes of simplification, the loads are usually lumped throughout the system. Figure 5-16 shows a simplified distribution system with two sources and seven loads.

The engineer is often engaged to design the original system or to recommend an economical expansion to the network. An expansion may involve additional housing or commercial developments, or it may be to handle increased loads within the existing area. In any case, the engineer is required to predict pressures throughout the network for various operating conditions, that is, for various combinations of sources and loads. The solution of such a problem must satisfy three basic requirements:

1. Continuity must be satisfied. That is, the flow into a junction of the network must equal the flow out of the junction. This must be satisfied for all junctions.

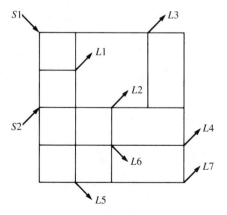

Figure 5-16   Pipe network

2. The head loss between any two junctions must be the same regardless of the path in the series of pipes taken to get from one junction point to the other. This requirement results because pressure must be continuous throughout the network (pressure cannot have two values at a given point). This condition leads to the conclusion that the algebraic sum of head losses around a given loop must be equal to zero. Here the sign (positive or negative) for the head loss in a given pipe is given by the sense of the flow with respect to the loop, that is, whether the flow has a clockwise or counter-clockwise sense.
3. The flow and head loss must be consistent with the appropriate velocity-head-loss equation.

Only a few years ago, these solutions were made by a trial-and-error hand computation, but recent applications using digital computers have made the older methods obsolete. Even with these advances, however, the engineer charged with the design or analysis of such a system must understand the basic fluid mechanics of the system to be able to interpret the results properly and to make good engineering decisions based on the results. Therefore, the method of solution first given by Cross (17) is presented as follows:

The flow is first distributed throughout the network so that the continuity requirement (requirement 1) is satisfied for all junctions. This first guess at the flow distribution obviously will not satisfy requirement 2 regarding head loss; therefore, corrections are applied. For each loop of the network, a discharge correction is applied to yield a zero net head loss around the loop. For example, consider the isolated loop in Fig. 5-17. In this loop, the loss of head in the clockwise sense will be given by

$$\sum h_{L_c} = h_{L_{AB}} + h_{L_{BC}}$$
$$= \sum k Q_c^n \tag{5-26}$$

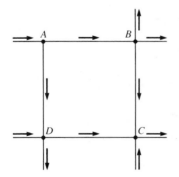

Figure 5-17   A typical loop of a network

The loss of head for the loop in the counterclockwise sense is

$$\sum h_{L_{cc}} = \sum kQ_{cc}^n \tag{5-27}$$

For a solution, the clockwise and counterclockwise head losses have to be equal or

$$\sum h_{L_c} = \sum h_{L_{cc}}$$
$$\sum kQ_c^n = \sum kQ_{cc}^n$$

As we noted, the first guess for flow in the network is undoubtedly in error; therefore, a correction in discharge, ΔQ, will have to be applied to satisfy the head loss requirement. If the clockwise head loss is greater than the counterclockwise head loss, ΔQ would have to be applied in the counterclockwise sense. That is, subtract ΔQ from the clockwise flows and add it to the counterclockwise flows:

$$\sum k(Q_c - \Delta Q)^n = \sum k(Q_{cc} + \Delta Q)^n \tag{5-28}$$

Expanding the summation on either side of Eq. (5-28) and including only two terms of the expansion, we obtain

$$\sum k(Q_c^n - nQ_c^{n-1}\Delta Q) = \sum k(Q_{cc}^n + nQ_{cc}^{n-1}\Delta Q)$$

Then solving for ΔQ, we get

$$\Delta Q = \frac{\sum kQ_c^n - \sum kQ_{cc}^n}{\sum nkQ_c^{n-1} + \sum nkQ_{cc}^{n-1}} \tag{5-29}$$

Thus if ΔQ as computed from Eq. (5-29) is positive, the correction is applied in a counterclockwise sense (add ΔQ to counterclockwise flows and subtract it from clockwise flows).

A different ΔQ is computed for each loop of the network and applied to the pipes. Some pipes will have two ΔQ's applied because they will be common to two loops. The first set of corrections usually will not yield the final desired result because the solution is approached only by successive approximations. Thus the corrections are applied successively until the corrections are negligible. Experience has shown that for most loop configurations, applying ΔQ as computed by Eq. (5-29) produces too large a correction. Fewer trials are required to solve for Q's if approximately 0.6 of the computed ΔQ is used.

EXAMPLE 5-8 For the given source and loads shown in Fig. A, how will the flow be distributed in the simple network, and what will be the pressures

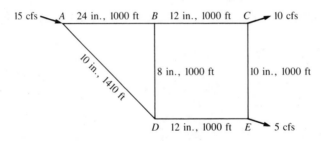

Figure A

at the load points if the pressure at the source is 60 psi? Assume horizontal pipes and $f = 0.012$ for all pipes.

SOLUTION   An assumption is made for the discharge in all pipes making certain that the continuity equation is satisfied at each junction. Figure B shows the network with assumed flows.

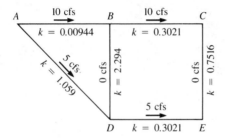

Figure B

The Darcy-Weisbach equation is used for computing the head loss; therefore, we have

$$h_f = f\left(\frac{L}{D}\right)\left(\frac{V^2}{2g}\right)$$

$$= 8\left(\frac{fL}{gD^5\pi^2}\right)Q^2$$

$$= kQ^2$$

where $k = 8\left(\frac{fL}{gD^5\pi^2}\right)$.

The loss coefficient, k, for each pipe is computed and shown in Fig. B. Next, the flow corrections for each loop are calculated as shown in the accompanying table. Since $n = 2$ (exponent on Q), $nkQ^{n-1} = 2kQ$. When the corrections obtained in the table are applied to the two loops, we get the pipe discharges shown in Fig. C. Then with additional iterations, we get the final distribution of flow as shown in Fig. D. Finally, the pressures at the load points are calculated (see page 276):

Loop ABC		
Pipe	$h_f = kQ^2$	$2kQ$
AB	$+0.944$	0.189
AD	-26.475	10.590
BD	0	0
$\sum kQ_c{}^2 - \sum kQ_{cc}^2 = -25.53$		$\sum 2kQ = \overline{10.78}$

$$\Delta Q = -25.53/10.78 = -2.40 \text{ cfs}$$

Loop $BCDE$		
Pipe	h_f	$2kQ$
BC	$+30.21$	6.042
BD	0	0
CE	0	0
DE	$-\ 7.55$	3.02
	$\overline{+22.66}$	$\overline{9.062}$

$$\Delta Q = 22.66/9.062 = 2.50 \text{ cfs}$$

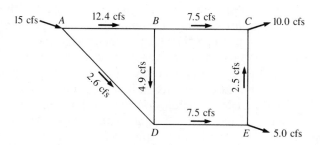

Figure C

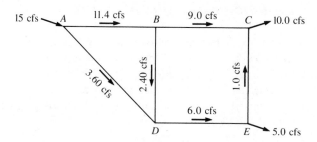

Figure D

$$p_C = p_A - \gamma(k_{AB}Q_{AB}^2 + k_{BC}Q_{BC}^2)$$

$$= 60 \text{ psi} \times 144 \text{ psf/psi} - 62.4(0.00944 \times 11.4^2 + 0.3021 \times 9.0^2)$$

$$= 8640 \text{ psf} - 1603 \text{ psf}$$

$$= 7037 \text{ psf}$$

$$= 48.9 \text{ psi}$$

$$p_E = 8640 - \gamma(k_{AD}Q_{AD}^2 + k_{DE}Q_{DE}^2)$$

$$= 8640 - 62.4(1.059 \times 3.5^2 + 0.3021 \times 6^2)$$

$$= 7105 \text{ psf}$$

$$= 49.3 \text{ psi} \qquad\qquad \blacksquare$$

5-6   Instruments and Procedures for Discharge Measurement

Direct and Indirect Methods of Flow Measurements

The methods of flow measurement can broadly be classified as either direct or indirect. Direct methods involve the actual measurement of the quantity of flow (volume or weight) for a given time interval. Indirect methods involve the measurement of a pressure change (or some other variable), which in turn is directly related to the rate of flow. Flow through *orifices, venturi meters,* and *flow nozzles* are all devices with which one employs indirect methods to measure the rate of flow in closed conduits. Still another indirect meter is the *electromagnetic* flow meter, which operates on the principle that a voltage is generated when a conductor moves in a magnetic field. We describe all these methods, as well as the *velocity-area integration* of flow measurement, in this section.

Direct Volume or Weight Measurements

One of the most accurate methods of obtaining liquid-flow rate is to collect a sample of the flowing fluid over a given period of time t. Then if the sample is weighed, the average weight rate of flow is W/t, where W is the weight of the sample. The volume of a sample can also be measured (usually in a calibrated tank), and from this the average volume rate of flow is given as Ψ/t, where Ψ is the volume of the sample.

Velocity-Area-Integration Method

If the velocity in a pipe is symmetrical, the distribution of the velocity along a radial line can be used to determine the volume rate of flow (discharge) in the pipe. The discharge is obtained by numerically or graphically integrating $V\,dA$ over the cross-sectional area of the pipe. Thus a velocity traverse across the flow section provides the primary data from which the discharge is evaluated. The velocity can be measured by a pitot tube or some other suitable velocity meter. We give one procedure for evaluating this discharge in the following paragraph.

From test data of V versus r, compute $2\pi Vr$ for various values of r; then when $2\pi Vr$ versus r is plotted, the area under the resulting curve (Fig. 5-18), will yield the discharge. This is so because $dQ = V\,dA = V(2\pi r\,dr)$, which is given by an elemental strip of area in Fig. 5-18. Hence the total area will yield the total discharge. This procedure involving the velocity-area-integration method is applicable to pipes when the velocity distribution is symmetrical with the axis of the pipe. However, even for unsymmetrical flows, it should be obvious that by summing $V\,\Delta A$ over a flow section, you can obtain the total flow rate. Such a procedure is commonly used to obtain the discharge in streams and rivers, as we noted in Chapter 4.

Orifice

A restricted opening through which fluid flows is an *orifice*, and if the geometric characteristics of the orifice plus the properties of the fluid are known, the orifice can be used to measure flow rates. Consider flow through the sharp-edged pipe orifice shown in Fig. 5-19. It is seen that the streamlines continue

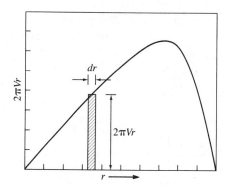

Figure 5-18   Graphical integration of $V\,dA$ in a pipe

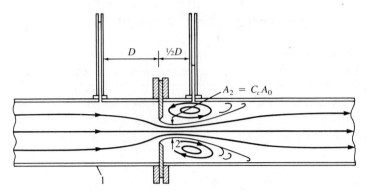

Figure 5-19   Flow through a pipe orifice

to converge a short distance downstream of the plane of the orifice; hence the minimum-flow area is actually smaller than the area of the orifice. To relate the minimum-flow area, A_j, often called the contracted area of the jet or *vena contracta*, to the area of the orifice A_o, we use the contraction coefficient, which is defined as

$$A_j = C_c A_o$$

$$C_c = \frac{A_j}{A_o}$$

Then, for a circular orifice, with diameter d,

$$C_c = \frac{(\pi/4)d_j^{\,2}}{(\pi/4)d^2} = \left(\frac{d_j}{d}\right)^2$$

Because d_j and d_2 are identical, we also have $C_c = (d_2/d)^2$. At low values of the Reynolds number, C_c is a function of the Reynolds number; however, at high values of the Reynolds number, C_c is only a function of the geometry of the orifice. Figure 5-20 shows the variation of C_c with d/D where D is the diam-

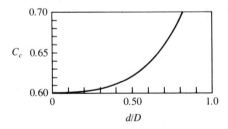

Figure 5-20   Contraction coefficient as a function of d/D for $\mathrm{Re}_d = 10^6$

eter of the pipe for a Reynolds number (Vd/v) of 10^6.* The discharge equation for the orifice is derived by writing the Bernoulli equation between sections 1 and 2 in Fig. 5-19 and then eliminating V_1 by means of the continuity equation $V_1 A_1 = V_2 A_2$. Then solving for V_2 and multiplying by the flow area, $C_c A_o$, we obtain the discharge equation

$$Q = \frac{C_c A_o}{\sqrt{1 - C_c^2 A_o^2 / A_1^2}} \sqrt{2g(h_1 - h_2)} \tag{5-30}$$

Equation (5-30) is the discharge equation for the flow of an incompressible inviscid fluid through an orifice. However, this is valid only at relatively high Reynolds numbers. For low and moderate values of the Reynolds number, viscous effects may be significant and an additional coefficient is applied to the discharge equation to relate the ideal to the actual flow. This is called the *coefficient of velocity* C_v; thus for viscous flow in an orifice, we have the following discharge equation:

$$Q = \frac{C_v C_c A_o}{\sqrt{1 - C_c^2 A_o^2 / A_1^2}} \sqrt{2g(h_1 - h_2)}$$

The product $C_v C_c$ is called the *discharge coefficient* C_d, and the combination $C_v C_c / (1 - C_c^2 A_o^2 / A^2)^{1/2}$ is called the *flow coefficient* K. Thus we have $Q = K A_o \sqrt{2g(h_1 - h_2)}$, where

$$K = \frac{C_d}{\sqrt{1 - C_c^2 A_o^2 / A_1^2}}$$

If Δh is defined as $h_1 - h_2$, the final form of the discharge equation for an orifice reduces to

$$Q = K A_o \sqrt{2g \, \Delta h} \tag{5-31}$$

If a differential pressure transducer is connected across the orifice, the transducer will sense a change in pressure equivalent to $\gamma \Delta h$. Therefore, in this application one simply uses $\Delta p / \gamma$ in place of Δh in Eq. (5-31) and in the parameter at the top of Fig. 5-21. Experimentally determined values of K as a function of d/D and Reynolds number based on orifice size, $4Q/\pi \, dv$, are given in Fig. 5-21. One type of problem is to determine Δh for a given discharge through an orifice in a given size of pipe. For such a problem Re_d is equal to $4Q/\pi \, dv$ and K is obtained from Fig. 5-21 (using the vertical lines and the bottom scale), and Δh is then computed from Eq. (5-31). When Q is to be determined, we use the

---

* These were obtained by using the values of K from Fig. 5-21 and values of C_v given by Lienhard (27) and then calculating to obtain C_c.

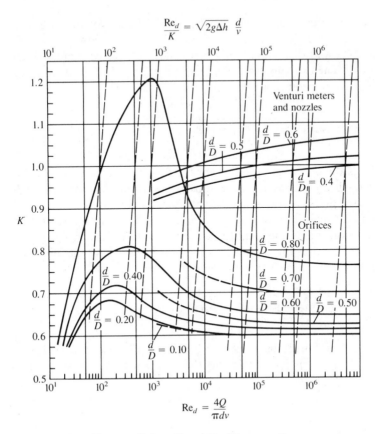

$$\frac{\mathrm{Re}_d}{K} = \sqrt{2g\Delta h}\ \frac{d}{\nu}$$

$$\mathrm{Re}_d = \frac{4Q}{\pi d \nu}$$

Figure 5-21   Flow coefficient K and Re_d/K versus the
Reynolds number for orifices, nozzles,
and venturi meters (20, 23)

top scale with the slanted lines to determine K for given values of d, D, Δh and
ν. With K, we can then solve for Q from Eq. (5-31).

The literature on orifice flow contains many discussions concerning the
optimum placement of pressure taps on both the upstream and downstream
side of the orifice. The data given in Fig. 5-21 are for "corner taps." That is, on
the upstream side, the pressure readings were taken immediately upstream of
the plate orifice (at the corner of the orifice plate and the pipe wall), and the
downstream tap was at a similar downstream location. However, pressure data
from flange taps (1 in. upstream and 1 in. downstream) and from the taps shown
in Fig. 5-19 all yield virtually the same values for K—the differences are no
greater than the deviations involved in reading Fig. 5-21.*

* For more precise values of K with specific types of taps, see the ASME report on fluid meters (20).

EXAMPLE 5-9   A 15-cm orifice is located in a horizontal 24-cm water pipe, and a water-mercury manometer is connected to either side of the orifice. When the deflection on the manometer is 25 cm, what is the discharge in the system? Assume the water temperature is 20°C.

SOLUTION   The discharge is given by Eq. (5-31): $Q = KA_o\sqrt{2g\,\Delta h}$. To either enter Fig. 5-21 or use Eq. (5-31), we will need to first evaluate Δh, the change in piezometric head in meters of fluid that is flowing. This is obtained by applying the equation of hydrostatics to the manometer shown as follows.

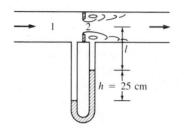

Writing the manometer equation from point 1 to point 2, we get

$$\Delta h = 0.25 \text{ m } (13.6 - 1)$$

$$= 3.15 \text{ m of water}$$

The kinematic viscosity of water at 20°C is $1.0 \times 10^{-6} \text{ m}^2/\text{s}$; so we now can compute $d\sqrt{2g\,\Delta h}/v$, the parameter needed to enter Fig. 5-21:

$$\frac{d\sqrt{2g\,\Delta h}}{v} = \frac{0.15 \text{ m } \sqrt{2(9.81 \text{ m/s}^2)(3.15 \text{ m})}}{1.0 \times 10^{-6} \text{ m}^2/\text{s}} = 1.2 \times 10^6$$

From Fig. 5-21 with $d/D = 0.625$, we read K to be 0.66 (interpolated). Hence

$$Q = 0.66 A_o\sqrt{2g\,\Delta h}$$

$$= 0.66 \frac{\pi}{4} d^2 \sqrt{2(9.81 \text{ m/s}^2)(3.15 \text{ m})}$$

$$= 0.66(0.785)(0.15^2 \text{ m}^2)(7.87 \text{ m/s})$$

$$= 0.092 \text{ m}^3/\text{s} \qquad\blacksquare$$

Venturi Meter

The orifice is a simple and accurate device for the measurement of flow; however, the head loss for the orifice is quite large. It is like an abrupt enlargement in a pipe: $h_L = (V_2 - V_1)^2/2g$. The venturi meter operates on the same

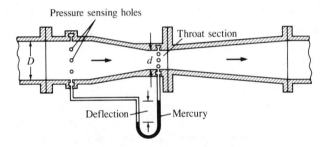

Figure 5-22   Typical venturi meter

principle as the orifice but with a much smaller head loss. The lower head loss results from streamlining the flow passage, as shown in Fig. 5-22. This streamlining eliminates any jet contraction beyond the smallest flow section; thus the coefficient of contraction has a value of unity, and the basic discharge equation for the venturi meter is

$$Q = \frac{A_2 C_d}{\sqrt{1 - (A_2/A_1)^2}} \sqrt{2g(h_1 - h_2)} \qquad (5\text{-}32)$$

$$= K A_2 \sqrt{2g\,\Delta h} \qquad (5\text{-}33)$$

The discharge equation for the venturi meter, Eq. (5-33), is the same as for the orifice, Eq. (5-31). However, K for the venturi meter approaches unity at high values of the Reynolds number and small d/D ratios. This trend can be seen in Fig. 5-21, where values of K for the venturi meter are plotted along with similar data for the orifice.

Electromagnetic Flow Meter

All the flow meters introduced to this point require that some sort of obstruction be placed in the flow. The electromagnetic flow meter neither obstructs the flow nor requires pressure taps, which are subject to clogging. Its basic principle is that a conductor that moves in a magnetic field produces an electromotive force. Hence liquids having a degree of conductivity will generate a voltage between the electrodes as in Fig. 5-23, and this voltage will be proportional to the velocity of flow in the conduit.

The main advantages of the electromagnetic flow meter are that the output signal varies linearly with the flow rate, and the meter causes no resistance to the flow. The major disadvantage is its high cost.*

---

* For a summary of the theory and application of the electromagnetic flow meter, see Shercliff (39).

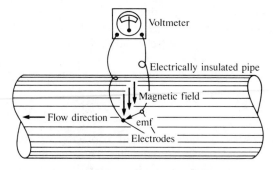

Figure 5-23 Electromagnetic flow meter

Ultrasonic Flow Meter

Another form of nonintrusive flow meter used in diverse applications is the ultrasonic flow meter. Basically, there are two different modes of operation for ultrasonic flow meters. One mode involves measuring the difference in travel time for a sound wave traveling upstream and downstream between two measuring stations. The difference in travel time is proportional to flow velocity. The second mode of operation is based on the Doppler effect. When an ultrasonic beam is projected into an inhomogeneous fluid, some acoustic energy is scattered back to the transmitter at a different frequency (Doppler shift). The measured frequency difference is related directly to the flow velocity.

Turbine Flow Meter

The turbine flow meter consists of a wheel with a set of curved vanes (blades) mounted inside a duct. The volume rate of flow through the meter is related to the rotational speed of the wheel, and this rotational rate is generally measured by a blade passing an electromagnetic pickup mounted in the casing. The meter must be calibrated for the flow conditions of interest. The turbine meter has an accuracy of better than 1% over a wide range of flow rates and operates with small head loss.

Vortex Flow Meter

The vortex flow meter consists of a cylinder mounted across the duct, which sheds vortices and gives rise to an oscillatory flow field. By proper design of the cylindrical element, the Strouhal number for vortex shedding ($S = nd/V_0$) will be constant for Reynolds numbers from 10^4 to 10^6. Over this flow range, the fluid velocity and volume flow rate are directly proportional to the frequency of oscillation, which can be measured by several different methods. An advantage of this meter is that it has no moving parts (reliability), but it does give rise to a head loss comparable to other obstruction meters.

Displacement Meter

The displacement meter works on the principle of displacing a piston or other mechanical part when flow passes through the meter. Thus the number of oscillations of the piston or disc (as in the nutating disc meter) can be monitored, which in turn will be proportional to the quantity of flow through the meter. Displacement meters are used extensively for measuring the quantity of water used by households or businesses from municipal water systems.*

Salt-Velocity Method

This method of discharge was first developed by Allen (1). The method, which is based on the increased electrical conductivity of a salt solution, is used in the following manner: A concentrated dose of salt solution is injected into the conduit at a given time and location in the conduit. Then by means of electrodes and associated instruments, the conductivity is recorded at a station farther downstream. The "instant" of passage of the salt solution is assumed to be at the center of gravity of the conductivity-time trace (the conductivity will increase and then decrease over a considerable length of time because of mixing due to turbulence). Thus the mean velocity can be determined from the length of travel and elapsed time of travel of the salt solution.

5-7 Forces and Stresses in Pipes and Bends

Forces on Bends and Transitions

BENDS Because of the change in momentum that occurs with flow around a bend, the momentum equation is used to calculate the forces acting on bends and transitions. The general momentum equation for steady one-dimensional flow is

$$\sum \mathbf{F}_{syst} = \sum_{c.s.} \mathbf{V} \rho \mathbf{V} \cdot \mathbf{A} \tag{5-34}$$

---

* For more complete information on displacement meters, as well as the other types of meters we have discussed, see the American Water Works manual on water meters (13).

Equation (5-34) is a vector form of the momentum equation using the control volume approach. Thus the subscript "syst" refers to everything inside the control volume, and the subscript "c.s." refers to the control surface. The $\sum \mathbf{F}$ term on the left-hand side of Eq. (5-34) includes all the external forces acting on the system (such as bend and water) within the control volume. Such forces could include forces of pressure, gravity, and the unknown force (usually acting through the pipe walls or anchor) to hold the bend or transition in place. $\mathbf{A}$ is the vector representation of area.

If Eq. (5-34) is written in scalar form and simplified for a single-stream application, such as a single stream of water flowing through a bend, we obtain

$$\sum F_x = \rho Q(V_{2x} - V_{1x}) \tag{5-34a}$$

$$\sum F_y = \rho Q(V_{2y} - V_{1y}) \tag{5-34b}$$

$$\sum F_z = \rho Q(V_{2z} - V_{1z}) \tag{5-34c}$$

Example 5-10 illustrates an application of Eq. (5-34).

EXAMPLE 5-10   A 1-m diameter pipe has a 30° horizontal bend in it, as shown in Fig. A, and carries water (10°C) at a rate of 3 m³/s. If we assume the pressure in the bend is uniform at 75 kPa gauge, the volume of the bend is 1.8 m³, and the metal in the bend weighs 4 kN, what forces must be applied to the bend by the anchor to hold the bend in place? Assume expansion joints prevent any force transmittal through the pipe walls of the pipes entering and leaving the bend.

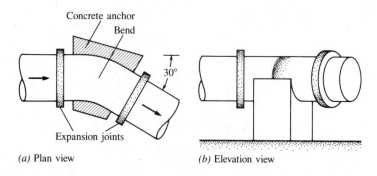

(a) Plan view   (b) Elevation view

Figure A

SOLUTION   Since only a single stream of water is involved in this problem, we can use Eqs. (5-34) for the solution. Consider the control volume shown in Fig. B, and first solve for the x component of force:

Then $\sum F_x = \rho Q(V_{2x} - V_{1x})$

$p_1 A_1 - p_2 A_2 \cos 30° + F_{\text{anchor},x} = 1{,}000 \cdot 3(V_{2x} - V_{1x})$

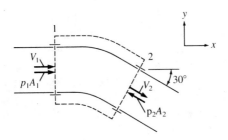

Figure B

where $p_1 = p_2 = 75,000$ Pa
$A_1 = A_2 = (\pi/4)D^2 = 0.785$ m^2
$V_{2x} = (Q/A_2)\cos 30° = 3.31$ m/s
$V_{1x} = Q/A_1 = 3.82$ m/s

$$F_{anchor,x} = 3,000(3.31 - 3.82) + 75,000 \times 0.785(0.866 - 1)$$
$$= -9,420 \text{ N}$$

Solve for F_y:

$$\sum F_y = \rho Q(V_{2y} - V_{1y})$$
$$F_{anchor,y} = 1,000 \cdot 3(-3.82\sin 30° - 0) - p_2 A_2 \sin 30°$$
$$= -35,170 \text{ N}$$

Solve for F_z:

$$\sum F_z = \rho Q(V_{2z} - V_{1z})$$
$$W_{bend} + W_{water} + F_{anchor,z} = 1,000 \cdot 3(0 - 0)$$
$$-4,000 - 1.8 \times 9,810 + F_{anchor,z} = 0$$
$$F_{anchor,z} = +21,660 \text{ N}$$

Then the total force that the anchor will have to exert on the bend will be

$$\mathbf{F}_{anchor} = -9,420\mathbf{i} - 35,170\mathbf{j} + 21,660\mathbf{k} \text{ N} \qquad \blacksquare$$

TRANSITIONS   The fitting between two pipes of different size is a *transition*. Because of the change in flow area and change in pressure, a longitudinal force will act on the transition. To determine the force required to hold the transition in place, the energy, momentum, and continuity equations are all applied. Example 5-11 illustrates the principles.

EXAMPLE 5-11   Water flows through the contraction at a rate of 25 cfs. The head loss coefficient for this particular contraction is 0.20 based on the velocity head in the smaller pipe. What longitudinal force (such as from an anchor) must be applied to the contraction to hold it in place? We assume the upstream pipe pressure is 30 psig, and expansion joints prevent force transmittal between the pipe and the contraction.

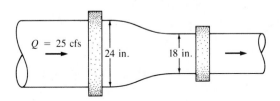

SOLUTION   Let the x direction be in the direction of flow, and let the control surface surround the transition as shown in the figure.

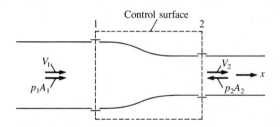

First solve for p_2 with the energy equation:

$$\frac{p_1}{\gamma} + \frac{V_1^2}{2g} + z_1 = \frac{p_2}{\gamma} + \frac{V_2^2}{2g} + z_2 + h_L$$

where $\dfrac{p_1}{\gamma} = 30 \times \dfrac{144}{62.4} = 69.2$ ft

$V_1 = \dfrac{Q}{A_1} = \dfrac{25}{(\pi/4)2^2} = 7.96$ ft/s

$V_2 = \dfrac{Q}{A_2} = \dfrac{25}{(\pi/4) \times 1.5^2} = 14.15$ ft/s

$z_1 = z_2$

$h_L = 0.20 \dfrac{V_2^2}{(2g)}$

Then $\dfrac{p_2}{\gamma} = 69.2 \text{ ft} + \dfrac{7.96^2}{2g} - \dfrac{14.15^2}{2g}(1 + 0.2)$

$$\frac{p_2}{\gamma} = 66.45 \text{ ft}$$

or $$p_2 = 4147 \text{ psf}$$

Now solve for the anchor force:

$$\sum F_x = \rho Q(V_{2x} - V_{1x})$$

$$p_1 A_1 - p_2 A_2 + F_{\text{anchor},x} = 1.94 \cdot 25(14.15 - 7.96)$$

$$F_{\text{anchor},x} = 1.94 \cdot 25(14.15 - 7.96) + 4147$$

$$\times \left(\frac{\pi}{4}\right) \cdot 1.5^2 - 30 \cdot 144 \cdot \left(\frac{\pi}{4}\right) \cdot 2^2$$

$$= -5,943 \text{ lb}$$

The anchor must exert a force of 5,943 lb in the negative x direction on the transition.

Note: In many cases, such as with a continuously welded steel pipe, the pipe walls are designed to carry this reaction, and no anchor block is required.

∎

Cavitation Effects

Cavitation occurs in flowing liquids when the flow passes through a zone in which the pressure becomes equal to the vapor pressure of the liquid, and then the flow continues on to a region of higher pressure. In the vapor pressure zone, vapor bubbles are formed (the liquid boils), and then when the liquid and bubbles enter the higher pressure zone, the bubbles collapse thereby producing dynamic effects that can often lead to decreased efficiency or equipment failure. Figure 5-24 shows three setups that could produce cavitation. In Fig. 5-24a, the high velocity flow through the venturi is accompanied by a reduced pressure, and if this pressure is as low as the vapor pressure of the liquid, cavitation will occur at point A inside the venturi section. In Fig. 5-24b, it is the combination of pipe elevation change, head loss along the pipe, and locally high velocity along the inside of the bend that creates the lowest pressure at point A in the pipe, and this is where cavitation would first occur.

In some systems, such as one using a pump (Fig. 5-24c), vapor pressure can arise because of deceleration of the water column if power to the pump is interrupted. In these cases, the pump will retard the flow, and a vapor cavity will form downstream of the pump. The vapor cavity will grow in size and then decrease. Just before the cavity vanishes, the columns of water on either side of the vapor pocket will be accelerating toward each other. Finally, when the vapor cavity disappears, the two columns of water will impact each other with

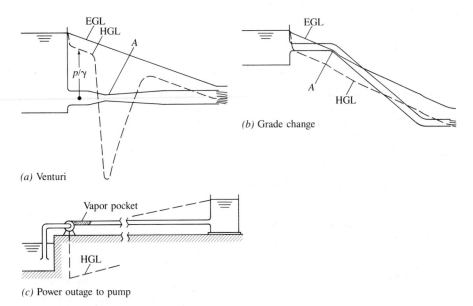

(a) Venturi

(b) Grade change

(c) Power outage to pump

Figure 5-24   Occurrence of cavitation

creation of dynamic pressures originating at the point of impact. These pressures can be large and can rupture the pipe and damage the pump (see Fig. 11-1, page 573).*

Because of the detrimental effects usually associated with cavitation, good hydraulic design practice will normally exclude any possibility of its occurrence. With a normal range of water temperature (40°F to 80°F), cavitation will occur if the pressure head gets as low as -33 ft (the hydraulic grade line is at an elevation 33 ft below the point in the question). Thus good design practice does not allow the pressure to be this low. In fact, to be conservative, the U.S. Bureau of Reclamation has recommended (46) that the pressure head throughout the pipe system should be greater than -10 ft. For important systems, special detailed analyses are required.[†]

Pipe Stress Due to Internal Pressure

In thin-walled pipes ($t/D < 0.1$), the formula for hoop tension stress due to pressure within the pipe is derived by considering a freebody of one half of a pipe section, as shown in Fig. 5-25. Taking a length (normal to page) of pipe L and applying the equation of equilibrium where σ is hoop stress in the

---

* For more details on the method of analyzing this phenomenon, see Chapter 11.

† For more information on cavitation in pipes, see Knapp (24) and Pearsall (33).

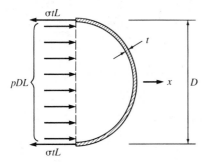

Figure 5-25   Freebody of a section of pipe

pipe wall to the section, we obtain

$$\sum F_x = 0$$

$$pDL - 2\sigma tL = 0$$

or
$$\sigma = \frac{pD}{2t}$$
(5-35)

In the derivation of Eq. (5-35), it was assumed there is a uniform distribution of stress across the wall because it was assumed to be thin. However, for thick-walled pipes, it can be shown (38) that the circumferential or hoop stress is given by

$$\sigma = \frac{pr_i^2}{r_o^2 - r_i^2}\left(1 + \frac{r_o^2}{r^2}\right)$$
(5-36)

where r = radius to a point in pipe wall
r_i = inside radius of pipe
r_o = outside radius of pipe

Temperature Stress and Strain

Temperature stresses develop when temperature changes occur after the pipe is installed and rigidly held in place. For example, if a pipe is restrained from expanding when the temperature changes $+ \Delta T°$, the pipe, in effect, would be subjected to a compressive longitudinal deflection of

$$\Delta L = \Delta T \alpha L$$
(5-37)

where α = coefficient of thermal expansion

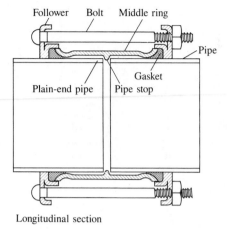

Longitudinal section

Figure 5-26   Dresser coupling expansion joint

Then the resulting effective longitudinal strain would be $\varepsilon = \Delta L/L = \Delta T\alpha$, and the resulting temperature stress would be

$$\sigma = E\varepsilon$$
$$= E\,\Delta T\alpha \qquad\qquad (5\text{-}38)$$

where E = elastic modulus

To eliminate the temperature stress, expansion joints are used, as shown in Fig. (5-26). These joints can be placed at regular intervals and must allow the pipe to expand a distance ΔL as given by Eq. (5-37), where L is the spacing between expansion joints.

External Loading

Pipes that are laid in a trench must be designed for external loading (soil forces) as well as for internal pressure. This is especially true for cases in which the internal pressure is low. Because the external loading on the pipe is in general not uniformly distributed around the circumference of the pipe, bending stresses develop within the pipe wall. These bending stresses are a function of the placement of the pipe (whether in a trench or under an open fill), as well as the supporting condition (bedding) under the pipe. Thus if the loading and bedding conditions were completely defined, it would be possible to solve for the resulting stresses in the pipe. However, because of the complexity of the usual underground pipe installations, a simpler more empirical design procedure

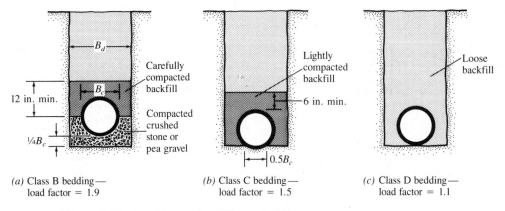

(a) Class B bedding —
   load factor = 1.9

(b) Class C bedding —
   load factor = 1.5

(c) Class D bedding —
   load factor = 1.1

Figure 5-27   Load factors for different bedding
conditions

is standard practice. Briefly, the procedure involves determining the design load
on the pipe, and then a pipe is chosen that will withstand the given loading
for a given bedding condition. In the following sections, we illustrate the
procedure involved in the design of pipes placed in trenches and backfilled
with soil.

Figure 5-27 shows a typical pipe installation with three different types of
bedding conditions. The load on the pipe is equal to the weight of the prism
of soil above the pipe in the trench minus the shear force between the trench
wall and soil prism. Marston and Spangler carried out a number of tests for
various pipe sizes, depths of cover, and different types of soil. Their research
was summarized by Spangler (40) in 1948. This work led to Marston's general
formula for soil load on pipe, which is $W = C\gamma_s B^2$, where W is the vertical load
acting on the pipe per unit of length of pipe (N/m or lb/ft), γ_s is the unit weight
of the soil (N/m^3 or lb/ft^3), B is the trench width or conduit width depending
on installation conditions, and C is a coefficient that is a function of the relative
fill height and soil type.

RIGID PIPE INSTALLATION   If the pipe is rigid (for example, concrete),
Marston's general formula is given as

$$W = C\gamma_s B_d{}^2 \tag{5-39}$$

where B_d = trench width

In Fig. 5-28, the coefficient C has been plotted as a function of the relative
height of soil cover, H/B_d.

FLEXIBLE PIPE INSTALLATION   If the pipe is flexible (for example, steel
or plastic) and the soil at the sides is well compacted, the side walls will carry

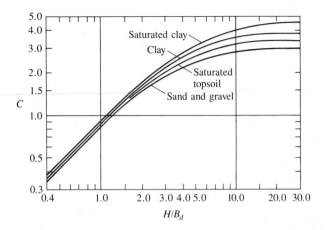

Figure 5-28   Load coefficients for various types of soil (4)

a significant portion of the total load; therefore, the load formula is modified as follows:

$$W = C\gamma_s B_d B_c \tag{5-40}$$

In Eq. (5-40), the coefficient C is obtained as before from Fig. 5-28, and B_c is the conduit diameter.

This procedure for load determination is applicable when pipes are laid in relatively narrow trenches ($B_d \leq 2B_c$). However, if the trench is greater than about twice the pipe diameter, the calculated load will be too large, and a different procedure must be used. If live loads (wheel loads from trucks or trains) are a factor, they, too, must be considered.*

Strength of Rigid Pipes

The strength of rigid pipe such as concrete or clay is determined by a three-edge bearing test such as shown in Fig. 5-29. These laboratory tests establish the basic load carrying capacity (strength) of the pipe (see Table 5-4, page 295, for representative strengths of concrete pipe). However, as we noted, the field strength of the pipe will depend on bedding conditions; therefore, the pipe strength based on the three-edge bearing test is modified by a load factor relating to the type of bedding. For example, if the three-edge bearing strength were 6500 lb/ft, the *field supporting strength* of this conduit with a class C type

---

* For these and other conditions, see the ASCE manual on sanitary storm sewers (4).

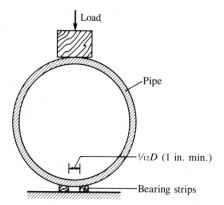

Figure 5-29   Three-edge bearing test

of bedding (see Fig. 5-27) would be $6500 \times 1.5 = 9750$ lb/ft. Besides the bedding factor, a factor of safety is usually applied to account for the variability of load distribution and other conditions not already accounted for. Thus the safe working strength of the conduit is given as

$$S_{safe} = \frac{S_{3\text{-edge}} \cdot F_{load}}{F_{safety}}$$ (5-41)

where S_{safe} = safe supporting strength of conduit
$\quad S_{3\text{-edge}}$ = three-edge bearing strength of conduit
$\quad F_{load}$ = bedding load factor
$\quad F_{safety}$ = factor of safety

The factor of safety is usually taken as 1.5 or greater.

5-8   Pipe Materials

The most common pipe materials are steel, ductile iron, concrete, asbestos cement, and plastic.

Steel

The type of steel in pipes used for civil engineering projects is generally of medium carbon content, which has high strength as well as high ductility. Because of these desirable physical qualities, steel pipe is used in a variety of applications from small pipes in household water systems to 12-ft (and larger) diameter penstocks or pipelines. The AWWA standard recommends (12) the use of a tensile stress for design of steel pipe equal to 50% of the yield point

Table 5-4 Crushing Strength of Concrete Pipe by the Three-Edge Bearing Method* [Adapted from Linsley, et al, Water-Resources Engineering, 3rd. ed. 1979 (43), McGraw-Hill Book Company, New York; used with permission of McGraw-Hill Book Company.]

Internal Diameter (in.)	Nonreinforced Concrete[†]			Reinforced Concrete[†] Ultimate Strength				
	Class I	Class II	Class III	Class I	Class II	Class III	Class IV	Class V
4	1500	2000	2400					
6	1500	2000	2400					
8	1500	2000	2400					
10	1600	2000	2400					
12	1800	2250	2600	—	1500	2000	3000	3750
15	2000	2600	2900	—	1875	2500	3750	4690
18	2200	3000	3300	—	2250	3000	4500	5620
21	2400	3300	3850	—	2625	3500	5250	6560
24	2600	3600	4400	—	3000	4000	6000	7500
27	2800	3950	4600	—	3375	4500	6750	8440
30	3000	4300	4750	—	3750	5000	7500	9380
33	3150	4400	4875	—	4125	5500	8250	10,220
36	3300	4500	5000	—	4500	6000	9000	11,250
39	—		—	—	4825	6500	9750	
42	—	—	—	—	5250	7000	10,500	13,120
48	—	—	—	—	6000	8000	12,000	15,000
54	—	—	—	—	6750	9000	13,500	16,880
60	—	—	—	6000	7500	10,000	15,000	18,750
66	—	—	—	6600	8250	11,000	16,500	20,620
72	—	—	—	7200	9000	12,000	18,000	22,500

* Data from 1977 Annual Book of ASTM Standards as follows: nonreinforced concrete, Specification C14; reinforced concrete, Specification C76, American Society for Testing and Materials, Philadelphia, Pa.
† All strengths in pounds per linear foot. Pipe strength in kilonewtons per meter can be obtained by multiplying the values in the table by 0.0146.

stress. The yield point stress ranges between about 25,000 and 45,000 psi depending on the grade of steel used.

Because steel is susceptible to corrosion, special coatings are used to provide corrosion resistance. Commonly used coatings are coal tar enamels, various kinds of polymers, plastics, cement mortar, and zinc (galvanized pipe). Special situations may require cathodic protection.*

---

* For more details on both protective coatings and electrical protection to resist corrosion, see AWWA manuals (10, 11) and NBS publication (45).

For drain systems such as highway culverts and sewers, the wall thickness needed to withstand hydraulic design pressures is small; therefore, such thin-walled pipes are very susceptible to collapse under the action of the imposed external soil loads. To prevent this failure, corrugated steel pipe was developed. Corrugated steel pipe is available in diameters from 4 in. to 144 in.

Steel will expand about 3/4 in. for every 100 ft of length for a temperature increase of 100°F; therefore, expansion joints are needed in many installations to prevent excessive temperature stresses. A common type of expansion joint is the Dresser coupling, as shown on page 291 in Fig. 5-26.*

Ductile Iron

Ductile iron pipe is manufactured in diameters from 4 in. to 48 in. From 4-in. to 20-in. diameters, the standard commercial sizes are available in 2-in. steps, and from 24-in. to 48-in. diameters, the sizes are available in 6-in. steps. Ductile iron is relatively resistant to corrosion and can withstand relatively large external loads (such as soil pressure); therefore, it is used extensively for water and sewer systems. For added protection against corrosion, the outside of the pipe is usually coated with an asphaltic compound and the inside, with either a cement lining or asphaltic coating. The pipe can be obtained in thickness to withstand working pressures up to 350 psi (2.41 MPa).

The most common pipe joints are either the push-on bell and spigot type or the mechanical flange type. The bell and spigot is sealed by a rubber gasket. Dresser couplings can also be used on ductile iron pipe.[†]

Ductile iron is generally more costly than steel; however, because of its rigid characteristic, it is less apt than steel to collapse under negative pressure.

Concrete

Some concrete pipe was first installed in the 1800s, but the greatest use began near the beginning of the twentieth century. Concrete pipe is now used for storm and sanitary sewers, highway and railroad culverts, and pressure pipes. For small culverts and low-pressure irrigation systems, the pipe is often unreinforced; however, for the larger sizes or higher-pressure systems, reinforcing is required.[‡]

---

* For more details on other types of expansion joints, as well as accepted procedures for design and installation of steel pipe, see AWWA manual no. 11 (12).

† For more details on ductile iron pipe, see the Cast Iron Pipe Association (16).

‡ For details on availability, strength, design, and installation of concrete pipe, see publications of the American Concrete Pipe Association (2, 3), AWWA manual no. M9 (9), and the ASTM manual on concrete pipe strength (8).

Asbestos Cement

This type of water pipe is composed of a nonmetallic mixture of asbestos, Portland cement, and silica, so it does not corrode in the usual sense. It is similar to ductile iron in that it is not as resistant to impact loading as steel. Asbestos cement water pipe is available in sizes from 3-in. to 36-in. diameter.

Plastic

The most common type of plastic pipe is polyvinylchloride (PVC). The main advantages of PVC pipe are its corrosion resistance, smoothness (less resistance to flow), and ease of field assembly. PVC pipe is used extensively for irrigation and sewer systems. Where plastic pipe must operate under high pressure, it is common to use the type reinforced with fiberglass for added strength.

5-9   Large Conduit Design

Scope and Use of Large Conduits

Large pressure conduits (6 ft and larger) are required in a variety of applications; however, the most typical cases are water-supply aqueducts for municipal or irrigation use, penstocks in hydropower installations, and outlet works associated with dams. In outlet works, the conduit is often a tunnel used to divert water around the dam during the project's construction. Then after completion of the dam, the tunnel is often used as part of the spillway system. Large conduits are usually made of steel or reinforced concrete. When the conduit is a tunnel, it is generally not economical to make it less than 6 ft in diameter because of the excessive cost of removal of material when excavating the tunnel. Tunnels in rock are constructed by two methods: drilling and blasting and boring. The tunnels may be unlined in sound rock, or they may be lined with reinforced concrete in fractured rock.

On some projects, there may be a combination of tunnel and pressure pipe in series depending on the specific topographic and geologic conditions of the project. Likewise, some tunnels may be lined with concrete, and others may be unlined, depending on the condition of the rock it is driven through. For example, the Apalachia project includes a conduit consisting of both tunnel and pressure pipes in series. The conduit is 8 mi (13 km) long and consists of about 2 mi of 22-ft diameter unlined tunnel, 4 mi of 18-ft concrete-lined tunnel, 1 mi of steel-lined tunnel, and several short lengths of 16-ft steel pipe in areas where the conduit was not tunneled.*

---

* For details of this project, see Goodhue (21) and Elder (18).

Table 5-5 Values of k_s for Different Conditions — Concrete Pipe (47)

Condition of Pipe	k_s (ft)	(mm)
New pipe, unusually smooth surface, steel forms, smooth joints	0.0001	(0.030)
Fairly new pipe, smooth pipe, steel forms, average workmanship, some pockets on concrete surface, smooth joints	0.0005	(0.15)
Granular or brushed surface in good condition, good joints	0.0010	(0.30)
Centrifugally cast concrete pipe	0.0010	(0.30)
Rough surface eroded by sharp materials in transit; marks visible from wooden forms or spalling of surface	0.002	(0.61)
Unusually rough wood form work, erosion of surface, poor alignment at joints	0.003	(0.91)

Head Loss in Large Conduits

The same basic theory for head loss in conduits presented in Sec. 5-3, pages 243–62 applies to large conduits. However, certain features of large conduits need further consideration in this section. The economics of the design of large conduits is very important, and economic optimization involves the cost required to achieve flow at reduced head loss (smoother conduit) as opposed to the cost of a rougher but larger conduit. Also, if power production is a factor, as in the case of penstocks or tunnels for hydropower plants, the increased revenue that can be obtained with less head loss must be considered. Because of the costs of large conduits, much more attention must be given to their resistance coefficients than for many of the installations that use smaller pipes. The basic approach for estimating head loss is to first estimate the equivalent sand roughness, k_s, and then with the relative roughness, k_s/D, and the Reynolds number, we can obtain the resistance coefficient f from Fig. 5-4, page 247, or Eq. (5-17), page 257. We discuss the special roughness conditions that must be considered for different kinds of large conduits as follows.

CONCRETE PIPE   The roughness of the pipe will of course depend on the surface finish of the pipe, which depends on the type of forms used in construction; however, other factors such as the degree of smoothness at the joints must also be considered. Sometimes concrete will erode with use, thereby causing the roughness to increase with age. Table 5-5 gives values of k_s for different conditions of concrete pipe.

STEEL PIPE   In steel pipe, the possibility of corrosion and incrustation of the surface with age must also be considered. Table 5-6 gives the values of k_s for different conditions of steel butt-welded pipe.

Table 5-6 Values of k_s for Different Conditions — Steel Butt-Welded Pipe (47)

Condition of Pipe	k_s (ft)	(mm)
New smooth pipe with centrifugally applied enamel on surface	0.0001	(0.03)
Hot asphalt dipped pipe	0.0003	(0.09)
Steel pipe with centrifugally applied concrete lining	0.0003	(0.09)
Light rust on surface	0.001	(0.30)
Heavy brush coated application of lining of asphalt or enamel	0.002	(0.61)
General tuberculation (1 to 3 mm in size) of surface	0.006	(1.8)
Severe tuberculation and incrustation	0.02	(6.1)

UNLINED TUNNELS   The roughness varies with the degree of excess overbreak and the kind of rock that determines the angularity of the roughness. Therefore, it is difficult to make definitive recommendations for k_s; however, some guidelines can be given based on measurements in existing tunnels. Analysis of 20 unlined tunnels (6) shows that the minimum k_s was about 0.5 ft, and the maximum k_s was about 3.0 ft. For tunnels less than 15 ft in diameter, the average k_s was about 0.85 ft; for tunnels greater than 15 ft in diameter, the average k_s was about 1.42 ft.

These guidelines are for tunnels that were drilled and blasted. Tunnels constructed using a boring machine have much smoother walls. For example, equivalent sand roughness values ($k_s's$) for granite quartzite and other hard rocks are about 0.04 ft, whereas for mudstone, it can be as low as 0.005 ft.

PROBLEMS

5-1   a. Shown below are the HGL and EGL for a pipeline. Indicate which is the HGL and which is the EGL.   (continued on next page)

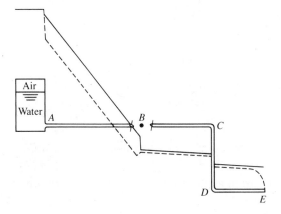

    b. Are all pipes the same size? If not, which is the smallest?

    c. Is there any region in the pipe where the pressure is below atmospheric pressure? If so, where?

    d. Where is the point of maximum pressure in the system?

    e. Where is the point of minimum pressure in the system?

    f. What do you think is located at the end of the pipe at point E?

    g. Is the pressure in the air in the tank above or below atmospheric pressure?

    h. What do you think is located at point B?

5-2   Water flows from reservoir A to B. The water temperature in the system is 10°C, the pipe diameter D is 1 m, and the pipe length L is 300 m. If $H = 16$ m, $h = 2$ m, and if the pipe is steel, what will be the discharge in the pipe? In your solution, sketch hydraulic and energy grade lines. What will be the pressure at point P halfway between the two reservoirs?

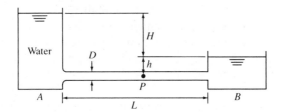

5-3   What horsepower must be supplied to the water to pump 2.5 cfs at 68°F from the lower to the upper reservoir? Assume the pipe is steel. Sketch the hydraulic and energy grade lines.

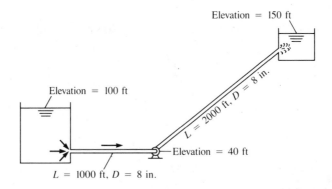

5-4   Irrigation water (20°C) is to be pumped through a 2-km long by 1-m diameter steel pipe from a river to an irrigation canal at a rate of 2.50 m³/s. The water surface elevation at the pump intake is 100 m, and in the canal, it is 150 m. What power should be supplied to the pump if the pump efficiency is 82% and the inlet and outlet head losses are nil?

5-5   Water flows from the reservoir through a pipe and then discharges from a nozzle as shown. What is the discharge of water? Also draw the HGL and EGL for the system.

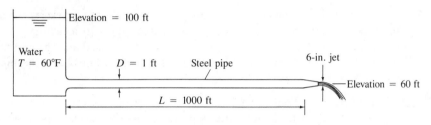

Elevation = 100 ft

Water
$T = 60°F$ $D = 1$ ft   Steel pipe   6-in. jet

Elevation = 60 ft

$L = 1000$ ft

PROBLEM 5-5

5-6   Points A and B are 3 mi apart along a 24-in. new cast-iron pipe carrying water. Point A is 30 ft higher than B. If the pressure at B is 20 psi greater than at A, determine the direction and amount of flow. $T = 50°F$.

5-7   Compare the head loss for water flow ($Q = 4$ cfs) in a 12 in., 3000-ft long steel pipe using the Darcy-Weisbach equation with Fig. 5-4, page 247, and using the Hazen-Williams formula.

5-8   Compare the head loss for water flow ($Q = 200$ cfs) in a 72 in., 10,000-ft long steel pipe using the Darcy-Weisbach equation with the Hazen-Williams formula and with the Swamee-Jain formula.

5-9   Consider flow in a 5-ft diameter steel pipe that is coated with a very smooth enamel finish. Assume water ($T = 60°F$) flows in it with a velocity of 15 ft/s. Determine the head loss per 1000 ft using
a. the Darcy-Weisbach equation
b. the Mannings equation
c. the Hazen-Williams equation

5-10   Consider the same pipe and flow conditions as in Prob. 5-9 except that the pipe is relatively smooth concrete. Determine the head loss with all the above equations plus the explicit equation for f (Swamee's equation).

5-11   A new steel pipe 24 in. in diameter and 2 mi long carries water from a reservoir and discharges it into air. If the pipe comes out of the reservoir 10 ft below the water level in the reservoir and the pipe slopes downward from the reservoir on a grade of 2 ft/1000 ft, determine the discharge.

5-12   What diameter cast-iron pipe is needed to carry water at a rate of 10 cfs between two reservoirs if the reservoirs are 2 mi apart and the elevation difference between the water surfaces in the reservoirs is 20 ft?

5-13   Referring to Fig. 5-1, page 242, if the pipe diameter is 2.0 ft, its length is 500 ft, and the pump is made to operate as a turbine with $Q = 32$ cfs, what horsepower will be generated by the turbine? Assume the upper reservoir water surface is at elevation 200 ft and the water surface elevation in the lower one is 100 ft. Assume the turbine efficiency is 85% and the pipe is steel. Also draw the HGL and EGL for this system.

5-14   The pressure at a water main is 300 kPa gauge. What pipe size is needed to carry water from the main at a rate of 0.025 m³/s to a factory 140 m from the main? Assume galvanized-steel pipe is to be used and the pressure required at the factory is 60 kPa gauge at a point 10 m above the main connection.

5-15   Two reservoirs with a difference in water surface elevation of 11 ft are joined by 45 ft of 1-ft diameter steel pipe and 30 ft of 6-in. steel pipe in series. The 1-ft line contains three bends ($r/D = 1$), and the 6-in. line contains two bends ($r/D = 4$). If the 1-ft and 6-in. lines are joined by an abrupt contraction, determine the discharge. $T = 60°F$.

5-16   a. Determine the discharge of water through the system shown.
b. Draw the hydraulic and energy grade lines for the system.
c. Locate the point of maximum pressure.
d. Locate the point of minimum pressure.
e. Calculate the maximum and minimum pressures in the system.

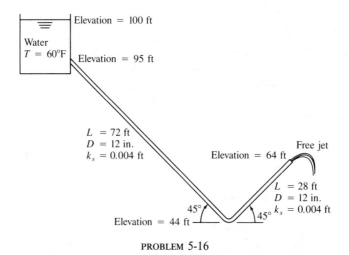

PROBLEM 5-16

5-17   Design a pipe system to supply water flow from the elevated tank to the reservoir at a discharge rate of 2.5 m³/s.

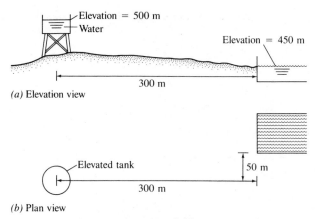

(a) Elevation view

(b) Plan view

PROBLEM 5-17

5-18   A pump that has the characteristic curve shown in the accompanying graph is to be installed in the system shown. What will be the discharge of water in the system?

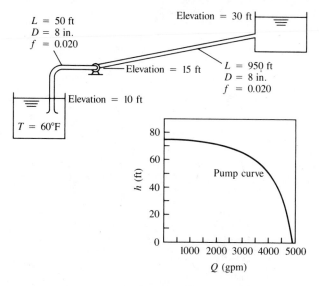

PROBLEM 5-18

5-19   Solve for the distribution of flow in the network of Example 5-8, pages 273–76, if the loads at points D, E, and C are 0 cfs, 5 cfs, and 20 cfs, respectively. The sole source is at point A. Also, what are the pressures throughout the system if the pressure and elevation at point A are 500 ft and 50 psi, respectively? The elevations at points B, C, D, and E are 450 ft, 430 ft, 440 ft, and 480 ft, respectively. Assume $f = 0.012$.

5-20   Assuming $f = 0.020$, determine the discharge in the pipes. Neglect minor losses.

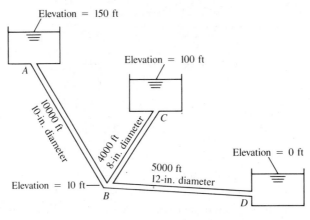

PROBLEM 5-20

5-21   Determine the water discharges in the pipes if they are all asphalted cast iron. The elevation of the center of the pipe at C is 50 ft.

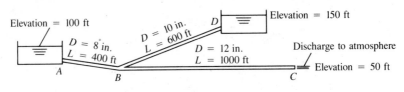

PROBLEM 5-21

5-22   With a flow of 20 cfs of water, find the head loss and the division of flow in the pipes from A to B. Assume $f = 0.030$ for all pipes.

$$
\begin{array}{c}
L = 3000 \text{ ft} \\
D = 14 \text{ in.}
\end{array}
$$

A • $\begin{array}{c} L = 2000 \text{ ft} \\ D = 24 \text{ in.} \end{array}$ $\begin{array}{|c|} \hline L = 2000 \text{ ft} \\ D = 12 \text{ in.} \\ \hline \end{array}$ $\begin{array}{c} L = 4000 \text{ ft} \\ D = 30 \text{ in.} \end{array}$ • B

$$
\begin{array}{c}
L = 3000 \text{ ft} \\
D = 16 \text{ in.}
\end{array}
$$

PROBLEM 5-22

5-23   This manifold is used to discharge heated effluent from a power plant into the Columbia River. There are ten discharge pipes spaced 10 ft apart (as shown), and the end pipe is to discharge water at a rate of 2.00 cfs. Determine the water surface elevation required in the reservoir and the total discharge.

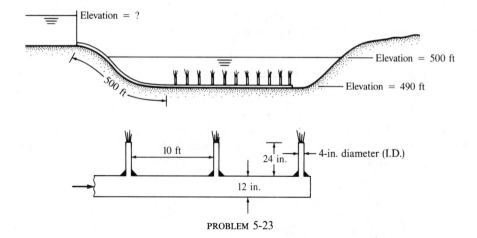

PROBLEM 5-23

5-24 Determine the distribution of flow for the given system. All pipes are 1 ft in diameter, and assume $f = 0.015$. Also, what is the pressure at load F if the pressure at source A is 60 psig?

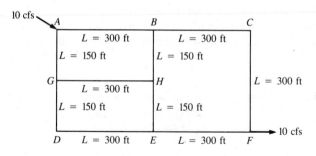

PROBLEM 5-24

5-25 A velocity traverse inside a 16-in. water pipe yields the data in the table below. What is the rate of flow in cubic feet per second and cubic feet per minute? What is the ratio of V_{max} to V_{mean}? Does it appear that the flow is laminar or turbulent?

y^* (in.)	0.0	0.1	0.2	0.4	0.6	1.0	1.5	2.0	3.0	4.0	5.0	6.0	7.0	8.0
V (ft/s)	0.0	7.2	7.9	8.8	9.3	10.0	10.6	11.0	11.7	12.2	12.6	12.9	13.2	13.5

* Distance from pipe wall.

5-26 What size of orifice is needed to produce a change in head of 8 m for a discharge of 2 m³/s of water in a 1-m diameter pipe.

5-27 An orifice is to be designed to have a change in pressure of 50 kPa across the orifice (measured with a differential pressure transducer) for a discharge of 3.0 m³/s of water in a 1.2-m diameter pipe. What orifice diameter will yield the desired result?

5-28   What water discharge is occurring in a 4-ft diameter horizontal pipe if a venturi meter in that pipe has a 2-ft throat diameter and a pressure differential of 10 psi between the upstream section and throat of the venturi meter?

5-29   A 6-in. orifice is placed in a 10-in. pipe, and a mercury manometer is connected to either side of the orifice. If a flow rate of water (60°F) through this orifice is 3 cfs, what will be the manometer deflection?

5-30   What throat diameter is needed for a venturi meter in a 200-cm horizontal pipe carrying water with a discharge of 10 m³/s if the differential pressure between the throat and upstream section is to be limited to 200 kPa at this discharge?

5-31   Water flows through a venturi meter that has a 30-cm throat. The venturi meter is in a 60-cm pipe. What deflection will occur on a mercury-water manometer connected between the upstream and throat sections if the discharge is 0.57 m³/s? Assume $T = 20°C$.

5-32   What compressive stress could develop in a 500-ft unreinforced concrete pipe that was laid in the winter ($T = 40°F$) between two rigid concrete structures (one on each end)? Assume the temperature in summertime can reach 80°F. Assume α and E are 6×10^{-6} ft/ft/°F and 4.5×10^{6} psi, respectively.

5-33   The pipe of Prob. 5-32 is to be 24 in. in diameter (I.D.), and it is to be installed so that a class C bedding condition can be assumed. What class of pipe should be installed in a 10-ft deep trench (3 ft wide) excavated in clay.

5-34   This 30° vertical bend in a pipe having a 2-ft diameter carries water at a rate of 31.4 cfs. If the pressure p_1 is 10 psi at the lower end of the bend where the elevation is 100 ft, and p_2 is 8 psi at the upper end where the

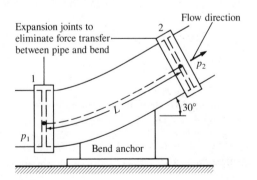

elevation is 103 ft, what will be the vertical component of force that must be exerted by the "anchor" on the bend to hold it in position? The bend itself weighs 300 lb, and the length L is 4 ft.

5-35   A 90° horizontal bend narrows from a 2-ft diameter upstream to a 1-ft diameter downstream. If the bend is discharging water into the atmosphere and the pressure upstream is 25 psi, what is the magnitude of the component of external force exerted on the bend in the x direction (the direction parallel to the initial flow direction) required to hold the bend in place?

5-36   A pipe 40 cm in diameter has a 135° horizontal bend in it. The pipe carries water under a pressure of 90 kPa gauge at a rate of 0.40 m³/s. What external force component in a direction parallel to the initial flow direction is necessary to hold the bend in place under the action of the water? What horizontal force component normal to the initial direction of flow is required to hold the bend?

5-37   This 130-cm overflow pipe from a small hydroelectric plant conveys water from the 70-m elevation to the 40-m elevation. The pressures in the water at the bend entrance and exit are 20 kPa and 25 kPa, respectively.

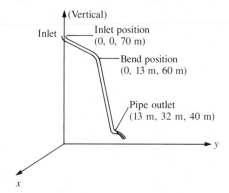

The bend interior volume is 3 m³, and the bend itself weighs 10 kN. Determine the force that a thrust block must exert on the bend to secure it if the discharge is 15 m³/s.

5-38   The pipe diameter D is 30 cm, d is 15 cm, and the atmospheric pressure is 100 kPa. What is the maximum allowable discharge before cavitation occurs at the throat of the venturi meter if $H = 5$ m?

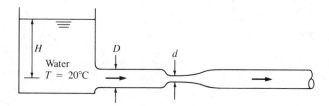

5-39   The sluiceway is steel lined and has a nozzle at its downstream end. What discharge may be expected under the given conditions? What force will be exerted on the joint that joins the nozzle and sluiceway lining?

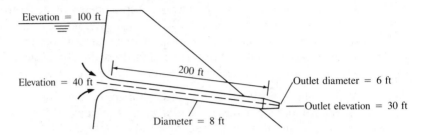

5-40   An 18-in. pipe abruptly expands to a 24-in. size. These pipes are horizontal, and the discharge of water from the smaller size to the larger is 25 cfs. What horizontal force is required to hold the transition in place if the pressure in the 18-in. pipe is 10 psi? Also, what is the head loss?

5-41   Determine the head loss per 1000 ft in this tunnel that is lined with concrete and is to have a water discharge of 1000 cfs.

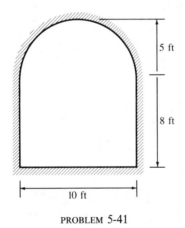

PROBLEM 5-41

5-42   Estimate the discharge of water in a tunnel 50% larger (linear dimension) than the one in Prob. 5-41 for a head loss per 1000 ft of 2 ft. Assume the tunnel is unlined, and make your own assumption(s) about the tunnel's degree of roughness.

REFERENCES

1. Allen, C.M., and E.A. Taylor. "The Salt Velocity Method of Water Measurement." *Trans. ASME*, 45 (1923), pp. 285–341.

2. Am. Concrete Pipe Assn. *Concrete Pipe Design Manual*. Am. Concrete Pipe Assn., Vienna, Vir., (1980).

3. Am. Concrete Pipe Assn. *Concrete Pipe Handbook*. Am. Concrete Pipe Assn., Vienna, Vir., (1980).

4. ASCE. *ASCE Manual on Construction of Sanitary Storm Sewers*." Manual no. 37, (1960).

5. ASCE. *Pipeline Design for Water and Wastewater*. ASCE, New York, (1975).

6. ASCE Task Force. "Factors Influencing Flow in Large Conduits." *Jour. Hydraulics Div.*, Proc. of ASCE, no. HY6, (November 1965).

7. ASHRAE. *ASHRAE Handbook, 1977 Fundamentals*. Am. Soc. of Heating, Refrigerating and Air Conditioning Engineers, New York, 1977.

8. *ASTM Standard Specification C76*. American Society for Testing and Materials, Philadelphia, Penn., (1977).

9. American Water Works Association. *Concrete Pressure Pipe*. Manual no. M9. AWWA, Denver, Colo., (1979).

10. AWWA. *Standard for Cement-Mortar Protective Lining and Coating for Steel Water Pipe 30-in. and Over — AWWA C205*. AWWA, New York (1962).

11. AWWA. *Standard for Coal-Tar Enamel Protective Coatings for Steel Water Pipe — AWWA Standard C203*. AWWA, New York (1962).

12. AWWA. *Steel Pipe Design and Installation*. AWWA Manual no. 11. AWWA, Denver, Colo. (1964).

13. AWWA. *Water Meters — Selection, Installation, Testing, and Maintenance*. AWWA Manual no. M6. AWWA, New York.

14. Beij, K.H. "Pressure Losses for Fluid Flow in 90° Pipe Bends." *J. Res. Nat. Bur. Std.*, 21 (1938).

15. Calhoun, Roger G. "A Statistical Roughness Model for Computation of Large-Bed-Element Stream Resistance." M.S. thesis, Washington State University, Pullman, Wash., 1975.

16. Cast Iron Pipe Res. Assn., *Handbook of Ductile Iron Pipe and Cast Iron Pipe*, 5th ed. Cast Iron Pipe Res. Assn., Oak Brook, Ill. (1978).

17. Cross, H. "Analysis of Flow in Networks of Conduits or Conductors." Univ. of Illinois Bulletin no. 286, (November 1936).

18. Elder, Rex. "Friction Measurements in Apalachia Tunnel." Trans. ASCE, 123, paper no. 2961, (1958), pp. 1249–74.

19. Eldridge, J.R. "An Analytical Method for Predicting Resistance to Flow in Rough Pipes and Open Channels." M.S. thesis, Washington State Universiity, Pullman, Wash., 1983.

20. *Fluid Meters — Their Theory and Applications*. ASME (1959).

21. Goodhue, H.W., R.L. Smart, and A.A. Meyer. "The Design of Recent TVA Projects: VIII. Apalachia and Oco No. 3." *Civil Engineering*, (October 1943), pp. 465–68.

22. Idel'chik, I.E. *Handbook of Hydraulic Resistance-Coefficients of Local Resistance and of Friction*. Trans. A. Barouch. Israel Program for Scientific Translations (1966).

23. Johansen, F.C. *Proc. Roy. Soc. London, Ser. A.*, 125 (1930).

24. Knapp, Robert T., J.W. Daily, and F.G. Hammitt. *Cavitation*. Inst. of Hydraulic Research, Univ. of Iowa, Iowa City, Iowa, 1979.

25. Kumar, Sushil, and John A. Roberson. "General Algorithm for Rough Conduit Resistance." *Jour. Hydraulics Div.*, ASCE, 106, no. HY11, Proc. Paper 15806 (November 1980), pp. 1745–64.

26. Laufer, John. "The Structure of Turbulence in Fully Developed Pipe Flow." *NACA Rept*, 1174 (1954).

27. Lienhard, J.H., and J.H. Lienhard IV. "Velocity Coefficients for Free Jets From Sharp-Edged Orifices." *Jour. Fluids Engineering, Transactions of ASME*, 106, (March 1984).

28. Linsley, Ray K., and J.B. Franzini. *Water-Resources Engineering*, 3d ed. McGraw-Hill, New York, 1979.

29. McNown, J.S. "Mechanics of Manifold Flow." *Trans. ASCE*, 119 (1954) pp. 1103–18.

30. Miller, D.S. *Internal Flow — A Guide to Losses in Pipe and Duct Systems*. British Hydromechanics Research Assoc., Cranfield-Bedford, England, 1971.

31. Moody, Lewis F. "Friction Factors for Pipe Flow." *Trans. ASME* (November 1944), p. 671.

32. Nikuradse, J. "Strömungsgesetze in rauhen Rohren." *VDI-Forschungsh.*, no. 361 (1933). Also translated in *NACA Tech. Memo* 1292.

33. Pearsall, I.S. *Cavitation*. Mills and Boon, London, 1972.

34. Rawn, A.M., F.R. Bowerman, and N.H. Brooks. "Diffusers for Disposal of Sewage in Sea Water." *Trans. ASCE*, 126, part III (1961), pp. 344–88.

35. Roberson, J.A., and C.K. Chen. "Flow in Conduits with Low Roughness Concentration." *Jour. Hydraulics Div.*, Am. Soc. Civil Eng., 96, no. HY4 (April 1970).

36. Roberson, J.A., and C.T. Crowe. *Engineering Fluid Mechanics*, 3d ed. Houghton Mifflin, Boston, 1985.

37. Schlichting, Hermann. *Boundary Layer Theory*, 6th ed. McGraw-Hill, New York, 1968.

38. Seely, F.B., and J.O. Smith. *Advanced Mechanics of Materials*, 2d ed. John Wiley & Sons, New York, 1959.

39. Shercliff, J.A. *Electromagnetic Flow-Measurement*. Cambridge University Press, New York, 1962.

40. Spangler, M.G. "Underground Conduits — An Appraisal of Modern Research." *Trans. ASCE*, 113 (1948).

41. Streeter, V.L., and E.B. Wylie. *Fluid Mechanics*, 7th ed. McGraw-Hill, New York, 1979.

42. Streeter, V.L. (ed.). *Handbook of Fluid Dynamics*. McGraw-Hill, New York, 1961.

43. Subramanya, K., and S.C. Awasthy. "Discussion of the Paper by Vigander." *Jour. Hydraulics Div.*, Proc. of ASCE, 96, no. HY12, (December 1970).

44. Swamee, P.K., and A.K. Jain. "Explicit Equations for Pipe-Flow Problems." *Jour. Hydraulic Div. ASCE*, 102 no. HY5 (May 1976).

45. "Underground Corrosion." Nat. Bur. of Stds. Circ. no. 579 (1957).

46. U.S. Bur. of Reclamation. *Design of Small Dams*, 2d ed. U.S. Govt. Printing Office, Washington, D.C., 1972.

47. US Bureau of Reclamation. "Friction Factors for Large Conduits Flowing Full." Engr. Monograph no. 7, US Bureau of Reclamation, U.S. Dept. of Interior, U.S. Govt. Printing Office, Washington, D.C., 1965.

48. Vigander, S., R.A. Elder, and N.H. Brooks. "Internal Hydraulics of Thermal Diffusers." *Jour. Hydraulics Div.*, Proc. of ASCE, 96, no. HY2, (February 1970).

49. Wright, S.J. "A Theory for Predicting the Resistance to Flow in Conduits With Nonuniform Roughness." M.S. thesis, Washington State University, Pullman, Wash., 1973.

6

Dworshak Dam on the North Fork of the Clearwater
River near Orofino, Idaho is a concrete gravity dam
with a height of 717 feet above bedrock. The gross
storage in the reservoir developed by the dam is
3,500,000 acre feet and the usable storage for power
and flood control is 2,000,000 acre feet. (Photo by
Kimberle A. Fennema)

Dams and
Reservoirs

Because of the very large size of certain dams (such as Grand Coulee, Hoover, and Aswan) one has a tendency to regard dams as isolated entities. Actually, virtually every dam is just one part of an overall water-resource project. Moreover, many projects have several objectives. For example, the Grand Coulee project in Washington was developed primarily for irrigation and hydroelectric power production, but both navigation and flood control are part of, and benefits of, the project. Thus Grand Coulee Dam should be viewed as one part of the comprehensive plan for utilizing the water resources of the Columbia River. All dams both large and small are designed and constructed to satisfy one or more water-resource objectives. The next section is a brief summary of the type of planning process that eventually leads to the constructed dam. Subsequent sections deal with the actual design and construction of dams. The last section in this chapter focuses on the role of the reservoir in the water resource scheme.

6-1   The Planning Process

Rivers are sources of energy (hydroelectric power) and water supply for municipalities and agriculture. Many rivers also serve as transportation arteries and are sources of recreation. Flooded rivers often wreak havoc, causing property damage and loss of life. Rivers are also often used for sewage disposal. As our population increases and demand for food, water, power, recreation, and land (often on flood plains) grows, we (that is, local and federal governments, private companies, and individual citizens) design and construct water-resource projects to control the rivers to our advantage. However, a use of the river that may be beneficial to some may be detrimental to others. Moreover, most water-resource projects cost a good deal of money. Careful planning should therefore be done to achieve optimum utilization of river basins as a whole as well as specific projects within them.

Definition of Planning

Planning means determining, in an orderly manner, the best way to accomplish a particular objective by evaluating various alternatives. For example, in the context of water resources, a problem may exist of not having enough water for the demands of a large city during drought periods. Careful planning should be done to bring about a solution to the problem. Planning involves evaluating several possible solutions. Some solutions entail building structures such as dams and supply pipes. Therefore, planning also involves designing these features and estimating their cost; cost comparisons often determine the "best" alternative.

Planning Phase

PRELIMINARY STUDY   Efficient planning of water-resource projects is done in several stages. First, the problem must be clearly defined and various alternatives for solutions to the problem are envisioned. A number of the alternatives may be either supported or rejected by use of rough data and shortcut analyses. These preliminary studies may be thought of as a coarse screening phase. In this phase the completely infeasible alternatives may be ruled out.

For example, let us return to the problem of the major city that does not have enough water to satisfy the demand during drought periods. Alternatives to this problem may include (1) water conservation programs, (2) constructing a dam on a river to provide storage from which water may be drawn during a drought, (3) a combination of water conservation and dam construction. Perhaps the water conservation can be implemented initially and a dam built eventually. In any event, thought will have to be given to the benefits and costs of each alternative project. In considering the construction of a dam, there may be several streams on which it could be built, and on any given stream, there may be several sites at which it could be located. The preliminary study allows all these different possibilities to be narrowed down to perhaps two or three of the best choices. By using rough data and simplified analyses in the reconnaissance study much time and money is saved over that which would be involved if a more thorough study were done on all the alternatives. The preliminary study will either result in a recommendation to cease further study (all alternatives are unsatisfactory for economic, environmental, or social reasons) or a recommendation will be made to proceed to the next phase, the feasibility study.

FEASIBILITY STUDY   During the feasibility study, a project or plan is developed to determine whether or not it should be implemented. The project or plan must be technically and economically feasible and environmentally acceptable. The preliminary study narrows the possible solutions down to a few reasonable alternatives, and the feasibility study establishes the scope, magnitude, and all essential features of these alternatives. Cost estimates are also refined so that a good assessment of cost versus benefits may be made for each alternative. Moreover, environmental issues are addressed and environmental impact statements (EIS) are prepared in this phase. Comparisons of the alternatives are then made, and the best project or plan is determined. Then a decision is made either to cease the study (too costly, or insurmountable environmental or social problems exist) or to proceed to the final phase.

FINAL STUDY   In the final phase of planning, detailed designs are made and plans and specifications are developed. The cost and benefit estimates may not change greatly from the feasibility phase to the final phase, though they may; for example, the cost of money (interest rates) may change significantly during that period. The possibility of major changes in the project should therefore be considered even after the final study has been done.

Planning Considerations of Projects Involving a Dam and Reservoir

Whether it be for municipal water supply, irrigation, or hydropower, several items must be considered in the planning and design of a dam and reservoir:

1. Hydrological data. Data of the stream that the dam is to be built on are analyzed to determine flood and drought flows and to determine the required capacity and operating procedure for the reservoir. Also, the required spillway capacity can be determined from the hydrologic data.
2. Geologic data. Some data of the geology of the area that the dam is to be built on will no doubt exist; however, on-site inspection, geologic mapping the drilling of exploratory holes, and collection of core-sample data by geologists are usually required. These data reveal the structural ability of the foundation material to withstand the loads that may act on it and indicate the leakage and erosion problems that may be encountered.
3. Reservoir data. A complete assessment of the area to be inundated by the reservoir must be made. This includes topographic maps, land ownership, land classification, and location of roads and public utilities. These data are used to estimate the cost of land acquisition and relocation of roads and utilities.

Many other factors, such as availability of materials and manpower, environmental aspects, sedimentation and dam safety must also be considered in the planning and design of a dam and reservoir.*

6-2   Types of Dams

Dams are classified according to the material (earth, rock, concrete) from which they are built (see Fig. 6-1) and according to their configuration and the way in which they resist the forces imposed on them. Thus a gravity dam is one in which the gravitational forces (such as the weight of the dam itself) are great enough to resist the overturning moment and sliding force of the hydrostatic forces imposed on it (Fig. 6-1c). Another type of gravity dam is the

---

* For more details on planning considerations, see Peterson (12) and the U.S. Bureau of Reclamation publication on the design of dams (20).

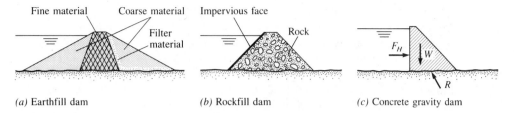

Fine material   Coarse material   Impervious face

Filter material   Rock

F_H W   R

(a) Earthfill dam   (b) Rockfill dam   (c) Concrete gravity dam

Figure 6-1   Earth dam, rockfill dam, and concrete gravity dam

buttress dam (Fig. 6-2a), in which reinforced concrete slabs constitute the face of the dam and are supported by vertical buttresses at intervals of 50 to 100 ft. In contrast, an arch dam is designed to transfer the imposed loads to adjacent rock walls on either side of the canyon it is located in (Fig. 6-2b). Both earthfill and rockfill dams are special types of gravity dams. We consider these types of dams in more detail in the following sections.

Earth Dams

INTRODUCTION   The first dams ever built by humans were made of earth. Biswas (1) notes that along the Nile River at about 2000 B.C. during the flood seasons some of the flood water was diverted through a canal to a natural storage depression. The flow in this canal was reportedly controlled by two earthen dams. On the Arabian peninsula in what is now Yemen, an earthfill dam was constructed sometime between 1000 B.C. and 700 B.C. Called Marib dam, it was 33 ft high and 1900 ft long (1). In Sri Lanka, an earth dam 70 ft high and 11 mi long was completed about 500 B.C. and it is still in use (14). Today, the most common type of dam is still the earth dam, which can be designed and built on almost any given site and foundation condition by using a wide range of earth materials available at the site. As for all dams, thorough field investigations are required for the design and construction of sound earth dams.

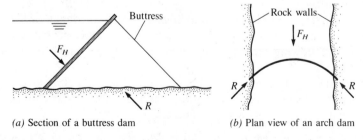

Buttress

F_H R

Rock walls F_H

R   R

(a) Section of a buttress dam   (b) Plan view of an arch dam

Figure 6-2   Buttress and arch dams

Economically, the earth dam is usually favored over the concrete gravity dam if suitable earth materials are available near the site and if the site is suitable for adequate spillway design and construction. It is not safe to have water spill directly over the top of an earth dam even when the spillway section is paved. Accepted design practice places the spillway structure on adjacent undisturbed ground or in a separate concrete gravity block. In earth dams, special attention must be given to the design, construction, and maintenance of the dam to resist internal erosion. If this is properly done, earth dams are as safe as any other type.

Some earthfill dams have been constructed by placing the earth in the dam by pumping the fill material to the site in a slurry and letting the water drain off. A dam constructed in this manner is called a *hydraulic fill dam*. The largest hydraulic fill dam in the United States is the Fort Peck Dam in Montana (126 million cu yd of fill material). However, stability problems can develop when constructing this type of dam. For example, a huge slide (5×10^6 cu yd) occurred during the construction of Fort Peck Dam (11). With the development of large earthmoving equipment, the cost of placing earth with earthmovers is competitive with the hydraulic process. Therefore, almost all earth dams are now constructed by using earthmoving equipment to place and compact the materials in the dam in layers so that a dense stable fill is produced. Such a dam is called a *rolled fill earth dam*.

The built-up section of an earthfill or rockfill dam is the *embankment*. The embankment must resist the hydrostatic forces of the water in the reservoir and must contain a section or zone impervious enough to prevent excessive seepage. If the soil material is relatively fine and abundant, a *homogeneous embankment* may be constructed. However, it is more common to find a variety of materials at the site of the dam. In this case, the usual practice is to design a *zoned embankment* (Fig. 6-1a). The finer material is compacted to produce a relatively impervious zone, and this is usually placed near the central part of the dam. The coarser material is placed upstream and downstream of the impervious core. This coarse material primarily provides stability to the dam. The interface zone between the fine material of the *core* and coarse material on the downstream side must be carefully designed and placed to eliminate the possibility of erosion of the fine material as seepage occurs after the reservoir is filled. Thus the interface zone should be a graded filter material that allows passage of seepage water but prevents dislodgment of fine particles of soil in the core. Other factors that should be considered in designing embankments are (1) adequacy of the foundation, (2) embankment stability, and (3) slope protection. We discuss these in more detail in the following sections.

ADEQUACY OF THE FOUNDATION   The safety of the embankment depends, in part, on the inherent shear strength of the foundation. If the foundation soils are weak, special steps must be taken. For example,

1. The slopes of the embankment may be flattened to distribute the load over a greater area.

2.   If the weak soils of the foundation are not too thick, they may be excavated.
3.   The embankment may be constructed at a slower pace than normal so that the weak soils will have time to consolidate without excessive differential settlement of the embankment. In this process, drains placed in the weak soils provide for removal of water and can help speed up the soil consolidation.

Whether the foundation is composed of weak or strong soils, the surface of the foundation material must be carefully prepared before the embankment material is laid down. This preparation includes removing all vegetation; digging out stumps and large roots; and stripping off sod, top soil, and any other organic material. Pockets of soft compressible soils should also be removed. When all the organic and soft material has been removed, stump holes and other pockets should be filled with soil and compacted by power tampers. The foundation is then plowed and rolled to compact the upper surface of the foundation. Sometimes a cutoff trench is excavated in the foundation material so that a contact is made between the core material and bedrock (see Fig. 6-3). Constructing a cutoff ensures that any old drains, pervious zones, or abandoned pipes are found and removed. However, if a cutoff trench is not used, then an inspection trench having a minimum width of 6 ft should be excavated to check for and remove undesirable features. The trench should then be backfilled and properly compacted.

If the foundation material is rock, it should be cleaned to remove all loose rock. Open joints in the rock should also be cleaned and filled with concrete. To prevent excessive seepage through fissures and crevices, grouting the foundation rock may be necessary.*

EMBANKMENT STABILITY   The design of the embankment should be based on the available soils, their water content, and the need for drying or wetting of the soils to achieve optimum conditions for compaction. Any large boulders in the soil should be removed because they will make rolling of the

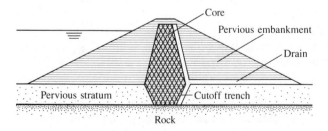

Figure 6-3   Cutoff trench and core of an earth dam

* For more details on foundation and abutment preparation, see Golzé (8) and USBR (20).

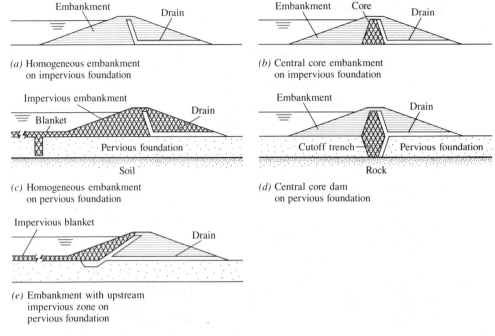

Figure 6-4   Different types of earth dam embankment (8)

soil difficult or impossible. It is impossible to prevent all seepage through any dam. A portion of an embankment dam will always be saturated and the permeability of the material will determine the rate at which water seeps through the embankment. At the downstream side of the core the fine materials of the core could be carried away from the core and into the coarser embankment material if measures are not taken to prevent this erosion. Thus filter zones should be provided on the downstream side of the core and a safe path (often in the form of special drains) should be provided to convey seepage water to the river channel downstream.

The embankment design depends on the type of soils available and on the objectives of the project. Thus each dam will have its unique design features. However, there are some general guidelines that are usually followed. Figure 6-4 illustrates the different features of embankments, and we discuss them here.

1.   A relatively impervious core is used to reduce seepage (Figs. 6-4*b* and 6-4*d*). The width of the core depends on the amount of impervious soils available and the method of placement of fill material. However, current practice (8) calls for a bottom width no less than one quarter of the net head between maximum pool level and minimum tailwater level. The top width of an impervious core should not be less than 10 ft to allow movement of equipment for placing and compacting the fill material (8).

2.   If the embankment rests on a pervious foundation, a cutoff trench may be used to reduce seepage (Figs. 6-4c and 6-4d). Upstream blankets may also be used to reduce seepage for embankments on pervious foundations where cutoff trenches would be very deep (Fig. 6-4e).

3.   Downstream control of seepage water is handled by means of pervious drainage blankets or by constructing most of the embankment of pervious material (Figs. 6-4a through 6-4e). The transition zone between the zone of fine soil and the coarser drainage blanket must be carefully designed and constructed to prevent the finer soils from being carried into the coarser drain material. Specifications for such a transition zone are given in reference (20).

The strength of the relatively impervious soils depends on their compacted density, and this in turn, depends on the water content and the kind and use of compaction equipment. Thus improving the strength of the soil by increasing or decreasing the water content is possible. Drying can be done by harrowing the fill in dry weather or by heating it in a specially designed mechanical dryer. Sprinkling of the borrow area is a common means of moistening a soil that is too dry.

In designing the slopes of the embankment, the usual practice is to choose slopes that have been found from past experience to be stable. These slopes are then checked for stability using the Swedish-slip-circle method. The factors involved in choosing the embankment slopes are:

1.   Character of soil materials available
2.   Foundation conditions
3.   Height of structure
4.   Possibility of rapid drawdown of the upstream pool

Speed of drawdown is important in embankment design because if the pool is drawn down rapidly, a relatively impervious embankment will not drain freely, and the pore-water pressure within the embankment will reduce the shearing resistance and increase the weight of the embankment material over what it would have been had it drained. The reduced shearing resistance may cause the embankment to fail by sliding. Therefore, if rapid drawdown of the reservoir can be expected, the upstream embankment must have a flatter slope, or it must be constructed of material that will drain rapidly. Table 6-1 shows recommended embankment slopes for selected materials and drawdown conditions.

SLOPE PROTECTION   Because of the erodibility of earth dams, special precautions must be taken to prevent erosion. The upstream slope of the embankment must be protected against rain and snowmelt runoff and against wave erosion. The most commonly used type of protection on the upstream face is dumped riprap, which consists of stones or rock fragments placed on a properly graded filter. The filter may be a specially placed blanket of sand and gravel, or it may be the upstream zone of a zoned embankment. The rock for the riprap should be hard, dense, and durable to resist wave action and normal

Table 6-1 Recommended Embankment Slopes for Small Earthfill Dams on Stable Foundations (20)*

Type of Dam	Drawdown Condition	Soil[†] Type	Upstream Slope	Downstream Slope
Homogeneous	Gradual[‡]	Silty gravel	$2\frac{1}{2}$:1	2 :1
		Sandy clays	3 :1	$2\frac{1}{2}$:1
		Inorganic clays	$3\frac{1}{4}$:1	$2\frac{1}{2}$:1
Modified homogeneous[§]	Rapid	Silty gravel	3 :1	2 :1
		Sandy clays	$3\frac{1}{2}$:1	$2\frac{1}{2}$:1
		Inorganic clays	4 :1	$2\frac{1}{2}$:1
Zoned with large core shell material of pervious sand, gravel, or rock	Gradual	Silty gravel	2 :1	2 :1
		Sandy clays	$2\frac{1}{2}$:1	$2\frac{1}{2}$:1
		Inorganic clays	3 :1	3 :1
Zoned with large core shell material of pervious sand, gravel, or rock	Rapid	Silty gravel	$2\frac{1}{2}$:1	2 :1
		Sandy clays	3 :1	$2\frac{1}{2}$:1
		Inorganic clays	$3\frac{1}{2}$:1	3 :1

* These recommendations were abstracted from the USBR for the design of dams less than 50 ft in height.
† Soil types are for embankments of homogeneous dams and cores of zoned dams.
‡ Rapid drawdown is defined as a drawdown rate of more than 6 in. per day.
§ A modified homogeneous dam is one that is entirely homogeneous except for a filter drain constructed at the base of the downstream part of the embankment.

Table 6-2 Thickness and Gradation Limits of Riprap on 3:1 Slopes (21)

Reservoir Fetch (mi)	Nominal Thickness (in.)	Maximum Size	Gradation, Percentage of Stones of Various Weights (lb)*		
			40 to 50 Percent Greater Than	50 to 60 Percent From — To —	0 to 10 Percent Less Than[†]
2.5 and less	30	2500	1250	75–1250	75
More than 2.5	36	4500	2250	100–2250	100

* Sand and rock dust shall be less than 5 percent, by weight, of the total riprap material.
† The percentage of this size material shall not exceed an amount that will fill the voids in larger rock.

weathering over a long period. The maximum size of rock and thickness of the riprap layer depends primarily on the size of waves that might be expected, and this in turn, depends on the reservoir *fetch*.* Table 6-2 gives recommended riprap thickness and rock sizes for small dams. The upstream slope protection should extend from the crest of the dam to several feet below the minimum water level.

---

* Fetch is the unobstructed overwater distance from the dam to the nearest land mass upwind from the dam. We discuss this topic in more detail in Sec. 6-5.

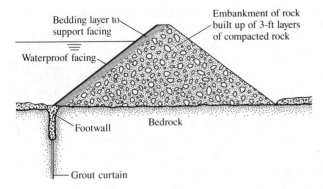

Figure 6-5 Rockfill dam

The downstream slope of the embankment requires less protection than the upstream slope. If the downstream zone of the embankment consists of rock or cobble fill, no special treatment of the slope is necessary. If the downstream embankment consists of fine materials (sand, gravel, or finer), common practice is to protect this surface with a 1- or 2-ft thick layer of rock or cobble. Grass turf also is often used for downstream slope protection where rainfall is adequate to maintain growth.*

Rockfill Dams

INTRODUCTION   A rockfill dam is an embankment that uses variable sizes of rock to provide stability and a thin membrane on its upstream face or a compacted earth core in the embankment for water tightness. The most commonly used membrane is a reinforced concrete slab (see Fig. 6-5). Other membranes are asphaltic concrete, steel plate, and wood timbers.

Rockfill dams with earth cores have essentially the same features as a central core earth dam except that the main embankment consists of rock instead of coarse soil. As in the central core earth dam, the filter zones on either side of the core must be carefully designed and placed to prevent erosion of the core.

The oldest rockfill-dam remains are found in Egypt at the Wadi el-Garawi about 30 km south of Cairo. An interdisciplinary group of engineers and scientists established the date of construction at approximately 2600 B.C., and they have documented it as a true rockfill dam with a central core of earth and rubble (7). The dam was planned to be 110 m long, and much of it was built to its design height of 14 m above its base. Unfortunately, however, the dam was never completed, probably because of a flood that washed out most of it

---

* For more details on earth dam design, see Golzé (8) and USBR (20).

during the final stages of construction. Apparently, no attempt was made to continue work thereafter.

The first modern rockfill dams were constructed in California in the 1800s during the California Gold Rush (6). Of the early dams built before 1900, only three were higher than 100 ft, and many of these early dams had a watertight facing made of timber. Between 1900 and 1932, 18 rockfill dams were constructed, and of these, 12 were faced with concrete. Most of the dams built in this period were more than 100 ft high, and one was 328 ft high. Since then, several rockfill dams more than 500 ft high have been built throughout the world. A rockfill dam is usually more economical to build in remote areas where rock is readily available, soil for an earth dam is not available, and bringing in materials for a concrete dam is too costly.

The early dams were constructed by dumping loose rock in lifts (layers) up to 25 ft thick. However, dams built in this way had considerable settlement with attendant damage to the upstream face. In recent years, most rockfill dams have been built by placing the rock in thin (no more than 3 ft thick) layers and compacting them with several passes of heavy-smooth drum, vibratory rollers.

THE FOUNDATION   Almost all rockfill dams are built on fairly solid rock foundations. As with earthfill dams, the foundation should be cleared of silt, clay, and organic material before construction of the rock embankment.

EMBANKMENT DESIGN FOR CONCRETE FACE DAMS   The upstream and downstream slopes of the embankment are constructed to have the angle of repose of the rock, that is, from about 1.3 horizontal to 1 vertical to 1.4 horizontal to 1 vertical. The rock is laid down in layers and each layer is compacted. The upstream face of the embankment requires special treatment to provide a suitable surface on which the impervious concrete face will rest. In current practice, the face supporting zone of the embankment consists of well-graded material with a maximum rock size of about 3 in. and a minimum of sand size (3). This zone is usually about 12 ft wide (horizontal measure) at the top of the dam and remains that width to the base for dams less than 300 ft high (3). For higher dams the width is made greater toward the base. For the 525-ft high Areia Dam in Brazil, the face supporting zone was increased to 30 ft at its base. The material in the face supporting zone is placed in layers about a foot thick and compacted with several passes of a vibratory roller. Then, after the entire embankment is in place, the upstream face is trimmed and compacted further with several passes of the roller, which is guided up and down the slope by means of a cable. In this manner, a sound foundation is prepared for support of the concrete slab face.

UPSTREAM BASE   At the upstream base of the dam, an anchor is required for the impervious face. Common practice is to excavate a portion of the foundation rock and pour a concrete wall (sometimes called a *footwall* or *toe slab*) to serve both as this anchor and as a grout cap for a grout curtain in the foundation (Fig. 6-5).

IMPERVIOUS FACING   The facing of modern concrete face dams consists of reinforced slabs with vertical joints but not horizontal joints (3). Thus the face of a 300-ft high dam might consist of slabs about 50 ft wide by about 500 ft long. The slab is usually about 1 ft thick at the top of the dam and, with distance down the slope, increases in thickness according to the formula $t = 1 + 0.003H$, where t is the slab thickness in ft, and H is the vertical distance in ft from the top of the dam to the location on the slope. The amount of steel rod reinforcing used in the face slab is usually about 0.4% of the concrete volume. The reinforcing runs both horizontally and vertically (up and down slope) through the slab. Because some settlement will occur in the embankment, the impervious face must be able to conform to the changes in the upstream surface of the dam. This is accomplished by the vertical contraction joints with water stops of rubber, plastic, or expandable metal between the slabs. The bottom slab of the face bears against the footwall. An expansion joint with a waterproof seal is also used here.*

Concrete Gravity Dams

FORCES ON THE DAM   Concrete gravity dams are designed so that the weight of the dam itself (the gravity force) is sufficient to resist overturning by the applied forces. The forces that must be considered in the design of the dam are the hydrostatic forces both upstream and downstream, hydrostatic uplift, the weight of the dam, earthquake forces, and ice forces.[†] Figure 6-6 shows how these forces are applied. We discuss each of these forces separately.

HYDROSTATIC FORCES   Because of the pressure of the water in the reservoir and in the downstream channel, hydrostatic forces will be exerted on the dam. In Fig. 6-6, the horizontal force per unit width is $F_{U,H}$. Here $F_{U,H} = \gamma h_U^2/2$, where γ is the specific weight of the water (62.4 lb/ft^3 for fresh water), and h_U is the vertical distance from the water surface to the base of the section of dam. The location of the line of action of this force is at 2/3 of the depth below the water surface. If the dam has a sloping face, there will be a vertical component of hydrostatic force; this is identified as $F_{U,V}$ (see Fig. 6-6). The magnitude of $F_{U,V}$ equals the weight of the water vertically above the sloping face of the dam, and its line of action is through the centroid of this

---

* For more details on the design and construction of rockfill dams, see Cooke (3), Creager (5), and Golzé (8).

[†] Forces due to temperature rise in a dam may also be significant if it is a large dam. For details, see Golzé (8).

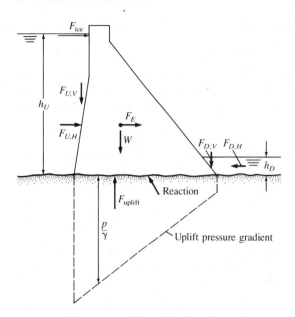

Figure 6-6   Forces acting on a section of a concrete
gravity dam

volume of water. Similar hydrostatic forces act on the downstream face of the
dam, as shown in Fig. 6-6.

HYDROSTATIC UPLIFT   After the reservoir is filled, water (under pressure) will seep into the pores of the concrete of the dam and through the pores
and fissures of the foundation rock. Once conditions of equilibrium have been
established (that is, once the seepage rate is constant), a pressure head gradient,
as is shown in Fig. 6-6, will be established in the pores of the concrete along
the base of the dam. The maximum head is at the heel (upstream limit) of the
dam, where $p/\gamma = h_U$, and the minimum head is at the toe of the dam and is
equal to h_D. Thus the magnitude of the hydrostatic uplift will equal the product
of average uplift pressure and the area of the base section. The line of action
of the uplift force will act through the centroid of the pressure prism at the
base of the dam. Customary practice is to reduce the uplift force by creating a
more impervious zone in the rock foundation by boring holes into the foundation rock and pumping cement grout into the holes. The grout then is forced
into the fissures and pores of the foundation. These grout holes usually are
spaced about every 10 ft along the length of the dam and are intended to create
a relatively impervious *grout curtain*. Therefore, when water seeps through this
curtain, the head loss is much greater across the curtain than in the rest of the
foundation material, thereby reducing the uplift pressure downstream of the
grout curtain. To relieve the uplift pressure even more, drains are usually installed between the zone just downstream of the grout curtain and the toe of

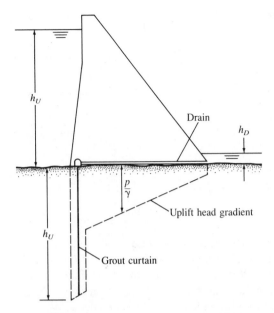

Figure 6-7 Uplift pressure on a dam with grout curtain and drains

the dam. Thus the uplift pressure distribution for a dam with a grout curtain and drains might appear as shown in Fig. 6-7. Even though the grout curtain and drains will greatly reduce the uplift pressure, engineers, assuming a more pessimistic scenario, often design dams to withstand greater pressure than this reduced value. Just how much hydrostatic uplift force is used depends on a given agency's or engineering firm's assumptions and practices.

WEIGHT OF DAM   In computing the gravity forces, one must include the weight of the concrete (usually assumed to equal 150 lb/ft^3) plus the weight of appurtenances such as gates and bridges. The resultant weight will act through the center of gravity of the entire mass.

EARTHQUAKE FORCES   When an earthquake occurs, the earth shakes (vibrates), as does the dam resting on the earth. From the standpoint of forces resulting from an earthquake, it is convenient to think of an inertial force due to the shaking of the dam. That is, the dam will be accelerated when the quake occurs so that an inertial force will act through the center of gravity of the dam and in a direction opposite to the direction of acceleration. Figure 6-8 shows a dam being accelerated to the left. The inertial force will act in the opposite direction (to the right in this case), and it will be equal to Ma, where a is the acceleration due to the earthquake, and M is mass. In the United States, the design value for the acceleration usually varies from 0.05g to 0.10g depending

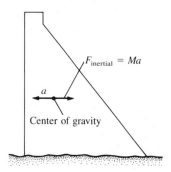

Figure 6-8 Inertial force due to acceleration of a dam
dam during an earthquake

on the area's susceptibility to severe earthquakes. Dams up to 50 ft high may be designed for earthquake forces using the simple procedure noted. If the dam is very high, however, the fundamental frequency of vibration of the dam may be in resonance with the vibration of the earthquake. Therefore, for very high dams, more sophisticated dynamic analyses are made for earthquake effects.*

Besides the inertial effects from an earthquake, the water pressure itself will be increased when the dam is accelerated in a direction toward the reservoir. A formula for this added pressure was developed by Zanger (23) and, for dams with a vertical upstream face, is given as

$$P_e = C\lambda\gamma h \tag{6-1}$$

$$\text{where } C = 0.365\left[\frac{y}{h}\left(2 - \frac{y}{h}\right) + \sqrt{\frac{y}{h}\left(2 - \frac{y}{h}\right)}\right] \tag{6-2}$$

and where λ = the earthquake intensity (the earthquake acceleration divided by the acceleration of gravity)

γ = specific weight of water

h = total depth of reservoir at section being studied

y = the vertical distance from the reservoir surface to the elevation in question (see Fig. 6-9)

According to the USBR (20), it may be shown analytically that the total horizontal force, on the face of the dam, V_e, above the elevation in question, and the total overturning moment, M_e, above that elevation are, respectively,

$$V_e = 0.726P_e y \tag{6-3}$$

$$M_e = 0.299P_e y^2 \tag{6-4}$$

---

* For a discussion of these analyses, see Golzé (8).

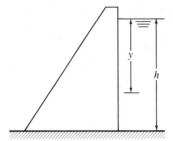

Figure 6-9 Definition sketch for terms in equations involving added water pressure from earthquakes

If the upstream face of the dam is not vertical, the added pressure due to the earthquake will be less than the value given by Eq. (6-1).*

ICE FORCES   In northern regions, where ice may become more than a foot thick, the force of ice against a dam may be large. The most severe situation occurs when the temperature of the ice is increasing and expanding. Golzé (8) indicates that the ice force may be taken as 10,000 lb/lineal ft of contact with the dam for ice thicknesses of 2 ft or more.

STABILITY ANALYSIS OF THE DAM   Several sophisticated methods are available for analyzing a dam for stability; however, we will describe only the simple gravity method. In the gravity method, a vertical slice of the dam is analyzed for stability, and it is assumed no forces are transmitted to or from this slice by adjacent elements.[†] The stability analysis checks

1. For resistance to overturning
2. For resistance to sliding
3. To make sure that allowable normal stresses in the concrete are not exceeded

We discuss the procedures and assumptions of each of these determinations.

RESISTANCE TO OVERTURNING   If the dam is too thin, it may not have enough weight to resist the action of the water pressure and may fail by tipping in the downstream direction about its toe. If this were to happen, the line of action of the resultant of the applied forces would lie outside the pivot point, as shown in Fig. 6-10. We might then conclude that a dam would be safe from overturning if a rule were adopted stating that the line of action of the resultant

---

* For the case of a sloping upstream face, see USBR (20).

† The more sophisticated methods of analyses take these forces into account. These forces are generally considered negligible for low dams (less than 50 ft high) but may be significant for high dams.

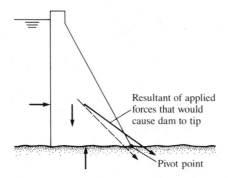

Figure 6-10 Consideration of forces causing overturning

should lie inside the toe of the dam (the broken line in Fig. 6-10). However, this may not be sufficient to guard against overturning. It can be shown that if the resultant acted just a short distance to the left of the pivot point, there would be tension in the concrete at the upstream face of the dam. But to assume unreinforced concrete can take a tensile stress is not prudent; therefore, good engineering practice requires that tensile stress not be allowed at the upstream face of the dam. By applying the flexure formula ($\sigma = F/A \pm Mc/I$) to a horizontal section at the base of the dam or on any other horizontal section, it can be easily shown that the resultant applied forces must lie within the middle third of the section being considered.

RESISTANCE TO SLIDING   The forces that tend to cause sliding are the pressure of the water on the face of the dam, horizontal earthquake forces, ice forces, and wave forces. If these forces are less than the resistance to shear at the base of the dam, or any other horizontal section through the dam, then the dam will not slide. The applied horizontal forces are determined by the methods we discussed in the previous section. The resistance to sliding is equal to the product of the normal force (vertical force for a horizontal shear plane) acting on the shear plane and the coefficient of friction f acting between the two surfaces at the shear plane.

Table 6-3 Representative Friction Factors for Foundation Material (20)

Material*	f
Sound rock, clean and irregular surface	0.8
Rock, some jointing and laminations	0.7
Gravel and coarse sand	0.4
Sand	0.3
Shale	0.3

* For silt and clay, testing is required.

The allowable friction factor, f, for rock is best determined by laboratory analyses. However, the values shown in Table 6-3 were given by the U.S Bureau of Reclamation (20) as a guide for preliminary analysis.

DETERMINATION OF MAXIMUM NORMAL STRESSES   To check whether maximum normal stresses in the concrete will be exceeded, one applies the flexure formula ($\sigma = F/A \pm Mc/I$) to determine the stresses acting on a horizontal plane through the dam at the upstream and downstream faces.

LOAD COMBINATIONS USED IN STABILITY ANALYSES   In the preceding sections, we discussed the kinds of loads that can be exerted on a dam, and we presented methods to determine the stresses that can result from these loads. The question now is whether all the loads should be applied to create the worst possible loading situation, or whether only some of them should be applied to represent a more probable situation. The worst possible situation might be one in which the reservoir is completely full, the maximum ice force is acting, and the maximum earthquake occurs. The worst possible situation is highly improbable; therefore, current design practice makes allowance for these situations by applying a lower factor of safety to them. The procedure is to categorize the possible loading situations according to *usual* load combination, *unusual* load combination, and *extreme* combination. Then, for example, factors of safety of 3.0, 2.0, and 1.0, respectively, are used to determine the concrete's maximum allowable compressive stress (19).

A typical usual load combination for a gravity dam is based on normal design reservoir surface elevation, representative ice load, and normal tailwater elevation. An unusual load combination might include maximum reservoir elevation, maximum ice load, and minimum tailwater elevation. The extreme load combination would be the usual load combination plus the loads resulting from the maximum credible earthquake for the region. The "maximum credible earthquake is one having a magnitude usually larger than any historically recorded event" (8).

To determine the allowable stresses, the factors of safety are applied as follows: The maximum allowable compressive stress for the concrete should be the specified compressive strength of the concrete divided by the safety factor (here the safety factor values are 3, 2, or 1 depending on the type of loading combination). According to Golzé (8), in no case should the compressive stress exceed 1500 psi for the usual load combination or 2250 psi for the unusual load combination. As noted earlier, usual practice is to design the dam so that tensile stresses do not occur in the concrete. More specifically, for the case of usual loading, tensile stresses are not allowed; however, for the cases of unusual or extreme loading, some tensile stresses are allowed (8).

THE DESIGN PROCESS   In designing a gravity dam, the usual procedure is to first choose a shape of the section of the dam based on previous experience with similar dams. Then, a stress and stability analysis of the structure is made

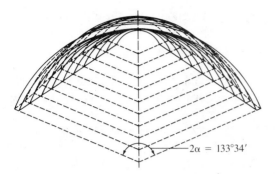

Figure 6-11 Arch dam geometry — plan view

to see if the structure passes the stability tests. If the dam is not stable or if stresses are too extreme, the section is reshaped to improve the design. Likewise, if the initial shape produces stresses much below the allowable limits or much safer than need be for stability, reshaping is also done to produce a more economical design. That process is continued until a satisfactory design is achieved.

Some guidelines for initial shaping of the section are (1) the upstream face is usually made vertical or near vertical, and (2) the downstream face usually has a slope of about 0.75 horizontal to 1.0 vertical.

Dams Categorized According to Special Features

Some types of dams have gotten their names from the distinguishing features that set them apart from the dams we have already discussed. These are arch dams and buttress dams.

ARCH DAM The arch dam is usually built in narrow canyons where the abutments are of massive sound rock so that the horizontal load acting on the dam (water pressure force) may be safely transferred to the canyon walls by arch action. A rule of thumb is that if the ratio of the width at the top of the dam to the height of the dam is less than 5, then an arch dam should be considered for the site (8). Figure 6-11 is a plan view showing the basic geometry of an arch dam. A dam like this is called a constant angle arch dam because the 2α is constant (approximately 133.6°).* Because the width of the canyon increases from the bottom of the dam to the top, the arch radius will also have

---

* It can be shown that for $2\alpha = 133.6°$, the volume of concrete needed for the arch is minimum. This assumes all the force is transferred to the canyon walls.

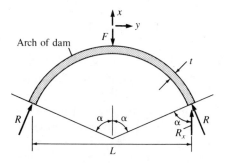

Figure 6-12   Forces on a single arch

to increase with height, as Fig. 6-11 shows. A simple approach to designing an arch dam is to divide it into several horizontal slices (separate arches) and to analyze each of these from the standpoint of hoop stress produced by the water pressure. For example, consider the arch shown in Fig. 6-12, where the width between the canyon walls is L. The horizontal component of water force acting downstream on one half of the arch will be resisted by one of the reaction components R_x, or

$$-\gamma h\, \Delta h\, \frac{1}{2} L + R_x = 0$$

where h is the depth of water above the arch, and Δh is the thickness of the horizontal slice of arch being analyzed.

But $R_x = R \cos (90° - \alpha)$

therefore,

$$R = \frac{\gamma h\, \Delta h (L/2)}{\cos (90° - \alpha)} \tag{6-5}$$

Also $R = \sigma t\, \Delta h$ $\tag{6-6}$

where σ is the normal stress in the concrete, and t is the thickness of the arch. Then, eliminating R from Eqs. (6-5) and (6-6) yields

$$t = \frac{\gamma h(L/2)}{\cos (90° - \alpha)\sigma} \tag{6-7}$$

By applying Eq. (6-7) from the crest of the dam to the bottom, we can determine how the thickness of the dam should vary from top to bottom. Of course, an extremely thin dam at the top is undesirable; therefore, the minimum top thickness is usually taken as about $L/60$, where L is the length of arc of the crest of the dam (4).

Equation (6-7) is derived by assuming the arch is a complete cylindrical shell; that is, no transverse stress is transmitted from one arch to the other. However, in the arch dam, this is not the case because as the load is applied (as the water surface in the reservoir rises), the upper arches deflect (the central portion moves downstream). The base of the dam, however, is fixed; therefore, because of this differential deflection, stress will always develop between the arches under full load. This all points to the fact that a much more sophisticated method of stress analysis is needed to design the most economical arch dam.

To actually design an arch dam, the engineer makes an approximate configuration of the dam based on the topography of the site and approximate formulas. Then, a stress analysis of the entire dam is made to reveal where unusually high or low stresses occur in the structure. This stress analysis is usually done using the finite element method. Once the stress analysis is made, changes in the original configuration are made to improve the stress distribution. Another stress analysis is done and any necessary further modifications are made. This process is repeated until a design is achieved that has (1) a reasonably uniform distribution of stress, (2) a compressive stress level throughout as near to the allowable limits as practicable, and (3) a minimum volume of concrete.

BUTTRESS DAM   A buttress dam is essentially a hollow gravity dam. Buttresses of reinforced concrete rest on the rock foundation and support a watertight sloping face of the dam. Figure 6-13 shows the general configuration of the nonoverflow section of a buttress dam. The facing of this dam is a flat slab. Other types of facings include the *round head deck* (Fig. 6-14) and the *multiple arch deck* (Fig. 6-15). The gravity part of the stability of this dam comes

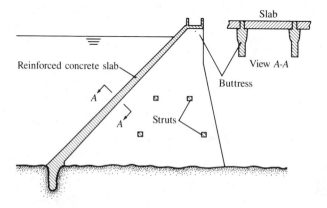

Figure 6-13   Nonoverflow section of a buttress dam

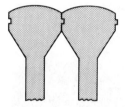

Figure 6-14 Section through round head deck of a buttress dam

Figure 6-15 Section through multiple arch deck of a buttress dam

from the dead load of the concrete in the buttresses and face and, more important, from the weight of the water above the sloping face. The spillway part of the dam consists of a reinforced downstream face that is also supported by the buttress (see Fig. 6-16).

The main advantage of the buttress dam is that it needs as little as 30 or 40% of the concrete needed for a solid gravity dam (4). However, this advantage is usually offset by the added labor costs of building forms and placing reinforcing steel. Moreover, in areas subject to freezing temperatures, the face slabs of buttress dams have been known to deteriorate from this type of weathering.

The flat-slab face is well suited for low buttress dams. For high buttress dams, however, it is difficult to satisfactorily transmit the large slab load to the buttress without creating serious stress concentrations. Therefore, the multiple arch or round head types are most often used on high head dams.

Figures 6-17 and 6-18 are examples of buttress dams built by the U.S. Bureau of Reclamation.

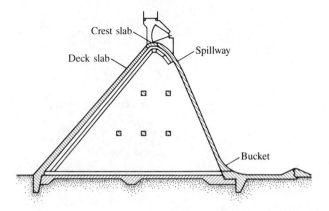

Figure 6-16 Spillway section of a buttress dam — elevation view

Figure 6-17   Bartlett Dam in Arizona — multiple arch
buttress (Courtesy of U.S. Bureau of Reclamation)

Figure 6-18   Stony Gorge Dam in California — slab
buttress (Courtesy of U.S. Bureau of Reclamation)

6-3 River Diversion

Whenever a dam is to be built across an existing river channel, the river must be diverted so that construction can be done. The manner in which the diversion is accomplished depends on the kind of dam being constructed, the character of the site, and the characteristics of the streamflow. We discuss several common methods of diversion and present the factors that favor one type of diversion over other types.

Two-Stage Diversion Using Cofferdams

In the construction of concrete gravity dams, the two-stage diversion is often used. During the first stage, a cofferdam is constructed across about half the channel, and the flow is diverted to the other half. Then the area inside the cofferdam is dewatered, and part of the dam is built inside this cofferdam (see Fig. 6-19a).

Part of the first-stage dam construction is often completed only up to a limited height, thus leaving a gap (Fig. 6-19b) in the dam that the river can flow through in the second stage of construction. A temporary diversion tunnel also may be included in the first-stage construction through which flow can be diverted in the second stage.

In the second stage, the first-stage cofferdam is removed and construction of part of the second-stage cofferdam is started. At this time, the flow is restricted to a small opening in the second-stage cofferdam (Fig. 6-19b). In a large river, cutting off the flow through the second-stage cofferdam is a critical operation because the flow velocity through the opening in the already constructed cofferdam frequently will be large and will increase as the opening becomes smaller. The process of stopping this flow is called the *closure* operation. Often this closure is effected by dumping large rock or concrete tetrahedrons into the restricted channel.* As the rocks or tetrahedrons are dumped, the bottom of the channel becomes progressively higher in elevation and the upstream water level rises. Finally an upstream level will be reached such that the flow will be diverted through a gap or diversion tunnel of the structure that was built inside the first-stage cofferdam. After the closure is completed, a more impervious blanket of fine material (small rocks and gravel) is placed on the upstream part of the closure section to reduce seepage through the large rocks or tetrahedrons of the closed-off section. When the seepage is reduced to a low level, the rest of the second-stage cofferdam can be completed and construction within the area of the second-stage cofferdam can be started (Fig. 6-19c).

---

* For example, 12-ton tetrahedrons were used in the closure operation of McNary Dam on the Columbia River.

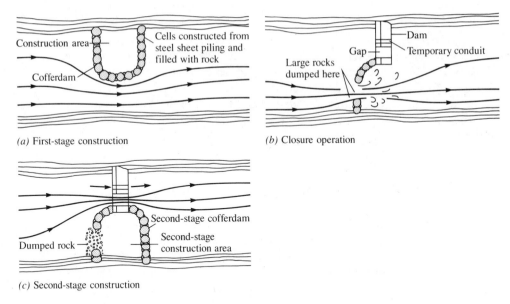

(a) First-stage construction

(b) Closure operation

(c) Second-stage construction

Figure 6-19   Two-stage diversion scheme

After construction is completed inside the second-stage cofferdam, the cofferdam is removed and flow is diverted through that part of the dam (spillway or powerhouse section). Then construction is completed on the part of the structure only partially completed during the first stage of construction. That is, the gap in the first-stage construction is filled, or the temporary diversion tunnel is plugged.

Cofferdams are often of the cellular type, in which linked steel piles are driven in a cellular pattern, as shown in Fig. 6-20, and each of the cells is filled with rock to provide stability against overturning. These cofferdams are also very durable in case they are overtopped during flood season. Figure 6-20 shows how the steel piles are linked.

Scheduling the different phases of construction of the cofferdams must be synchronized with the normally expected variation in streamflow. For example, the first cofferdam will normally be built during a low-flow period, and then the stream will be diverted to the completed structure during another low-flow season. Thus depending on the stream characteristics, the designer may have to think in terms 1-yr time increments between low-flow seasons.

The designer also will have to decide how high to design the cofferdams. Normally, they are not made so high that they would never be overtopped by the design flood for the dam. The designer must balance the added cost of a very high cofferdam against the damage that would result from its overtopping. Damages might include the costs due to work stoppage and cleanup. For major dams, designing the cofferdam so that it will withstand a 20-yr flood is common.

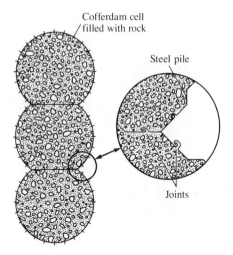

Cofferdam cell filled with rock

Steel pile

Joints

Figure 6-20   Details of cellular cofferdam — plan view

Tunnel Diversion Using Cofferdams

If the dam is to be built in a very narrow canyon, the usual case for arch dams, then the two-stage diversion described above may not be suitable because of the limited working space in the canyon. A common procedure is to excavate a tunnel through one abutment and then build cofferdams upstream and downstream of the damsite. When the cofferdams are closed, the water is diverted through the tunnel, and then construction of the dam can take place inside the cofferdams (see Fig. 6-21).

In designing arch dams, it is fairly common to include a tunnel as part of the spillway. In these cases, part of the spillway tunnel can usually be used as part of the diversion tunnel. The spillway tunnel often has an intake at an elevation near the crest of the dam, which would be too high for diverting flow during construction. To use the spillway tunnel, the typical procedure is to construct a

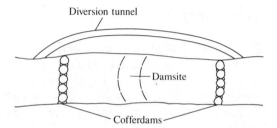

Diversion tunnel

Damsite

Cofferdams

Figure 6-21   Diversion in a narrow canyon

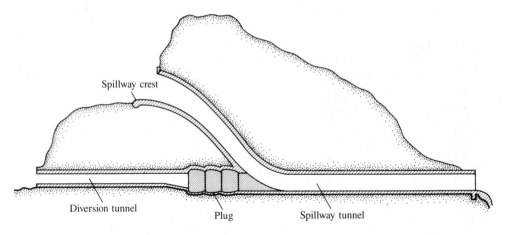

Figure 6-22   Typical arrangement for using part of a
spillway tunnel in a diversion scheme

separate diversion tunnel, which is ultimately connected to the spillway tunnel at
the lower level. When the dam is completed, the diversion tunnel is plugged, as
shown in Fig. 6-22.

Flumes

If a dam is to be built on a fairly small river in a wide canyon, it may
be possible to divert the flow away from the construction area by means of a
separate channel, or *flume*. A flume, which is often made of a steel or timber
frame with a timber lining, usually diverts the flow around the damsite or over
a low block in the dam. As construction proceeds, the flume can be moved to
another area. Figure 6-23 shows a diversion flume used on the construction of
the Canyon Ferry damsite in Montana. In Fig. 6-23, also note the steel sheet
piling used to guide the flow into the flume. This is the same type of piling
used to construct the cellular cofferdams.

Diversions for Earth Dams

In the case of concrete dams, if the cofferdam is overtopped, some
cleanup work will be required and some repairs may have to be made; however,
the cost for this work should not be exorbitant. If cofferdams for an earth dam
were overtopped, it is possible that all the work done to that date could be
wiped out; therefore, the diversion scheme is much more important for earth
dams than for concrete dams. That is, the cost of overtopping a cofferdam
protecting an earth embankment is usually much more than that for a concrete
dam. Therefore, a much lower risk is usually assumed when designing the diver-

Figure 6-23   Flume diversion

sion scheme of an earth dam; it must be designed to accommodate very large floods. A thorough study of the meteorology of the region and the hydrology of the stream will help schedule critical operations such as closure to reduce the risk of failure.

6-4   Dam Safety

Because the sites for future dams are usually less suitable (canyons not as narrow, weaker foundations) than the sites already used, the potential for problems and even failure is perhaps greater now than before. However, as more and more experience is gained in the construction and operation of dams, engineers are finding better ways to handle potentially serious problems. Some of this experience comes from studying dams that have failed. This part of the text examines the various problems that can lead to the failure of a dam and methods of guarding against the occurrence of such problems are discussed.

In Section 6-2, we addressed the safety of the concrete gravity dam, which is designed so that it will have a certain factor of safety with respect to sliding, overturning, and development of maximum stress. Also, if earthfill dams are well compacted and their slopes are designed and constructed to be consistent with the strength (found from laboratory tests) of the materials from which they are built, then their embankments should be safe against failure from sliding or excessive settlement. These kinds of failures are fairly obvious and it is fairly easy to control the factors involved in these failures. Most failures have to do with foundations and seepage problems (9). Problems with spillways (inadequate capacity or erosion of outlet works), erosion of embankments, and massive slides upstream into the reservoir also have contributed to or caused failures in dams. These common causes of failures can be prevented by careful analyses and proper design and construction. We now discuss measures that can be taken to prevent these failures.

Piping

All earth dams with a central core of compacted earth have some seepage through them. If seepage occurs without dislodging and removing soil particles, no damage will result. However, if soil particles are washed away in the seepage, severe problems may develop. *Piping* is the term given to concentrated leaks that erode surrounding material (soil particles) along the path of leakage. The flow passage enlarges and leakage increases until a serious problem develops and failure possibly occurs.

Piping was the cause of failure of the Teton Dam in Idaho on June 5, 1976. This was a central-core zoned earthfill dam having a height of 305 ft, and the dam failed during the initial filling of the reservoir. More than a score of people died in the ensuing flood, and property damage was estimated at $400 million. Piping was first observed during the morning, and by noon, internal erosion had progressed so far that the top of the dam collapsed.*

Piping can develop in earth dams where the filter between fine soil and coarse material in the embankment is not properly designed to prevent the movement of the fine particles. Thus careful design *and* careful construction of the filter zones are required for safety against piping. All filters should be constructed from screened sands and gravels (9).

Piping can also occur where the fill material joins the abutment material or where it joins a solid structure such as an outlet conduit. If the fill material is not carefully placed and hand tamped next to surfaces of discontinuity, the less consolidated material may be the starting point for piping. One way of reducing the possibility of piping around buried conduits is to construct collars around the conduit, as Fig. 6-24 shows.

In theory, the collars tend to block any flow passages along the exterior of the conduit, and the distance the seepage water must travel is increased by the collars. However, proper compaction is difficult to achieve around these collars, and some authorities discount their effectiveness.

---

* For more details on the cause of the dam's failure, see Jansen (9).

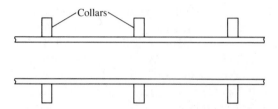

Figure 6-24   Collars around conduit to eliminate piping

When earth dams are built on compressible alluvial soils, the possibility exists that differential settlement of the embankment will occur as the reservoir is filled and the embankment and foundation material become saturated. If differential settlement of the embankment occurs, cracks may develop in it, which may lead to excessive leakage and piping. Sherard (15) has indicated that certain gradations of soil materials are especially susceptible to cracking.

In small earth dams, burrowing animals (muskrats and ground squirrels) can be a souce of piping problems. Therefore, regular inspections should be made of the embankment at low pool levels and if burrows are found, action must be taken to fill them and to prevent new ones from forming.

Piping can be detected by observing the leakage that may be occurring from the downstream side of the dam. If the leakage is in the form of clear water, piping is not occurring. If, however, the leakage is muddy, piping is to be suspected, and one must assume the dam is in danger of imminent failure. Piping that develops along cracks that extend to the exterior of the dam can sometimes be stopped by trenching along the crack and backfilling with compacted impervious soil (15).

Once piping occurs, rapid remedial action is usually called for. First, the reservoir pool should be drawn down as fast as possible by discharging water through outlet conduits. After the pool level has been drawn down, repairs can be made to the defective parts of the embankment or foundation. In the design of the dam, the outlet works should be large enough so that the reservoir can be drawn down quickly.

Foundation Problems

The weight of the dam itself must be carried by the foundation; therefore, the designer must be aware of the character of foundation material, including how impervious it is to seepage. Information about the foundation is obtained by geologic investigations that may include core drilling, geophysical tests, and geologic mapping of the area. If foundation exploration is deficient, the designer cannot anticipate possible difficulties, and the dam may develop problems relating to differential settlement, sliding, and leakage. If the foundation is composed of water soluble rock, the foundation could also develop solution cavities, become weak, and eventually cause dam failure. This condition contributed to the failure of the St. Francis Dam in California in 1928. This was a 205-ft high concrete dam (9). The best way to avoid foundation problems is to conduct a thorough geologic survey of the site and to make certain that the resulting information is studied by competent geologists and engineers so that a complete understanding of the foundation is obtained. Once that has been achieved, a safe design can be executed, or the site may be found unsuitable for economical development.

Slides on Reservoir Slopes

When a reservoir is filled or drawn down after filling, physical changes (saturation of base material, development of excessive pore pressure, added weight to reservoir foundation) of the earth in and around the reservoir can lead to slides when unstable materials exist. Slides can produce waves that completely overtop the dam, as occurred at the Vaiont Dam in Italy in 1963. In this particular disaster, about 315 million cu yd of earth slid into the reservoir and caused a 300-ft wave of water to wash over the crest of the 869-ft high dam. About 2600 people died from the flood wave that traveled down the canyon and devastated everything in its path (9). The basic concrete structure of the dam itself was not damaged!

Geologic investigations should be made of the reservoir area to locate potential slide hazards. According to Jansen (9), "the potential for landslides may exist in nearly any kind of rock, some slates and schists are notoriously susceptible to movement." If a slide hazard exists, special precautions may have to be taken to alleviate or design against the event. An alternative may include changing the site of the dam.

Spillway Problems

Dams may be overtopped if the spillway capacity is not great enough to carry the flood flow. This is especially serious for earthfill dams because of their susceptibility to erosion during overtopping. The most common cause for inadequate spillway capacity is underestimating the peak flow rate or volume of the design flood. Only competent, experienced hydrologists should be given the responsibility of determining the design flood. The design flood should be carefully chosen in the light of potential hazards resulting from dam failures. In Chapter 2, (Sec. 2-7, page 78) we provide more details on design floods.

Failure of the spillway itself has occurred in several dams. One common cause of spillway failure has been inadequate design of the stilling basin. In the early years of dam construction in the Midwest, this type of failure occurred on several concrete gravity dams where the spillway aprons and cut-off walls did not adequately prevent erosion of the foundation material. Erosion at the end of chute spillways has also caused failures. Once a structure such as the end section of a chute spillway is undermined and it starts to drop into the eroded cavity, the process is irreversible. Accelerated erosion proceeds up the slope under the channel until the entire spillway section is destroyed. To prevent this kind of failure, the stilling basin must be large enough to dissipate the energy of the high-velocity water from the chute, adequate wing walls must be built at the end of the chute to eliminate possible undermining, and riprap of adequate size and gradation must be used in the stilling basin itself to prevent erosion of the base material.

Chute spillways have also failed from other causes such as earth slides obstructing the channel and misalignment of joints of the spillway floor sections, which can lead to cavitation and eventual destruction of the spillway.

Design and Inspection

In the preceding sections, we have briefly referred to the necessity for thorough foundation exploration by engineering geologists, design by competent professional engineers, and careful construction practices. The philosophy, organization, and action required for *safe* design and construction of a dam has been most effectively summarized by Jansen (9):

> Safety of dams requires consideration of more than the technical factors. Looking at the organization, for example, one thing which must be assured is that all voices are heard. Ideas may come from within the organization — from nearly any level — or from outside. The latter includes, of course and particularly, consultants. One of the greatest hazards in the engineering of major structures is the exclusion of the ideas of those who may have valuable contributions to make. The management of any organization must exert special effort to assure that this does not happen.
>
> In case histories of projects gone wrong, the dominance of single decisionmakers — sometimes authorities whose reputations for expertise were well earned — is not uncommon. Even experts can make mistakes, and probably the worst is to assume that an expert's judgment need not be questioned by those qualified to question.
>
> Another consideration, especially in large organizations composed of many compartments, is to assure that information flows among the units. The many ideas essential to good engineering must be shared freely across the internal boundaries. The integration of separate efforts should be continuous throughout the evolution of designs, rather than simply gluing together individual final products. This means that designing must start with a general perspective and then focus on the individual parts — not vice versa.
>
> There must be recognition of the inseparable relationship of design and construction. These functions are best considered as a single

process. Design is not completed until construction is accomplished. Designers and construction engineers have to work in concert during design and while the dam is being built so that the site conditions disclosed can be weighed against design objectives. Any necessary modifications in design during this period should be a collaborative effort of designers, geologists, and construction engineers.

The vital relationship between the engineer and the geologist needs continuing emphasis. They must work as closely together as must the dam and its foundation. . . .

The tendency to think of averages in the engineering of dams must be resisted. Failures occur where the dam or its foundation is weakest, not where it is in average condition. The design must focus on the potential weaknesses. Exploration and testing will necessarily depend on sampling techniques, with results varying sometimes over a wide range. Natural materials available for construction may exhibit average characteristics that meet requirements; yet, they may be judged totally unacceptable when their variations are considered.

The variability of natural conditions does not encourage unreserved faith in standard guidelines or "cookbook" approaches to dam engineering. No matter how many exploratory holes are drilled or how many samples are tested, the reservoir site may still contain surprises — and these may appear at any time during the life of the structure.

Flexibility is the key. Rigid criteria are useful only as long as conditions match the underlying assumptions. They fail when deviations are not perceived or when the latitude and the judgment are not available to make the necessary adjustments. The trouble with a "cookbook" is that some of its users may come to think that it contains all the recipes. Design by the book is especially hazardous in an organization insulated from professional interchange.

Once the dam is completed, surveillance of the dam and reservoir is required to guard against potential hazards over the years of operation. Surveillance includes

1. *Periodic visual inspection by experienced engineers.* These inspections will reveal uneven settlement, cracks, discoloration, or increase of seepage and embankment sloughing.
2. *Monitoring of instruments.* The design of all major dams includes installing instruments that will reveal changes within the dam. For example, sensors should be included in gravity dams to allow detection of structural and foundation movement as the reservoir is filled, and piezometers should be included in earth dams to reveal any changes in the pressure field within the dam and its foundation. Any anomaly that shows up in the pressure field should be a warning of possible abnormal leakage problems.
3. *Interpretation of information.* Information obtained from field inspection and measurements from instruments will be virtually worthless unless intelligently analyzed by experienced people. Observed data must be analyzed for deviations from reading to reading and for slow trends that may have

subtle meanings. The implications of long-term changes are sometimes overlooked, for example, serious piping may take years to develop. Thus a change in pressure gradient (by observing piezometers) may reveal the progressive development of seepage channels that have less resistance to flow than a well-compacted embankment.

Employment of Qualified Personnel

This aspect of dam safety is succinctly presented by Golzé (8):

> The use of qualified personnel is another basic element. The design must not only be by competent engineers but should be supervised by registered or licensed engineers for the State in which the structure is to be located. In like manner, construction is a professional undertaking which, again, for maximum performance and safety, should be supervised and directed by graduate registered or licensed engineers. The operation of completed facilities should be done by men skilled by years of experience working under the supervision of one or more licensed engineers qualified in this particular field.
>
> In connection with the use of qualified personnel which may either be persons in the employ of the owner or persons employed by firms of private engineers, independent consultants should be available at all stages of design and construction of the facility to advise and counsel the engineers in charge. In connection with the periodic and special inspections of the structure, independent consultants should likewise be employed. Their advice to the dam owners concerning the physical condition and the efficiency of his facility brings to bear a professional judgement supported by years of experience.

Governmental Controls

Besides the safeguards we have already presented, which are generally under the control of the owners and designers of the project, state and federal statutes dictate certain actions that must be taken to ensure a safe project. All major hydropower projects must be licensed by the federal government, and all these (currently more than 400 projects) must be inspected periodically by the Federal Energy Regulating Commission (FERC). These inspections are made at individual dams at least every five years, and if deficiencies are found, the owners of the project must advise FERC within 30 days of the corrective measures they plan to take.

Many states also have enacted laws that exercise control over the design, construction, maintenance, and operation of dams and reservoirs.*

---

* For more information on statutes governing the safety of dams, see Golzé (8). For more details on all aspects of dam safety, see Jansen (9).

6-5   Reservoirs

A reservoir is a manmade lake or structure used to store water. Inherent in the definition of a reservoir is that people have the major control over the use of water in it.

For one type of reservoir, for example, elevated tanks used in municipal water supply systems, the inflow to the reservoir is completely controlled, but the outflow is primarily dictated by consumers' needs and desires. For a system like this, the source of supply to the reservoir may be a river from which water may be withdrawn as needed. In this kind of system, the reservoir makes it possible to use pumps of moderate size pumping into the reservoir at a fairly constant rate. During peak demand periods, the reservoir is drawn down, for during these periods, the outflow rate is much more than the inflow provided by the pumps. If reservoirs were not included in this system, many more pumps or pumps having a much greater range of discharge would be needed. The reservoir also supplies water needs during emergencies such as power failures, fires, or pipe ruptures.

Another type of reservoir, for example, one created by damming a stream, has an uncontrolled inflow* but a largely controlled outflow. Natural occurrences and environmental factors are important in the design and operation of this type of reservoir. For example, the water available for storage is totally a function of the natural streamflow that empties into it. Moreover, because the streamflow downstream of the reservoir will be altered by operation of the reservoir, changes in the stream environment (often detrimental) will usually occur.

Other questions that have to be answered in the design and construction of a reservoir on a stream are

1.   What height of dam will be needed to yield the desired objectives of the reservoir?
2.   Will leakage from the reservoir be a problem?
3.   Will evaporation significantly affect the yield from the reservoir?
4.   Will there be any problems regarding stability of the earth around the reservoir (slides) when the reservoir is filled and operated?
5.   Will incoming sediment be a problem?
6.   What is involved in preparing the area of the reservoir (tree clearing, soil removal)?

Other concerns that relate more to social problems are

1.   Relocation of utilities and transportation facilities such as roads and railroads.

---

* This assumes the reservoir is on a stream that has no other reservoirs or other kind of control upstream.

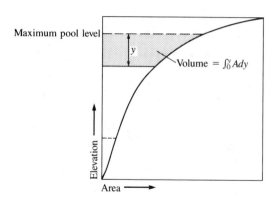

Figure 6-25   Area versus elevation for a reservoir

2. Purchase of land for the reservoir and condemnation of property owned by people who resist selling.*
3. Downstream pollution problems.
4. Development of recreation sites around the reservoir.
5. Impact on natural resources and mitigation measures required.

Some of these topics will be discussed in more detail in the following sections.

Reservoir Capacity

Reservoir capacity is the volume of water that can be stored in the particular reservoir. In the case of manmade tanks, it is simply the inside volume below the maximum water surface level in the tank. Likewise, in a reservoir behind a dam, it is the volume in the reservoir below the *normal maximum* pool level. This can be calculated by using a topographic map of the region. First, the area inside different elevation contours within the reservoir is measured, then a curve of area versus elevation can be constructed as shown in Fig. 6-25. At any given elevation, the increment of storage in the reservoir at that elevation will be $A\,dy$, where dy is a differential depth. Then the total storage below maximum pool level to any depth will be given by $\int_0^y A\,dy$. In other words, the shaded area in Fig. 6-25 will be the total storage between the maximum pool level and depth y.

The *normal maximum pool* level is the maximum possible level in the reservoir when the spillway gates are closed (level at the top of the gates). However, if there are no gates on the spillway (uncontrolled spillway), the normal maximum pool level is assumed to be at the crest of the spillway. The minimum

---

* Condemnation proceedings apply only to government projects and to certain private power projects.

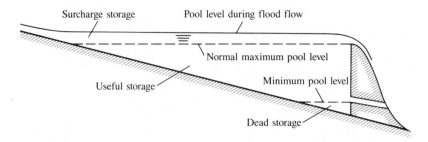

Figure 6-26   Storage relations for a reservoir with an uncontrolled spillway

pool level is usually taken as the lowest level at which flow can be released from the reservoir. This is the invert elevation of the lowest outlet pipe, as shown in Fig. 6-26. The *useful storage* in a reservoir is the storage between normal maximum pool level and minimum pool level. The storage below minimum pool level is called *dead* storage. When water is being discharged over the spillway, the reservoir pool may actually rise above the level of normal *maximum* pool level. This storage above normal maximum pool level, called surcharge storage or flood storage, is important in routing a flood through the reservoir (see Sec. 4-5, page 217).

The foregoing discussion pertains to the storage available because of the volume above ground surface in the reservoir. Besides this obvious storage, some water (as much as 2 or 3% of useful storage in some cases) will be stored in the soil and rocks of the banks of the reservoir. This stored water is called *bank storage*. If the bank material is quite porous, bank storage may be useful; it will be released when the pool is drawn down. However, if the bank material is relatively impervious and if the reservoir is normally drawn down rapidly, the water in bank storage may not drain into the reservoir fast enough to be useful.

How to Determine What Storage Capacity Is Needed

When an elevated tank, such as in a city water system, is used to augment peak flow demands, one can determine the needed pumping capacity and storage volume for a given demand by a simple numerical calculation. Example 6-1 illustrates the procedure.

E X A M P L E   6 - 1   Table A shows the average water demand for each hour of a common 24-hr working day of a small city.

A pump takes water from a well and delivers it to a reservoir from which the water system is supplied. Based on the demand data, what reservoir volume

Hour (ending at)	1 A.M.	2	3	4	5	6	7	8	9	10	11	12N
Q (gpm)	800	800	900	1000	1200	1425	1900	2200	2000	1575	1600	1700

Hour (ending at)	1 P.M.	2	3	4	5	6	7	8	9	10	11	12M
Q (gpm)	1500	1300	1400	1600	1800	2300	2100	1500	1100	900	800	800

Table A

is required and what pump capacity is needed if the pump is to operate continuously for the 24-hr period?

SOLUTION

(1) Hour (ending at)	(2) Q_{system}	(3) Q_{pump}	(4) Q_{reserv}	(5) $\sum Q_{reserv} \Delta t$ (gals)
1 A.M.	800	1,425	−625	
2	800	1,425	−625	
3	900	1,425	−525	
4	1,000	1,425	−425	
5	1,200	1,425	−225	
6	1,425	1,425	0	0 (reserv full)
7	1,900	1,425	475	28,500
8	2,200	1,425	775	75,000
9	2,000	1,425	575	109,500
10	1,575	1,425	150	118,500
11	1,600	1,425	175	129,000
12N	1,700	1,425	275	145,500
1 P.M.	1,500	1,425	75	150,000
2	1,300	1,425	−125	142,500
3	1,400	1,425	−25	141,000
4	1,600	1,425	175	151,500
5	1,800	1,425	375	174,000
6	2,300	1,425	875	226,500
7	2,100	1,425	675	267,000
8	1,500	1,425	75	271,500 (peak withdrawal)
9	1,100	1,425	−325	252,000
10	900	1,425	−525	220,500
11	800	1,425	−625	183,000
12M	800	1,425	−625	145,500

Table B

First, determine the average pumping rate. It will be equal to the average of the demand rate:

$$Q_{\text{pump}} = \frac{\sum_{i=1}^{24} Q_i}{24}$$

where Q_i is the demand discharge as given in Table A.

$$Q_{\text{pump}} = \frac{34{,}200 \text{ gpm-hr}}{24} = 1425 \text{ gpm}$$

The required storage is obtained by setting up Table B. Column 1 is the end time of the period in question, column 2 is the system demand discharge, column 3 is the pumping rate, and column 4 is the difference between the system demand rate and the pumping rate ($Q_{\text{system}} - Q_{\text{pump}}$). Therefore, column 4 gives the rate of reservoir filling if the values in this column are negative, or the rate of reservoir emptying if the values are positive. Assume the reservoir is full at 6 A.M. after the night of low demand when the reservoir is being filled. Then, by summing the $Q_{\text{reserv}} \Delta t$ from that time throughout the day (column 5), it is found that the greatest value of $\Sigma Q_{\text{reserv}} \Delta t$ occurs at 8 P.M., and the value of $\Sigma Q_{\text{reserv}} \Delta t$ at that time is the maximum volume of water that had to be drawn from the tank to satisfy the demand. Thus the storage required is 271,500 gallons.   ■

One common method of storage requirement analysis for a reservoir on a natural stream uses the mass diagram of flow in the stream entering the reservoir. That is, by constructing a curve of accumulated volume of flow versus time and analyzing it, one can determine the necessary storage to yield a desired dry-period flow, or given a certain volume of storage, one can determine what dry-period flow can be achieved. Consider the mass diagram of a stream as shown in Fig. 6-27. The ordinate is the total volume of flow passing the given station starting from an initial time (in this case, October 1, 1974, the beginning of the 1975 water year).* By definition, the discharge Q is volume rate of flow, or $d\Psi/dt$; therefore, the slope of the mass curve at any point on the curve is the river discharge at that time. Thus the steepest parts of the curve indicate flood periods, and the flattest parts indicate low-flow or dry periods. To determine storage needs, one can scan the streamflow records for all years of record and pick out the severest period so far as drought conditions are concerned. That period is then plotted as a mass curve and analyzed. In the next paragraph, we give a detailed description of how such an analysis is made.

---

* The volume unit used for the ordinate of Fig. 6-27 is sec-ft-mo or sfm which is the abbreviation for a flow of 1 cfs for a period of 1 month. Thus for a 30 day month 1 sfm = 1 ft^3/s × 30 da/mo × 24 hr/da × 3600 sec/h = 2.592 × 10^6 ft^3.

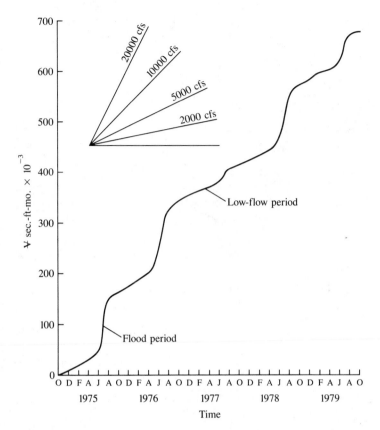

Figure 6-27   Mass diagram for the Salmon River
at White Bird, Idaho, from October 1974 to
October 1979

Figure 6-28 is the mass curve for the Salmon River for one of the driest
periods of record, and it is to be analyzed to determine what reservoir storage
would be needed to produce a minimum rate of flow of 6000 cfs downstream
of the reservoir for that period. In the upper-left corner of Fig. 6-28, a chart
shows the slopes that represent discharges from 2000 to 15,000 cfs. Now, focusing on the mass-flow chart of Fig. 6-28 in the June–July period of 1976, we
can see that in June the curve was very steep ($Q = 45,000$ cfs); then, from August
1976 to April 1977, it flattens out (in February 1977, the discharge was only
about 4000 cfs). During this period, it is obvious that water from storage in
the reservoir is needed to augment the flow. To determine the amount of storage
needed to maintain the flow at 6000 cfs, we draw a line with a slope representing
6000 cfs tangent to the mass curve of the 1976 summer season, as shown in
Fig. 6-28. The point of tangency (point a) of that line is the time that the natural

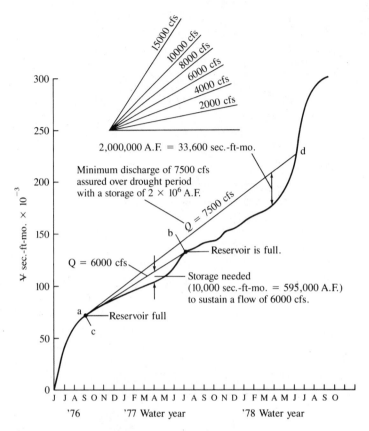

Figure 6-28 Storage-discharge relations for Salmon
River at White Bird, Idaho, for 1977 and 1978
water years

streamflow went from a discharge greater than 6000 cfs to a discharge less than
6000 cfs (the slope of the mass curve drops below 6000 cfs). Then, as long as
the natural streamflow slope is less than the slope of the 6000 cfs line (the
6000 cfs is the discharge being released from the reservoir), it is obvious that
the discharge leaving the reservoir is greater than that coming in so that stored
water is being used up. The maximum amount of stored water needed is given
by the maximum vertical spread between the 6000 cfs line and the mass curve,
as Fig. 6-28 shows. In this case, the maximum spread between two curves is
10,000 s-ft-mo. If the reservoir were full when releases from the reservoir were
started (point of tangency), the lowest pool level would occur where the greatest
spread between the two lines (6000 cfs line and mass curve) occurs, and after
that instant, the reservoir would begin to fill again. When the two curves cross

at point *b* (in June 1977), the reservoir would again be full, and thereafter excess water would be discharged over the spillway or released to the river through outlet works. This sort of analysis can be done for the next low-flow period (1978 water year) to see what storage would be needed, and so forth. Thus if the object is to get through a drought period with a constant flow rate of 6000 cfs, we can carry out this kind of analysis to determine the amount of storage needed, and a dam could be designed and built to the required height to provide that storage capacity in the reservoir.

Of course, we can easily turn the problem around and pose the problem this way: If *y* acre-ft of storage is available, what flow could be assured through the dry period? Assume we are going to analyze the 1977–1978 period. Then, by moving the vertical line representing the storage available between line *cd* and the mass curve until line *cd* reaches its minimum slope, the slope of *cd* then is the discharge that could be assured for that period for the given storage. In this case, a flow of 7500 cfs would be assured with a storage of 2,000,000 acre-ft. Note also that reservoir withdrawal would occur over a 2-yr period because of the drought's duration.

We have presented the foregoing procedure for graphical solutions of the problem. However, it is easy to see that all the quantities and concepts can be set up on a computer for a much more rapid and accurate solution. The solution procedure was for the simplest of problems; in actuality, one would probably not require that the outflow be constant during the drought period, and one would also have to consider evaporation and perhaps leakage as other withdrawals from the reservoir. If the period of record of streamflow is short, it may also be necessary to generate long-term synthetic streamflow records that could be analyzed for the low-flow periods. The generation of synthetic records involves applying statistical methods and is part of the science of *stochastic hydrology*. One of the weakest features of the yield analysis using a mass curve of historical records is that the sequence of inflows can never be expected to recur. As a result, it is impossible to assign any probability to the expected yield. In Chapter 2, beginning on page 49, we provide some further considerations on the determination of yield based on return period.

EXAMPLE 6-2   For the conditions shown in Fig. 6-28, what flow could be assured for this period if a storage of 1 million acre-ft were available to draw on?

SOLUTION   First, convert 1 million acre ft (A.F.) to s-ft-mo. (sfm) to have storage units consistent with Fig. 6-28.

$$1 \text{ sfm} = 1 \text{ ft}^3/\text{s} \times 30 \text{ da/mo} \times 24 \text{ hr/da} \times 3600 \text{ s/hr}$$
$$= 2.592 \times 10^6 \text{ ft}^3$$
$$= 59.504 \text{ A.F.}$$

Thus 1 sfm for a 30-day month is equivalent to 59.504 A.F. Then, 1×10^6 A.F. = $(1 \times 10^6 \text{ A.F.})/(59.504 \text{ A.F./sfm}) = 16{,}806$ sfm. With a storage of 16,806 sfm, it is found by trial and error that the greatest drawdown in the reservoir occurs around April 1978, and that water during the drought period is first drawn from the reservoir around September 1976. The sustained flow for this period is found to be about 6600 cfs.   ∎

Storage Allocations

INTRODUCTION   The storage in a reservoir is rarely allocated for just one use. Even though a dam may be built primarily for power production, consideration will almost always have to be given to operation of the reservoir to mitigate floods, to mitigate damage to fish and wildlife, and to be compatible to a degree to recreational activities on the stream. To satisfy these various demands for the reservoir storage, special terms have been adopted for each of these storage requirements. They are storage for *instream flow requirements* (fish and other aquatic life, recreation, wildlife, and navigation), *flood control, irrigation,* and *power production.*

CONFLICTING STORAGE DEMANDS   Storage requirements often conflict with one another. For example, for maximum power production, keeping the reservoir at full pool level is necessary to maintain the highest head possible. However, if the reservoir is to mitigate flood damage, it would have to be drawn down considerably over time before floods are expected so that a significant portion of the reservoir is available to store the flood waters and reduce the flood peaks. Another conflict can arise between power interests and recreational activity on the river below the reservoir. In many power systems, hydropower is used to supply peak load demand. Thus on many streams, the flow varies greatly throughout the day. Highly varying flows from such an operation can be a severe annoyance to fishermen and boaters on the river, and in several cases, lives have been lost because people have been stranded in places (such as on islands) where they could not escape the rising water.

The most pronounced conflict often develops between instream flow needs and irrigation requirements. On many watersheds, most or all of the water may be allocated for irrigation. In that case, the stream below the irrigation diversions will be depleted, and much or all of the fish and other aquatic life will be eliminated, navigation and recreational boating will be made impractical, and other forms of wildlife usually will be adversely affected. A severe situation like this may develop on streams with little or no storage initially developed on it. Thus dam and reservoir projects are often developed for these streams to help restore some of the benefits that existed before excessive withdrawals for irrigation. The design of such a project should include storage allocations for irrigation, instream requirements, and perhaps flood control.

When a project is being planned, decisions must be made about the allocation of water for the various demands. If this cannot be worked out amicably between the parties that have the various interests, legal action is often taken to resolve the matter. In any case, by the time the final design of the dam and reservoir are being carried out, there should be storage allocations for various needs. That is, specific volumes within the reservoir should be reserved for specific purposes. For example, the U.S. Bureau of Reclamation will assign specific volumes in the reservoir for *flood control, joint use,* and *conservation* (19). The top part of the reservoir is reserved for flood control, and the next level is reserved for joint use. The joint use part is assigned for flood control during a period of the year and for conservation during a different period of the year. The part of the reservoir below the joint use part is reserved for conservation, by which is meant all water used for irrigation, power production, municipal and industrial water supply, fish and wildlife, recreation, navigation, and for water quality enhancement.

Once the different storage capacities have been allocated, reservoir operating procedures must be developed. These procedures include the way and time of year the reservoir is to be drawn down to provide maximum flood control, the way the water is to be released when flood flows arrive, the way the water is to be released for instream uses, the power and irrigation demands, and so on. Obviously, much care and analysis is needed to develop operating plans for a multiple-use project. Often, linear program formulations systems analysis is used to develop the operating procedure that will optimize the system.* We present additional information on storage allocation for flood control in Chapter 9.

Wind-Generated Waves, Setup, and Freeboard

INTRODUCTION   Whenever wind blows over an open stretch of water, waves develop, and the mean level of the water surface may change. The latter phenomenon, called *setup* or *wind tide*, is significant only in relatively shallow reservoirs. When a dam is being designed, the crest of the dam must be made higher than the maximum pool level in the reservoir to prevent overtopping of the dam as the wind-generated waves strike the face of it. The additional height given to the crest of the dam to take care of wave action, setup, and possibly settlement of the dam (if it is earthfill) is called *freeboard*. In the next two sections, we discuss the factors that control setup and wave height.

SETUP   Consider the basin of water shown in Fig. 6-29. The solid line depicting the water surface is the case when no wind is blowing; the water surface

---

* For a more thorough discussion of reservoir system operation, see Louchs (10) and Toebes (16).

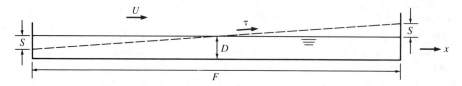

Figure 6-29   Definition sketch for setup

is horizontal. When the wind is blowing, a shear stress acts on the water surface, and because of this, the surface will tilt, as shown by the broken line in the figure. If one considers just the static forces relating to this problem, determining the amount of setup S is easy.* Consider the forces acting on the water in the basin in the x direction. Assume the basin has a length F in the direction of the wind velocity U, and assume the dimension normal to the page is ℓ. The forces in the x direction will be the shearing force produced by the wind and the hydrostatic forces on the ends of the basin. The force balance equation is

$$\sum F_x = 0$$

$$\frac{\gamma_w(D-S)^2\ell}{2} - \frac{\gamma_w(D+S)^2\ell}{2} + \tau_0 F\ell = 0 \tag{6-8}$$

but τ_0 can be given as

$$\tau_0 = \frac{C_f \rho_{\text{air}} U^2}{2} \tag{6-9}$$

where C_f is the average shear stress coefficient. Solving Eq. (6-8) for S yields

$$S = \frac{\gamma_a}{\gamma_w} \frac{C_f}{8} \frac{V^2 F}{gD} \tag{6-10}$$

where γ_a = specific weight of air
γ_w = specific weight of water
g = acceleration due to gravity

If one assumes γ_a/γ_w and C_f are essentially constant, Eq. (6-10) can be expressed as

$$S = K \frac{V^2 F}{gD} \tag{6-11}$$

---

* Actually the problem is more complicated than this because once the shear stress acts, circulation will start; however, this simple derivation illustrates the role of the primary variables.

Analysis of work by Saville (13) shows that a reasonable K value is 2.025×10^{-6} for reservoirs; therefore, Eq. (6-11) can be written as

$$S = 2.025 \times 10^{-6} \frac{V^2 F}{gD} \tag{6-12}$$

where V = velocity in ft/s (or m/s)
$\qquad D$ = average depth of the reservoir in ft (m)
$\qquad F$ = fetch in ft (m)

EXAMPLE 6-3   A reservoir is oval shaped with a length of 10 mi and a width of 5 mi. If the wind blows in a direction lengthwise to the reservoir with a velocity of 80 mph, what will be the setup if the average depth is 20 ft?

SOLUTION   The fetch will be 10 mi or 52,800 ft, and the wind velocity is 117.3 ft/s; therefore, the setup will be

$$S = 2.025 \times 10^{-6} \times \frac{V^2 F}{gD}$$

$$= 2.025 \times 10^{-6} \times \frac{(117.3 \text{ ft/s})^2 \times 52,800 \text{ ft}}{32.2 \text{ ft/s}^2 \times 20 \text{ ft}}$$

$$= 2.28 \text{ ft} \qquad\qquad \blacksquare$$

HEIGHT OF WIND WAVES   Starting in 1950, the U.S. Army conducted tests to evaluate the wave heights that might be expected in a reservoir with a given fetch and wind speed. *Fetch* is the open water distance (in the direction of the wind velocity) upwind of the point in question. Thus if one were to determine the fetch for the dam on the reservoir shown in Fig. 6-30, and with the given wind direction, the fetch, F, would be as shown.* The results of the U.S.

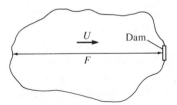

Figure 6-30   Definition sketch for fetch

---

* Saville showed that if the reservoir is fairly narrow in a direction normal to the wind direction, then the effective fetch would be less than the simple definition given here. For the method of computing effective fetch for long narrow reservoirs see Saville (13).

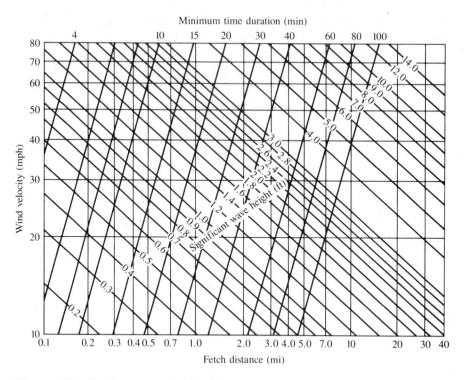

Figure 6-31 Significant wave height, H_s, as a function of wind speed and fetch (13)

Army studies were published by Saville in 1963 (13). A particularly useful chart developed from that study allows one to determine wave height as a function of wind velocity and fetch. Figure 6-31 is an adaptation of the chart developed by Saville. By entering the chart with wind velocity (given on the ordinate) and fetch in miles (abscissa), one can read off significant wave height (lines that slope downward to the right). Focusing on the lines that slope upward to the right, one can find the minimum time duration that the wind must blow to develop that particular wave height. Actually, an entire spectrum of waves is developed in a storm, and the heights indicated in Fig. 6-31 are just the larger ones; they are called *significant waves*. In Saville's study, the significant wave height is the average height of the highest one third of the waves. Of a total sample of waves, it was also found that the significant wave will be exceeded by waves of greater height 13% of the time. Waves having heights larger than the significant wave height will be exceeded by fewer than 13% of the waves depending on the given height. The relative wave height (the given wave height to the significant wave height) as a function of distribution is given in Table 6-4. Thus a wave height 1.67 times the significant wave height would have only 0.4% of

Table 6-4 Wave Heights Distributions (13)

Total Number of Waves in Series Averaged to Compute Specific Wave Height, H, (%)	Ratio of Specific Wave Height, H, to Significant Wave Height, H_s (H/H_s)	Waves Exceeding Specific Wave Height, H (%)
1	1.67	0.4
5	1.40	2
10	1.27	4
20	1.12	8
25	1.07	10
30	1.02	12
$33\frac{1}{3}$	1.00	13
40	0.95	16
50	0.89	20
75	0.75	32
100	0.62	46

waves exceeding its height. Table 6-4 can be used with Fig. 6-31 to choose the freeboard needed for a given dam. The designer must choose what wave height to design for. One may wish to be conservative and choose the $H/H_s = 1.67$. If the freeboard were designed for this, only about 0.4% of the waves would splash to the top of the dam. A design based on a lesser wave height would allow more waves to splash to the top of the dam, which might be acceptable if a drainage system were designed to carry the excess water away without erosion. At first glance, one might be tempted to choose a freeboard based on the very largest wave that might be expected to reach the top of the dam, but proper drainage design might allow some splashing of the larger waves over the top of the dam.

Once the wave height has been determined, the amount of freeboard chosen depends on the amount of wave *runup* on the face of the dam. Wave *runup, R,* is the difference between maximum elevation attained by wave runup on a slope and the static water elevation at the toe of the slope. The concept of

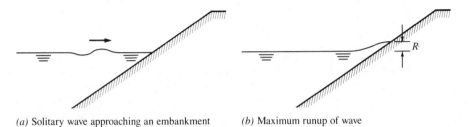

(*a*) Solitary wave approaching an embankment   (*b*) Maximum runup of wave

Figure 6-32   Wave runup

runup is best perceived by visualizing a solitary wave approaching an embankment, as shown in Fig. 6-32a. When the wave reaches the embankment, it will break and run up the slope, as shown in Fig. 6-32b.

The amount of runup for a wave of given size has been shown by Saville (13) to be a function of the slope of the embankment, the degree of roughness of the embankment, and the relative steepness of the wave, H_0/L_0 (see Fig. 6-33), where H_0 is the wave height in ft or m, and L_0 is the wave length. Here L_0 is given as

$$L_0 = 0.159gT^2 \tag{6-13}$$

where T is the wave period in seconds, and g is the acceleration due to gravity in ft/s^2 or m/s^2. T is a function of fetch and wind speed:

$$T = \frac{0.429U^{0.44}F^{0.28}}{g^{0.72}} \tag{6-14}$$

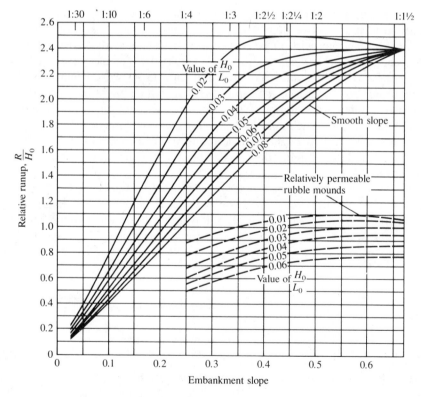

Figure 6-33   Wave runup ratios versus wave steepness and embankment slopes (13)

In Eq. (6-14), U is in ft/s or m/s, and F is in ft or m. The preceding formulas and Figs. 6-1, 6-33, and Table 6-4 are for deep water waves; that is, for depths greater than about one half the wave length (13). Therefore, if the reservoir depth is greater than $0.50 \, (0.159gT^2)$ (from Eq. 6-13), one can reliably use these charts and formulas. However, if shoal water occurs before the embankment or dam face, the shallow wave case exists, and one must use different relations.*

FREEBOARD DETERMINATION   Freeboard is calculated by summing the amount of setup and the runup that might be expected and adding another increment for intangible effects. These intangibles could include settlement of the embankment. Example 6-4 gives the process of calculating freeboard.

EXAMPLE 6-4   Considering the wind conditions and reservoir size of Example 6-3, what freeboard should be used for an earth dam having an upstream slope of 1 V to 3.0 H? Assume the upstream embankment is faced with well-designed riprap. Further assume the freeboard is to be based on the wave height that will be exceeded in height by only 2% of the waves. The average reservoir depth is 100 ft, and the dam is 200 ft high.

SOLUTION   The freeboard will be equal to setup plus runup plus allowance for settlement of the embankment plus an amount for contingencies.

First determine runup: Runup is a function of the wave height and slope and roughness of embankment. The significant wave height H_s is obtained from Fig. 6-31 and for the conditions of this example ($F = 10$ mi and $V = 80$ mph), is found to be 10.6 ft. It is also found from Fig. 6-31 that the wind must blow for about 75 min for this wave height to develop. From Table 6-4, page 359, the wave height that will be exceeded by only 2% of the waves is equal to $1.40H_s$ or $H = 1.40 \times 10.6 = 14.8$ ft.

Next, we must determine the wave length so that we may determine runup with the use of Fig. 6-33. From Eq. (6-13),

$$L = 0.159gT^2$$

where $T = 0.429U^{0.44}F^{0.28}/g^{0.72}$
$U = 80$ mph $= 117.3$ ft/s
$g = 32.2$ ft/s^2
$F = 10$ mi $= 52,800$ ft
$T = 0.429 \times 117.3^{0.44} \times 52,800^{0.280}/32.2^{0.72} = 6.02$ s

Then $L = 0.159 \times 32.2 \times 6.02^2 = 186$ ft

---

* For information on the shallow wave relation, see Saville (13). Other design data on waves and runup may be found in the U.S. Army Corps of Engineers manual on shore protection (17).

$$\frac{H}{L} = \frac{14.8 \text{ ft}}{186 \text{ ft}} = 0.080$$

From Fig. 6-33, for a value of $H/L = 0.080$ and a slope of 0.33, it is estimated that $R/H \approx 0.50$, (extrapolation of Fig. 6-33) or

$$R = 0.50 \times 14.8 = 7.40 \text{ ft}$$

Setup: From Eq. (6-12), we have

$$S = \frac{2.025 \times 10^{-6} V^2 F}{gD}$$

or, for this example,

$$S = \frac{2.025 \times 10^{-6} \times 117.3^2 \times 52,800}{32.2 \times 100}$$

$$= 0.46 \text{ ft}$$

Settlement: Assume a 1% settlement in the embankment, so settlement freeboard would be 2.0 ft.
Contingencies: For other uncertainties, allow 1.5 ft.

$$\text{Total freeboard} = 7.40 + 0.46 + 2.0 + 1.5 = 11.36 \text{ ft} \qquad ■$$

Sedimentation in Reservoirs

INTRODUCTION   All streams carry sediments that originate from erosion processes in the basins that feed the streams. In some streams, the average rate of sediment inflow to the reach will equal the rate of outflow; the stream will be in equilibrium. However, if a dam is constructed across the stream and a reservoir is produced, the velocity in the reservoir will be negligible so that virtually all the sediment coming into the reservoir will settle out and be trapped. Therefore, the reservoir should be designed with enough volume to hold the sediment and still operate as a water storage reservoir over the project's design life. For large projects, the design life is often considered 100 years.

Sediment carried in a stream is classified as either *bed load* or *suspended load*. The bed load consists of the coarsest fractions of the sediment (sands and gravels), and it rolls, slides, and bounces along the bottom of the stream. The finer sediments are suspended by the turbulence of the stream. When the sediment enters the lower velocity zone of the reservoir, the coarser sediments will be deposited first, and it is in this region that a delta will be formed (see Fig. 6-34). The finer sediments will be deposited beyond the delta at the bottom of the reservoir.

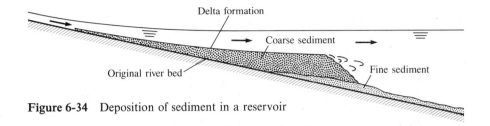

Figure 6-34 Deposition of sediment in a reservoir

Problems relating to sedimentation so far as design and operation of the reservoir are concerned are

1. Estimating the rate of accumulation of sediment in the reservoir so that enough volume can be reserved for the sediment accumulation.
2. Including the anticipated sediment accumulation in the operating plan of the reservoir. For example, because the initial phases of delta formation will be in the higher elevation zone of the reservoir, more reservoir storage will be lost in that zone than in zones at lower elevations. Thus flood-control storage may be reduced at a much faster rate than instream flow storage if the storage for flood control were initially allocated in the upper part of the reservoir. As sediment accumulates, it may be desirable to adjust the storage allocation for the different classes of storage.
3. Planning for location of recreational sites. For example, a boat dock would not be included in an area where a delta would likely form in a few years.

In the next two sections, we address the process of estimating the rate of flow of sediment into the reservoir and the volume it may be expected to occupy.

SEDIMENT YIELD   The total sediment outflow from a watershed or drainage basin measured in a specified period is the sediment yield. Often, the yield is expressed in terms of tons per acre per year. The engineer designing a reservoir must estimate the average sediment yield for the basin supplying the reservoir to determine at what rate the reservoir will fill with sediment. The estimate of sediment yield is usually based on one or more of the following procedures, which we list from highest to lowest expected accuracy:

1. Obtain records of measured sediment discharge at the reservoir site or near the site. These records should cover ten or more years to establish a reasonable dependable average yield. The methods of measuring sediment discharge are given by Vanoni (22).
2. Look for measured sediment discharge records on basins that have characteristics similar to the basin under consideration. Start a sediment discharge measuring program at the reservoir site as soon as possible, and compare short-term records for the site with similar records from basin(s) having similar characteristics. If a good correlation between the records for the two basins exists, the long-term averages from the other basin can possibly be used for preliminary estimates to design the reservoir.

Table 6-5 Sediment Yields Based on Reservoir Surveys* (22)

(1) Reservoir and Location	(2) River Basin	(3) Period of Record (yr)	(4) Reservoir Volume (acre-ft)	(5) Reservoir Volume (acre-ft per sq mi)
Sardis near Sardis, Miss.	Little Tallahatchie River	20.7	91,900	59.0
Bodeman near Brunswick, Ind.	Tributary of West Creek	13.0	52	19.0
Lake Tandy near Hopkinsville, Ky.	Little River	52.5	770	130.0
Kiser Lake near St. Paris, Ohio	Mosquito Creek	14.5	3,330	380.0
Upper Hocking #2 near Lancaster, Ohio	Hunter's Run	5.0	57	30.0
Lake Dante near Wagner, S. Dak.	Tributary to Choteau Creek	17.9	261	90.0
John Martin near Caddoa, Colo.	Arkansas River	15.2	423,000	22.0
Altus near Altus, Okla.	North Fork Red River	12.6	157,000	62.0
Bellerud Pond near Adams, N. Dak.	Dark River	13.0	17	15.0
Walnut Cove near Walnut Cove, N.C.	Dan River	9.0	970	2.5
Mission Lake near Horton, Kans.	Mission Creek	30.2	1,900	229.0
Guernsey near Guernsey, Wyo.	Platte River	30.3	74,000	14.0[§]

* Randomly selected from the Summary of Reservoir Sediment Deposition Surveys made in the United States through 1960, United States Department of Agriculture, Miscellaneous Publication No. 964, Appendix A, Feb., 1965.

[†] Computed from Conservation Needs Inventories of Individual States—usually available by inquiry to state offices of Soil Conservation Service, United States Department of Agriculture.

[‡] Estimated.

[§] Based on sediment-contributing area only.

Watershed Land Use[†] (%)			Mean Annual Quantities		
(6)	(7)	(8)	(9)	(10)	(11)
Cropland	Idle Pasture	Forest	Precipitation (in.)	Runoff (in.)	Sediment Yield (tons per sq mi)
30	35	30	52	20	1,100
55	30	5	36	10	280
55	10	30	47	18	1,700
65	20	10	38	12	4,900
65	20	10	36	13	950
60	35	—	22	0.5	260
15	80	5	15	0.3	850
60	30	5	24	0.9	1,300
75	15	5	18	0.7	110
25	10	65	50	19	240
70	20	5	32	4.4[‡]	740
15	80	—	8–24	1.3	210

3. Study the records of sedimentation in existing reservoirs throughout the region. Base the estimate of sediment yield on basins having similar characteristics. Since about 1940, the U.S. Army Corps of Engineers, the U.S. Soil Conservation Service, and the U.S. Bureau of Reclamation have been measuring the amount of sedimentation in reservoirs throughout the U.S. These data are forwarded to the Federal Interagency Committee on Sedimentation, which compiles the data and distributes it to all offices of these three agencies. The designer may find information by contacting any of these agencies. Table 6-5 shows a summary of sediment yields for several basins based on reservoir surveys.

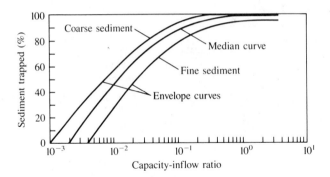

Figure 6-35   Trap efficiency curve (2)

TRAP EFFICIENCY   Once the sediment reaches the reservoir, most or all of it will be "trapped" in the reservoir if it has a long enough period to settle out. Thus most sediment of sand size and larger will be trapped in a new reservoir. However, some of the fine suspended sediment may pass through the reservoir if the reservoir is relatively small in comparison to the average discharge of the stream that feeds it. Brune (2) studied 44 reservoirs and developed a relationship between trap efficiency and the capacity-inflow ratio, as shown in Fig. 6-35. Here, the capacity-inflow ratio is storage volume in the reservoir divided by the mean annual volume of water flow into the reservoir. Obviously, as the reservoir fills up with sediment (capacity-inflow ratio decreases), the velocity of water flow through the reservoir increases because the cross-sectional flow area decreases; therefore, a smaller percentage of sediment will be trapped, as Fig. 6-35 shows. Thus one may use trap efficiency curves to estimate the useful life of a reservoir.

SEDIMENT WEIGHT   Because sediment yield is given in weight (tons/year), it is necessary to know the unit weight of the sediment deposit to estimate the actual volume it will occupy when it gets into the reservoir. Coarse sediments usually do not change volume much after they are deposited in a reservoir; however, fine sediments may compress considerably with time. Moreover, if the reservoir is drawn down to a level where the fine sediments are exposed and they dry out, they will consolidate much more than if never dried. Even when covered with water after drying, they will not expand appreciably. An empirical formula developed by Lane and Koelzer and presented in Vanoni (22) can be used to approximate the specific weight of sediment deposits in terms of age of deposit T in years and its initial specific weight, γ_1, taken after a year of consolidation:

$$\gamma = \gamma_1 + B \log_{10} T \tag{6-15}$$

Table 6-6 Constants in Eq. 6-15 for Estimating the Specific Weight of
Reservoir Sediments* (22)

	Sand		Silt		Clay	
Reservoir Operation	γ_1	B	γ_1	B	γ_1	B
Sediment always submerged or nearly submerged	93	0	65	5.7	30	16.0
Normally a moderate reservoir drawdown	93	0	74	2.7	46	10.7
Normally considerable reservoir drawdown	93	0	79	1.0	60	6.0
Reservoir normally empty	93	0	82	0.0	78	0.0

* All values given in lb/ft³.

In Eq. (6-15), γ_1 and the coefficient B (both given in weight per unit volume)
are functions of the type of sediment (sand, silt, or clay) and the manner in
which the reservoir is operated (this reflects the effects of drying of the sediment
deposit). Table 6-6 gives values of γ_1 and B in lb/ft³ for various sediments and
conditions.

PROBLEMS

6-1   Using the flexure formula, prove that the resultant force acting on a
gravity dam must pass within the middle one third of the base of the dam to
ensure that none of the concrete is in tension along the base.

6-2   Determine the normal stresses at the heel and toe of this concrete dam
using the simple flexure formula for solving for the stress. Consider weight
of dam, 2/3 hydrostatic uplift, and hydrostatic force on the face of the dam
in your analysis.

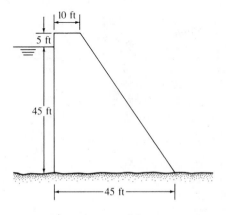

PROBLEM 6-2

6-3   For the dam of Prob. 6-2, what would be the stresses if earthquake forces were also included? Consider only horizontal acceleration toward reservoir.

6-4   Consider the dam and conditions of Prob. 6-3. Is the dam stable against sliding? Assume the dam rests on sound rock.

6-5   Using simple arch analysis, design an arch for an arch dam. This particular arch is to be at elevation 1000 ft, and the width of the canyon wall at this elevation is 1970 ft. The maximum pool level for the reservoir is to be 1270 ft. Assume the allowable compressive stress in the concrete is 2000 psi.

6-6   Figure A shows the plan view of Wasco reservoir east of Portland, Oregon. Figure B shows the area-capacity curves for the reservoir. The

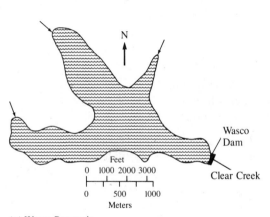

(a) Wasco Reservoir

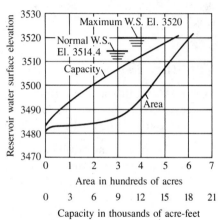

(b) Area-Capacity curves
for Wasco Reservoir

| Month | Flow (cfs) | | | | Pan Evaporation (in.) |
	1975	1976	1977	1978	
O	13	6	4	14	0.5
N	67	20	18	55	0.2
D	30	15	25	41	0.0
J	62	30	23	56	0.0
F	23	10	15	37	0.0
M	26	12	20	29	0.5
A	39	14	24	42	3.0
M	56	20	19	61	4.0
J	57	10	15	45	5.0
J	9	5	8	12	7.0
A	5	3	6	7	6.0
S	7	2	8	6	4.0

storage reserved for conservation (irrigation) is between elevation 3488 ft and 3514 ft. The data in the table are the mean monthly flows for the creeks discharging into the Wasco reservoir and pan evaporation estimates during a drought period. For this reservoir, what outflow could be assured during the drought period?

6-7   Determine the wave height and runup for Wasco Dam (refer to Prob. 6-6 for information about the reservoir). Wasco Dam is a 50-ft high earth dam with an upstream slope of 1 V to $2\frac{1}{2}$ H. The reservoir side of the embankment is surfaced with 24-in. riprap.

6-8   Design a nonoverflow section of a gravity dam at a site where the design earthquake intensity is assumed to be 0.1 g. Assume the normal maximum pool elevation in the reservoir is 400 ft, the surface of the rock foundation is at elevation 100 ft, and the normal downstream water surface elevation is 150 ft. Make your own assumptions for other data needed in your design.

6-9   For a site chosen by your instructor, select the type of dam most appropriate for the site and give reasons for your choice.

6-10   The following table gives the anticipated average hourly water demand for a small city for a particular time of year. A reservoir is to be designed to even out the pumping rate from the well source to the reservoir. If the pumping rate is to be constant throughout the day and if it is assumed the reservoir is full when the demand is greater than Q_{avg} in early morning, what reservoir volume is required and what will be the pumping rate?

Hour (ending at)	1 A.M.	2	3	4	5	6	7	8	9	10	11	12N
Demand Q (gpm)	800	800	900	1000	1200	1425	1900	2200	2000	1575	1600	1700

Hour (ending at)	1 P.M.	2	3	4	5	6	7	8	9	10	11	12M
Demand Q (gpm)	1500	1300	1400	1600	1800	2300	1800	1500	1200	1000	900	800

6-11   Prove that $2\alpha = 133.57°$ is the arch angle that minimizes the volume required in a rib of an arch dam. *Hint*: First write an equation for the volume of a given rib in terms of r and α using Eq. (6-7), page 331. Then the volume equation will be expressed in terms of α, which can be differentiated and set equal to zero, which can then be solved to obtain the α for minimum volume.

6-12   For a dam chosen by your instructor, determine the following information about the dam:
a. What type of dam is it?
b. What earthquake intensity was used in design?

   c. Is there a grout curtain in the foundation?
   d. Are there drains at the base of the dam?
   e. Are there any other drains?
   f. Was uplift considered in the design? To what extent?
   g. Were ice forces considered in the design?
   h. What is the foundation rock?
   i. Was any other type of dam considered for the site?
   j. Is a powerhouse a part of the project? Is so, what type, number, and size (power generating capacity) of turbines are installed in the powerhouse?
   k. What are the basic dimensions of the dam?
   l. Have there been any leakage problems? If so, what are the details about it? How were they remedied?
   m. How was the river diverted during construction?
   n. Have there been any problems relating to the reservoir, such as slides or abnormal sedimentation?
   o. Are any monitoring instruments installed in the dam to detect dam or foundation movement?
   p. What is the magnitude of the design flood? Was it based on the probable maximum storm?

6-13   Lake Okechobee in Florida has an average depth of 7 ft and is about 37 mi long and 30 mi wide. If winds of 40 mph were to blow over the lake, what would be the setup?

6-14   For a 1000-acre reservoir site in your region of the country (specific location to be chosen by you or your instructor), estimate the sediment yield (tons/year) for the site. What volume in acre-feet is this equivalent to?

REFERENCES

1. Biswas, A.K. *History of Hydrology*. North-Holland Publishing, Amsterdam, 1970.
2. Brune, G.M. "Trap Efficiency of Reservoirs." *Trans. Am. Geophysical Union*, 34, no. 3, Washington, D.C. (1953), pp. 407–18.
3. Cooke, J.B. "Progress in Rockfill Dams." *Jour. of Geotechnical Engineering*, ASCE, 110, no. 10 (October 1984).
4. Creager, W.P., and J.D. Justin. *Hydroelectric Handbook*. John Wiley & Sons, New York, 1950.
5. Creager, W.P., J.D. Justin and J. Hinds. *Engineering for Dams*. John Wiley & Sons, New York, 1945.
6. Galloway, J.D. *Trans. ASCE*, 104, Paper no. 2015 (1939).
7. Garbecht, G., "Sadd-el-Kabara: The World's Oldest Large Dam." *International Water Power and Dam Construction* (July 1985).
8. Golzé, Alfred R. *Handbook of Dam Engineering*. Van Nostrand Reinhold, New York, 1977.

9. Jansen, R.B. *Dams and Public Safety.* U.S. Dept. of Interior, U.S. Govt. Printing Office, Washington, D.C., 1980.

10. Louchs, D.P., J.R. Stedinger, and D.A. Haith. *Water Resource Systems Planning and Analysis.* Prentice-Hall, Englewood Cliffs, N.J., 1981.

11. Middlebrooks, T.A. "Fort Peck Slide." *Trans. ASCE,* 107 (1942).

12. Peterson, M.S. *Water Resource Planning and Development.* Prentice-Hall, Englewood Cliffs, N.J., 1984.

13. Saville, T., E.W. McLendon, and A.L. Cochran. "Freeboard Allowances for Waves in Inland Reservoirs." *Trans. ASCE,* 126 (1963), Part IV, pp. 195–226.

14. Schuyler, J.D. *Reservoirs for Irrigation.* John Wiley & Sons, New York, 1905.

15. Sherard, J.L., R.J. Woodward, S.F. Gizienski, and W.A. Clevenger. *Earth and Earth-Rock Dams.* John Wiley & Sons, New York, 1963.

16. Toebes, G.H., and A.A. Shepherd (eds.). Proc. of the Nat. Workshop on *Reservoir Systems Operations.* ASCE, 1979.

17. U.S. Army Corps of Engineers, *Shore Protection Manual,* 4th ed. U.S. Army Waterways Experiment Station, Coastal Engineering Research Center, U.S. Govt. Printing Office, Washington, D.C., 1984.

18. U.S. Bureau of Reclamation. *Design of Arch Dams.* U.S. Govt. Printing Office, Washington, D.C., 1977.

19. U.S. Bureau of Reclamation. *Design of Gravity Dams.* U.S. Govt. Printing Office, Washington, D.C., 1976.

20. U.S. Bureau of Reclamation. *Design of Small Dams.* U.S. Govt. Printing Office, Washington, D.C., 1960.

21. U.S. Bureau of Reclamation. *Design of Small Dams,* 2d ed. U.S. Govt. Printing Office, Washington, D.C., 1977.

22. Vanoni, Vito A. (ed.) *Sedimentation Engineering.* ASCE, 1975.

23. Zanger, C.N. Engineering Monograph No. 11, *Hydrodynamic Pressures on Dams Due to Horizontal Earthquake Effects.* U.S. Bureau of Reclamation (May 1952).

Strontia Springs Diversion Dam near Denver, Colorado. (Courtesy of Harza Engineering Co., Chicago, Illinois)

Hydraulic Structures

The design of hydraulic structures is based on the application of fundamental principles in fluid mechanics and basic ideas from the applied areas of closed conduit and open channel hydraulics. It is therefore not necessary that you be exposed to an exhaustive list of hydraulic structures to be adequately prepared for hydraulic design. Carefully chosen structures can both introduce the most common applications and illustrate basic principles and areas requiring special emphasis in design. In this chapter, we address traditional areas of design as well as introduce some recent technological advances in modern application where more design sophistication is required.

7-1   Functions of Hydraulic Structures

Although hydraulic structures have a broad range of applications, most structures fall into one of several broad categories. For example, the most common type of hydraulic structure is a pipe or an open channel that conveys water from one point to another. Because this type of structure is so widespread, extensive design methods have been developed based on the analysis and systematic collection of both laboratory and field-generated experimental data. Pipes and channels, however, are usually not addressed under the subject of hydraulic structures, and in this book, we discuss them in separate chapters.

Spillways, another type of conveyance structure, transport water around, through, or over dams, usually at high speed. Adapting the basic prismatic closed conduits and open channels to specialized applications associated with hydroelectric power plants and pumping installations leads to structures that require careful attention to acceleration, deceleration, and turning of the flow. How the direction and speed of the flow is modified greatly influences the efficiency and operational characteristics of hydraulic machinery such as pumps and turbines. Intake structures, approach channels, draft tubes, and other integral components of power and pumping projects should be designed to carefully change flow velocity to optimize plant efficiency and operation.

Hydraulic structures are also used to convey materials along with the water. Modern urban sewage systems and the coal industry use pipes to economically transport solids with the flow. To pass fish by dams or pumping plants, special structures must be designed.

Other major categories of hydraulic structures involve the control and measurement of flow and energy dissipation. Control is usually accomplished with various types of valves and gates, but it can also be accomplished with bypass structures such as surge tanks, which control flow and pressure by providing additional storage in the system. Potential energy associated with hydroelectric projects is usually extracted and transformed into useful electrical energy. In many instances, however, it is neither possible nor economical to transform this energy, and the energy must be dissipated. This frequently is associated with high velocity flows in spillways and chutes in which the energy must be dissipated in a controlled fashion before releasing the flow into an unprotected stream bed or canal. High velocity flow can produce unwanted phenomena such as vibration and cavitation unless the designer produces a design that either prevents their occurrence or, if they do occur, mitigates their effects.

7-2   Dam Appurtenances

Associated with dams are many types of hydraulic structures that offer a wide variety of applications. Although the major structures usually consist of spillways and energy dissipators, diversion works, low-level outlet works, and intake structures are also important.

Spillways

INTRODUCTION   A spillway is nearly always required to pass flow by a dam. In the case of storage or hydroelectric dams, where large flows pass through hydraulic turbines, spillways may be used infrequently to pass floods. For diversion dams, where the diversion represents a small portion of the total flow, the spillway may operate continuously.

The safe operation of spillways is the main objective in design. Failure of the spillway to perform its design function can lead to failure of the dam with possible property damage and loss of life. The determination of the design flood flows (see Chapter 2) is critical, particularly for earthfill and rockfill dams, which cannot withstand overtopping. Because dams raise the water level in a stream, spillways usually must be designed for high velocity flow, since this additional potential energy is transformed into kinetic energy. Not only the spillway must be designed to withstand these velocities but also the terminal structure, or dissipator, which must release flow at a small enough velocity and produce conditions so that the dam will not be endangered by excessive downstream erosion.

Spillways for dams impounding water in reservoirs having little storage require a capacity that will nearly equal the peak of the inflow flood to the reservoir. However, if much of the flood volume entering the reservoir can be temporarily stored, the spillway need not be designed for the flood peak. This situation we address in Chapters 4 and 9.

Several spillway configurations are available to the designer. Besides the basic types of spillways, the designer may choose to use two spillways. One spillway, usually termed a service spillway, can be used to pass frequently occurring smaller floods. Another type of structure, termed an auxiliary spillway, can be designed to pass the larger floods. Auxiliary spillways are particularly adaptable to sites with natural saddles, where flow can be discharged into a natural waterway remote from the dam and where less stringent safety requirements may prevail.

Figure 7-1   Spillway of Mangla Dam in Pakistan
(Courtesy of Harza Engineering Co., Chicago, Illinois)

Many separate features contribute to the overall function of spillways (Fig. 7-1). An approach channel frequently delivers flow to some type of control structure, which determines the amount of flow through the spillway. The control structure usually consists of either an overflow device such as a weir or curved crest or a submerged opening such as an orifice. The control structure can be fitted with gates or valves to control the rate of flow. Except for arch dams, where flow can plunge directly from the dam, spillway control structures usually connect directly with a channel that conveys the flow to the downstream river channel. Some type of energy dissipation structure is built at the end of the channel to enable the flow to be returned to the river without causing unacceptable scour.

OVERFLOW SPILLWAYS   Overflow spillways can be classified into two categories: (1) a straight drop design that consists of a control section but no downstream channel to convey the flow, such as flow over the top of an arch dam, and (2) a scheme that provides a continuous channel shaped to form an S-shape, or ogee spillway, such as on a concrete gravity dam.

The crest section for an uncontrolled ogee spillway is commonly shaped to conform with the underside of water flowing over a weir. This will produce a pressure distribution on the crest boundary that is close to atmospheric for the selected flow rate. The head above the spillway crest is the design head H_D. Flow rates less than the design flow will produce pressures on the spillway face above the atmospheric level, whereas flow rates greater than the design flow

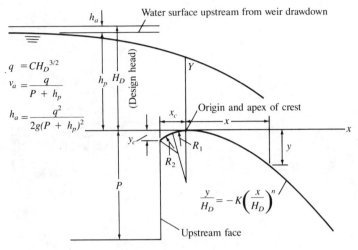

$q = CH_D^{3/2}$

$v_a = \dfrac{q}{P + h_p}$

$h_a = \dfrac{q^2}{2g(P + h_p)^2}$

Figure 7-2   Crest configuration for overflow
spillway (20)*
* For vertical upstream face and $h_a = 0$, $K = 0.50$, $n = 1.872$,
  $R_1 = 0.53H_D$, $R_2 = 0.235H_D$, $x_c = 0.283H_D$, and
  $y_c = 0.127H_D$

will cause subatmospheric pressures. Subatmospheric pressures increase the flow-passing capability of the spillway, but for high heads can lead to cavitation, which if sufficiently severe, will lead to vibration and surface erosion.

Analytical methods have been developed that allow one to compute the nappe coordinates, but these methods require sophisticated computation, and spillway design is frequently based on experimental measurements made by the U.S. Bureau of Reclamation (21). Figure 7-2 shows the suggested spillway shape for a vertical upstream face.

The discharge over an ungated spillway is controlled primarily by the head on the crest. The weir equation (Eq. 4-38, page 210) is also used for spillway flow:

$$Q = C\sqrt{2g}LH^{3/2} \tag{7-1}$$

where Q is volumetric flow rate, C is the dimensionless coefficient of discharge, L is the crest length perpendicular to the flow, and H is the total head on the crest, including the approach velocity head, h_a. The coefficient C is larger for a spillway than for a weir because the head H is referenced to the spillway crest (the top of the trajectory), whereas the head for the weir equation is referenced to the spring point of the jet (the elevation of the top of the weir plate). The coefficient C depends on the approach depth, actual shape of the crest, and upstream face slope. Figure 7-3a gives the variation of C_D, the value of C when H equals the design head H_D, with the relative upstream depth P/H_D. Here P is the height of the spillway crest with respect to the channel bed.

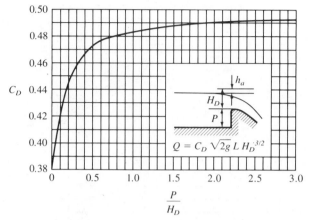

(a) Discharge coefficient C_D versus P/H_D for design head

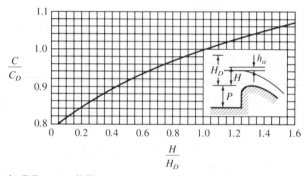

(b) C/C_D versus H/H_D

Figure 7-3 Coefficient of discharge for ogee crests with vertical faces (20)

The ratio of the coefficients C/C_D will vary from about 0.8 for small H/H_D to about 1.06 for H/H_D of 1.5 (see Fig. 7-3b). A spillway crest becomes more efficient at heads that exceed the design head, but this comes at the expense of subatmospheric pressures on the spillway, which leads to concerns about cavitation damage.

Overflow spillways frequently use undershot radial gates for releases over the dam, as shown in Fig. 7-4. The presence of the gate, particularly for small openings, produces orifice-type flow, which produces a parabolic trajectory above that given by the underside of a free-flowing nappe. Adherence of gate flow to the nappe profile results in reduced pressures on the spillway face. These reduced pressures can be avoided if the spillway is broadened to adhere to the parabolic trajectory; this, however, will reduce the discharge capacity for the ungated design condition. Some reduction of the magnitude of the reduced

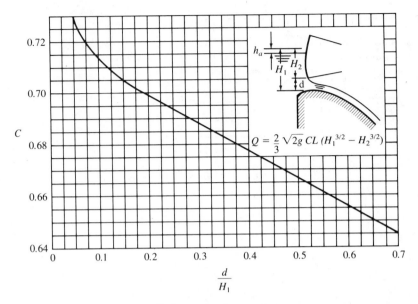

Figure 7-4   Coefficient of discharge for flow under gates (20)

spillway pressure for gated flows can be achieved by moving the gate sill downstream from the crest.

The governing equation for gated flows can be derived by integrating the flow rate over the orifice opening to produce

$$Q = \frac{2}{3}\sqrt{2g}CL(H_1^{3/2} - H_2^{3/2})$$

(7-2)

where C is a coefficient of discharge, and H_1 and H_2 are total heads to the bottom and top of the gate opening. The coefficient C is a function of geometry and the ratio d/H_1 (Fig. 7-4), where d is the gate aperture.

Piers placed on spillways to furnish structural support for the gates not only reduce the effective flow-passing length of the spillway crest by the width of the piers but also cause local flow contractions that further reduce the effective length, particularly if the nose of the pier is not rounded. The reduction, which depends on the total head on the crest and the shape of the upstream pier nose, is fairly small, ranging up to 4% for square-nosed piers that have a head on the crest equal to the opening between piers.

SHAFT SPILLWAYS   Shaft spillways (Fig. 7-5) are often selected for projects where space limitations or topographic features do not permit installing other types of spillways. They are also often used with embankment dams because it is not safe for standard chutes to be constructed on the embankment.

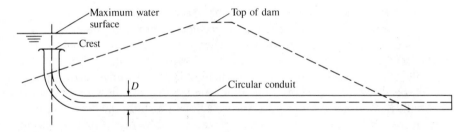

Figure 7-5 Typical shaft spillway through abutment
of dam

The major components of a shaft spillway are a circular crest section, a vertical shaft, an elbow in a vertical plane, a tunnel section, and a terminal structure. For large structures, the various components are usually constructed from concrete, with the tunnel and much of the vertical curve tunneled through rock.

There are three types of possible flow regimes in vertical shaft spillways. For low flows, a continuous volume of air persists through the flow passages and the flow rate is governed by the relationship for weir flow at the crest (Eq. 7-1): $Q = C\sqrt{2g}LH^{3/2}$. The head on the weir H is measured from the crest (apex of the trajectory). The length L is the circumference of the crest measured at the spring point (see Fig. 7-6). The coefficient C is a function of both H/R_s and P/R_s, where R_s is the radius of the spring point.*

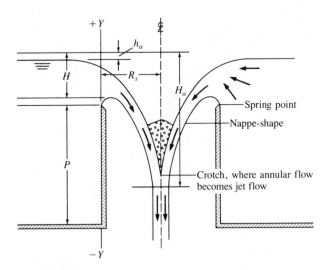

Figure 7-6 Definition sketch for circular weir (20)

---

* This functional relationship can be found in the USBR publication on the design of gravity dams (19).

For H/R_s above approximately 0.45, the weir becomes partially submerged. As H/R_s approaches unity, the entrance becomes completely submerged, and the second type of flow, orifice control, occurs, where Q is proportional to $H_a^{1/2}$. The third type of flow will occur when the downstream tunnel becomes full and full-pipe flow relationships apply, where Q is proportional to $H_p^{1/2}$ (H_p is the full head for full-pipe flow). The flow rate for the pipe flow regime varies only with $H_p^{1/2}$ rather than $H^{3/2}$, as for weir flow. For full-pipe flow, relatively large increases in head produce small increases in flow rate. For this reason, the designer must be extremely careful in estimating the maximum discharge. Thus this type of spillway frequently requires an auxiliary spillway to pass large floods. Figure 7-7 illustrates the rating curve for a shaft spillway. It is common practice to avoid designing for full-pipe flow, shaft spillways are usually designed so that the flow depth in the downstream tunnel is not more than 75% of the diameter. This allows air to enter from the downstream portal, thus preventing subatmospheric pressures in the tunnel. Terminal structures for shaft spillways frequently use a deflector bucket that throws the flow beyond the structure.

Shaping the crest of a shaft spillway follows the procedures outlined for ogee spillways in which the crest is shaped to follow the underside of flow

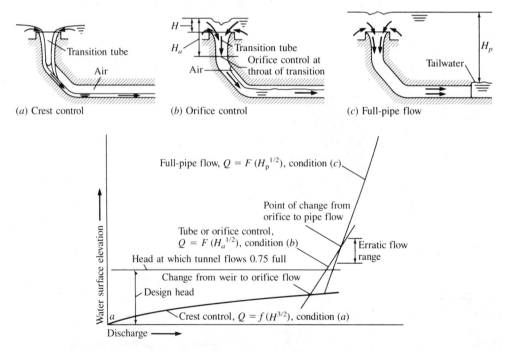

Figure 7-7   Characteristics of a rating curve for a morning glory spillway (20)

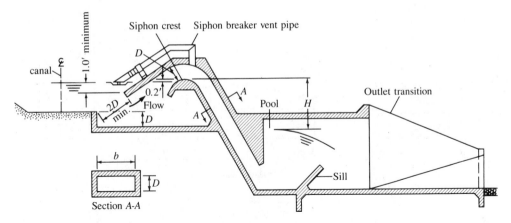

Figure 7-8 Profile of a siphon spillway (20)

over a circular sharp-crested weir. A feature of circular weir flow, however, is that for larger flows, lower nappe shapes lie below those for low flows. Thus a crest designed for a high flow can experience subatmospheric pressures for low flows. Crest profiles are therefore usually designed so that the head on the crest is only about 0.3 times the crest radius R_s.

SIPHON SPILLWAYS   Siphon spillways are often used in low-head small capacity installations, where it is desirable to keep the reservoir level within a modest range of fluctuation, or in larger installations, where it is used as a service spillway and large floods are carried by an auxiliary spillway. A siphon spillway, shown in Fig. 7-8, is a closed conduit consisting of an inlet, an upward sloping section, a crown with a level crest section, a downward sloping section, and an outlet. When air has been evacuated from the siphon, the full potential of the difference in head between the reservoir and the outlet is used in producing flow. The elevated crown section, operating under subatmospheric pressure, should not be constructed so high above the reservoir level that pressures are reduced to the vapor pressure, thus producing cavitation. Pressure heads at the crown should not be lower than −20 ft of water. A siphon breaker is designed to allow air to enter the system so that flow can be stopped when the reservoir level is below the crown. The outlet is frequently designed to rise in the flow direction to prevent air from entering the downstream end, thus preserving the siphon action.

The inlet section is sufficiently submerged to prevent ice and debris from entering the siphon and to prevent the formation of air-entraining vortices that would interfere with the siphon action. Alternating priming and depriming of the siphon, which would significantly change flow rates, can be prevented by properly designing the vent pipe. Siphon spillways are sometimes chosen because of the automatic operation provided without the need for any special moving parts. Closer flow regulation can be achieved with several smaller siphons rather than a single large one.

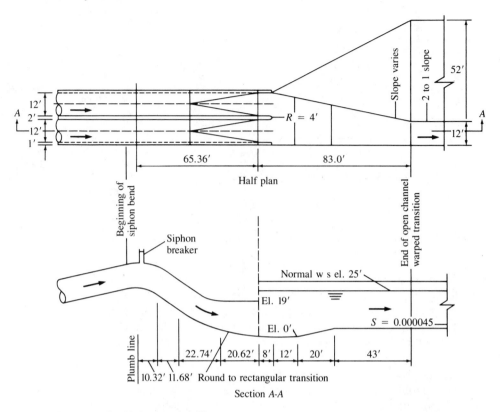

Figure 7-9 Siphon outlet (3)

Another application for siphons occurs at the outlets of pump discharge lines discharging into canals, as shown in Fig. 7-9. This type of installation eliminates the need for a large gate to prevent backflow and is used where there is concern that gates might become inoperable because of an earthquake or plugging by debris. The system consists of the siphon and a siphon breaker. The siphon breaker is a valve that can be opened to allow air to enter the siphon. Thus when a power outage occurs due to an earthquake or other emergency, the siphon breaker is automatically opened, which in turn, prevents backflow.

Terminal Structures

Flow in a spillway must be either deflected or decelerated before being released to the downstream channel if erosion of the streambed and sidebanks could result in danger to the dam itself. Deceleration of flow in a terminal structure is normally accomplished by designing the structure to create a hydraulic jump. In this process, the high level of kinetic energy is reduced through dissipation and by a transfer to potential energy represented by the higher-depth subcritical flow downstream from the jump. In Chapter 4, page 196, we developed the basic relationship between the depths in a hydraulic jump for a rectangular channel, which is given as

$$\frac{y_2}{y_1} = \frac{1}{2}(\sqrt{1 + 8\text{Fr}_1{}^2} - 1) \tag{7-3}$$

where y_1 and y_2 are the upstream and downstream depths, and Fr_1 is the upstream Froude number.

The head loss through the jump for a rectangular channel is given by

$$h_L = \frac{(y_1 - y_2)^3}{4y_1y_2} \tag{7-4}$$

The energy equation for the jump, given by

$$\frac{V_1{}^2}{2g} + y_1 = \frac{V_2{}^2}{2g} + y_2 + h_L \tag{7-5}$$

can be rearranged and made dimensionless by dividing by the upstream kinetic energy, $V_1{}^2/2g$, to give

$$\frac{h_L}{V_1{}^2/2g} + \frac{y_2 - y_1}{V_1{}^2/2g} + \frac{V_2{}^2/2g}{V_1{}^2/2g} = 1 \tag{7-6}$$

Equation (7-6), plotted in Fig. 7-10 on the next page, illustrates the fate of the initial upstream kinetic energy, which is equal to the sum of the energy loss in the jump, the increase in potential energy through the jump, and the down-stream kinetic energy. Equation (7-6) shows clearly the increase in relative energy loss as Fr_1 increases. Also note that for very high Fr_1, little residual kinetic energy remains downstream from the jump.

Hydraulic structures are designed so that hydraulic jumps occur under controlled conditions in a structure called a stilling basin. Although the energy associated with high velocity spillway flow will eventually be dissipated to the level of energy in the receiving channel, where and how this energy is dissipated is of utmost importance in controlling erosion. The positioning of a hydraulic jump on an unobstructed horizontal surface is very sensitive to the close match of sequent depths defined by Eq. (7-3). If the downstream depth matches the sequent depth y_2, the hydraulic jump will occur as desired on the apron. If, however, the downstream depth is less than y_2 (and it has to be only moderately less), the jump will occur downstream from the apron (a swept-out jump), and the river will become exposed to high scouring velocities. If on the other hand, the downstream depth is greater than y_2, the jump will become submerged. Although a submerged jump is preferable to a swept-out jump, much of the initial

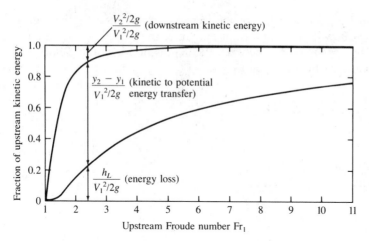

Figure 7-10   Distribution of upstream kinetic energy
as a function of Froude number

kinetic energy remains in the form of a submerged jet, which alone can result
in considerable downstream scour.

A carefully designed stilling basin will not only improve the dissipation
characteristics of a hydraulic jump, it will also shorten its length and stabilize
the position of the jump so that it is not sensitive to fluctuations in tailwater
level. This latter attribute makes the design safer. Stilling basins, Fig. 7-11, are

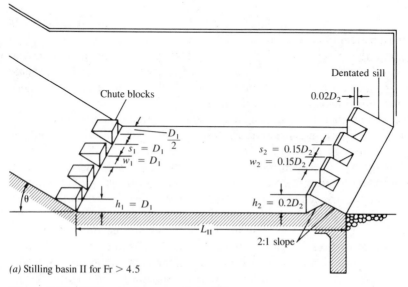

(a) Stilling basin II for Fr > 4.5

Figure 7-11   Stilling basin proportions (20)

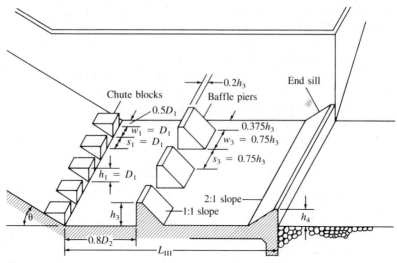

(b) Stilling basin III for Fr > 4.5

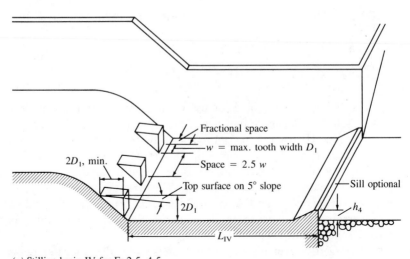

(c) Stilling basin IV for Fr 2.5–4.5

often furnished with appurtenances such as chute blocks, baffle blocks, and end sills. Chute blocks, located at the entrance to the stilling basin, help initially to spread some of the water in the vertical plane. The resisting force of the baffle blocks on the flow (reaction to the drag on the block) helps the downstream hydrostatic pressure force create and stabilize the jump. Because of their blunt shape, cavitation on the blocks can be a concern when velocities are high. The sill at the end of the basin lifts the flow away from the downstream bed and produces a return current that deposits bed material immediately downstream from the stilling basin.

Because stilling basin block arrangements are difficult to design analytically, their design must be based on experimental methods. Standard designs have been developed through both observations of existing installations and a systematic series of model studies. Three types of stilling basins developed by the U.S. Bureau of Reclamation are shown in Fig. 7-11.

The dimensions D_1 and D_2 in Fig. 7-11 take the same value as y_1 and y_2 (Eq. 7-3), respectively. The lengths L_{II}, L_{III}, and L_{IV} are given as follows:

$$L_{II} = D_2[4.0 + 0.055(Fr_1 - 4.5)] \quad \text{for } 4.5 < Fr_1 < 10.0 \qquad (7\text{-}7)$$

$$L_{II} = 4.35D_2 \qquad\qquad\qquad\qquad \text{for } 10.0 < Fr_1$$

$$L_{III} = D_2[2.4 + 0.073(Fr_1 - 4.5)] \quad \text{for } 4.5 < Fr_1 < 10.0$$

$$L_{III} = 2.8D_2 \qquad\qquad\qquad\qquad \text{for } 10.0 < Fr_1$$

$$L_{IV} = D_2[5.2 + 0.40(Fr_1 - 2.5)]$$

For the type III stilling basin, the dimensions h_3 and h_4 are given by

$$h_3 = D_1[1.30 + 0.164(Fr_1 - 4.0)]$$
$$h_4 = D_1[1.25 + 0.056(Fr_1 - 4.0)]$$

The type II basin is designed for use on high spillways for Froude numbers greater than 4.5. The chute blocks and end sill help reduce the basin length by 33%. The type III basin reduces the length by 60% with the addition of chute blocks, baffle piers, and an end sill. This structure is also used for Froude numbers greater than 4.5, but its use is restricted to small spillways where the upstream velocity is less than 50–60 fps. The type IV basin is used for Froude numbers between 2.5 and 4.5, and thus is used primarily on canal structures and diversion dams. This basin attenuates some of the wave action prevalent at low Froude numbers.

Submerged bucket dissipators are used when the tailwater depth is too great for the formation of a hydraulic jump. Both solid and slotted bucket types can be used, as Fig. 7-12 shows. The bucket dissipator generates one eddy within the bucket and another downstream from the bucket that moves riverbed material toward the lip of the bucket. Flow leaves the slotted bucket at a flatter angle, and in the process, produces a surface roller less violent than for the solid bucket. This results in smoother downstream flow. However, the slotted bucket operates satisfactorily over a smaller range of tailwater than the solid bucket. The slotted bucket is more conducive to sweep-out at low tailwater and will create a diving jet that produces objectionable scour at high tailwater.

Sometimes it is convenient to direct spillway flow into the river without passing through a stilling basin. This is accomplished with a deflector bucket

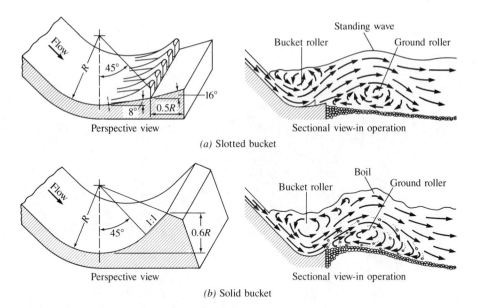

(a) Slotted bucket

(b) Solid bucket

Figure 7-12   Submerged buckets (20)

designed so that the jet strikes the riverbed a safe distance from the spillway and dam. This type of spillway is often called a flip bucket or ski jump spillway. The flow follows the equation for projectile dynamics, given by

$$y = x \tan \theta - \frac{x^2}{K[4(d + h_v) \cos^2 \theta]} \tag{7-8}$$

where x, y, d, and θ are defined in Fig. 7-13. K is a factor less than unity, having an average value of about 0.9, that accounts for energy losses due to turbulence and air resistance in the jet before the jet strikes the riverbed, and h_v is the velocity head at the lip. Figure 7-14 shows a deflector bucket in operation.

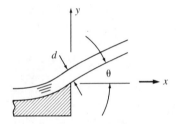

Figure 7-13   Definition sketch for deflector bucket

Figure 7-14   Deflector bucket in operation at end of
a tunnel spillway, Lucky Peak Dam, Idaho
(Bechtel Group, Inc.)

Gates and Valves

Two major classes of gates are (1) spillway gates, installations to control
spillway discharge and (2) conduit gates to completely shut off flow in penstocks
and outlet conduits, for maintenance, repair, or inspection. Spillway gates are
usually designed as vertical lift, radial, or drum gates.

Vertical lift gates (Fig. 7-15) are rectangular, usually made of steel and span
a section of conduit or channel between two vertical piers. The gates slide in
guides fabricated into the piers and are raised by a hoist. Smaller gates slide
on the guides, whereas large installations use wheels mounted on the gate (fixed-
wheel gate) or a roller-train that operates independently of both the gate and
the guides to reduce friction. The use of wheels or rollers greatly reduces the
size of the lifting mechanism.

The design of vertical lift gates considers the following static loading conditions: hydrostatic water load, gate weight, friction, and force transmitted by
the hoist. In addition, because of the undershot orifice flow inherent in this

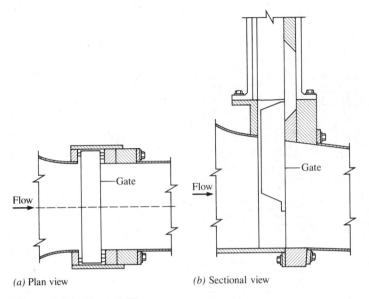

(a) Plan view   (b) Sectional view

Figure 7-15 Vertical lift gate

design, dynamic forces can be produced under the gate, which can produce either additional uplift or downpull forces on the underside of the gate that can have both steady and unsteady components. The unsteady forces can lead to severe vibration, which should be considered in designing spillway gates, but this vibration has been found to be of more consequence in the higher head installations of outlet works.

Radial gates (Fig. 7-16) consist of a cylindrical skin segment connected by radial arms to a trunnion pin. The trunnion pin is at the center of the cylindrical segment so that all hydrostatic pressure loads pass through the pin. Thus the hoist capacity must overcome only the gate weight, seal friction, and pin resistance. Counter weights are sometimes used to balance the gate weight and further reduce hoist capacity. Note in Fig. 7-16, the discharge coefficient is given in terms of the angle of the gate face and the location of the gate seat relative to the crest; therefore, it is more general than the coefficient and formula given in Fig. 7-4.

Drum gates (Fig. 7-17) are hinged at the downstream end of a recessed hydraulic chamber, are essentially hollow, and when in the open position, fit flush with the spillway crest. The gate's position is changed by controlling the water level in the chamber. Controls for filling the chamber are located in the adjacent piers and frequently are designed to provide automatic operation.

Besides these three types of gates, modest reservoir storage can be gained for smaller installations during low-flow periods by using temporary structures such as needles, flashboards, and stoplogs (Fig. 7-18). Needles consist of nearly

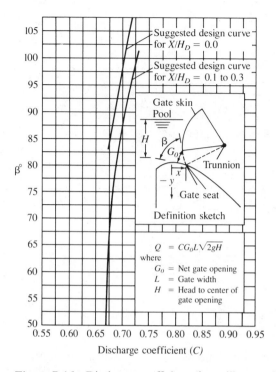

Figure 7-16 Discharge coefficients for spillway radial gates (17)

vertical members whose lower end rests in a slot along the spillway crest and whose upper end is supported by a horizontal beam. Flashboards are relatively low structures consisting of horizontal boards (usually wood) supported by pipes or rods inserted into the spillway crest. Flashboards are sometimes designed to fail when a designated head is exceeded. Stoplogs are also horizontal members but are supported by slotted piers. Depending on their size, stoplogs can be removed either by hand or by a hoist.

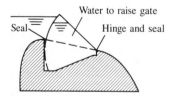

Figure 7-17 Drum gate

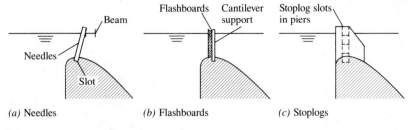

(a) Needles   *(b)* Flashboards   *(c)* Stoplogs

Figure 7-18   Temporary crest gates

Outlet Works

Outlet works are normally provided at most dams so that water can be released for the various purposes for which the dam was constructed. Flow may be passed through or around the dam to the downstream river for satisfying instream flow requirements, or it may be diverted from the river for irrigation or domestic water supply.

Outlet works in concrete dams (see Fig. 7-19) primarily consist of a conduit through the dam that includes a converging entrance protected by a trashrack, an emergency gate, a valve or gate to control the flow rate, and a terminal structure frequently designed as an energy dissipator. The elevation of the intake is influenced by considerations such as providing the design discharge at minimum reservoir. The option of withdrawing water from selected levels to

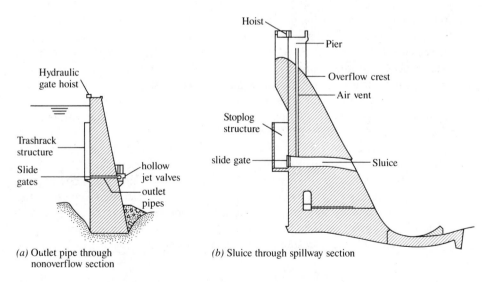

(a) Outlet pipe through
nonoverflow section

(b) Sluice through spillway section

Figure 7-19   Typical outlet works installations

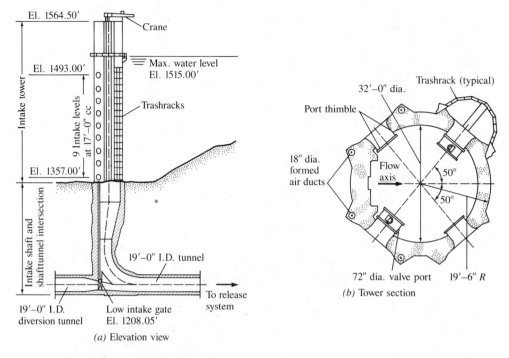

El. 1564.50'

Crane

Max. water level
El. 1515.00'

El. 1493.00'

Intake tower

9 Intake levels
at 17'-0" cc

Trashracks

El. 1357.00'

Intake shaft and
shafttunnel intersection

19'-0" I.D. tunnel

19'-0" I.D.
diversion tunnel

Low intake gate
El. 1208.05'

To release
system

(a) Elevation view

Trashrack (typical)

32'-0" dia.

Port thimble

18" dia.
formed
air ducts

Flow
axis

50°

50°

72" dia. valve port   19'-6" *R*

(b) Tower section

Figure 7-20 Selective withdrawal intake tower,
Castaic Dam, California (2)

meet temperature and water quality standards may require a special intake
with openings at several elevations.* Outlets for evacuating the reservoir, sometimes called sluices, will be placed as low as possible, making allowances for
sediment deposition. If downstream requirements are different with respect to
water temperature, separate outlets may be required at different levels if the
reservoir is stratified; for example, fish may require cold water during the
spawning season, whereas irrigation may require warm water to promote faster
plant growth.

Intake structures can vary from a simple submerged opening on the upstream face of a dam to an elaborate free-standing tower (Fig. 7-20). Outlet
works conduits through dams are usually designed for high velocities because
of the availability of a substantial head to produce these high velocities and
the economic advantage that accompanies the smaller conduit sizes made possible by high velocity. The high velocity, however, requires that the designer
carefully consider the hydraulic characteristics of the intake flow passages.

---

* For more information on this, see the U.S. Army Corps of Engineers publication on selection
withdrawal intake structures (18).

Nonstreamlined entrances cause flow separation from the boundaries that lead to many undesirable flow conditions whose effects are accentuated by high velocities. These include: pressure and velocity fluctuations that can lead to structural vibration, cavitation, noise, and energy loss.

Inlet transition shapes having elliptical curves are frequently designed to follow the shape of a free jet. Although this shape provides good flow conditions, other simple circular shapes can also be used provided they do not cause flow to separate from the boundaries. Flows in outlet works must be closely regulated so that downstream water needs can be satisfied. This regulation is achieved with a gate or valve designed to function throughout the entire range of flow; thus the gate or valve must be able to operate at any gate opening. For maintenance of the conduit, additional gates are also required upstream from the operating gates or valves. These gates frequently are closed or opened only under zero flow conditions created by closing the downstream operating gate. This simplifies the upstream gate's design, since adverse operating conditions usually occur for partial gate openings where the large separation zone behind the gate produces excessive turbulence. However, when the downstream valve is inoperable, the gate must be able to be closed. The regulating valve may be located at the downstream end of the conduit (see Fig 7-19a), which allows for more uniform conduit flow — since the throttling effect of the control valve would otherwise produce high velocity flow in the zone downstream from the control — and which tends to minimize vibration and cavitation effects within the conduit — since most energy will be dissipated downstream from the conduit. However, the static pressure within the conduit is large, requiring substantial structural strength. Depending on the characteristics of the site, regulating valves or gates can be installed on the upstream faces of dams or in chambers within the dam itself.

Many types of gates and valves are available to the designer for operation or emergency use. Slide or fixed-wheel gates are often used for controlling flow at intakes and interior locations in the conduit, whereas regulating the downstream end is usually accomplished, particularly under high head, with needle or hollow jet valves like the Howell-Bunger valve (Fig. 7-21).

Fish Passage Facilities

Many rivers and streams are used by anadromous fish such as salmon or steelhead trout to move between the spawning areas in upstream watersheds and to the ocean where they spend much of their adult life. Extremely important in planning a project involving dam construction on these rivers is preserving the fisheries resource. High dams are incompatible with fish passage. Besides costly but often ineffective passage facilities, the long, deep reservoirs upstream from high dams produce velocities so slow they can delay the movement of downstream migrants to where the young fish do not reach the ocean in sufficient numbers to support the run. In the early years of dam construction in

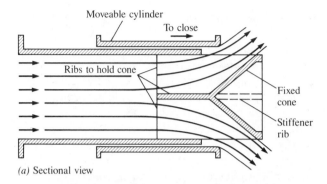

(a) Sectional view

(b) Shop assembly photo
(Courtesy Allis Chalmers, Inc.)

Figure 7-21   Howell-Bunger valve

the United States, many high dams, such as Coulee Dam on the Columbia River in Washington, were constructed without fish passage facilities and thus eliminated fish runs upstream from the dam. Except for Chief Joseph Dam, immediately downstream from Coulee Dam, all other dams on the U.S. portion of the Columbia River have fish passage facilities. Even so, runs are greatly reduced from historical highs, and efforts are now under way to enhance the number of fish in the river. One example of where the economic value of the anadromous fish run has been emphasized is the Fraser River in British Columbia, which despite its tremendous potential for power generation, has experienced no dam construction (except on some of its tributaries).

Facilities built to permit upstream passage, called fish ladders (Fig. 7-22), comprise a series of pools connected to one another with weirs and submerged orifices. The fish usually pass through the orifices as they move in the upstream

Figure 7-22   Fish-passing facilities at Rocky Reach
Dam, Washington (Courtesy Chelan County Public
Utility No. 1)

direction from pool to pool. Fish moving downstream either pass through the
turbines, over the spillway, or are diverted into a collection device for safe passage by the dam. Although fish may suffer only a 10% mortality rate in passing
through the turbines, predators in the tailrace area increase the mortality of
the temporarily disoriented fish. If the fish must pass a series of dams before they
reach the ocean, the cumulative effect of the dams can decimate the run. Recent
efforts have been to divert fish from the intakes of turbines, such as the traveling
screen shown in Fig. 7-23.* Most of the downstream migrants swim in the top
15 ft of the river. Therefore, as the fish are swept into the intake of the power
plant, they are near the top of the intake structure. As the water passes through
and around the traveling screen near the upper part of the intake, the fish are
diverted in the downstream direction along the top of the screen (from *A* toward
B in Fig. 7-23) and from there upward into the gate well (toward *C*). From the
gate well, part of the flow and the fish are carried in a separate channel (fish
bypass gallery) to the river downstream from the dam.

During migration season, fisheries agencies require minimum spill to assist
downstream migration. If spill is excessive, however, and the tailwater area is

---

* The screen is in the form of a closed loop supported on a frame. Rollers on each end of the frame
allow it to operate like a belt on pulleys. In the configuration shown in Fig. 7-23, the screen is driven
in a clockwise sense at a speed of about 0.1 ft/s.

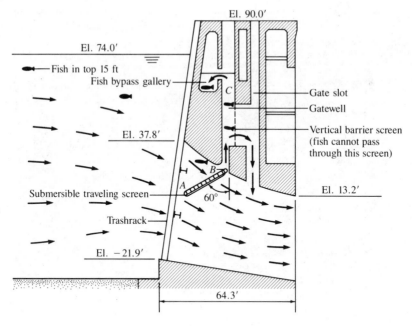

Figure 7-23 Powerhouse intake traveling screen,
Bonneville Second Powerhouse, Columbia River

deep, entrainment of air in the flow can lead to excessively high dissolved nitro-
gen levels in the water, which in turn, can cause physical injury to the fish, further
increasing their mortality.

7-3   Hydroelectric Facilities

Intake Structures

Intake structure design for hydroelectric plants is frequently affected
by nonhydraulic considerations such as geologic formations that dictate location and orientation, which may not produce the best flow conditions. Moreover,
conventional design practice usually requires separate structures for spillways
and power plants. This results in many intake structures constructed to the side
of the main river channel (Fig. 7-24), causing flow to turn before entering the
intake. Because of space limitations, powerhouses may be oriented parallel to
the flow, causing flow to turn into the intakes. This change in direction causes
separation from the banks of the intake channel, leading to the formation of
eddies and, possibly, if the submergence is insufficient, vortices over the intakes.
Vortices should be prevented from forming over intakes because they lead to

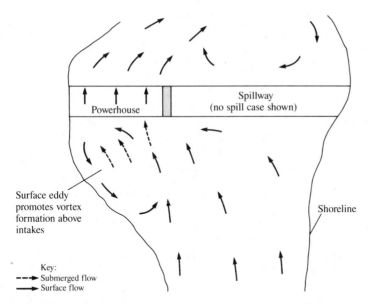

Figure 7-24   Effect of powerhouse location on vortex formation*

* Most of the surface eddy would be eliminated if the
  approach flow was perpendicular to intake face.

inefficient approach conditions, can entrain air, which reduces the discharge capacity, and may lead to damaging vibrations in the rotating machinery.

Other conditions associated with the intake channel that affect hydraulic conditions are the magnitude of the velocity and the divergence or convergence of the channel, as shown in Fig. 7-25. High velocities can cause vortices and aggravate the effects of flow separation. Converging channels in the downstream direction, particularly if the channel is straight, tend to produce good flow conditions, whereas diverging channels tend to produce separation and poor flow behavior.

Although submergence is important in determining the extent of vortex formation, the geometry of the intake structure also influences the character of flow. Most intake designs will draw flow at some depth below the water surface. This design feature inherently tends to lead to stagnated and circulating flow above the intakes that can produce vortices. This condition can partially be overcome if the intake roofs curve up to the free water surface (Fig. 7-26); this withdraws the surface water and inhibits vortex formation. If the intake roof, as shown in Fig. 7-26, does not extend to the surface, the slope on the upstream face of the intake structure becomes an important consideration. A wall that leans in the downstream direction promotes vortex activity. A vertical wall is usually preferred because an upstream sloping wall, although having good vortex-inhibiting

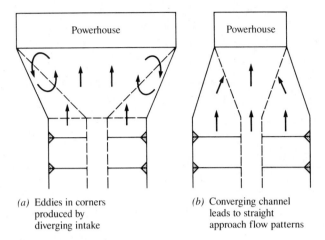

(a) Eddies in corners
produced by
diverging intake

(b) Converging channel
leads to straight
approach flow patterns

Figure 7-25   Effect of intake channel divergence or
convergence on vortex formation

characteristics, also inhibits trash removal. Protuberances, such as piers projecting out from the intake face, should be avoided, because circulating flow tra-
versing the face of the intake may trigger vortices in the lee of the projecting
piers. Short vertical walls parallel to the intake face and suspended from the
piers (Fig. 7-26) and piercing the water surface, are effective in reducing vortex
formation.

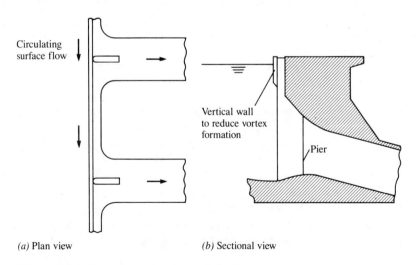

Circulating
surface flow

Vertical wall
to reduce vortex
formation

Pier

(a) Plan view

(b) Sectional view

Figure 7-26   Powerhouse intake configuration

Trashracks

Trashracks are installed at intakes to prevent large debris from being carried into the turbines. They consist primarily of vertical or nearly vertical steel bars supported by steel beams (see Fig. 7-23). Headloss through the trashracks depends on the bar thickness and spacing. Although some attempts have been made to reduce headlosses through the trashrack by streamlining the individual rack elements and the main supporting members, trashracks usually are not streamlined. This lack of streamlining, when combined with poor approach flow, can lead to vibration and possible failure. Vibration is of particular concern in pumped storage projects where reverse flow at the intake during the pumping cycle can separate from the flow boundaries and produce velocities through the racks that far exceed those experienced during the generation cycle.

Draft Tubes

Any residual velocity retained by flow entering the tailrace downstream from a hydraulic turbine represents a reduction in the available head for producing power. High specific-speed turbine runners, which are usually selected to reduce equipment and construction costs, result in high velocity flow leaving the runner. If this flow were immediately released to the stream, a substantial fraction of the total energy would be wasted in the case of a low-head plant. To recover some of this energy, the flow is expanded carefully and gradually in a diffuser called a draft tube.

The simplest type of draft tube is a conical expansion. When used with vertical axis turbines, however, the conical diffuser is also vertical, thus requiring excavations well below tailwater level. To overcome the high cost of this excavation, conventional draft tube design consists of an elbow-type configuration, as Fig. 7-27 shows. Draft tubes are usually designed by the turbine manufacturer as part of the turbine system. Experience, laboratory testing, and an understanding of the principles of flow separation are required for effective draft tube design.

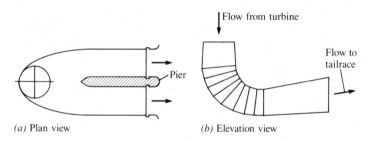

(a) Plan view   *(b)* Elevation view

Figure 7-27   Typical draft tube for a Francis turbine

The underlying objective is to expand the cross-sectional flow area at a rate gradual enough to prevent separation, thus preventing vibration in the system and minimizing loss of energy.

Tailraces

The open channel zone immediately downstream from the draft tube is called the tailrace. Its purpose is to convey water back into the stream with a minimum of energy loss, with flow patterns not detrimental to other parts of the project, and to isolate the powerhouse from the effects of hydraulic components such as spillways.

Tailrace energy losses can be minimized by orienting the plant so that flow from the draft tubes is already oriented in the stream direction. Examples of two plants associated with one project (Rock Island, Columbia River), one having undesirable flow orientation and the other, desirable, are shown in Fig. 7-28. Rock Island Dam, the first dam built (in 1933) on the Columbia River, has the initial powerhouse constructed on the left bank (right side of Figure 7-28) in such a way that flow from the draft tubes has to move laterally past the face

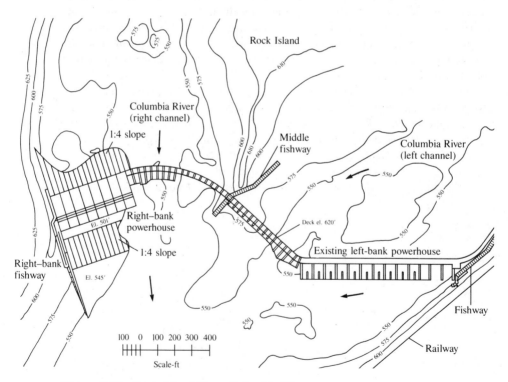

Figure 7-28 Rock Island Dam, Columbia River

of the plant before reentering the main river channel. This type of flow, called spatially varied flow, is inefficient, particularly when the tailwater depths are low. The second powerhouse at Rock Island, shown on the right bank, is oriented so that flow in the tailrace is aligned directly with the main channel flow, which minimizes energy loss.

Other hydraulic considerations in designing tailraces involve protection of the channel bed and adjoining embankments from velocities and waves that could produce scour. Other important aspects of tailrace design are creating flow patterns that produce currents that will attract migrating adult fish and direct them toward fish ladders and eliminating eddies that trap downstream migrants. Dividing walls are used to separate the tailrace from the spillway stilling basins to prevent wave action and adverse currents from propagating their effects upstream through the draft tubes to the turbines.

7-4 Pump Intake Structures

An important hydraulic design consideration associated with most pump installations is the configuration of the pump intake structure. Flow approaching the inlet of a pump impeller should have near axial flow with some radial component and uniform flow patterns to assure efficient and safe operation. Radial flow entering the pump bell is a good indication that these flow patterns will exist. Approach flow that is nonuniform or consists of circumferential velocity components (swirl), as shown in Fig. 7-29, can lead to severe asymmetrical loading on the pump impeller, pump vibration, and cavitation. Many pumps fail when forced to operate under these adverse conditions. Even if pumps do not fail, poor approach conditions produce inefficiency, vibration, and noise. All vortices, including both surface and subsurface types, should be

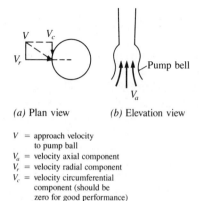

(a) Plan view   *(b)* Elevation view

V = approach velocity
to pump ball
V_a = velocity axial component
V_r = velocity radial component
V_c = velocity circumferential
component (should be
zero for good performance)

Figure 7-29 Approach flow patterns to a pump

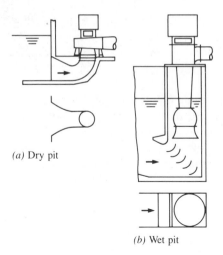

(a) Dry pit

(b) Wet pit

Figure 7-30   Typical pump installations

eliminated. Surface vortices may entrain air from the surface, resulting in a reduction in capacity as well as the other detrimental factors mentioned.

Pump installations can be classified as either a dry-pit (flow is delivered to the pump suction port through a conduit) or a wet-pit (fluid surrounds the pump casing or suction pipe) installation, as shown in Fig. 7-30. Flow patterns in pump intake structures are complex and cannot be computed analytically. Thus designs are based either on published standards where design is similar to previously documented installations or on reduced-scale model studies. Wet-pit installations in the United States often are designed in accordance with the Hydraulic Institute standards (11). Figures 7-31 and 7-32 show pit dimensions recommended by the Hydraulic Institute.*

7-5   Cavitation in Hydraulic Structures

When a liquid has its temperature increased under constant pressure or its pressure reduced under constant temperature, a state is eventually reached where bubbles, or cavities, begin to appear and grow in the liquid. The growth rate is moderate if dissolved gases are diffusing into the bubble but can be explosive if the growth is primarily a result of vaporization of the liquid into the cavity. The explosive growth rate is called boiling when it results from increased temperature and cavitation when produced by a hydrodynamically induced pressure reduction.

---

* A comprehensive report on pump intake structures has also been prepared by Prosser and is published by the British Hydromechanics Research Association (16).

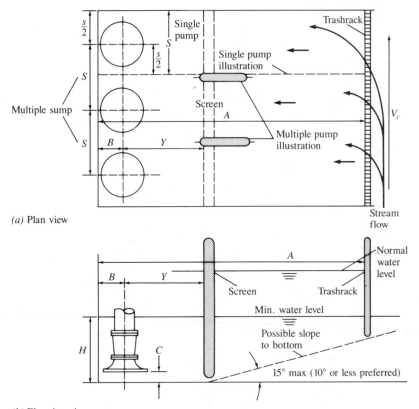

(a) Plan view

(b) Elevation view

Figure 7-31 Proportions for pump sumps (Adapted from *Hydraulic Institute Standards*, 14th ed. © 1983 by the Hydraulic Institute, 712 Lakewood Center North, 14600 Detroit Av., Cleveland, OH 44107)

Cavitation is a speed-related phenomenon in which locally increased velocities are associated with pressure reductions. Its occurrence in a flow field usually causes reduced performance or potential damage to the flow surfaces. Early observations of cavitation were noted in the propellers of steam-driven vessels, particularly, in war ships, where achieving higher speeds was a design objective. Other cavitation problems began to appear in pump and turbine installations. High velocities associated with the rotating elements of these hydraulic machines will produce cavitation when the ambient pressure is too low. Hydraulic structures have also exhibited several failures created by cavitating conditions. These failures generally result from some combination of high velocity flow and boundary discontinuities either into or away from the flow that produces locally low pressures. In particular, spillways, stilling basins, and gate slots require that separated flows, conduit misalignment, and boundary roughness be carefully considered.

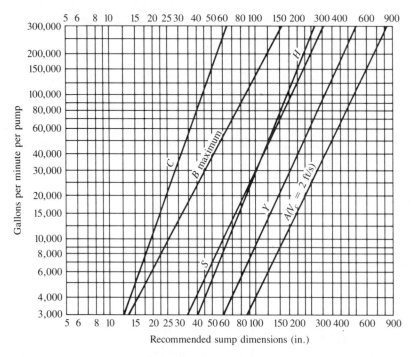

Figure 7-32   Pump sump dimensions for Fig. 7-31
(Adapted from *Hydraulic Institute Standards*, 14th ed.
© 1983 by the Hydraulic Institute, 712 Lakewood
Center North, 14600 Detroit Av., Cleveland, OH
44107)

Causes of Cavitation

For cavitation to occur, there must be a source of low pressure, and there must be nuclei (gas bubbles or other foreign material that carries gas) in the flow. Although nuclei are important in laboratory experiments, most hydraulic structures operate with water of relatively uniform quality, and the availability of nuclei is generally sufficient to allow cavitation to develop. To complete the cavitation process, the cavity, formed under low pressure, will eventually be exposed to high pressure again at some downstream location, whereupon it collapses. The collapse mechanism is largely responsible for the erosion of material, excessive noise, damaging vibration, and other adverse effects of cavitation.

Control of Cavitation

One obvious way to prevent cavitation is to eliminate the source of low pressure. For example, a lower setting of the turbine in a hydropower plant will increase the pressure and inhibit cavitation from forming. Designs that eliminate eddy formation are equally important in reducing sources of low pressure. Limiting the velocity to a low value will also reduce the level of cavitation. Frequently, it is neither possible nor desirable to design hydraulic structures that will be free from cavitation. In these cases, where erosion of material is the primary concern, the effects of cavitation can be mitigated in several different ways:

1. The boundaries may be designed so that the cavities collapse well out in the flow field where high pressures due to the collapse cannot act on the boundary.
2. Cavitation-resistant materials, such as stainless steel, or fiber-reinforced concrete can be used to greatly retard the damage rate, even though the system is still in a cavitating regime.
3. Introducing air into the flow changes the type of cavitation from vaporous to gaseous, which eliminates the explosive growth and complete collapse of the cavities, and thus reduces the damaging effects.

Cavitation Index

To predict cavitation, the hydraulic engineer uses a dimensionless parameter that correlates the forces preventing cavitation (ambient pressure) with those causing cavitation (dynamic pressure). A cavitation index is defined as follows:

$$\sigma = \frac{p_0 - p_v}{\rho V_0^2/2} \qquad (7\text{-}9)$$

where σ is the cavitation index, p_0 is the ambient pressure, p_v is the vapor pressure, ρ is the mass density of the fluid, and V_0 is the velocity of the fluid relative to the body. High values of σ correspond to low cavitation potential. As σ decreases, either from a decrease in ambient pressure or an increase in velocity, a value for σ is eventually reached where cavitation begins. This cavitation inception value is designated as σ_i. The variation in pressure between two points in a flow field can be expressed in the conventional way as follows:

$$C_p = \frac{p - p_0}{\rho V_0^2/2} \qquad (7\text{-}10)$$

where C_p is the pressure coefficient, and p is the pressure at a point on the body. In the particular case where the pressure p is lowered to the point of cavitation inception, it attains a value close to p_v. For this special case,

$$\sigma_i = -C_p \qquad (7\text{-}11)$$

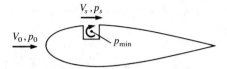

Figure 7-33   Role of eddies in cavitation

If σ is greater than σ_i, no cavitation will occur. If σ is equal to or less than σ_i, cavitation will occur.

Separated Flows

We discussed the role of eddies in producing cavitation earlier. An example of flow separation is shown in Fig. 7-33, where a slot has been installed in a streamlined strut. In this example, an eddy is set up in the slot so that the pressure in the center of the eddy is reduced from the reference pressure p_0 by both the geometry of the strut and the presence of the eddy.

The velocity V_s, adjacent to the slot, can be expressed in terms of the upstream velocity V_0 and the pressure coefficient C_p by

$$V_s = V_0(1 - C_p)^{1/2} \tag{7-12}$$

By assuming the velocity distribution in the slot eddy is approximated by that of a forced vortex (12), the pressure at the center of the eddy, p_{min}, is given by

$$p_{min} = p_s - \rho \frac{V_s^2}{2} \tag{7-13}$$

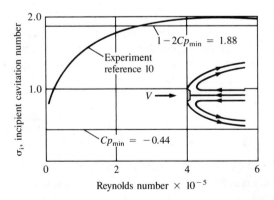

Figure 7-34   Incipient cavitation number for discs (12)

where p_s is the pressure at the edge of the slot. Combining Eqs. (7-11), (7-12), and (7-13) leads to the following prediction for the critical cavitation index, σ_i:

$$\sigma_i \text{ (separated flow)} = 1 - 2C_p \tag{7-14}$$

Experiments show that Eq. (7-14) predicts σ_i quite well for high Reynolds numbers (Fig. 7-34). However, at low Reynolds numbers, the V_s used in Eq. (7-13) is reduced from the theoretical value because of the larger viscous effects in the boundary layer. Smaller V_s values yield smaller values of σ_i. The σ_i values given by Eq. (7-14) should therefore be considered the upper limit for σ_i.

EXAMPLE 7-1   The velocity $V_A = 10$ ft/s when $V_0 = 5$ ft/s. If the converging boundary shown in the sketch is in a horizontal plane, determine the value of V_0 that will just begin to cause cavitation in the slot B.

SOLUTION   For point A, the pressure coefficient is

$$C_p = 1 - \frac{V_A^2}{V_0^2}$$

$$= 1 - \left(\frac{10}{5}\right)^2 = -3$$

$$\sigma_{i_B} = 1 - 2C_p = 1 - 2(-3) = 7$$

$$= \frac{p - p_v}{\rho V_0^2/2}$$

$$7 = \frac{20(144) - 37}{1.94 V_0^2/2}$$

$$V_0 = 20.5 \text{ fps} \qquad \blacksquare$$

Cavitation in Spillways

For many years, the design of spillways has used crest profiles that minimize low pressures. Spillways associated with high dams, however, produce high velocities that combine with moderate roughness or other irregularities, such as construction joints on the surface, to produce cavitating conditions.

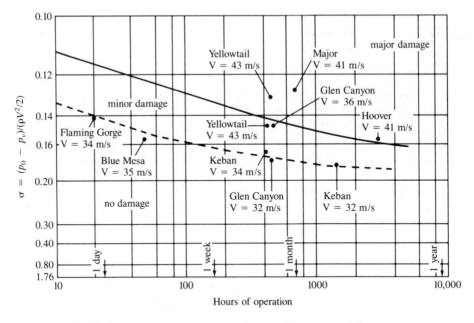

Figure 7-35   Cavitation damage in spillway tunnels (9)

Serious damage occurred in the high-velocity tunnel spillways at Hoover Dam (8) and Yellowtail Dam (4). The 1984 flood on the Colorado River led to similar damage at Glen Canyon Dam and again, though to a lesser degree, at Hoover Dam. Aeration slots have since been added to mitigate the effects of cavitation at these structures.

One of the requirements in designing for cavitation is to make an estimate of the σ-value that will prevent excessive damage. Damage is not just a function of σ but also of the time of operation. Experience with U.S. Bureau of Reclamation structures (21), shown in Fig. 7-35, illustrates this.

7-6   Experimental Design

Introduction

Although hydraulic structures often have simple shapes, such as round straight pipes, many structures have unique boundaries that defy an analytically based design or a design based on a previously constructed geometrically similar structure. To predict many of the flow details associated with these structures is virtually impossible without directly observing the performance of the actual structure. Even the performance of a simple, round pipe cannot be predicted from first principles for turbulent flow and requires experimentally determined coefficients. Pretesting cannot, of course, be done with large structures; thus the

Figure 7-36   Boundary Dam Model (Courtesy of the
Albrook Hydraulics Laboratory, Washington State
University, Pullman, Washington)

design of most important structures is guided by the results gathered from tests
conducted on a reduced-scale model, such as shown in Fig. 7-36.

The prediction of full-scale (prototype) performance from model tests is
based on widely used laws of similitude.* In this section, we review the important considerations related to hydraulic structures and introduce several areas
not addressed in a first exposure to similitude applications.

The fundamental scaling relationships are often developed by either dimensional analysis or use of a physical interpretation that develops the same scaling
relationships by maintaining equal ratios of forces in the model and prototype.
For example, in a hydraulic structure, the most important forces generating a
flow pattern and pressure distribution are gravity and viscous forces. In fluid

---

* These are described in elementary texts on fluid mechanics; see, for example, Roberson (14).

mechanics, acceleration is commonly referred to as the inertia force per unit mass. The formulation of equal ratios of model and prototype inertia forces to viscous forces and of equal ratios of inertia forces to gravity forces produces dimensionless numbers referred to as the Reynolds and Froude numbers, respectively.

Reynolds and Froude Scaling Parameters

The Reynolds number Re is defined by

$$\text{Re} = \frac{VL}{v} \tag{7-15}$$

where V and L are, respectively, a characteristic velocity and length, and v is the kinematic viscosity.

The Froude number Fr is given by

$$\text{Fr} = \frac{V}{\sqrt{gL}} \tag{7-16}$$

where g is the acceleration due to gravity.

Before the Froude and Reynolds numbers are used for developing scaling ratios, we first must develop the general scaling ratio for force from Newton's second law written for both model and prototype as follows:

$$F_m = M_m a_m \tag{7-17}$$

$$F_p = M_p a_p \tag{7-18}$$

where F is force, M is mass, a is acceleration, and the subscripts m and p, respectively, refer to model and prototype.

Dividing Eq. (7-17) by Eq. (7-18) produces

$$F_r = M_r a_r \tag{7-19}$$

where the subscript r refers to the ratio: model to prototype. Eq. (7-19) yields

$$F_r = (\rho_r L_r^3)\left(\frac{L_r}{T_r^2}\right) \tag{7-20}$$

$$= \rho_r \frac{L_r^4}{T_r^2}$$

where ρ_r is the fluid mass density ratio between model and prototype. Similarly, T_r is the time ratio, and L_r is the length ratio.

Eq. (7-20) may be written in terms of a fluid velocity ratio, since $V_r = L_r/T_r$. Accordingly,

$$F_r = \rho_r L_r^2 V_r^2 \qquad (7\text{-}21)$$

If maintaining *equal Reynolds numbers* in model and prototype is necessary, the velocity ratio is

$$V_r = \frac{v_r}{L_r} \qquad (7\text{-}22)$$

This combines with Eq. (7-21) to yield for the force ratio

$$F_r = \frac{\mu_r^2}{\rho_r} \qquad (7\text{-}23)$$

If the simulated flow system requires *equal Froude numbers* in model and prototype, the velocity ratio is

$$V_r = \sqrt{g_r L_r} \qquad (7\text{-}24)$$

This leads to other scaling ratios for force (F_r), volumetric flow rate (Q_r), and pressure (p_r) where g_r is assumed to be 1:

$$F_r = \rho_r L_r^3 \qquad (7\text{-}25)$$

$$Q_r = L_r^{5/2} \qquad (7\text{-}26)$$

$$p_r = \rho_r L_r \qquad (7\text{-}27)$$

If both the Reynolds and Froude numbers were to be kept the same in model and prototype, Eqs. (7-22) and (7-24) lead to the following requirement:

$$L_r = v_r^{2/3} \quad \text{(assuming } g_r = 1\text{)} \qquad (7\text{-}28)$$

Eq. (7-28) is difficult to satisfy because it has only one degree of freedom. The choice of a length scale L_r immediately dictates the fluid viscous properties. On the other hand, choice of a model fluid determines the length scale. It is rarely convenient, or even possible, to satisfy this requirement in simulating flow in hydraulic structures. Thus considerable attention must be given to assessing the relative importance of maintaining model and prototype Reynolds and Froude number equality.

It is usually not possible to operate the model at prototype Reynolds number. Fortunately, however, conducting reduced scale tests at a Reynolds number high enough to guarantee acceptable similitude but less than the prototype value is usually sufficient. This procedure is permissible because Reynolds

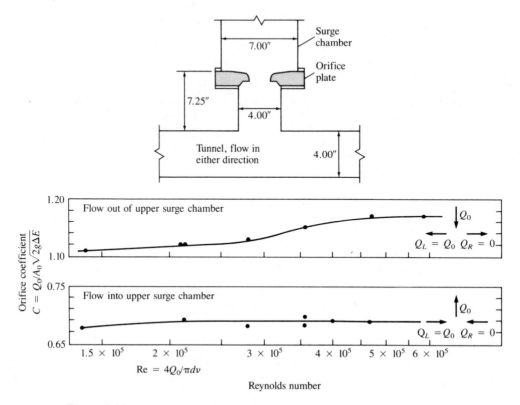

Figure 7-37 Variation of orifice coefficients with Reynolds number for a surge tank orifice

number effects most often are significant only up to some limiting value of the Reynolds number; above this, forces and other parameters tend to be independent of the Reynolds number. For example, Fig. 7-37, which is a plot of discharge coefficient versus Reynolds number for flow through a model of a surge tank orifice, illustrates that the coefficient does not vary significantly with Reynolds number above a value of 5×10^5 for flow out of the surge tank and 2×10^5 for flow into the surge tank.

If gravity forces are important in shaping the flow patterns, such as often occurs in flow situations where there is an interface between fluids of different densities, the Froude number should be made equal in model and prototype. The most commonly occurring case with hydraulic structures is a water-air interface. However, the simulation of all these free-surface flows does not require Froude number equality; for example, wave motion or rapidly varied flow such as over a spillway crest require Froude number equality, but flow in a low-velocity canal may not. Requirements for similitude are not always straightforward, so design and laboratory engineers must show judgment and have experience in designing a test program.

One special type of gravity-dominated flow involves two liquids of slightly different densities. This flow could involve a freshwater-saltwater interface in an estuary, a sewage outfall discharging into the ocean, or a discharge of heated water from a power plant into a cooler body of water. For this last case, formulating the inertia to gravity force ratio produces a dimensionless number, termed a densimetric Froude number, that must have the same model and prototype values. The densimetric Froude number has the form

$$\text{Fr} = \frac{V}{\sqrt{\left(\frac{\Delta\rho}{\rho}\right)gL}} \tag{7-29}$$

where $\Delta\rho$ is the difference in mass densities between the two fluids, and ρ is the mass density of one of the fluids. Maintaining both the ordinary Froude number and the densimetric Froude number equal in model and prototype is often necessary and leads to scaling ratios similar to those for the ordinary Froude number. In this case $(\Delta\rho/\rho)_r = 1$.

Modeling procedures frequently involve other variations, depending on the objectives of the study, such as movable beds, scale distortion, and hydroelastic modeling. Movable beds are mostly used when scour prediction or sediment deposition is a major objective of the study. The simulation of prototype scour in a model is not always easy, and the bed material is chosen after considering Reynolds number, size, and density of the material. A movable bed model should be verified by comparing values such as velocity or river stage with prototype data. A distorted model is one in which the vertical scale is larger than the horizontal scale; for example, a model having a vertical scale of 1:20 and a horizontal scale of 1:100 would have a distortion ratio of 5. Models are distorted for many reasons, including achieving higher model Reynolds numbers, better simulation of bed movement in movable bed models, and reducing overall model size.

Hydroelastic Modeling

Hydroelastic modeling is done when flow-induced vibrations are important and the hydraulic structure moves as a result of fluid-induced forces. The movement of the structure influences the magnitude and frequency of the forces and there is continuous interaction between the structure and the fluid forces. For this type of model, the elastic properties should be simulated for proper reproduction of the hydroelastic forces. A common application occurs in the study of large gates, where high speed flow under a partially open gate may produce fluctuating high amplitude pressure forces under the gate. Gates of this type are frequently suspended with steel beams that act as stiff springs when subjected to a variable force. Proper modeling of this system involves simulation of the beam stiffness and natural frequency.

EXAMPLE 7-2   The sketch shows a diffuser-port check valve for an outfall structure that is to be studied in a 1:4 scale model under unsteady conditions produced by wave motion. Assume operation of the flap of the check valve is governed by the following expression relating load to deflection ($\Delta \propto FL^3/EI$), where Δ is deflection, F is force on the flap consisting of both hydrodynamic drag and net flap weight (W), E is elastic modulus, and I is the moment of inertia. If $E_r = 1$ (same material in model and prototype), and choosing $a_r = 1$ (this infers that the Froude number is the same in model and prototype), determine W_r, I_r, V_r, T_r (time ratio), and f_r (frequency ratio). How can the W_r and I_r requirements be satisfied? Assume water in both model and prototype.

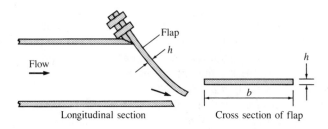

Longitudinal section   Cross section of flap

SOLUTION

$$F_r = \rho_r L_r^3 a_r = L_r^3$$

$$W_r = F_r = L_r^3 = (1:4)^3 = 1:64$$

$$a_r = \frac{L_r}{T_r^2} = 1 \quad \text{or} \quad T_r = L_r^{1/2} = (1:4)^{1/2} = 1:2$$

$$V_r = \frac{L_r}{T_r} = L_r^{1/2} = 1:2$$

$$f_r = \frac{1}{T_r} = \frac{1}{(L_r)^{1/2}} = \frac{1}{1:2} = 2:1$$

The moment of inertia $I \propto bh^3$, where h is thickness of the material

$$\therefore \Delta_r = \frac{W_r L_r^3}{E_r I_r} = \frac{L_r^3 \cdot L_r^3}{1 \cdot I_r} = L_r$$

$$I_r = L_r^5 = b_r \cdot h_r^3 = L_r h_r^3$$

$$\therefore h_r = L_r^{4/3} = 1:6.35$$

Note that h_r is less than L_r (1:4). If E_r was chosen to be L_r, then h_r would equal L_r. Using the computed value of h_r, additional weight must be added to the flap of the model check valve.   ■

7-7   Diffusers for Wastewater

The Purpose of Diffusers for Wastewaters

Coastal cities often discharge their wastewater (usually treated) underwater far enough from shore so that the wastes become sufficiently diluted with the ambient fluid to render the resulting mixture harmless. The design of such a system requires detailed information about the body of water into which the wastewater is discharged. Such information will include salinity and temperature of the ambient fluid and data on coastal currents. Also one must have information about the bottom characteristics (depth variation and whether the bottom is composed of mud, rock, or sand) of the ocean, bay, or lake in the vicinity in which the diffuser is to be located. With this information and with discharge rate and temperature of the wastewater one can design a diffuser to dispose of the wastewater. Because of the need for large dilutions of the wastewater (in the order of 100 to 200), it is usually necessary to discharge it at considerable depth. Depths from 100 to 250 ft have been used for disposing of wastewater from large coastal cities.

In this section, we present the basic fluid mechanics associated with the mixing process that produces the dilution and then the procedure for designing the diffuser will be presented.

Fluid Mechanics of a Jet Discharging into an Ambient Fluid of Same Density

ROUND JET   Consider a jet of fluid from a pipe or orifice being discharged into an ambient fluid of the same density such as shown in Fig. 7-38. If the Reynolds number of the jet is great enough, turbulent mixing between the jet and the ambient fluid will occur. Labus (13) has shown that fully turbulent mixing will occur if the Reynolds number of the round jet ($w_0 D/v$) is greater than about 4000. If the jet is wastewater, the wastewater will become diluted as the jet mixes with the ambient water. And as mixing occurs, the velocity of the resulting mixture decreases with distance (z) from the origin of emission. Fig. 7-38 shows how the velocity w is distributed downstream of the origin. Note that the velocity not only decreases with z but also with radial distance x from the centerline of the jet. If w_m is the maximum velocity for any given distance z downstream of the origin, w_m is given as

$$w_m = 6.2 w_0 \left(\frac{D}{z} \right) \tag{7-30}$$

Equation (7-30) is valid for $z/D > 10$.

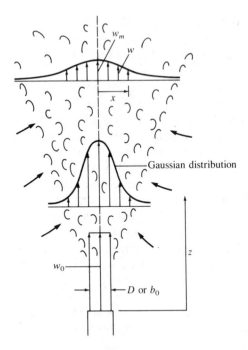

Figure 7-38 Definition sketch for jet discharging into ambient fluid of equal density

Several researchers (1, 15) have shown that for any value of z, the velocity will reduce with increasing values of x according to a Gaussian distribution, as shown in Fig. 7-38. The formula for w as a function of w_m and x is

$$w = w_m e^{-8.7(x/z)^2} \qquad (7\text{-}31)$$

The mean dilution of the wastewater as given by Fischer (10) is $\bar{S} = 0.282\,(z/D)$. Here, the dilution, $\bar{S}$, is the local volume flux divided by the initial discharge. Thus for a jet having an initial diameter of 1 ft, for example, the average dilution 100 ft downstream of the origin of the jet would be 28 (the mixture of the wastewater and ambient fluid at $z = 100$ ft would be 1/28 as concentrated with contaminant as the raw wastewater).

PLANE JET   A plane jet (also called a two-dimensional jet) occurs when fluid is discharged from a slot instead of a round orifice or pipe. Then, in theory, the plane jet is infinitely long (dimension normal to the plane of the page in Fig. 7-38), and its width is identified as b_0 instead of D. The formulas for w_m, w, and $\bar{S}$, obtained from Fischer (10), for the plane jet are given as

$$w_m = 2.41 w_0 \left(\frac{b_0}{z}\right)^{1/2} \qquad (7\text{-}32)$$

$$w = w_m e^{-74(x/z)^2} \tag{7-33}$$

$$\bar{S} = 0.50 \left(\frac{z}{b_0} \right)^{1/2} \tag{7-34}$$

Fluid Mechanics of a Plume

A plume is flow that is mainly driven by the buoyancy resulting from the density difference between the fluid of a jet and the ambient fluid. Consider the case shown in Fig. 7-39, where fluid is discharged from a pipe into an ambient fluid. Assume the fluid from the pipe has lower density than the surrounding fluid. The mixing action will initially be driven by the momentum of the jet, but Fischer points out that in a relatively short distance, the buoyant effects will control mixing, and the resulting flow pattern is called a *plume*.

The significant buoyancy variable is called the initial specific buoyancy flux, B, where

$$B = g \left(\frac{\Delta\rho}{\rho} \right) Q = g'Q \quad \text{(for round plume)} \tag{7-35}$$

$$B = g \left(\frac{\Delta\rho}{\rho} \right) q = g'q \quad \text{(for plane plume)}$$

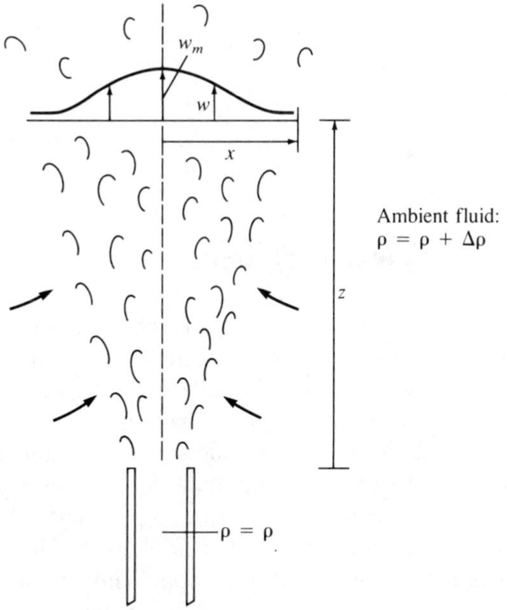

Figure 7-39 Definition sketch for a buoyant plume in ambient fluid of constant density

$$g' = g\left(\frac{\Delta\rho}{\rho}\right) \tag{7-36}$$

where g = acceleration due to gravity

$\quad\Delta\rho$ = difference in density between fluid discharged from pipe and ambient fluid

$\quad\rho$ = density of fluid discharged from pipe

$\quad Q$ = discharge from pipe

$\quad q$ = discharge per unit length for a plane plume

For a simple plume (round or plane) shown in Fig. 7-39, formulas are available for determining the velocity, concentration, and dilution. The formulas for the round plume are

$$w_m = 4.7\left(\frac{B}{z}\right)^{1/3} \tag{7-37}$$

$$w = w_m e^{-100(x/z)^2} \tag{7-38}$$

$$\bar{S} = \frac{0.15B^{1/3}z^{5/3}}{Q} \tag{7-39}$$

The corresponding equations for the plane plume are

$$w_m = 1.66B^{1/3} \tag{7-40}$$

$$w = w_m e^{-74(x/z)^2} \tag{7-41}$$

$$\bar{S} = \frac{0.38B^{1/3}z}{q} \tag{7-42}$$

Design of Diffuser for Sewage Effluent

GENERAL CONSIDERATIONS   The primary objective in discharging sewage to a large body of water is to achieve enough dilution to eliminate its health hazard. Dilution is achieved through turbulent mixing. Usually the wastewater after primary or secondary treatment is carried by a pipe to the body of water it is to be discharged into, and then it is discharged from orifices or nozzles into the ambient water. Usually the wastewater is emitted from multiple ports along the discharge pipe (see a manifold system presented in Chapter 5 page 265) to achieve much more dilution than if it were discharged from a single port. Thus a line of waste-water jets issuing from the main outfall pipe approximates the conditions of a plane jet. Therefore, a plane jet formula is used to define the degree of dilution. Figure 7-40 shows how a typical outfall sewer system might appear. The part of the system the jets are emitted from is called a *diffuser*.

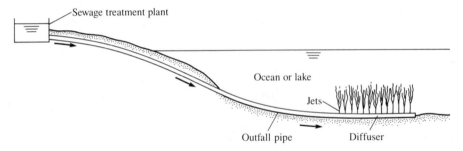

Figure 7-40 Typical outfall pipe and diffuser system

OUTFALL PIPE AND DIFFUSER DESIGN The design process is an iterative one. Many trials involving location, length, and diameter of outfall pipe;
length of diffuser; spacing of ports; and size of ports are made before the final
design is achieved. Various aspects of the design process include determining

1. The distance between shoreline and diffuser to prevent transport of sewage
   back to shore area by ocean currents
2. The depth of diffuser to ensure desired dilution will be achieved
3. The size of outfall and diffuser pipes so that adequate velocities are maintained to prevent deposition of soilds
4. The size of diffuser diameter and port diameters to maintain reasonably
   uniform discharge of effluent from each port

We now discuss how each of these is considered in the design process.

After initial dilution, which should be in the order of 100 or more, the sewage
effluent is further diluted by mixing due to currents in the ambient fluid. Over
time, bacterial decay produces further beneficial effects. The initial dilution is the
most important design objective, but even this initially diluted sewage should not
be allowed to drift back to beach areas. If onshore currents are present, the
diffuser will have to be placed far enough from the beach to provide the time and
additional mixing to render the effluent unobjectionable. Thus knowledge of the
prevailing ocean currents for all seasons of the year is needed to assist in siting
the diffuser. Outfall pipes for major cities (Seattle, Los Angeles, San Diego) range
from 1 to 5 mi in length (10).

Initial dilution is achieved by jet and plume mixing. Assuming we have an
effluent less dense than the ambient fluid and the ambient fluid is of uniform
density, the typical flow pattern for a vertically directed effluent is shown in
Fig. 7-41. The cross section of the diffuser pipe in Fig. 7-41 includes a section
through one of the ports. In a typical diffuser several hundred ports may be
spaced evenly along the length of the diffuser. Spacing usually ranges from 10
to 20 ft. Even though initial mixing is from jet action, Fischer (10) shows that
the predominant mixing results from buoyant effects; therefore, the formula for
the buoyant plume is used to predict dilution. The formula given by Fischer is

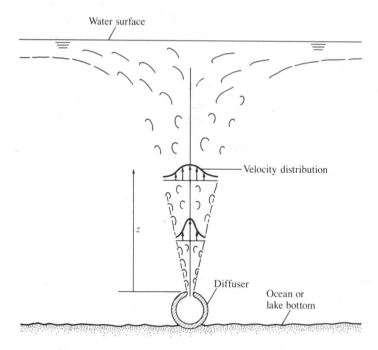

Water surface

Velocity distribution

z

Diffuser

Ocean or
lake bottom

Figure 7-41 Jet discharging from a diffuser pipe

Eq. (7-42). In slightly different form, it is

$$\bar{S} = \frac{0.38g'^{1/3}z}{q^{2/3}}$$ (7-43)

If we wanted to achieve a dilution of at least 200, for example, for a given g' and q, we could determine the minimum z value (approximate depth of diffuser) to achieve that dilution.

EXAMPLE 7-3 What height above a diffuser (z) is required for a flow (q) of 0.01 m²/s (0.10 ft²/s) and a $\Delta\rho/\rho$ of 0.025 to achieve a dilution of 200?

SOLUTION

$$\bar{S} = \frac{0.38g'^{1/3}z}{q^{2/3}}$$

Then $$z = \frac{\bar{S}q^{2/3}}{(0.38 \times g'^{1/3})}$$

where $g' = \dfrac{g\,\Delta\rho}{\rho} = 9.81(\text{m/s}^2) \times 0.025 = 0.25$ m/s².

or $\qquad z = \dfrac{200 \times (0.01)^{2/3}}{(0.38 \times 0.25^{1/3})}$

$\qquad\quad = 38.8$ m ■

As noted by Fischer, the discharge value of 0.01 m²/s and $\Delta\rho/\rho = 0.025$ are typical values used for outfall diffuser systems where wastewater is discharged into the ocean. A depth of 45 m for the diffuser itself is fairly typical.

When a density gradient exists in the ambient fluid, as would be the case where the temperature decreases with depth, the plume may reach a maximum height (less than the elevation of the water surface) and spread out laterally (see Fig. 7-42). For this condition, with a linear variation in density, z_{max} is a function of the density gradient and the other significant variables and as determined by Wallace (22), it is given as

$$z_{max} = 3.6(g'q)^{1/3}\left\{\frac{-g}{\rho}\frac{d\rho_a}{dz}\right\}^{-1/2} \tag{7-44}$$

where $d\rho_a/dz$ is the density gradient of the ambient fluid. For this condition, the minimum dilution in the plume is given as

$$S = \frac{0.24g'^{1/3}z_{max}}{q^{2/3}} \tag{7-45}$$

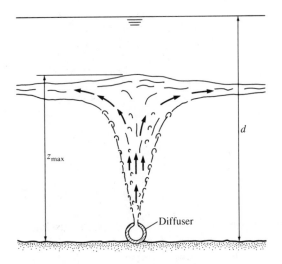

Figure 7-42 Plume in an ambient fluid with temperature gradient

EXAMPLE 7-4   Determine z_{max} and the minimum dilution in a plume for effluent discharging at a rate of 0.015 m²/s from a diffuser 60 m below the ocean surface. The $\Delta\rho/\rho = 0.025$ and the temperature of the ocean varies linearly from 19°C at the surface to 16°C at the bottom. For these temperatures and for a salinity of 3.4%, the densities at 19°C and 16°C will be 1024.28 kg/m³ and 1025.00 kg/m³.

SOLUTION   First calculate g'.

$$g' = g\,\Delta\rho/\rho = 9.81(\text{m/s}^2) \times 0.025 = 0.245 \text{ m/s}^2$$

Now compute the density gradient term of Eq. (7-44), $(1/\rho)(d\rho_a/dz)$.

$$\frac{1}{\rho}\frac{d\rho_a}{dz} = \frac{1}{\rho_{avg}}\left(\frac{\rho_{surface} - \rho_{60m}}{60 \text{ m}}\right)$$

$$= \frac{1}{[(1024.28 + 1025.0)/2]}\left(\frac{1024.28 - 1025.00}{60}\right)$$

$$= -1.17 \times 10^{-5} \text{ m}^{-1}$$

Then from Eq. (7-44),

$$z_{max} = 3.6(g'q)^{1/3}\left(\frac{-g}{\rho}\frac{d\rho_a}{dz}\right)^{-1/2}$$

$$= 3.6(0.245 \text{ m/s}^2 \times 0.015 \text{ m}^2/\text{s})^{1/3}$$

$$\times (-9.81 \text{ m/s}^2 \times -1.17 \times 10^{-5} \text{ m}^{-1})^{-1/2}$$

$$= 52 \text{ m}$$

From Eq. (7-45), the dilution will be

$$S = 0.31g'^{1/3}\frac{z_{max}}{q^{2/3}}$$

$$= \frac{0.31(0.245 \text{ m/s}^2)^{1/3}52}{(0.015 \text{ m}^2/\text{s})^{2/3}}$$

$$= 166 \qquad\blacksquare$$

The diffuser pipes and ports are sized to achieve about the same discharge from each port along the diffuser. To do this, the designer has the option of changing the port diameters and the diffuser pipe diameter. Usually the port spacing is kept constant. To give an idea of port size and spacing already in use,

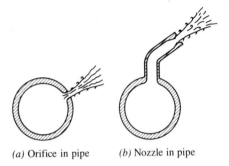

(a) Orifice in pipe   *(b)* Nozzle in pipe

Figure 7-43 Two different types of diffuser ports

Fischer (10) notes that the diffuser at Whites Point No. 4 (Sanitation Districts of Los Angeles County) has port diameters ranging from 2 to 3.6 in. with a spacing of 6 ft, whereas the diffuser for the city of San Diego has port diameters ranging from 8.0 to 9.0 in. with a spacing of 48 ft. Most other large installations fall within these ranges of port size and spacing. In designing the diffuser port sizes, port spacing and diffuser pipe size are initially chosen, and the discharge through each port is calculated using the method for manifold design (see Sec. 5-5, page 265). As one proceeds upstream, the diffuser pipe will be made larger in two or three steps to maintain approximately the same velocity in the main diffuser pipe. Moreover, because one wishes to keep the discharge from every port at about the same rate, the port sizes will usually be decreased in several steps as one proceeds upstream in the design process. If one is designing a diffuser for discharging into ocean water, the difference in density of the effluent and the ambient salt water must be taken into account as one calculates the head at each port from one port to the next. This density-difference effect is relevant only if the diffuser lies on a sloping bottom. That is, if the diffuser slopes downward in the direction of flow, the pressure head across the port (due to density difference) at an upstream port will be greater than at a downstream port by an amount $(\Delta\rho/\rho)\,\Delta z$, where $\Delta\rho$ is the difference in density between the ambient fluid (seawater) and the effluent, ρ is the density of the effluent, and Δz is the change in elevation between the upstream port and downstream port.

The actual ports in the diffuser can be simple orifices with rounded corners in the side of the diffuser pipe, as shown in Fig. 7-43*a*, or they can be in the form of a nozzle-riser configuration, as shown in Fig. 7-43*b*. Experience indicates that the ports should usually be sized so that the summation of the port exit areas will be from one third to two thirds of the outfall pipe cross-sectional area. In Fig. 7-43, the ports are not directed vertically upward, which is done to produce less interaction between the jets of adjacent ports, thus effecting slightly more mixing and dilution than if they were aligned vertically. When the jets are directed horizontally, the trajectory of a jet is longer before reaching the surface than one directed vertically, thereby providing a longer mixing length and greater dilution.*

7-8   Thermal Effluent Diffusers for Power Plants

Modern thermal electric power plants require cooling water for efficient operation of the condensers. The cooling water may be in a closed loop system, where cooling towers or ponds cool the water after it passes through the power plant, or the power plant may be located near an ocean or lake, where cooling water may be drawn directly from the body of water and then discharged back into it after passing through the condensers of the power plant. Thus the "disposal" of the heated water is similar to disposal of the wastewater. That is, the buoyant effect due to temperature difference between the effluent and ambient water will produce mixing in the plumes as we have described. However, there are many differences between the design and operation of a wastewater effluent system and a power plant effluent system:

1. The discharge in the thermal plant cooling system is often of an order of magnitude greater than that of the wastewater system.
2. The buoyancy in the thermal system is of an order of magnitude less than that of the wastewater system.
3. The required dilution for the thermal plant is of an order of magnitude less than that of the wastewater system.
4. In the thermal plant, the same large discharge drawn from the ocean or lake (coolant) is discharged back into the ocean or lake (heated effluent). Thus short circuiting of the effluent into the intake may be a possibility for the thermal plant, but this situation does not exist for the wastewater system.

Because of these differences, the thermal effluent system generally modifies the near-shore circulation. Thus the formulas for dilution and velocity presented in the preceding section serve only as gross approximations to the physical phenomenon. Current design practice uses hydraulic modeling to obtain a better prediction of the circulation and dilution in the prototype.[†]

PROBLEMS

7-1   An ungated spillway is to deliver 150,000 cfs with a crest elevation of 1000 ft. The upstream reservoir elevation is not to exceed elevation 1030 ft.

---

* For more details on diffuser system design and for effects of cross-currents and density stratification, see Fischer (10).

[†] For details of this hydraulic modeling, see Fischer (10).

The streambed is at elevation 800 ft. Determine the length of the crest required.

7-2   Develop the rating curve for the spillway in Prob. 7-1.

7-3   For Prob. 7-1, determine the shape of the crest.

7-4   If the spillway in Prob. 7-1 were to have radial gates, choose the number of gates and their major dimensions. Assume: the gate radius is $0.85H_D$, the gate trunnion is to be positioned $0.5H_D$ above the crest, and the gate lip is to rest on the spillway crest in the closed position.

7-5   Show that the trajectory for flow downstream from a slightly open vertical gate (located at the end of a horizontal channel) follows the equation

$$y = -\frac{x^2}{4H}$$

where x and y are, respectively, coordinates in the horizontal and upward directions, and H is the head on the gate opening.

7-6   Discuss the effect of gate lip angle and location of the gate sill on reducing the potential for cavitation on spillway surfaces downstream from partially open radial gates.

7-7   Design a spillway, including the crest configuration and type of gates, piers, and so on for an ogee spillway of a concrete gravity dam that is to be 300 ft high and 2000 ft long at the top. The design flood for the spillway is 200,000 cfs. Assume the bottom width of the dam is 1500 ft.

7-8   What discharges would occur over the spillway of Prob. 7-7 if the head on the spillway is 1/4 and 1/2 the design head?

7-9   Letting the configuration of Fig. 7-16, page 390, be the same as shown in Fig. 7-4, page 378, compare the discharge obtained from Fig. 7-4 with that from Fig. 7-16 for a gate opening $d = 5$ ft, and $H_1 = 30$ ft. Assume the design head, H_D, for the spillway is 40 ft and the radius of the radial gate is 0.85 H_D. Let the pivot point of the gate be at an elevation $H_D/2$ above the crest, and let the lip of the gate be right at the crest when the gate is closed.

7-10   Determine the basic configuration for a stilling basin for the spillway of Prob. 7-7. Make your own assumptions about data that may be missing.

7-11   Design either a shaft or siphon spillway for conditions given by your instructor.

7-12   For the conditions of Prob. 7-1, design a stilling basin if the tailwater elevation is at elevation 840 ft.

7-13   Derive Eq. (7-8), page 387.

7-14   The vertical lift gate shown in Fig. 7-15, page 389, is known as a downstream seal type of gate. This type of gate can be difficult to close under flow because of hydraulic forces under the gate. Draw a free-body diagram of the gate just before closure (say a 4-in. opening) to illustrate why this may or may not be a problem. Discuss design features that will allow this problem to be overcome.

7-15   Design a system of supporting pipes and flashboards that will fail when the head reaches 3 ft above a spillway crest.

7-16   Conduct a study to determine what measures are now being taken in your geographical location to assure compatibility of preservation of the fisheries resource and hydropower development.

7-17   Show by means of a sketch how reverse flow through the intake shown below can result in high velocities during the pump-back cycle of a pumped storage hydroplant.

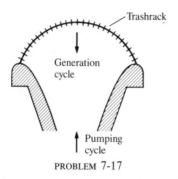

PROBLEM 7-17

7-18   Show how a draft tube reduces energy losses by drawing the hydraulic and energy grade lines for flow through the draft tube shown in Fig. 7-27, page 399.

7-19   Design a pump pit for a three-pump installation to handle a total flow of 60,000 gpm. Use Fig. 7-32, page 404.

7-20   A model of this sudden expansion energy dissipator indicated that cavitation damage could be expected in the prototype in both the expansion and the downstream pipe. Suggest at least two modifications to the design that could reduce the cavitation potential.

PROBLEM 7-20

7-21   Sketch a few streamlines for flow passing this submerged body. The body is solid and all cross sections have a circular shape. Which zones do you think could first begin to cavitate as the cavitation index is lowered? (Refer to the letters.)

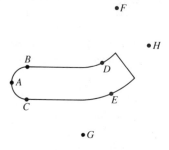

PROBLEM 7-21

7-22   Experience with tunnel spillways has shown that damage frequently occurs near the downstream end of the vertical curve. Give reasons why damage does not occur as frequently at points A, B, and C.

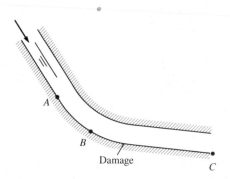

PROBLEM 7-22

7-23   A 24-in. diameter pipe converges into an 8-in. pipe as shown. A gate slot is located in the smaller pipe. Assuming a forced vortex forms in the gate slot (point A), estimate the flow rate that will cause incipient cavitation in the gate slot.

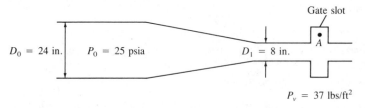

PROBLEM 7-23

7-24   Why does cavitation often occur on rough surfaces located downstream from smooth surfaces?

7-25   Derive Eqs. (7-12) and (7-14), pages 406 and 407.

7-26   This object is submerged in water flowing at 100 ft/s and a pressure of 5195 lb/ft². C_p for most of the underside of the body is constant at $+0.1$. C_p for the upper side varies linearly from 0 to -1.2 for the first third of the length, is constant at -1.2 for the middle third, and varies back to 0 for the downstream third. Is any part of the object subject to vapor pressure and thus cavitation?

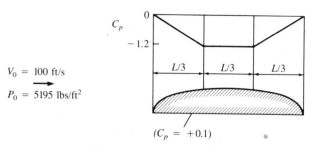

$V_0 = 100$ ft/s

$P_0 = 5195$ lbs/ft²

PROBLEM 7-26

7-27   A 1:25 scale model of a spillway is tested. If the discharge in the model is 0.1 m³/s, to what prototype discharge does this correspond? If it takes 1 min for a particle to float from one point to another in the model, how long would it take a similar particle to traverse the corresponding path in the prototype?

7-28   A newly designed dam is to be modeled in the laboratory. The prime objective of the general model study is to determine the adequacy of the spillway design and to observe the water velocities, elevations, and pressures at critical points of the structure. The reach of the river to be modeled is 1200 m long, the width of the dam (also the maximum width of the reservoir upstream) is to be 300 m, and the maximum flood discharge to be modeled is 5000 m³/s. If the maximum laboratory discharge is limited to 0.90 m³/s and the floor space available for the model construction is 50 m long and 20 m wide, determine the largest feasible scale ratio (model/prototype) for such a study.

7-29   Model studies yielded a minimum C_p on a boundary surface of -0.20. Then, for flow past the prototype surface, would you expect cavitation to occur if $V_0 = 50$ ft/s, $p_0 = 5.0$ psig, and with a water temperature of 60°F? Would cavitation occur if $V_0 = 70$ ft/s and $p_0 = 10.0$ psig?

7-30   A vertical roller gate is suspended on a vertical stem that can be assumed to act like a spring. The gate weighs 40,000 lb, and its spring constant

K equals 155,000 lb/in. If a length scale of 1:16 is chosen for a model, and assuming the acceleration ratio is unity (this is the same as saying the Froude numbers are the same in model and prototype) and water will be used in both model and prototype, find
a. The required model mass
b. The required model spring constant
c. The prototype frequency if 10 hz was measured in the model

7-31   The modeling of an outfall diffuser discharging heated effluent into a river requires that the heat transfer be simulated as well as the hydromechanics. The solution of a differential equation equating the net heat leaving the surface of the river to the change in heat stored in the river water can be used to show that model temperatures at any scaled time will be the same as in the field if the ratio

$$\left(\frac{Kt}{\gamma C_p y}\right)_r = 1$$

where K is the heat transfer coefficient, t is time, γ is specific weight, C_p is specific heat of water, and y is depth. If the model and prototype fluids are both water, $\gamma_r = 1$ and $C_{p_r} = 1$, and

$$K_r = \frac{y_r}{t_r}$$

A horizontal length ratio $L_r = 1:100$ and a density ratio $(\Delta\rho/\rho)_r = 1$ were chosen for a particular model study.
a. If the heat transfer coefficient ratio $K_r = 1:3$, what vertical scale ratio should be chosen?
b. What should the model flow rate Q_m be to simulate a flow of 100,000 cfs in the prototype?
c. If the river is very wide (hydraulic radius = river depth) what should n_r (Manning's n ratio) be?

7-32   A model of a venturi meter has a length ratio of 1:6. Water at 20°C flows in the prototype; water at 90°C flows in the model. If the prototype throat diameter is 100 cm and the throat velocity is 10 m/s, what model discharge will simulate these conditions? What serious problems can you envision with this type of model?

7-33   A round jet of wastewater is discharging upward from near the bottom of a bay that is 400 ft deep. The jet is 1 ft in diameter, and the discharge rate is 30 cfs. What are the velocities along the axis of the jet at distances of 100 ft and 200 ft from the origin of the jet ($z = 100$ ft and $z = 200$ ft)? What are the velocities at these same z-distances but at a radial distance of 20 ft from the axis of the jet? The density of the wastewater is 1.94 slugs/ft³, and the density of the water in the bay is 1.99 slugs/ft³.

7-34   A round jet of water is discharging upward from near the bottom of a bay 100 ft deep. The jet is 1 ft in diameter, and the discharge rate is 30 cfs. What is the velocity at the points (shown by Xs) in the diagram below? Assume the density of the jet is the same as the density of the water in the bay.

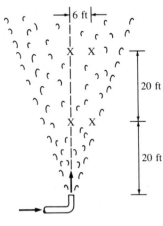

PROBLEM 7-34

7-35   The outfall diffuser system for Honolulu, Hawaii, for a certain period of the year as described in Fischer (10) has the following characteristics:
a. Wastewater discharge $= 7.00 \text{ m}^3/\text{s}$
b. Wastewater temperature $= 75°F$
c. Surface seawater density $= 1023.0 \text{ kg/m}^3$
d. Seawater density at diffuser depth $= 1024.0 \text{ kg/m}^3$
e. Outfall pipe diameter $= 84$ in. (2.13 m)
f. Diffuser depth $= 70$ m
g. Diffuser length $= 1030$ m
h. Number of diffuser ports $= 285$
i. Diameter of diffuser ports: varies from $d = 7.6$ cm to $d = 9.0$ cm
Determine the maximum height of plume and minimum dilution for these conditions.

REFERENCES

1. Albertson, M.L., Y.B. Dai, R.A. Jensen, and H. Rouse. "Diffusion of Submerged Jets." *Trans. ASCE*, 115 (1950).
2. Amorocho, J., A.F. Babb, and J.A. Ross. "Hydraulic Model Investigations of the Castaic Dam High Intake Tower," *Report to State of California Department of Water Resources*, University of California, Davis, Calif., June 1971.
3. Babb, A.F., and W.K. Johnson. *Performance Characteristics of Siphon Outlets*. Proc. of the Jour. of Hydraulic Div., ASCE, (November 1968).

4. Borden, R.C., D. Colgate, J. Legas, and C.E. Selander. *Documentation of Operation, Damage, Repair, and Testing of Yellowtail Dam Spillway.* U.S. Bureau of Reclamation, Report no. REC-ERC-71-23 (May 1971).

5. Bradley, J.M., and A.J. Peterka. "Hydraulic Design of Stilling Basins, High Dams, Earth Dams, and Large Canal Structures (Basin II)." *Journal of the Hydraulic Division, ASCE*, 83, no. HY5, Proc. Paper 1402 (October 1957).

6. Bradley, J.M., and A.J. Peterka. "Hydraulic Design of Stilling Basins: Short Stilling Basin for Canal Structures, Small Outlet Works, and Small Spillways (Basin III)." *Journal of the Hydraulic Division, ASCE*, 83, no. HY5, Proc. Paper 1403 (October 1957).

7. Bradley, J.M., and A.J. Peterka. "Hydraulic Design of Stilling Basins: Stilling Basin and Wave Suppressors for Canal Structures, Outlet Works, and Diversion Dams (Basin IV)." *Journal of the Hydraulic Division, ASCE*, 83, no. HY5, Proc. Paper 1404 (October 1951).

8. Bradley, J.M. *Study of Air Injection into the Flow in the Boulder Dam Spillway Tunnels, Boulder Canyon Project.* U.S. Bureau of Reclamation, Division of Research, Report no. HYD 186 (October 1945).

9. Falvey, H.T. "Predicting Cavitation in Spillway Tunnels." *Water Power and Dam Construction* (August 1982).

10. Fischer, H., J. List, R. Koh, J. Imberger, and N. Brooks. *Mixing in Inland and Coastal Waters.* Academic Press, New York, 1979.

11. Hydraulic Institute Standards, 14th ed. Cleveland, Ohio, 1983.

12. Johnson, V. "Mechanics of Cavitation." *Journal of the Hydraulic Division, ASCE* (May 1963).

13. Labus, T.L., and E.P. Symons. *Experimental Investigation of an Axisymmetric Free Jet With an Initially Uniform Velocity Profile.* NASA TN D-6783, 1972.

14. Roberson, J.A., and C.T. Crowe., *Engineering Fluid Mechanics*, 3d ed. Houghton Mifflin, Boston, Mass., 1985.

15. Rosler, R.S., and S.G. Bankoff. "Large-Scale Turbulence Characteristics of a Submerged Water Jet." *AICE Journal*, 9, no. 5 (September 1963).

16. *The Design of Pumps and Intakes*, first ed. British Hydromechanics Research Association. Cranfield, Bedford, England (July 1977).

17. U.S. Army Corps of Engineers. EM 1110-2-1603, 1961.

18. U.S. Army Corps of Engineers. "Proceedings: CE Workshop on Design and Operation of Selective Withdrawal Intake Structures." Misc. paper HL-86-3, Nat. Technical Information Service, Springfield, Vir. (May 1986).

19. U.S. Bureau of Reclamation. *Design of Gravity Dams.* Denver, Colo. 1976.

20. U.S. Bureau of Reclamation. *Design of Small Dams*, 2d ed. U.S. Govt. Printing Office, Washington, D.C., 1977.

21. U.S. Bureau of Reclamation. *Part VI—Hydraulic Investigations; Bulletin 3, Studies of Crests for Overfall Dams.* Boulder Canyon Final Reports, Denver, Colo. 1948.

22. Wallace, R.B., and S.J. Wright. "Spreading Layer of Two-Dimensional Buoyant Jet." *Journal of Hydraulic Division, ASCE*, 110, no. 6 (June 1984).

8

Interior of powerhouse of Lower Granite Dam on
Snake River near Pullman, Washington (Photo by
Kevin G. Coulton)

Hydraulic
Machinery

8-1   An Introduction to Pumps and Turbines

Use of Pumps

Water lifting devices were probably the first machines to be built by man, and today only the electric motor surpasses the pump for numbers of machines in use throughout the world. In hydraulic engineering, we are primarily interested in pumps for irrigation, flood control, water supply, wastewater, and thermal power plant cooling systems. The design of the pump is primarily dictated by the discharge rate and head to be developed by the pump. Another design consideration is the clarity of the water to be pumped; that is, is it clear water from a lake or well, or is it wastewater that may contain sediment particles, debris, or corrosive products? In this chapter, we present various aspects of the problems associated with pump design and pump station design.

Use of Turbines

Hydraulic turbines are used to convert the power of flowing water into usable electrical or mechanical power. The turbine design is dictated by the head on the turbine and the discharge through the turbine.

In low-head plants (6–100 ft) with moderate to high discharge, the propeller type of turbine is most often used. Some propeller type turbines have adjustable blades to effect higher efficiencies over a wider range of flow conditions; these are called Kaplan turbines. Kaplan turbines are typically used on run of the river hydroelectric plants, as found on the lower Columbia and Ohio rivers in the United States or on the Danube and Rhine rivers in Europe. A typical section of a Kaplan turbine and flow passages is shown in Fig. 8-1a. A Kaplan turbine runner used in a hydroelectric plant on the Danube River is shown in Fig. 8-1b.

For medium-head (90–1500 ft) power plants, the Francis type of turbine is usually used. Here the water approaches the turbine impeller in a radial direction and leaves the impeller parallel to the propeller's axis. In the Francis turbine, the impeller blades are fixed, but vanes that guide the water into the impeller are adjustable so that high efficiencies can be realized over a wide range of discharge. Francis type turbines are found in the power plants of Grand Coulee Dam in Washington, Shasta Dam in California, and Hoover Dam in Nevada. Figure 8-2 shows a typical Francis turbine installation.

Water completely fills the flow passage of the Francis and propeller type turbines. These installations allow maximum use of head between the upstream and downstream pool levels. They also can accommodate a relatively large discharge of water.

For high-head hydropower installations ($H > 1500$ ft), the impulse turbine is used. In this type of turbine, water is most often conveyed to the turbine

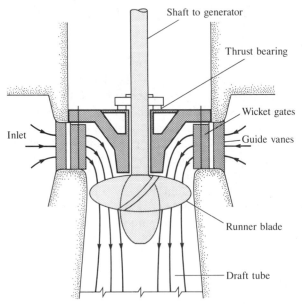

Shaft to generator

Thrust bearing

Wicket gates

Inlet

Guide vanes

Runner blade

Draft tube

(a) Schematic of section of unit in place.

Figure 8-1   Kaplan turbine (Courtesy of Voith Hydro Inc.)

(b) A Kaplan turbine runner

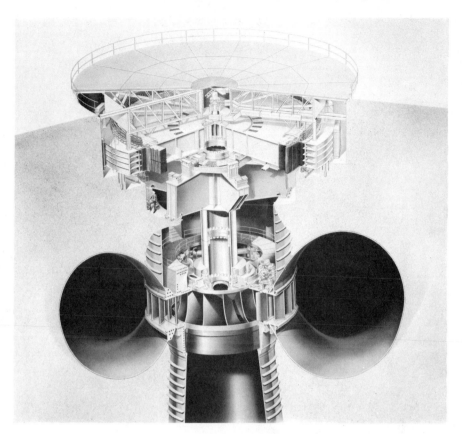

Figure 8-2   Schematic view of Francis-type turbine used in Grand Coulee Dam (Courtesy of Voith Hydro Inc.)

through a pipe called a penstock and then through a nozzle from which a high velocity jet of water issues. This jet impinges on curved vanes (buckets) placed on the periphery of the turbine wheel, thus causing the wheel to rotate and to generate power (see Fig. 8-26). In that application, the axis of the wheel is horizontal. To develop more power from a single wheel, a wheel having a vertical axis with multiple jets is used (see Fig. 8-3). Although the impulse turbine is the logical choice of turbine type for high-head installations, it is also suitable for many lower head sites if discharge is relatively small. Thus many small-scale hydropower plants use impulse turbines. The primary advantages of the impulse turbine are its simplicity and ease of maintenance. One of its disadvantages is that the impeller must be placed so that it is always above the highest level of the downstream pool; therefore, in run of the river plants, much head would be wasted when the river discharge is low and the downstream pool level is at low elevation.

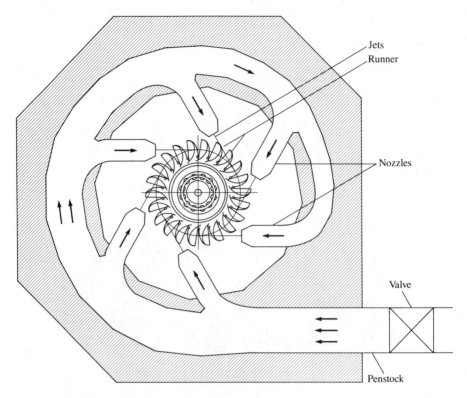

Figure 8-3   Plan view of vertical-axis-impulse-turbine
runner with six jets

8-2   Dimensionless Parameters for Turbomachines

When designing turbomachines, the designer must apply the basic
equations of fluid mechanics to shape the vanes and flow passages of the machine so that separation-free flow will occur. Through this as well as through
previous experience and testing a final design for a pump or turbine is achieved.
Once a pump or turbine is produced, performance tests are made to determine
the machine's actual operating characteristics. The tests yield relationships between dimensionless parameters for that particular machine. In the following
paragraphs, we develop by means of dimensional analysis the relevant dimensionless parameters used with turbomachines.*

---

* As shown in Roberson (12), these parameters could also be developed by approaching the problem
from the fundamental theory of lift and drag of an airfoil; however, for the sake of brevity, we use
the dimensional analysis approach.

In applying dimensional analysis to the problem, we first must identify the significant variables. In the case of pumps, the variables of primary interest to the user are Δp, Q, P and η, where Δp is pressure increase produced by the pump, Q is discharge, P is power supplied to the pump, and η is efficiency. Other significant variables are n, speed of rotation, D, diameter of pump impeller, and ρ, the mass density of the fluid being pumped. Next, we must identify the dependent and independent variables. Of the above, the variables n, D, Q, and ρ are all identified as independent variables. Thus Δp, P, and η are the dependent variables. However, in dimensional analysis of a group of variables, only one dependent variable is allowed. We therefore do an analysis first with one of the dependent variables, then with another one, and so on. Let us do the first analysis with Δp as the dependent variable. The corresponding functional equation is

$$\Delta p = f(\rho, n, D, Q) \qquad (8\text{-}1)$$

where $[\Delta p]^* = F/L^2$
$\quad\quad\quad [\rho] = FT^2/L^4$
$\quad\quad\quad [n] = T^{-1}$
$\quad\quad\quad [D] = L$
$\quad\quad\quad [Q] = L^3/T$

Letting ρ, n, and D be the variables that we use to combine with other variables to form dimensionless groups, first, we nondimensionalize the equation with respect to the force dimension, F. So doing we obtain

$$\frac{\Delta p}{\rho} = f(n, D, Q) \qquad (8\text{-}2)$$

Next, we nondimensionalize it in time, T:

$$\frac{\Delta p}{\rho n^2} = f\left(D, \frac{Q}{n}\right) \qquad (8\text{-}3)$$

Finally, we make the functional equation completely dimensionless by combining powers of D with the remaining groups of variables:

$$\frac{\Delta p}{\rho n^2 D^2} = f\left(\frac{Q}{nD^3}\right) \qquad (8\text{-}4)$$

In working with pumps, we often focus on the change in head, ΔH, rather than Δp. Therefore, in Eq. (8-4), if we let $\rho = \gamma/g$ and $\Delta p = \gamma \Delta H$ and cancel out the

---

* The brackets around the variable mean dimensions of.

γ's we obtain

$$\frac{\Delta H}{n^2 D^2/g} = f\left(\frac{Q}{nD^3}\right) \tag{8-5}$$

The dimensionless parameter on the left-hand side of Eq. (8-5) is the *head coefficient*, C_H, and the parameter inside the parentheses on the right-hand side is the *discharge coefficient*, C_Q. Thus

$$C_H = \frac{\Delta H}{n^2 D^2/g} \quad \text{and} \quad C_Q = \frac{Q}{nD^3}$$

where $C_H = f(C_Q)$ \hfill (8-6)

By applying dimensional analysis to the variables P, ρ, n, D, and Q it can be shown that

$$\frac{P}{\rho D^5 n^3} = f\left(\frac{Q}{nD^3}\right) \tag{8-7}$$

or \qquad $C_P = f(C_Q)$ \hfill (8-8)

where C_P is defined as the *power coefficient*. Similarly, it can be shown that $\eta = f(C_Q)$.

Summarizing, the dimensionless parameters used in similarity analyses of pumps are as follows:

$$C_H = \frac{\Delta H}{D^2 n^2/g} \tag{8-9}$$

$$C_P = \frac{P}{\rho D^5 n^3} \tag{8-10}$$

$$C_Q = \frac{Q}{nD^3} \tag{8-11}$$

where C_H and C_P are functions of C_Q for a given type of pump.

8-3   Axial-Flow Pumps

Figure 8-4 is a set of curves of C_H, C_P, and η versus C_Q for a typical axial-flow pump. Dimensional curves (head, power, and efficiency versus Q for a constant speed of rotation) from which Fig. 8-4 was developed are shown in Fig. 8-5. Curves like those shown in Figs. 8-4 and 8-5 characterize the pump's performance, so they are often called *characteristic curves* or *performance curves*. These curves are obtained by experiment.

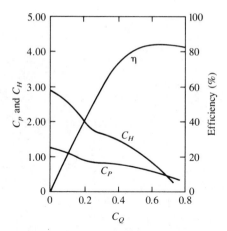

Figure 8-4   Dimensionless performance curves for a typical axial-flow pump (15)

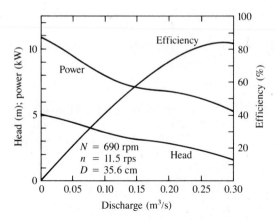

Figure 8-5   Performance characteristics of a typical axial-flow pump (15)

Performance curves are used to predict prototype operation from model tests or the effect of change of speed of the pump. Two examples of these applications follow.

EXAMPLE 8-1   For the pump represented by Figs. 8-4 and 8-5, what discharge in cubic meters per second will occur when the pump is operating against a 2-m head and at a speed of 600 rpm? What power in kilowatts is required for these conditions?

SOLUTION   First compute C_H. Here,

$$D = 35.6 \text{ cm} \quad \text{and} \quad n = 10 \text{ rps}$$

Then, $C_H = \dfrac{2 \text{ m}}{(0.356 \text{ m})^2(10^2 \text{ s}^{-2})/(9.81 \text{ m/s}^2)} = 1.55$

With a value of 1.55 for C_H, a value of 0.38 for C_Q is read from Fig. 8-4. Hence, Q is calculated as follows:

$$C_Q = 0.38 = \frac{Q}{nD^3}$$

or $Q = 0.38(10 \text{ s}^{-1})(0.356 \text{ m})^3$

$$= 0.171 \text{ m}^3/\text{s}$$

From Fig. 8-4, the value of C_P is 0.80 for $C_Q = 0.38$, then,

$$P = 0.80 \; \rho D^5 n^3$$
$$= 0.80(1.0 \text{ kN} \cdot \text{s}^2/\text{m}^4)(0.356 \text{ m})^5(10 \text{ s}^{-1})^3$$
$$= 4.57 \text{ km} \cdot \text{N/s} = 4.57 \text{ kJ/s}$$
$$= 4.57 \text{ kW} \qquad\qquad\qquad \blacksquare$$

EXAMPLE 8-2 If a 30-cm axial-flow pump having the characteristics shown in Fig. 8-4 is operated at a speed of 800 rpm, what head ΔH will be developed when the water pumping rate is 0.127 m³/s? What power is required for this operation?

SOLUTION   First compute

$$C_Q = \frac{Q}{nD^3}$$

where $Q = 0.127 \text{ m}^3/\text{s}$

$$n = \frac{800}{60} = 13.3 \text{ rps}$$

$$D = 30 \text{ cm}$$

Then, $C_Q = \dfrac{0.127 \text{ m}^3/\text{s}}{13.3 \text{ s}^{-1}(0.30 \text{ m})^3}$

$$= 0.354$$

Now, enter Fig. 8-4 with a value of $C_Q = 0.354$, and read off a value of 1.60 for C_H and a value of 0.80 for C_P. Then,

$$\Delta H = \frac{C_H D^2 n^2}{g} = \frac{1.60(0.30 \text{ m})^2(13.3 \text{ s}^{-1})^2}{(9.81 \text{ m/s}^2)}$$

$$= 2.60 \text{ m}$$

and
$$P = C_P \rho D^5 n^3 = 0.80(1.0 \text{ kN} \cdot \text{s}^2/\text{m}^4)(0.30 \text{ m})^5(13.3 \text{ s}^{-1})^3$$

$$= 4.56 \text{ kW}$$ ∎

In practical applications, axial-flow pumps are best suited for relatively low heads and high rates of flow. Hence, pumps used for dewatering lowlands, such as those behind dikes, are almost always of the axial-flow type. Figure 8-6 shows a typical setup for an axial-flow pump. For larger heads, radial- or mixed-flow machines are more efficient.

8-4   Radial- and Mixed-Flow Pumps

Figure 8-7 shows the type of impeller used for many radial-flow pumps. These pumps are also called *centrifugal pumps.* Liquid from the inlet pipe enters the pump through the eye of the impeller, then travels outward between the vanes of the impeller to the edge of the impeller, where the fluid enters the casing of the pump and is then conducted to the discharge pipe. The principle of the radial-flow pump is different from that of the axial-flow pump in that

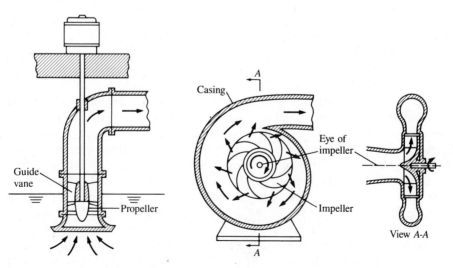

Figure 8-6 Axial-flow pump   **Figure 8-7** Centrifugal pump

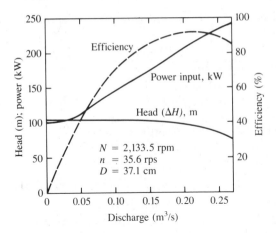

Figure 8-8 Typical performance curves for a centrifugal pump (3)

the change in pressure largely results by rotary action (pressure increasing outward like that of a rotating tank of water). Additional pressure increase is produced in the radial-flow pump when the high velocity flow leaving the impeller is reduced in the expanding section of the casing.

Although the basic designs of the radial- and axial-flow pumps are different, it can be shown that the same similarity parameters (C_Q, C_P, and C_H) apply for both types. Thus the methods we discussed for relating size, speed, and discharge in axial-flow machines also apply to the radial-flow machine.

The major practical difference between the axial- and radial-flow pumps, so far as the user is concerned, is in the performance characteristics of the two pumps. Figure 8-8 shows the dimensional performance curves for a typical radial-flow pump operating at a constant speed of rotation, and Fig. 8-9 shows the dimensionless performance curves for the same pump. Note that at shutoff flow, the power required is less than for flow at maximum efficiency. Normally, the motor to drive the pump will be chosen for conditions of maximum pump efficiency. Hence, it can be seen that the flow can be throttled between the limits of shutoff condition and the normal operating condition without any chance of overloading the pump motor. This is not the case for an axial-flow pump, as seen in Fig. 8-4. In that case, when the pump flow is throttled below maximum efficiency conditions, the required power increases with decreasing flow, thus leading to the possibility of overloading at low-flow conditions. For large installations, special operating procedures are followed to avoid overloading; for example, a valve in a bypass from the pump discharge back to the pump inlet can be adjusted to maintain a constant flow through the pump. However, for small-scale applications, it is often desirable to have complete flexibility in the flow control without the complexity of special operating procedures. In this latter case, a radial-flow pump offers a distinct advantage.

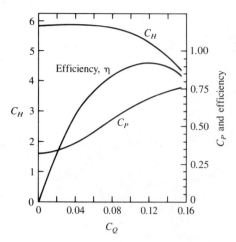

Figure 8-9 Dimensionless performance curves for a centrifugal pump, from data given in Figure 8-8 (3)

Figure 8-10 Cutaway view of a single-suction, single-stage, horizontal-shaft radial pump (Courtesy of Ingersol Rand Co.)

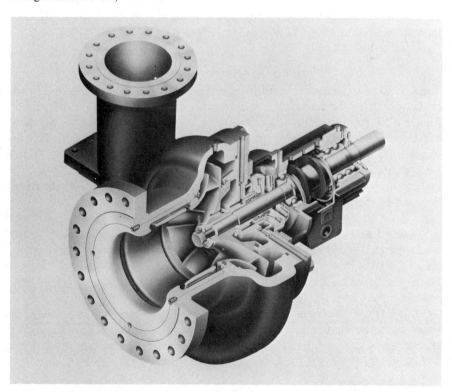

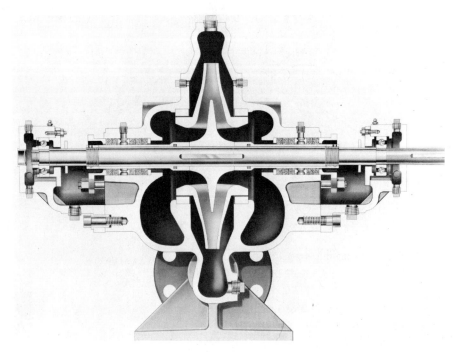

Figure 8-11   Sectional view of double-suction, single-stage, horizontal-shaft split-casing pump (Courtesy of Worthington Pump Co.)

Radial-flow pumps are manufactured in sizes from 1 hp or less and heads of 50 or 60 ft to thousands of horsepower and heads of several hundred feet. Figure 8-10, page 443, shows a cutaway view of a single-suction, single-stage, horizontal-shaft radial pump. Another common design has flow entering the impeller from both sides (*double-suction impeller*), as shown in Figs. 8-11 and 8-12, which is equivalent to two single-suction impellers placed back to back and made as a single casting. This arrangement gives balanced end thrust on the shaft of the impeller.

EXAMPLE 8-3· The pump having the characteristics given in Fig. 8-8, when operated at 2133.5 rpm, is to be used to pump water at maximum efficiency under a head of 76 m. At what speed should the pump be operated, and what will be the discharge for these conditions?

SOLUTION   Since the diameter is fixed, the only change that will occur will result from the change in speed (assuming negligible change due to viscous effects). The C_H, C_P, C_Q, and η for this pump operating at maximum efficiency against a ΔH of 76 m are the same as these for operating at maximum efficiency

Figure 8-12   Overall outside view of a double-
suction, single-stage, horizontal-shaft split-casing pump
(Courtesy of Worthington Pump Co.)

with a speed of 2133.5 rpm, since both operating conditions correspond to the
point of maximum efficiency in Fig. 8-8. Thus we can write

$$(C_H)_N = (C_H)_{2133.5 \text{ rpm}}$$

Here N refers to the speed of rotation with $\Delta H = 76$ m. The graph of Fig. 8-8
indicates that $\Delta H = 90$ m and $Q = 0.225 \text{ m}^3/\text{s}$ at maximum efficiency for
$N = 2133.5$ rpm. Thus

$$\frac{76 \text{ m}}{N^2} = \frac{90 \text{ m}}{(2133.5)^2}$$

$$N^2 = (2133.5)^2 \frac{76}{90}$$

$$N = 2133.5 \left(\frac{76}{90}\right)^{1/2} = 1960 \text{ rpm}$$

Using $(C_Q)_{1960} = (C_Q)_{2133.5 \text{ rpm}}$ and solving for the ratio of discharge, we have

$$\frac{Q_{1960}}{Q_{2133.5}} = \frac{1960}{2133.5} = 0.919$$

$$Q_{1960} = 0.207 \text{ m}^3/\text{s}$$ ∎

EXAMPLE 8-4   The pump having the characteristics shown in Figs. 8-8 and 8-9 is a model of a pump that was actually used in one of the pumping plants of the Colorado River Aqueduct (3). For a prototype that is 5.33 times larger than the model and operates at a speed of 400 rpm, what head, discharge, and power would be expected at maximum efficiency?

SOLUTION   From Fig. 8-9, we pick off values of 0.115, 5.35, and 0.69 for C_Q, C_H, and C_P, respectively, for the maximum efficiency condition. Then for $n = (400/60)$ rps and $D = 0.371 \times 5.33 = 1.98$ m, we solve for P, ΔH, and Q:

$$P = C_P \rho D^5 n^3$$

$$= 0.69(1.0 \text{ kN} \cdot \text{s}^2/\text{m}^4)(1.98 \text{ m})^5 \left(\frac{400}{60} \text{ s}^{-1}\right)^3$$

$$= 6200 \text{ kW}$$

$$\Delta H = \frac{C_H D^2 n^2}{g} = \frac{5.35(1.98 \text{ m})^2(400/60 \text{ s}^{-1})^2}{(9.81 \text{ m/s}^2)}$$

$$= 95.0 \text{ m}$$

$$Q = C_Q n D^3$$

$$= 0.115 \left(\frac{400}{60} \text{ s}^{-1}\right)(1.98 \text{ m})^3 = 5.95 \text{ m}^3/\text{s}$$ ∎

8-5   Performance Characteristics Under Abnormal Operating Conditions

The performance curves we have presented up to now are for pumps operating at a speed and discharge to yield the highest efficiency. However, there are cases, such as during a power outage, when the pump will operate abnormally for a few seconds. For these occurrences, it is often desirable to predict the performance of the pump as well as the discharge and pressure in the entire flow system as a function of time. These analyses use the method of characteristics,* and this in turn, requires the input of boundary conditions such as the head versus discharge at the pump. Thus the performance characteristics (ΔH and Q for a wide range of pump speed) are required. Under abnormal conditions, the pump speed may reduce to zero and then reverse so that it operates as a turbine. When operating as a turbine, the pump is said to have a negative speed. Likewise, the discharge will be positive when operating normally, or negative when operating as a turbine when the flow reverses. The performance characteristics for these operating conditions can be presented as shown in Fig. 8-13, page 448. All quantities on this graph are expressed as a percentage of the value at the point of best efficiency.

8-6 Specific Speed

A pump's performance is given by the values of its power and head coefficients (C_P and C_H) for a range of values of the discharge coefficient C_Q. Certain types of machines are best suited for certain head and discharge ranges. For example, an axial-flow machine is best suited for low heads and high discharges, whereas a radial-flow machine is best suited for higher heads and lower discharges. The parameter used to pick the type of pump (or turbine) best suited for a given application is specific speed n_s. Specific speed is obtained by combining C_H and C_Q in such a manner that the diameter D is eliminated:

$$n_s = \frac{C_Q^{1/2}}{C_H^{3/4}} = \frac{(Q/nD^3)^{1/2}}{[\Delta H/(D^2 n^2/g)]^{3/4}} = \frac{nQ^{1/2}}{g^{3/4}\,\Delta H^{3/4}} \tag{8-12}$$

Thus specific speed relates different types of pumps without reference to size.

When the maximum efficiencies of different types of pumps are plotted against n_s (Fig. 8-14a), it is seen that certain types of pumps have higher efficiencies for certain ranges of n_s. In fact, in the range between the completely axial- and radial-flow machine, there is a gradual change in impeller shape to accomodate the particular flow conditions with maximum efficiency (see Fig. 8-14b–d).

The specific speed traditionally used for pumps in the United States is defined as $N_s = NQ^{1/2}/\Delta H^{3/4}$. Here, the speed N is in revolutions per minute, Q,

---

* For details of these analyses, see Chaudhry (1).

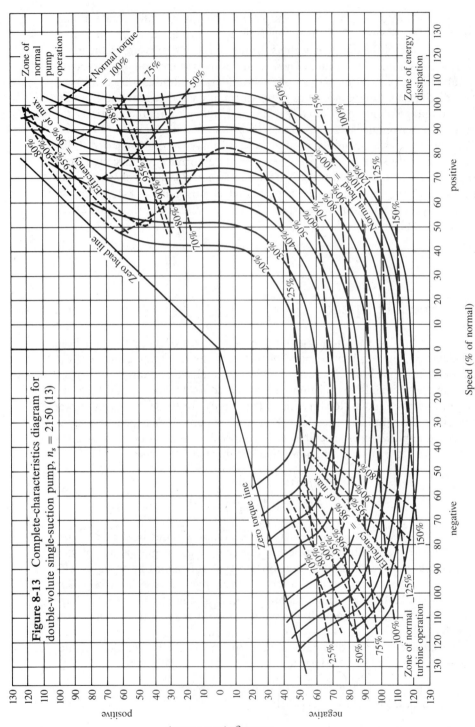

Figure 8-13 Complete-characteristics diagram for double-volute single-suction pump, $n_s = 2150$ (13)

Speed (% of normal)

Discharge (% of normal)

the discharge per suction inlet, is in gallons per minute, and ΔH is in feet. This form is not dimensionless, and its values are much larger than those found for n_s (the conversion factor is 17,200). Most texts and references published before the introduction of the SI system of units use the traditional definition for specific speed.

8-7 Multistage Pumps

Because n_s varies inversely with $\Delta H^{3/4}$, n_s will obviously decrease as the head ΔH increases. It may also be seen (Fig. 8-14) that the efficiency is small for low values of n_s. If only one stage (one impeller) is used to pump water with a very large head difference, the efficiency would be quite low. Therefore, to maintain the n_s near values for which the efficiency is high multistage pumps are manufactured. That is, the stages are in series, wherein the discharge from the

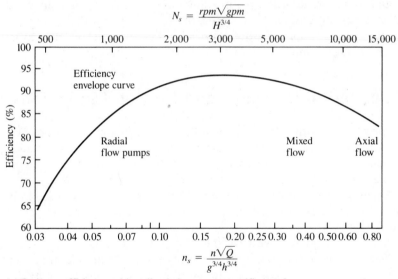

(a) Optimum efficiency and impeller designs versus specific speed n_s

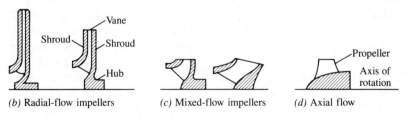

(b) Radial-flow impellers   (c) Mixed-flow impellers   (d) Axial flow

Figure 8-14   Optimum efficiency and impeller designs versus specific speed

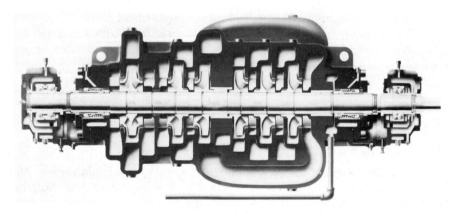

Figure 8-15   Cutaway view of multistage pump
(Courtesy of Worthington Pump Co.)

first stage (first impeller) discharges directly into the suction side of the second impeller and so on. In multistage pumps, all impellers are on a single shaft and enclosed in a single housing. By dividing the total head by the number of impellers, the n_s can be kept at the optimum value. Figure 8-15 shows a cutaway view of a multistage pump.

8-8   Cavitation in Pumps and Suction Limitations

The pressure within the water as it flows through a pump is lowest on the suction side of the pump. And as the water flows past the impeller blades, more extreme local low pressure zones exist in regions of greatest blade curvature. Thus if the pressure on the suction side of the pump is too low, cavitation may occur. Figure 8-16 shows the pitting on an impeller that occurred from cavitation. Cavitation can also reduce the efficiency of the pump or cause severe vibration problems.

In Chapter 7, page 405, we gave the cavitation index as $\sigma = (p_0 - p_v)/(\rho V_0^2/2)$. The index in this form is useful for application to hydraulic structures. For pumps, the index is changed in the following ways. First, instead of $\rho V_0^2/2$ in the denominator, the pressure produced by the pump, Δp, is used. Next, the ambient pressure, p_0, is now the total head ($p_i/\gamma + V_i^2/2g + z_i$) in the intake pipe of the pump converted to pressure. However, by using the energy equation, this total pressure can be given as the pressure at the source (reservoir, for example) minus the pressure loss due to friction between the source and inlet minus the pressure change, since the pump is located at an elevation different from the source. Thus the cavitation index for a pump can be given as

$$\sigma = \frac{p_0 - \gamma h_L - \gamma(z_i - z_0) - p_v}{\Delta p} \tag{8-13}$$

Figure 8-16 Cavitation damage to impeller of a centrifugal pump

Because most pump data are given in terms of head rather than Δp, we divide the numerator and denominator of the right-hand side of Eq. (8-13) by γ to obtain

$$\sigma = \frac{p_0/\gamma - h_L - (z_i - z_0) - p_v/\gamma}{\Delta H} \qquad (8\text{-}14)$$

Moreover, because the vapor pressure, p_v, is usually given in terms of absolute pressure, it is customary to express p_0 similarly. The definition sketch shown in Fig. 8-17 shows how the variables of Eq. (8-14) relate to a physical situation and gives representative values for p_0 and p_v. The numerator on the right-hand side of Eq. (8-14) is the net positive suction head (NPSH), and the value of σ when significant cavitation is first observed is the critical value, σ_c.* Through experimental tests, the pump manufacturer will determine σ_c values for different pumps, and these values are usually made available to those who buy the pumps. It is also common for manufacturers to convert the σ_c values to NPSH values so that NPSH curves (required NPSH versus N_s), or at least critical NPSH values for the rated conditions (maximum efficiency condition), are available.

---

* Cavitation can be produced in an experimental test program by operating the pump at a given speed and then gradually lowering the pressure on the suction side of the pump (for example, the head could be lowered in the reservoir shown in Fig. 8-17) until cavitation occurs.

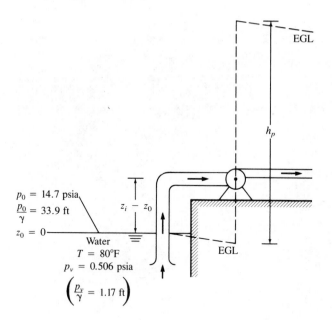

Figure 8-17 Definition sketch

EXAMPLE 8-5 Tests on a given centrifugal pump yielded a value for σ_c of 0.075 when the pump was operating at maximum efficiency. For this maximum efficiency condition, $h_p = 140$ ft and $Q = 1.35$ cfs. If the pump is to be installed in a setting such as shown in Fig. 8-17, what is the maximum value of $(z_i - z_0)$ that can be used with cavitation-free operation? Assume it will be operating at maximum efficiency. Further assume the intake-pipe diameter is 8 in. and the total head loss coefficient (sum of loss coefficients for inlet, pipe friction, bend, and so on) has a value of 1.5.

SOLUTION   From Eq. (8-14), we have

$$\frac{p_0}{\gamma} - h_L - (z_i - z_0) - \frac{p_v}{\gamma} = \sigma \, \Delta H$$

or $$(z_i - z_0) = \frac{p_0}{\gamma} - h_L - \frac{p_v}{\gamma} - \sigma \, \Delta H$$

But $$h_L = \frac{KV^2}{2g} = 1.5 \frac{(Q^2/A^2)}{2g} = 0.35 \text{ ft}$$

Assume $T = 80°F$ and standard sea-level atmospheric pressure ($p_0 = 14.7$ psia and $p_0/\gamma = 33.9$ ft). Also $p_v = 0.506$ psia (from Appendix Table A-4, page 648) from which $p_v/\gamma = 1.17$ ft.

Then $(z_i - z_0) = 33.9 - 1.17 - 0.35 - 0.075 \times 140$

$$= 21.9 \text{ ft}$$ ∎

Because most centrifugal pumps used in a given range of n_s have about the same shape and performance characteristics, certain general limitations on the flow conditions on the suction side of the pump may be established to prevent cavitation. These limitations are published by the Hydraulic Institute (6) and are given in terms of maximum ΔH versus n_s for different suction lifts or suction heads. For example, a chart for single-suction mixed-flow and axial-flow pumps is shown in Fig. 8-18. Example 8-6 illustrates the use of the chart.

EXAMPLE 8-6   An axial-flow pump is to be used to lift water from a main irrigation canal to a smaller irrigation canal at a higher level. If the total head (elevation difference plus head losses in the pipe) is to be 11 m, and if the total suction head is to be 2 m (the impeller is below the water level in the main canal), what is the safe upper limit of specific speed? Would it be safe to operate a pump at a speed of 1200 rpm and with a discharge of 0.5 m³/s under these conditions?

SOLUTION   We enter Fig. 8-18 with a total head of 11 m and a total suction head of 2 m and read a value of n_s of 0.51. Thus the safe upper limit of specific speed is 0.51.

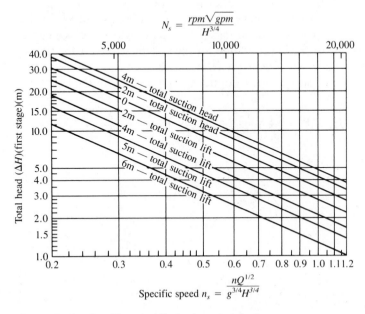

$$N_s = \frac{rpm\sqrt{gpm}}{H^{3/4}}$$

Specific speed $n_s = \dfrac{nQ^{1/2}}{g^{3/4}H^{3/4}}$

Figure 8-18   Specific speed limitations for single-suction mixed-flow and axial-flow pumps* (6)
* Pumping clear water, 30°C at sea level

By definition, we have

$$n_s = \frac{nQ^{1/2}}{g^{3/4} \Delta H^{3/4}}$$

Hence, for $N = 1200$ rpm or $n = 20$ rps, $Q = 0.50$ m³/s, and $\Delta H = 11$ m. We compute n_s as follows:

$$n_s = \frac{20(0.50)^{1/2}}{(9.81)^{3/4}(11)^{3/4}} = 0.42$$

The n_s computed here is less than the allowable n_s; thus the stated operating conditions are *within the safe range*.   ■

8-9   Pumps Operating in a Pipe System

In Chapter 5, we considered several problems involving head loss for a given discharge or vice versa. In this chapter, we have considered how the head developed by a pump is related to the discharge through the pump. We now link the two to see what head and discharge will prevail when a given pump is operated in a given pipe system. The solution (that is, the flow rate for a given system) is obtained when the system equation (or curve) of head versus discharge is solved simultaneously with the pump equation (or curve) of head versus discharge. The solution of these two equations (or the point where the two curves intersect) will yield the operating condition for the system. Consider the flow of water in the system shown in Fig. 8-19. When the energy equation is written from the reservoir water surface to the outlet stream, we obtain the equation:

$$\frac{p_1}{\gamma} + \frac{V_1^2}{2g} + z_1 + h_p = \frac{p_2}{\gamma} + \frac{V_2^2}{2g} + z_2 + \sum K_L \frac{V^2}{2g} + \frac{fL}{D} \frac{V^2}{2g}$$

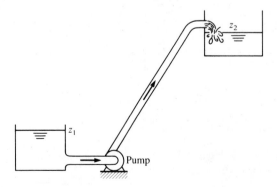

Figure 8-19   Pump and pipe combination

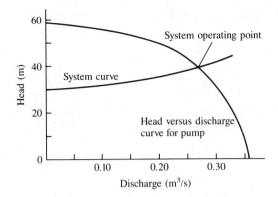

Figure 8-20   Pump and system curves

This equation simplifies to

$$h_p = (z_2 - z_1) + \frac{V^2}{2g}\left(1 + \sum K_L + \sum \frac{fL}{D}\right) \tag{8-15}$$

Hence, for any given discharge, a certain head h_p must be supplied to maintain that flow. Thus we can construct a head versus discharge curve, called the *system curve*, as shown in Fig. 8-20. We also plot the ΔH–Q curve for the pump producing the flow in Fig. 8-20.

As the discharge increases in a pipe, the head required for flow also increases. However, the head produced by the pump decreases as the discharge increases. Thus the two curves will intersect, and the operating point is at the point of intersection — that point where the head produced by the pump is just the amount needed to overcome the head loss in the pipe and the elevation change.

If we have more than one pump in the system, we simply use the composite ΔH–Q curve (which we introduce in the next section) with the system curve to obtain a solution.

E X A M P L E   8 - 7   What will be the discharge in this water system if the pump has the characteristics shown in Fig. 8-20? Assume $f = 0.015$.

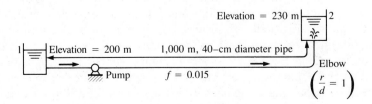

SOLUTION   First, write the energy equation from water surface to water surface.

$$\frac{p_1}{\gamma} + \frac{V_1{}^2}{2g} + z_1 + h_p = \frac{p_2}{\gamma} + \frac{V_2{}^2}{2g} + z_2 + \sum h_L$$

$$0 + 0 + 200 + h_p = 0 + 0 + 230 + \left(\frac{fL}{D} + K_e + K_b + K_E\right)\frac{V^2}{2g}$$

where $K_e = 0.5$, $K_b = 0.35$, and $K_E = 1.0$ (from Table 5-3, page 256). Hence,

$$h_p = 30 + \frac{Q^2}{2gA^2}\left[\frac{0.015(1000)}{0.40} + 0.5 + 0.35 + 1\right]$$

$$= 30 + \frac{Q^2}{2 \times 9.81 \times [(\pi/4) \times 0.4^2]^2}\,(39.3) = 30\ \text{m} + 127\ Q^2\ \text{m}$$

Now, we make a table of Q versus h_p to give values to produce a system curve that will be plotted with the pump curve. When the system curve is plotted on the same graph as the pump curve, it is seen (Fig. 8-20) that the operating condition occurs at $Q = 0.27\ \text{m}^3/\text{s}$.

Q, m³/s	Q^2, m⁶/s²	$127Q^2$	$h_p = 30\ \text{m} + 127Q^2\ \text{m}$
0	0	0	30
0.1	1×10^{-2}	1.3	31.3
0.2	4×10^{-2}	5.1	35.1
0.3	9×10^{-2}	11.4	41.4

8-10 Pumps Operated in Combination

Parallel Installation

When pumps are used to supply a system in which the demand may vary considerably, installing several pumps but using only those needed to satisfy the demand at any particular time is customary. Therefore, to determine the discharge in the system, one must construct pump performance curves that represent the multiple operation. For example, it is common practice to have three or more pumps discharging into a common manifold pipe that supplies water to a distribution system, as shown in Fig. 8-21. For such an installation, we say the pumps are installed in parallel. In this setup, the pressure throughout the length of the manifold into which the pumps are discharging will be essentially constant. Thus the ΔH across all the pumps will be virtually the same no matter how many pumps are operating. Then whatever the ΔH may be, each pump will be discharging a rate of flow consistent with the ΔH–Q curve for that

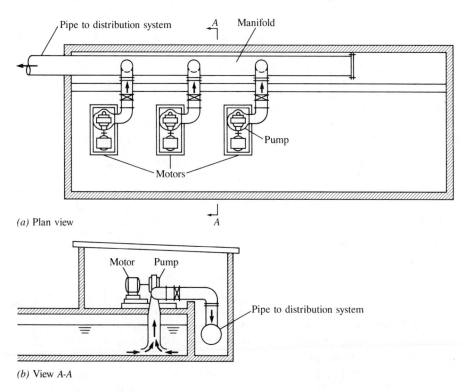

(a) Plan view

(b) View A-A

Figure 8-21 Parallel pump installation

particular pump, and the total discharge from all the pumps will be the sum of those discharges. One can develop a ΔH–Q curve for multiple pump operation by taking different H's and determining the corresponding $\sum Q$ for each ΔH. If the pumps have identical performance characteristics, then the $\sum Q$ would simply be nQ, where n is the number of pumps in operation. The ΔH–Q curve for two pumps having the performance curve shown in Fig. 8-20 operating in parallel would be as shown in Fig. 8-22.

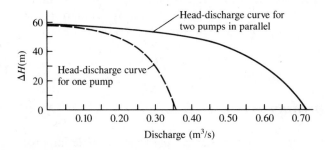

Figure 8-22 Performance curve for two identical pumps in parallel

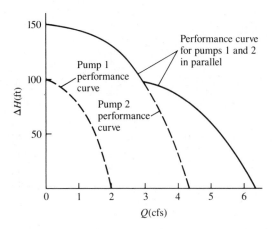

Figure 8-23   Performance of two dissimilar pumps in parallel

If two pumps having dissimilar performance curves were operated in parallel, one would still sum the discharges for the given ΔH as noted. For example, if we had pumps with performance curves indicated by the broken lines in Fig. 8-23, the composite performance curve for these two pumps would be as given by the solid line in Fig. 8-23. Note that pump 2 is of no use in augmenting the discharge if ΔH is greater than its shutoff head ($\Delta H = 100$ ft in this case). The composite curve is constructed with the assumption that a check valve prevents flow from passing back through pump 1 when the ΔH exceeds 100 ft. If there were no check valve, there would be a negative contribution from pump 1 at the higher heads. Thus one must be careful about installing and operating pumps in combination having dissimilar performance curves; otherwise, the combined operation may not be any better than the operation of a single pump.

SERIES OPERATION   Although less common than parallel pump installations, the series pump installations are sometimes desirable. Figure 8-24 shows two pumps in series. For a series installation, the discharge will be the same through each pump, so the total head for the combined operation will be the summation of ΔH for each pump having the given Q. Thus the composite performance curve for the two pumps of Fig. 8-20 would be as shown in Fig. 8-25.

8-11   Hydraulic Turbines

Impulse Turbine

In the impulse turbine, a jet of water issuing from a nozzle impinges on vanes (sometimes called buckets) of the turbine wheel (often called runner),

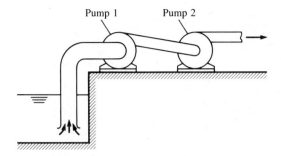

Figure 8-24   Two pumps in series

thus producing power as the runner rotates (see Fig. 8-26). Figure 8-27 shows a runner for the Henry Borden hydroelectric plant in Brazil. The impulse turbine is sometimes called the Pelton wheel, after the man who developed the main features of the modern impulse turbine. The impulse turbine is especially suited for application in high-head plants where the head is typically 500 ft or more. Installations with heads of 1000 to 2000 ft are not uncommon. If the discharge is small, impulse turbines can also be effectively used with lower heads. Recently, impulse turbines have been installed in several small-scale hydroelectric plants following the increase in power costs in the past decade. Before the increased costs in power, the small-scale hydropower plants were often not economically feasible.

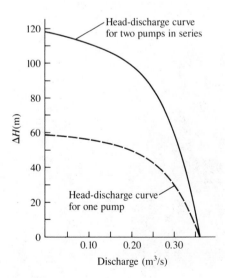

Figure 8-25   Performance curve for two identical pumps in series

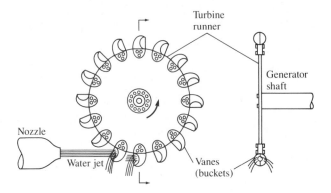

Figure 8-26   Impulse turbine wheel

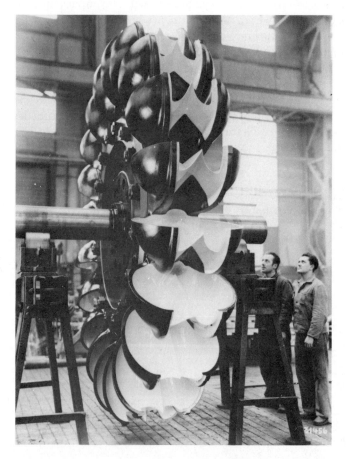

Figure 8-27   Spare runner for the Henry Borden
power plant in Brazil. (Courtesy of Voith Hydro Inc.)

Reaction Turbine

In contrast to the impulse turbine, where a jet under atmospheric pressure impinges upon only one or two vanes at a time, the flow in a reaction turbine is under pressure and this flow completely fills the chamber in which the turbine runner is located (see Figs. 8-1 and 8-2, pages 434–35). In the reaction turbine, the runner consists of several vanes attached to a hub. As flow passes through the runner, the vanes change the direction of flow, thus producing a force on the vanes by a change in momentum. The force causes the runner to rotate, and power is generated. Because flow in a reaction turbine continuously acts on all the vanes (in contrast to the impulse turbine where only one or two buckets are acted on by the jet at any time), a runner of given size will develop more power per unit head than the impulse turbine.

Reaction turbines can be designed to operate under quite low heads (only 3 or 4 ft). But some have been installed to operate efficiently at heads of more than 1000 ft. The runner of a medium- to high-head turbine has a different shape than a low-head turbine. For example, the most effective type of runner for a low-head (3 ft $< H <$ 100 ft) turbine is a propeller or *axial*-flow type shown in Fig. 8-1. The propeller blades may be fixed; however, for higher efficiencies

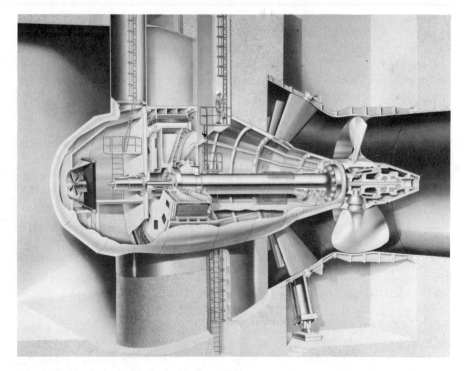

Figure 8-28 Schematic view of bulb turbine
(Courtesy of Voith Hydro Inc.)

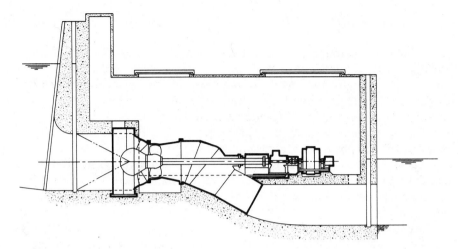

Figure 8-29 Schematic view of an S-turbine (Voith Hydro, Inc.)

over a wider range of load, the adjustable blade or Kaplan type of turbine is favored. The conventional configuration for the axial-flow turbine is a vertical turbine shaft that drives the generator above the turbine, as shown in Fig. 8-1. However, other innovative designs of axial-flow machines have different configurations. The axis of the *bulb*-type turbine has a horizontal orientation, and the generator is in a bulb-shaped housing just upstream of the turbine runner, as shown in Fig. 8-28. The advantage of this type of turbine is that cost savings can be achieved because the exit flow passages are simpler to construct, and the power-house is essentially eliminated because the generators are in the flow passages within the dam. Another variation of the axial-flow machine is the S–turbine, as shown in Fig. 8-29.

The reaction turbine in Fig. 8-2 is called a Francis turbine, after the man who perfected it. In the Francis turbine, the water before entering the turbine runner is prerotated by the spiral *scroll case* and by guide vanes. Then, as the water passes through the runner, the action of the vanes of the impeller turns the water so that it leaves the impeller in a direction essentially parallel to the axis of the runner and without rotation. The water is then conveyed to the downstream *tail race* through a *draft tube*.

Impulse Turbine Theory

FORCE ON BUCKET   Figure 8-30 shows the bucket with a jet of water impinging on it and being deflected. The absolute velocity of the jet is V_j, and the absolute velocity of the bucket is V_B. Therefore, if the momentum equation is applied to the control volume as shown and we let it move with the bucket,

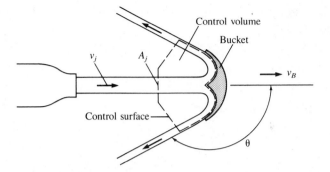

Figure 8-30 Deflection of jet by vane

it can be shown that the force on the bucket will be

$$F_{Bx} = -\rho(V_j - V_B)A_j[(V_j - V_B)_{2x} - (V_j - V_B)_{1x}] \tag{8-16}$$

where $(V_j - V_B)$ is the velocity of the jet relative to the bucket, and the subscripts $2x$ and $1x$ refer to the x components of the relative jet velocity exiting and entering the control volume, respectively. $(V_j - V_B)_{2x} = (V_j - V_B) \cos \theta$. Thus, for a given $(V_j - V_B)$, F_B will be maximized if $\theta = 180°$. Then Eq. (8-16) becomes

$$F_B = 2\rho(V_j - V_B)A_j(V_j - V_B) \tag{8-17}$$

The quantity $(V_j - V_B)A_j$ represents the discharge of water turned by a single bucket. It is less than the discharge issuing from the nozzle of the turbine. However, when we consider the discharge turned by the whole runner it can be appreciated that it will be the same as the discharge from the nozzle. Physically this can be explained by the fact that as a given bucket intercepts the jet from the nozzle not only is it deflecting that flow but there will also be a small cylinder of water being turned by the bucket just ahead of the given bucket (see Fig. 8-26). Therefore, the total force acting on the buckets will be

$$F_B = 2\rho V_j A_j(V_j - V_B) \tag{8-18}$$

POWER   The power developed will be the product of the speed of the bucket and the force acting on it, or

$$P = V_j F_B = 2\rho V_j V_B A_j(V_j - V_B)$$
$$= 2\rho A_j(V_j^2 V_B - V_j V_B^2)$$

Note that if $V_B = 0$ or if $V_B = V_j$, the power will be zero. There must be an optimum bucket speed between these two limits that will maximize the power.

This speed can be determined by differentiating P with respect to V_B and setting the differential equal to zero:

$$\frac{dP}{d(V_B)} = 2\rho A_j(V_j^2 - 2V_jV_B) = 0 \tag{8-19}$$

Then, solving Eq. (8-19) for V_B yields

$$V_B = \frac{1}{2} V_j \tag{8-20}$$

and $$P = Q\gamma\left(\frac{V_j^2}{2g}\right) \tag{8-21}$$

Thus it has been shown that the maximum power will be developed when the bucket speed is one half the jet speed. And Eq. (8-21) shows that the power thus developed is the same as the maximum theoretical power because $V_j^2/2g$ is the total head in the jet.

PRACTICAL CONSIDERATION   From a practical standpoint, the jet is usually turned less than 180° because of interference of the exiting jet with the incoming jet. Experience indicates that the optimum θ should be about 165°. Experience also shows that the desired speed ratio, V_B/V_j, is about 0.45 instead of 0.5. This is due to the hydraulic friction between the jet and the bucket surfaces. The efficiency of large impulse turbines is near 90% (2).

E X A M P L E 8 - 8   What power in kilowatts can be developed by the impulse turbine shown if the turbine efficiency is 85%? Assume the resistance coefficient f of the penstock is 0.015 and the head loss in the nozzle itself is nil. What will be the angular speed of the wheel assuming ideal conditions ($V_j = 2V_{bucket}$), and what torque will be exerted on the turbine shaft?

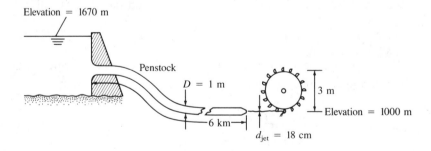

S O L U T I O N   First determine the jet velocity by applying the energy equation from the reservoir to the free jet before it strikes the turbine buckets.

$$\frac{p_1}{\gamma} + \frac{V_1{}^2}{2g} + z_1 = \frac{p_j}{\gamma} + \frac{V_j{}^2}{2g} + z_j + h_L$$

where $p_1 = 0$
$z_1 = 1670$ m
$V_1{}^2/2g = 0$
$p_j = 0$
$z_j = 1000$ m
$\gamma = 9810$ N/m^3 at 10°C (assumed)

The penstock water velocity is

$$V_{\text{penstock}} = \frac{V_j A_j}{A_{\text{penstock}}} = 0.0324 V_j$$

Then $h_L = \frac{fL}{D} \frac{V^2}{2g} = \frac{0.015 \times 6000}{1} (0.0324)^2 \frac{V_j{}^2}{2g} = 0.094 \frac{V_j{}^2}{2g}$

Now solving the energy equation for V_j yields

$$V_j = \left(\frac{2g \times 670}{1.094}\right)^{1/2}$$

$$= 109.6 \text{ m/s}$$

The gross power is

$$P = Q\gamma \frac{V_j{}^2}{2g} = \frac{\gamma A_j V_j{}^3}{2g}$$

or $P = \dfrac{9810(\pi/4)(0.18)^2(109.6)^3}{2 \times 9.81}$

$$= 16{,}760 \text{ kW}$$

The power output of turbine is

$$P = 16{,}760 \times \text{efficiency} = 14{,}245 \text{ kW}$$

The tangential bucket speed will be $\frac{1}{2}V_j$: therefore,

$$V_{\text{bucket}} = \frac{1}{2} \, 109.6 \text{ m/s} = 54.8 \text{ m/s}$$

or $r\omega = 54.8$ m/s

Thus $\qquad \omega = \dfrac{54.8 \text{ m/s}}{1.5 \text{ m}} = 36.53 \text{ rad/s}$

The wheel speed is

$$N = (36.53 \text{ rad/s}) \frac{1 \text{ rev}}{2\pi \text{ rad}} \, 60 \text{ s/min} = 349 \text{ rpm}$$

$$\text{Power} = T\omega$$

Thus $\qquad T = \dfrac{\text{Power}}{\omega} = 14{,}245 \text{ kW}/36.53 \text{ rad/s} = 390 \text{ kN·m}$ ■

Reaction Turbine Theory

CHARACTERISTICS OF THE REACTION TURBINE   In contrast to the impulse turbine, where a jet under atmospheric pressure impinges on only one or two vanes at a time, the flow in a reaction turbine is under pressure, and this flow completely fills the chamber in which the impeller is located (see Fig. 8-31). There is a drop in pressure from the outer radius of the impeller, r_1, to the inner radius r_2. This is also different from the impulse turbine, where the pressure is the same for the entering and exiting flow. The original form of the reaction turbine, first extensively tested by J.B. Francis, had a complete radial-flow impeller (Fig. 8-32). That is, the flow passing through the impeller had velocity components only in a plane normal to the axis of the runner. More recent impeller designs such as the mixed-flow and axial-flow types are still called reaction turbines.

TORQUE AND POWER RELATIONS FOR THE REACTION TURBINE   We will use the angular-momentum equation to develop formulas for the torque and power.* The segment of turbine runner shown in Fig. 8-32 depicts the flow conditions that occur for the entire runner. We can see that guide vanes outside the runner itself cause the fluid to have a tangential component of velocity around the entire circumference of the runner. Thus the fluid will have an initial amount of angular momentum with respect to the turbine axis when it approaches the turbine runner. As the fluid passes through the passages of the runner, the runner vanes effect a change in the magnitude and direction of velocity. Thus the angular momentum of the fluid is changed, which produces a torque on the runner. This torque drives the runner, which in turn, generates power. To quantify the above, we let V_1 and α_1 represent the incoming velocity

---

* The angular momentum approach, yielding equations of the same form, is also applicable for radial-flow pumps.

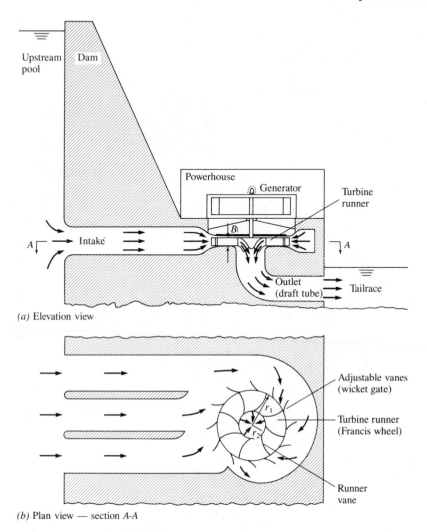

(a) Elevation view

(b) Plan view — section *A-A*

Figure 8-31   Schematic view of reaction turbine
installation

and angle of the velocity vector with respect to a tangent to the runner, respectively. Similar terms at the inner-runner radius are V_2 and α_2.

To obtain the torque on the turbine shaft, the angular-momentum equation
is applied to a control volume. Then for steady flow, we have

$$\sum \mathbf{M} = \sum_{cs} (\mathbf{r} \times \mathbf{V}) \rho \mathbf{V} \cdot \mathbf{A}$$

or $$\mathbf{T}_{shaft} = \sum_{cs} (\mathbf{r} \times \mathbf{V}) \rho \mathbf{V} \cdot \mathbf{A} \qquad (8\text{-}22)$$

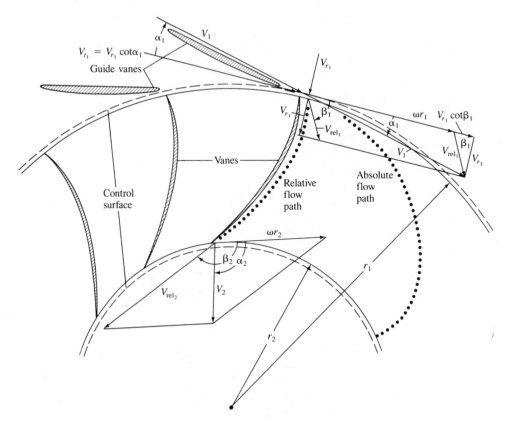

Figure 8-32   Velocity diagrams for a Francis-type runner

For the velocities of Fig. 8-32, Eq. (8-22) becomes

$$T = (-r_1 V_1 \cos \alpha_1)\rho(-Q) + (-r_2 V_2 \cos \alpha_2)\rho(+Q)$$
$$= \rho Q(r_1 V_1 \cos \alpha_1 - r_2 V_2 \cos \alpha_2)$$

The power from this turbine will be $T\omega$, or

$$P = \rho Q\omega(r_1 V_1 \cos \alpha_1 - r_2 V_2 \cos \alpha_2) \qquad (8\text{-}23)$$

Equation (8-23) shows that the power production is a function of the direction of the flow velocities entering and leaving the impeller, α_1 and α_2.

It may be noted that even though the pressure varies within the flow in the reaction turbine, it does not enter into the expressions we have derived using the angular-momentum equation. The reason it does not appear is that the outer and inner control surfaces that we chose are concentric with the axis about which we are evaluating the moments and angular momentum. The pressure forces acting on these surfaces all pass through the given axis; therefore, they do not produce moments about the given axis.

VANE ANGLES   The head loss in a turbine will be less if the flow enters the runner with a direction tangent to the runner vanes than if the flow approaches the vane with an angle of attack. In the latter case, separation will occur with consequent head loss. Thus vanes of an impeller designed for a given speed and discharge and with fixed guide vanes will have a particular optimum blade angle β_1. However, if the discharge is changed from the original design condition, the guide vanes and impeller vane angles will not "match" the new flow conditions. Most turbines for hydroelectric installations are made with movable guide vanes on the inlet side to effect a better match at all flows. Thus α_1 is increased or decreased automatically through governor action to accommodate fluctuating power demands on the turbine.

To relate the incoming-flow angle α_1 with the vane angle β_1, we first assume the flow entering the impeller will be tangent to the blades at the periphery of the impeller. Likewise, the flow leaving the stationary guide vane is assumed to be tangent to the guide vane. Now, we will consider both the radial component and the tangential components of velocity at the outer periphery of the wheel ($r = r_1$) in developing the desired equations. We can easily compute the radial velocity, given Q and the geometry of the wheel, by the continuity equation:

$$V_{r_1} = \frac{Q}{2\pi r_1 B}$$

where B = height of turbine blades.

The tangential (tangent to the outer surface of the runner) velocity of the incoming flow is given as

$$V_{t_1} = V_{r_1} \cot \alpha_1 \qquad (8\text{-}24)$$

However, in relation to the flow through the runner, this same tangential velocity is equal to the tangential component of relative velocity in the runner, $V_{r_1} \cot \beta_1$, plus the velocity of the runner itself, ωr_1. Thus the tangential velocity when viewed with respect to the runner motion is

$$V_{t_1} = r_1 \omega + V_{r_1} \cot \beta_1 \qquad (8\text{-}25)$$

Now, on eliminating V_{t_1} between Eqs. (8-24) and (8-25), we have

$$V_{r_1} \cot \alpha_1 = r_1 \omega + V_{r_1} \cot \beta_1 \qquad (8\text{-}26)$$

Equation (8-26) can be rearranged to yield

$$\alpha_1 = \text{arccot}\left(\frac{r_1\omega}{V_{r_1}} + \cot \beta_1\right) \tag{8-27}$$

EXAMPLE 8-9   A Francis turbine is to be operated at a speed of 600 rpm and with a discharge of 4.0 m³/s. If $r_1 = 0.60$ m, $\beta_1 = 110°$, and the blade height B is 10 cm, what should be the guide vane angle α_1 for a nonseparating flow condition at the runner entrance?

SOLUTION

$$\alpha_1 = \text{arccot}\left(\frac{r_1\omega}{V_{r_1}} + \cot \beta_1\right)$$

where $r_1\omega = 0.6$ m × 600 rpm × 2π rad/rev × $\dfrac{1}{60}$ min/s = 37.7 m/s

$$V_{r_1} = \frac{Q}{2\pi r_1 B} = \frac{4.00 \text{ m}^3/\text{s}}{2\pi \times 0.6 \text{ m} \times 0.10 \text{ m}} = 10.61 \text{ m/s}$$

$$\cot \beta_1 = -0.364$$

Then, $\alpha_1 = \text{arccot}(3.55 - 0.364)$

$$= 17.4° \qquad\blacksquare$$

PERIPHERAL SPEED OF TURBINE RUNNER AS A FUNCTION OF ΔH In our discussion on the impulse turbine, we showed that the bucket speed must be about 0.5 of the speed of the jet that drives the turbine for maximum efficiency. For reaction turbines, a similar relationship exists. That is, design experience shows that the ratio of the peripheral speed of the runner to $\sqrt{2g\,\Delta H}$ must be in a particular range of values to yield maximum efficiency. This ratio is identified by the symbol ϕ or

$$\phi = \frac{u}{\sqrt{2g\,\Delta H}} \tag{8-28}$$

where $u = r\omega = \pi DN/60$
ΔH = change in head across the turbine.

The denominator of the term on the right-hand side of Eq. (8-28) is often called the *spouting velocity*, and ϕ is called the *speed ratio*. The ranges of speed ratios for maximum efficiency for the different types of turbines are impulse turbine, 0.43–0.47; Francis turbine, 0.5–1.0; propeller turbine, 1.5–3.0.

Equation (8-28) along with representative values of ϕ for different types of turbines shows that the speed of a given turbine will be a function of ΔH. Because turbines are usually designed to operate at constant speed (which is necessary to maintain constant 60-cycle current), the optimum speed and size of turbine are obviously closely linked for most efficient operation. For example, if we were considering a turbine for a medium-head hydropower site, we would undoubtedly use a Francis turbine, and the average ϕ value would be about 0.75. Then, when this value is substituted into Eq. (8-28), we would have

$$0.75 = \frac{\pi DN/60}{\sqrt{2g\,\Delta H}}$$

or
$$D = \frac{60 \times 0.75 \sqrt{2g\,\Delta H}}{\pi N} \tag{8-29}$$

Thus D is essentially fixed for a given N and ΔH. The actual speed N for a specific installation must be a synchronous speed for the generator. For a 60-cycle frequency, it can be shown that the speed in rpm is given as

$$N = 7200/n \tag{8-30}$$

where n is the number of poles in the generator and must be an even integer. In Europe and South America, where the frequency is 50 cycles/s, the speed is given as $N = 6000/n$.

TURBINE OPERATION WITH VARIABLE LOAD   For a turbine and generator to supply power under varying load demands at a constant speed, control gates must be adjusted to increase or decrease the discharge to match the load. The flow conditions for a given gate opening will be somewhat different from the flow conditions for other gate openings; therefore, efficiency and overall performance characteristics will vary somewhat with the gate opening. Figure 8-33 shows power and efficiency versus gate opening for a typical reaction turbine operating at a speed of $N = 138.6$ rpm. In Fig. 8-33, data are given for various heads so that such information might be available in an installation where the head might also vary, such as when a reservoir water surface elevation changes over time.

TURBINE SPECIFIC SPEED   As in the case of pumps, *specific speed* is a dimensionless parameter that is useful in selecting the proper type of turbine for a particular set of ΔH and Q. It is also useful for correlating results of cavitation tests on turbines. For pumps, the specific speed was defined as

$$n_s = \frac{nQ^{1/2}}{g^{3/4}\,\Delta H^{3/4}} \tag{8-31}$$

However, for turbines, we are more interested in the power of the turbine rather than Q. Therefore, the turbine n_s is expressed in terms of power, P, by

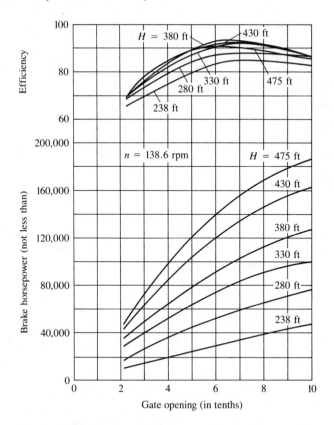

Figure 8-33   Power and efficiency versus gate opening for a turbine (13)

multiplying and dividing Eq. (8-31) by $\gamma^{1/2} \Delta H^{1/2}$ to obtain

$$n_s = \frac{nQ^{1/2}\gamma^{1/2} \Delta H^{1/2}}{g^{3/4} \Delta H^{3/4}\gamma^{1/2} \Delta H^{1/2}} \tag{8-32}$$

But $Q\gamma \Delta H = P$, so Eq. (8-32) can be expressed as

$$n_s = \frac{nP^{1/2}}{\gamma^{1/2}g^{3/4} \Delta H^{5/4}} \tag{8-33}$$

This is the dimensionless form of n_s. A dimensional form of specific speed has customarily been used by the hydraulic turbine industry, and this is given as

$$N_s = \frac{NP^{1/2}}{\Delta H^{5/4}} \tag{8-34}$$

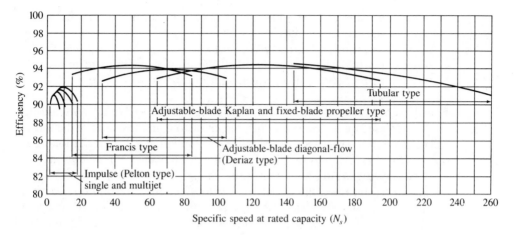

Figure 8-34 Typical peak efficiencies of various
turbines in relation to specific speed

where N = rotational speed in rpm
P = power in horsepower
ΔH = head on the turbine in feet.

For a given Q and ΔH, the actual speed is directly proportional to specific
speed, as can be seen in Eq. (8-34). Moreover, higher values of speed result in
reduced diameters (see Eq. 8-29) and weights of the generator and turbine, as
well as the reduced size and cost of powerhouse. Therefore, to take advantage
of these possible lower costs, the design engineer always desires to choose
turbines with high specific speeds even if there is some sacrifice in efficiency.

Figure 8-34 shows typical peak efficiencies for various types of turbines as
a function of specific speed. A plot like this is helpful in determining the type of
turbine to use for a particular hydropower site.

CAVITATION IN TURBINES Like pumps, turbines are also susceptible
to cavitation. The region of the flow where cavitation is most likely to occur is
on the downstream side of the turbine impeller blades. The cavitation index
for turbines is the same as for pumps (Eq. 8-14); however, the elevation of the
turbine refers to the tailrace water surface elevation, and the head loss is assumed
to be negligible. Thus the cavitation index is given as

$$\sigma = \frac{p_0/\gamma - p_v/\gamma - (z_t - z_0)}{\Delta H}$$
(8-35)

where p_0 = absolute atmospheric pressure
p_v = absolute vapor pressure of the water

z_t = elevation of the downstream side of the turbine above the water surface in the tailrace*

z_0 = elevation of tailrace water surface

ΔH = net head across the turbine (head change from upstream of turbine to the downstream end of draft tube)

For a given turbine operating with a given ΔH and speed, one can sense that as z_t is increased or p_0 decreased, the pressure acting on the blades of the turbine will decrease, and eventually one would reach a point at which cavitation would occur. Thus lower values of σ indicate a greater tendency for cavitation.

Because cavitation susceptibility also changes with the speed of the impeller (greater speed means greater relative velocities and less pressure on the downstream side of the impeller), the critical sigma values can be correlated to specific speed for different types of turbines. The actual σ_c values are obtained experimentally. Results of experimental tests are shown in Fig. 8-35.

If one considers a given water temperature (also given p_v), one can solve for z_t (allowable elevation of turbine) for given ΔH and atmospheric pressure, p_0. For standard atmospheric pressure and 80°F water temperature, such computations yield the plots shown in Fig. 8-36, which is similar to Fig. 8-18 for pumps.

EXAMPLE 8-10   Select the type, speed, and size of turbine for a site where the net head is 330 ft and $Q = 4300$ cfs.[†] Determine also the elevation of the turbine with reference to the water surface in the tailrace. Assume the turbine will drive a 60-cycle generator.

SOLUTION   From Fig. 8-36 for a net head, ΔH, of 330 ft, we see that N_s is about 45 for the normal practice of the 1960s. We also see from this figure and Fig. 8-34 that the turbine should be a Francis type. From Fig. 8-34, we can expect the turbine to have a peak efficiency of about 94%. Then the maximum developed power for the turbine will be

$$P = \frac{Q\gamma\,\Delta H \times \text{efficiency}}{550}$$

$$= \frac{4300 \text{ ft}^3/\text{s} \times 62.4 \text{ lb/ft}^3 \times 330 \text{ ft} \times 0.94}{550}$$

$$= 151{,}330 \text{ hp}$$

Because $N_s = NP^{1/2}/H^{5/4}$, one can solve for an approximate value of N:

---

* The *tailrace* is the channel (canal or natural stream) that the flow from the draft tube discharges into.

† The head and discharge values are those for one of several turbines installed at Grand Coulee Dam in Washington state.

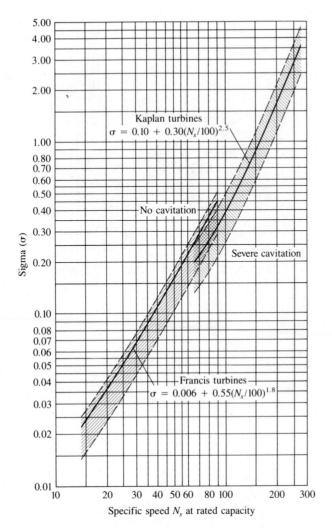

Figure 8-35 Sigma values for various specific speeds*
[From Davis and Sorensen (4), *Handbook of Applied
Hydraulics*, 3rd ed., McGraw Hill, 1969. Used with
permission.]
* Shaded bands show approximate range of critical sigma of
Francis and Kaplan turbines; curves indicate minimum plant
sigma recommended as a guide for preliminary selection
purposes.

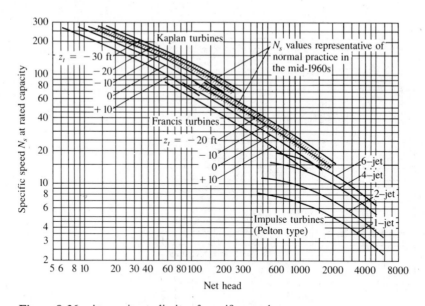

Figure 8-36 Approximate limits of specific speed
for various turbines and net heads* [From Davis and
Sorensen (4), *Handbook of Applied Hydraulics*, 3rd ed.
McGraw Hill, 1969. Used with permission.]
* Effect of draft head z_t on maximum allowable N_s of Kaplan
and Francis-turbines and effect of number of jets on N_s of
impulse turbines are illustrated. Shaded bands show range of
N_s values representative of present normal practice for Kaplan
and Francis turbines.

$$N = \frac{N_s H^{5/4}}{P^{1/2}}$$

$$= \frac{45 \times (300)^{5/4}}{(151,330)^{1/2}} = 162.7 \text{ rpm}$$

However, it is necessary that the turbine operate at a synchronous speed for the
generator that will be obtained from Eq. (8-30) or $N = 7200/n$. Using 46 poles
(the number of poles that yields the synchronous speed closest to but not exceeding $N = 162.7$ rpm), we have

$$N = \frac{7200}{46} = 156.5 \text{ rpm}$$

Then a new value of N_s is calculated based on this speed:

$$N_s = \frac{156.5 \times (151,330)^{1/2}}{(330)^{5/4}} = 43.3$$

With an N_s of 43.3 and a ΔH of 330 ft, we see from Fig. 8-36 that z_t (the elevation of turbine relative to the elevation of the water surface in the tailrace) should be about -10 ft. To determine the approximate diameter of the Francis wheel, we use Eq. (8-34):

$$D = \frac{60 \times 0.75\sqrt{2g \times 330}}{\pi \times 156.5}$$

$$= 13.3 \text{ ft}$$ ■

Small-Scale Hydropower Systems

In the 1970s and early 1980s, many small-scale hydropower systems were designed and developed. These became economically feasible when the cost of power generation increased markedly because of the increased cost of fuel for thermal power plants. The basic principles for these small-scale plants are the same as for conventional hydropower developments. However, because of their small scale, design and development costs have to be minimized. Some reduction in cost is achieved by purchasing standardized turbines and other equipment from manufacturers. It is also possible to use centrifugal pumps, operated in reverse direction, as turbines.*

8-12   Pump Turbines

A *pump turbine* is a hydraulic machine that operates as a pump in one mode of operation and as a turbine in a different mode. Almost all pump turbines are reversible; that is, the machine operates as a pump when rotating in one direction and as a turbine when rotating in the reverse direction. Pump turbines may be of the Francis, mixed-flow, or axial-flow type depending on the head the machine is designed for.

Use of Pump Turbines

Pump turbines are used in pumped storage installations to level the peaks and valleys of a typical electrical utility load curve. For example, a typical electrical utility will have high loads during the day and low loads at night. If a pumped storage installation is used in the system, then during periods of low demand, some of the "excess" energy can be used to pump water (using the

---

* For more on small-scale hydropower systems, see Warnick (18), Haroldsen (5), Kittredge (8), Mayo (10), and Shafer (14).

pumping mode of the pump turbine) from a low elevation to a storage reservoir at higher elevation. Thus energy is stored in the reservoir. Then during periods of high demand, the water from the high reservoir is run back through the pump turbine using the turbine mode to generate power. Pumped storage installations are especially desirable when used with a power plant that is most effectively operated at constant power for long periods, such as a nuclear power plant. The pumped storage part of the system does not produce any additional energy — it simply transfers energy produced during lower demand periods to high demand periods. The overall efficiency (including mechanical, electrical, and hydraulic losses) for a modern pumped storage system is about 75%.

Head Ranges for Different Types of Pump Turbines

The propeller or axial-flow type of pump turbine is generally designed to operate under fairly low heads (3 to 100 ft) and is often designed with adjustable blades so that it may be operated at high efficiency over wide ranges of head and discharge. This latter aspect is especially suited for use in tidal power installations.

For intermediate heads (35 to 300 ft), the mixed-flow type of impeller with adjustable blades is most often used. The radial-flow or Francis type runner with fixed vanes is used with high heads. Single-stage Francis type pump turbines have been designed and built for heads from 75 to 2000 ft. If heads greater than about 2000 ft are encountered, a more complex multistage machine must be used.*

Cavitation in Pump Turbines

Pump turbines are generally more susceptible to cavitation than a turbine or pump with the same specific speed. Therefore, to avoid cavitation, pump-turbine impellers have to be placed at an elevation lower than the tailwater elevation. The lower elevation produces a higher pressure on the impeller, which suppresses the onset of cavitation.

8-13   Viscous Effects

In earlier sections, we developed similarity parameters to predict prototype results from model tests, but we neglected to discuss the viscous effects in model pumps. Although viscous effects are usually small, they are not negligible, especially if the model is quite small.

---

* For more details on pump turbines, see Webb (19).

To minimize the viscous effects in modeling pumps, the Hydraulic Institute standards (6) recommend that the size of the model be such that the model impeller is not less than 30 cm in diameter. These same standards state that "the model should have complete geometric similarity with the prototype, not only in the pump proper, but also in the intake and discharge conduits."

Even with complete geometric similarity, one can expect the model to be slightly less efficient than the prototype. An empirical formula proposed by Moody that is used for estimating prototype efficiencies of radial- and mixed-flow pumps and turbines from model efficiencies is

$$\frac{1 - e_1}{1 - e} = \left(\frac{D}{D_1}\right)^{1/5} \tag{8-36}$$

where e_1 is the efficiency of the model, and e is the efficiency of the prototype.

EXAMPLE 8-11   A model having an impeller diameter of 45 cm is tested and found to have an efficiency of 85%. If a geometrically similar prototype has an impeller diameter of 1.80 m, estimate its efficiency when it is operating under conditions dynamically similar to those in the model test ($C_{Q \text{ model}} = C_{Q \text{ prototype}}$).

SOLUTION   We apply Eq. (8-36) with the condition that $e_1 = 0.85$ and $D/D_1 = 4$. Then

$$e = 1 - \frac{1 - e_1}{(D/D_1)^{1/5}}$$

$$= 1 - \frac{0.15}{1.32} = 1 - 0.11 = 0.89$$

The efficiency of the prototype is estimated to be 89%. ∎

8-14   Other Types of Pumps

The pumps and turbines we have discussed so far in this chapter are all classified as turbomachines. In turbomachines, the exchange of energy is accomplished by means of hydrodynamic forces developed between a moving fluid and the rotating and stationary parts of the machine. For example, in the axial-flow pump, the lift force of the rotating blades of the impeller produces the pressure increase of the pump.

Another entirely different class of pump is the positive displacement type. All positive displacement pumps have parts that interact in such a way that definite volumes of fluid are conveyed in the desired pumping direction essentially in proportion to the speed of operation of the pump. One of the simplest

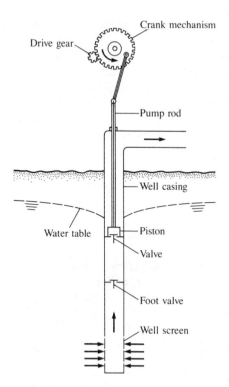

Figure 8-37   Reciprocating piston pump

positive displacement pumps is the reciprocating piston pump shown in Fig. 8-37. In this pump, as the pump rod and piston are raised by the crank mechanism, the valve in the piston is closed so that the piston draws water into the well and up through the well pipe (well casing). On this upstroke, the foot valve remains open. On the downstroke of the piston, the foot valve closes, but the valve in the piston opens. Thus one can see that for each cycle of the crank that drives the piston a definite volume of water will be "lifted" from the well and through the outlet pipe. If the speed of the crank is doubled, the rate of pumping would also be doubled (neglecting leakage past seals of the piston). The work required to pump the water can be expressed in terms of the essentially static force applied to the piston to lift the water times the distance through which the piston acts when water is being lifted.

Many other types of positive displacement pumps have configurations different from that of the simple piston pump. Several of these pumps are the *gear pump, two-lobe rotary pump,* and *screw pump.*

Besides the broad categories of turbomachines and positive displacement pumps, *jet pumps* and *hydraulic rams* have limited but important use in special situations.

Descriptions of the aforementioned pumps are given under separate headings below.

Gear Pump

Figure 8-38 is a section through a spur-gear pump. The gears rotate in the direction indicated, and these gears have very close clearance with the casing of the pump. Where the gear teeth contact, they form a tight liquid seal. Thus as the gears rotate, liquid flows in between the gear teeth on the suction side in very much the same way that liquid is drawn into the cylinder of a piston pump when the piston is on the suction stroke. As the gears rotate, the liquid is trapped between the teeth and the casing and is carried around to the discharge side of the pump, where the liquid is forced out as the teeth of the gears mesh together.

Gear pumps are just one class of rotary pumps that are used for pumping various kinds of liquids over a wide range of pressure, viscosities, and temperatures. Several applications of rotary pumps are

1.   Chemical processing
2.   Food handling
3.   Tank truck loading and unloading
4.   Machine tool coolants
5.   Pressure lubrication
6.   Hydraulic power transmission
7.   General transfer of liquids

The efficiency depends on the viscosity of the liquid being pumped, but it may be as high as 70% for low viscosity liquids. Rotary pumps can be designed to develop pressures up to 5000 psi, and some have capacities as high as 5000 gpm.

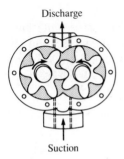

Figure 8-38   Spur-gear pump

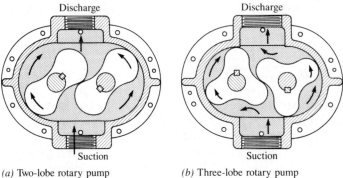

(a) Two-lobe rotary pump   (b) Three-lobe rotary pump

Figure 8-39   Lobe pumps

Lobe Pumps

Figure 8-39 shows two-lobe and three-lobe rotary pumps. These pumps operate exactly like gear pumps. However, because of the smaller number but larger volume of "chambers" that produce the pumping action, there may be more of a pulsating flow from the lobe pump than from the gear pump.

Screw Pumps

Screw pumps are similar to gear and lobe pumps in that pumping occurs as the elements of the pump rotate and mesh. In the screw pump, the liquid is carried between screw threads on one or more rotors and is displaced axially as the screws rotate.

Disadvantages of Rotary Pumps

The main disadvantages of rotary pumps are their cost is quite high because of the precise machining required to produce close tolerances, and they are not suited for pumping liquids that have abrasives in them. Because of the close clearances in rotary pumps, liquids containing abrasives (such as sand) will usually cause rapid wear of the surfaces.

Jet Pumps

Jet pumps derive their pumping action from a high velocity jet of fluid that then becomes entrained with the fluid it is pumping. The high momentum

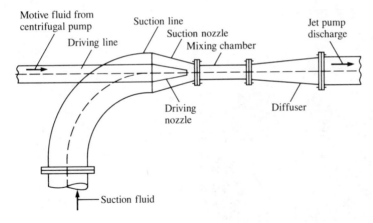

Figure 8-40   Jet pump

of the jet is converted to pressure in a diffuser. Liquid jet pumps are sometimes also called *eductors*. Figure 8-40 shows the essential features of a jet pump. There are many advantages of the jet pump, for example, it is self priming; it has no moving parts; and it can be made from any machinable materials, glass, and fiberglass. The main disadvantage of the jet pump is its relatively low efficiency. The entrainment process inherent in its operation produces large head losses that account for this low efficiency. Despite its low efficiency, it has several uses, including

1.   Deep-well pumping
2.   Bilge pumping on ships
3.   Providing circulation in rearing tanks of fish hatcheries (absence of moving mechanical parts do not injure fish)
4.   Chemical processing mixing
5.   Pumping out wells, pits, sumps where there is an accumulation of sand or mud

The deep well application is illustrated in Fig. 8-41. The jet pump and centrifugal pump act as a two-stage pumping unit. In the pumping process, the jet pump near the bottom of the well produces enough pressure so that the pressure on the suction side of the centrifugal pump is well above the vapor pressure of the liquid. Thus the centrifugal pump provides the remaining necessary head to yield the desired results. Without the jet pump, the centrifugal pump alone at the surface of the ground would not be able to pump water from a well more than about 30 ft deep because the water would vaporize when the suction pressure reaches the vapor pressure of the water (equivalent to about − 33 ft of head at normal temperatures). The jet pump is well suited for this kind of application because it can be designed to be a relatively compact unit that can be easily installed in a well. A typical commercial deep well unit has a 1-in. pressure pipe

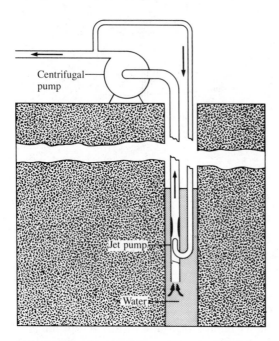

Figure 8-41   Jet pump in combination with a
centrifugal pump for pumping water from a well

with a $1\frac{1}{4}$-in. discharge pipe and, depending on the depth of the well, several
available nozzle and diffuser combinations (7).

Hydraulic Ram

The hydraulic ram was first developed in England in about 1800. It uses
a relatively large flow of water under low head to pump a much smaller amount
of water to a much higher elevation. Figure 8-42 shows the essential features of
a hydraulic ram. Valve W is the waste valve, and valve C is a check valve.
Assuming the cycle of operation starts with zero velocity in the drive pipe with

Table 8-1 Data on Selected Hydraulic Rams in the United States

Location	Discharge (cfs)		Head (ft)		Strokes
	to ram	to reservoir	drive pipe	pump head	per minute
U.S. Naval Coaling Station, Bradford, Rhode Island	1.29	0.52	37	84	130
Seattle Water Works, Seattle, Washington	1.63	0.55	49	131	65

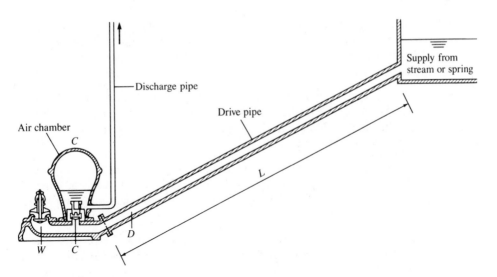

Figure 8-42 Hydraulic ram

valve W open and valve C closed, flow starts past valve W and, because of continuity, water also starts moving in the drive pipe. The flow through valve W and the drive pipe will accelerate until the velocity past the valve is so great that valve W closes quickly. The valve closure is initiated because the drag on the valve overcomes the weight of the valve that tends to keep it open. Once valve W closes, it produces a water hammer pressure in the drive pipe and in the body of the ram. This pressure will be large enough to open the check valve C, and some water will be forced into the air chamber. The pressure in the air chamber increases and further compresses the air. More water goes into the air chamber and, because of this increase of pressure, flow occurs in the discharge pipe. After a short time, the water hammer pressure in the drive pipe and ram subsides so that valve C closes. A short time later, the pressure in the drive pipe and ram is further relieved (the relief wave of water hammer starts) and valve W opens. Once valve W is opened, the cycle repeats itself.

In the early 1900s, many hydraulic rams were used for municipal water supplies as well as for individual farms. Table 8-1 gives data for two of these early rams. A study of some of these early rams (11) indicated that if the drive pipe was three times as long as the pumping head, it would be long enough to develop water hammer pressures to operate satisfactorily.

PROBLEMS

8-1   If the pump having the characteristics shown in Fig. 8-4, page 439, has a diameter of 40 cm and is operated at a speed of 1000 rpm, what will be the discharge when the head is 3 m?

8-2   If a pump geometrically similar to the one characterized in Fig. 8-5, page 439, is operated at the same speed (690 rpm) but is twice as large, $D = 71.2$ cm, what will be the water discharge and the power demand when the head is 10 m?

8-3   For a pump having the characteristics given in Figs. 8-4 or 8-5, what water discharge and head will be produced at maximum efficiency if the pump diameter is 24 in. and the angular speed is 1100 rpm? What power is required under these conditions?

8-4   A pump has the characteristics given in Fig. 8-4. What discharge and head will be produced at maximum efficiency if the pump size is 50 cm and the angular speed is 45 rps? What power is required when pumping water under these conditions?

8-5   For a pump having the characteristics shown in Fig. 8-4, plot the head-discharge curve if the pump is 14 in. in diameter and is operated at a speed of 900 rpm.

8-6   For a pump having the characteristics shown in Fig. 8-4, plot the head-discharge curve if the pump diameter is 60 cm and the speed is 690 rpm.

8-7   If the pump having the characteristics in Fig. 8-8, page 442, is operated at a speed of 30 rps, what will be the shutoff head?

8-8   If the pump having the characteristics in Fig. 8-9, page 443, is 40 cm in diameter and is operated at a speed of 25 rps, what will be the discharge when the head is 50 m?

8-9   If the pump having the characteristics in Fig. 8-8 is doubled in size but halved in speed, what will be the head and discharge at maximum efficiency?

8-10   For a pump having the characteristics shown in Fig. 8-9, plot the head-discharge curve if the pump diameter is 1.52 m and the speed is 500 rpm.

8-11   If a pump having the characteristics given in Figs. 8-8 or 8-9 is operated at a speed of 1500 rpm, what will be the discharge when the head is 160 ft?

8-12   If the pump having the performance curve shown is operated at a speed of 1500 rpm, what will be the maximum possible head developed?

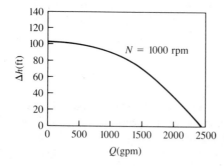

8-13   What type of pump should be used to pump water at a rate of 12 cfs and under a head of 25 ft? Assume $N = 1500$ rpm.

8-14   For most efficient operation, what type of pump should be used to pump water at a rate of 0.30 m³/s and under a head of 8 m? Assume $n = 25$ rps.

8-15   What type of pump should be used to pump water at a rate of 0.40 m³/s and under a head of 70 m? Assume $N = 1100$ rpm.

8-16   What type of pump should be used to pump water at a rate of 12 cfs and under a head of 600 ft? Assume $N = 1100$ rpm.

8-17   You want to pump water at a rate of 1.0 m³/s from the lower to the upper reservoir shown in the figure. What type of pump would you use for this operation if the impeller speed is to be 600 rpm?

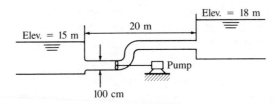

PROBLEM 8-17

8-18   The pump used in the system shown has the characteristics given in Fig. 8-5, page 439. What discharge will occur under the conditions shown, and what power is required?

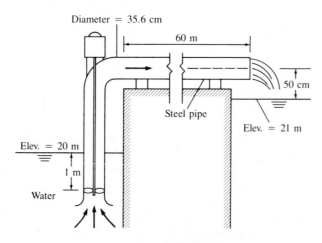

PROBLEMS 8-18, 8-19

8-19   If the conditions are the same as in Prob. 8-18 except that the speed is increased to 900 rpm, what discharge will occur, and what power is required for the operation?

8-20   An axial-flow pump is to be used to lift water against a head (friction and static) of 5.0 m. If the discharge is to be 0.40 m³/s, what maximum speed in revolutions per minute is allowed if the suction head is 1.5 m?

8-21   What is the specific speed for the pump operating under the conditions given in Prob. 8-18? Is this a safe operation with respect to the susceptibility to cavitation?

8-22   What is the specific speed for the pump operating under the conditions of Prob. 8-19? Is this a safe operation with respect to the susceptibility to cavitation?

8-23   For the conditions of Prob. 8-17, would the pump operate withoout cavitation?

8-24   A pump having the characteristics given in Fig. 8-8, page 442, pumps water from a reservoir at an elevation of 366 m to a reservoir at an elevation of 450 m through a 36-cm steel pipe. If the pipe is 610 m long, what will be the discharge through the pipe?

8-25   The pump delivers water from the large reservoir to the tank through the 800-ft long, 2-ft diameter steel pipe. The performance curve for the pump is also shown. The pump is started when the water surface elevation in the tank is 100 ft and continues to operate until the water surface elevation is 200 ft.

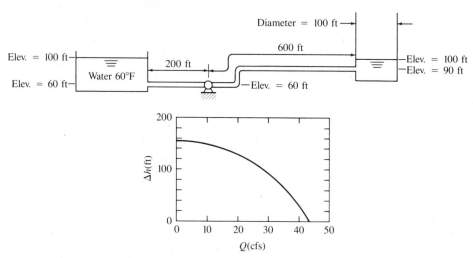

a. Where will the point of minimum pressure occur in the pipe? What is the magnitude of the minimum pressure?

b. What power must be supplied to the pump when the water surface elevation in the tank is 100 ft (a minute or so after the pump is turned on) if the pump efficiency is 70%?

c. Estimate the time required to fill the reservoir to the 200-ft level.

8-26   If two pumps like the one given in Prob. 8-25 had been installed and operated in parallel for the conditions of Prob. 8-25, what would be the initial discharge?

8-27   If two pumps like the one given in Prob. 8-25 had been installed in series for the conditions of Prob. 8-25, what would be the initial discharge?

8-28   Two pumps having the performance curve shown are operated in series in the 18-in. diameter steel pipe. When both are operating, estimate the time to fill the tank from the 150-ft level to the 200-ft level. Estimate the maximum pressure in the pipe during the filling phase. Where will this maximum pressure occur? What would have been the initial discharge if the pumps had been installed in parallel?

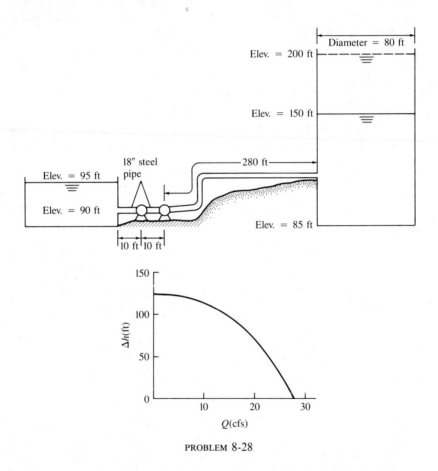

PROBLEM 8-28

8-29   The pump of Prob. 8-12 is used to pump water from reservoir A to reservoir B. The pump is installed in a 2-mi long, 12-in. pipe joining the two reservoirs. There are two bends in the pipe ($r/D = 1.0$), and two gate valves are open when pumping. When the water surface elevation in reservoir B is 30 ft above the water surface in reservoir A at what rate will water be pumped?

8-30 Work Prob. 8-29 but have two pumps like that of Prob. 8-12 operating in parallel.

8-31 Work Prob. 8-29 but have two pumps like that of Prob. 8-12 operating in parallel and have an 18-in. pipe instead of a 12-in. pipe.

8-32 The pump in the system shown discharges water into a wye connection and thus into the two pipes shown. The levels in the three reservoirs remain constant. With all regulating valves open, head losses in each of the three pipes can be expressed as $H_L = KQ^2$, where H_L is the head loss in feet, K is a constant of proportionality, and Q is in cfs. The magnitudes of the constants for the three lines are pipe A: $K = 5$; pipe B: $K = 20$; pipe C: $K = 30$. The pump has the following head-discharge characteristics:

Q (cfs)	0	1	2	3	4
H (ft)	200	198	182	145	75

If all valves in the system are open, determine the discharge in each of the three pipes. What is the horsepower output of the pump?

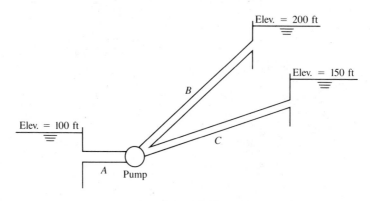

PROBLEM 8-32

8-33 A penstock 1 m in diameter and 10 km long carries water from a reservoir to an impulse turbine. If the turbine is 83% efficient, what power can be produced by the system if the upstream reservoir elevation is 650 m above the turbine jet, and the jet diameter is 16.0 cm? Assume $f = 0.016$, and neglect head losses in the nozzle. What should the diameter of the turbine wheel be if it is to have an angular speed of 360 rpm? Assume ideal conditions for the bucket design ($V_{bucket} = \frac{1}{2}V_j$).

8-34   Assume the characteristic curves shown in Fig. 8-8, page 442, are for a single-stage, single-suction centrifugal pump. This pump is to be installed in the system shown below. Do you think the pump will cavitate? Assume head loss in the system (from reservoir to reservoir) is negligible. Show computations to justify your answer.

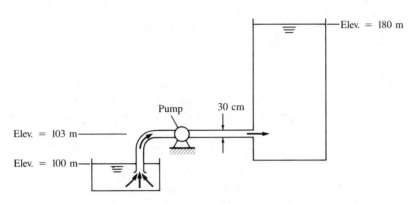

PROBLEM 8-34

8-35   A Francis turbine is to be operated at a speed of 60 rpm and with a discharge of 4.0 m³/s. If $r_1 = 1.5$ m, $r_2 = 1.20$ m, $B = 30$ cm, $\beta_1 = 85°$, and $\beta_2 = 165°$, what should α_1 be for nonseparating flow to occur through the runner? What power and torque should result with this operation?

8-36   A Francis turbine is to be operated at a speed of 120 rpm and with a discharge of 113 m³/s. If $r_1 = 2.5$ m, $B = 0.90$ m, and $\beta_1 = 45°$, what should α_1 be for nonseparating flow at the runner inlet?

8-37   a. For a given Francis turbine, $\beta_1 = 60°$, $\beta_2 = 90°$, $r_1 = 5$ m, $r_2 = 3$ m, and $B = 1$ m. What should α_1 be for a nonseparating flow condition at the entrance to the runner when the discharge rate is 126 m³/s and $N = 60$ rpm?
b. What is the maximum attainable power with these conditions?
c. If you were to redesign the turbine blades of the runner, what changes would you suggest to increase the power production if the discharge and overall dimensions are to be kept the same?

8-38   Select the type, speed, and size of turbine for a site where the net head is 600 ft and $Q = 10$ cfs.

8-39   Select the type, speed, and size of turbine for a site where the net head is 200 ft and $Q = 1000$ cfs.

8-40   Select the type, speed, and size of turbine for a site where the net head is 50 ft and $Q = 3000$ cfs.

REFERENCES

1. Chaudhry, M.H. *Applied Hydraulic Transients*. 2nd ed. Van Nostrand Reinhold, New York, 1987.
2. Creager, W.P., and J.P. Justin. *Hydroelectric Handbook*. John Wiley & Sons, New York, 1950.
3. Daugherty, Robert L., and Joseph B. Franzini. *Fluid Mechanics with Engineering Applications*. McGraw-Hill, New York, 1957.
4. Davis, C.V., and K.E. Sorensen (eds.). *Handbook of Applied Hydraulics*, 3d ed. McGraw Hill, New York, 1969.
5. Haroldsen, R.O., and F.B. Simpson. "Micro Hydropower in the U.S.A." *Water Power and Dam Construction*, 33, no. 33 (1981).
6. *Hydraulic Institute Standards*, 12th ed. Hydraulic Institute, New York, 1969.
7. Karassik, I.J. (ed.). *Pump Handbook*. McGraw Hill, New York, 1976.
8. Kittredge, C.P. "Centrifugal Pumps Used as Hydraulic Turbines." Paper no. 59-A-136, *Trans. ASME* (1959).
9. Knapp, R.T., J.W. Daily, and F.G. Hamitt. *Cavitation*. Institute of Hydraulic Research, Iowa City, Iowa, 1979.
10. Mayo, H.A., Jr., and W.G. Whippen. *Small Scale Hydro/Centrifugal Pumps as Turbines*. Allis-Chalmers, Inc., York, Pa., 1981.
11. Mead, D.W. *Hydraulic Machinery*. McGraw-Hill, New York, 1933.
12. Roberson, J.A., and C.T. Crowe. *Engineering Fluid Mechanics*, 3d ed. Houghton Mifflin, Boston, Mass., 1985.
13. Rouse, H. (ed.). *Engineering Hydraulics*. John Wiley & Sons, New York, 1950.
14. Shafer, L., and A. Agostinelli. "Using Pumps as Small Turbines." *Water Power and Dam Construction*, 33, no. 11 (1981).
15. Stepanoff, A.J. *Centrifugal and Axial Flow Pumps*, 2d ed. John Wiley & Sons, New York, 1957.
16. Strohmer, F., and E. Walsh. "Appropriate Technology for Small Turbines." *Water Power and Dam Construction*, 33, no. 11, (1981).
17. U.S. Department of Energy. *Microhydropower Handbook*. U.S. Govt. Printing Office, 1983.
18. Warnick, C.C. *Hydropower Engineering*. Prentice-Hall, Englewood Cliffs, N.J., 1984.
19. Webb, Donald E. *Pump Turbine Schemes*. The American Society of Mechanical Engineers, New York, 1979.

Linda, California, during the flood of February 1986.
The North Fork of the Feather River flows from top
center, and the Yuba River flows from the middle
right. (Courtesy of the State of California Department
of Water Resources)

9

Flood Control

9-1   Flooding and Historical Control

People have had to cope with floods throughout time. As long as people were nomadic, floods were only an inconvenience, requiring that camp be moved to higher ground during the wet season. In some parts of the world, this practice is still prevalent. In the Sudd there are large swamps created by the Nile River in Sudan and herds of cattle are moved to higher ground during the months when the river flow is high and vast areas are flooded. As the water recedes in the spring and summer, more grass is exposed, and the herds are moved back into the area to graze. The culture produced by this annual flooding has persisted for centuries.

In more populated areas, floods produce consequences far more serious than inconveniences. The attraction of the *flood plain* — the flat land along rivers or streams that is periodically flooded — is natural. Year after year, rivers have brought both water and sediment downstream. During flood flows, both the rate of flow and the concentration of sediment carried by the flow are higher. As the river rises, overtops its banks, and flows out of the *main channel* and over the flood plain, part of the transported sediment is deposited. The net result is an accumulation of fertile soil, which was early recognized for its agricultural importance. Thus early settlers farmed the flood plains, frequently building homes and buildings there. Because flood plains are fertile, easy to farm, and easy to build on, people have continually endeavored to occupy them permanently. In early times, people struggled with flood problems, often attempting to build local levees to keep flood waters from reaching their homes.

Because of the convenience of the river as a navigation route, communities developed along the river banks, providing a center from which agricultural products could be shipped and at which other raw materials or products could be landed for sale and distribution. The economic investment in flood plains has therefore grown throughout the world, and consequently annual damages produced by floods have almost continually increased.

As population increases, flood-control measures, such as the construction of levees and flood-control reservoirs, becomes a role of government. In the United States, the U.S. Army Corps of Engineers has general responsibility and authority for the development of flood-control projects (16). Even with the provision of some form of flood protection, the damages from flooding increase annually. This phenomenon is due both to the increasing investment in the flood plains and to the tendency of more people to move onto the flood plain after protection from flooding has been provided. Table 9-1 shows annual flood damages and deaths in the United States and indicates that the annual damage, though erratic, is increasing.

Hoyt (13) developed an annual flood index by summing flooding depths that occurred at index points on several key U.S. rivers each year. Thus a large flood index generally indicates a large amount of flooding. Hoyt showed that the amount of flooding as well as actual damages increased between 1903 and 1940 and concluded that of the increase in annual flooding damages during that

Table 9-1 Annual Deaths and Flood Damages in the United States (U.S. National Weather Service)

Year	Lives Lost	Property Damage (thousands of 1985 dollars)	Year	Lives Lost	Property Damage (thousands of 1985 dollars)
1903	178	800,000	1945	91	1,200,000
1904	0	100,000	1946	28	430,000
1905	2	170,000	1947	55	1,400,000
1906	1	6,000	1948	82	1,100,000
1907	7	200,000	1949	48	460,000
1908	11	130,000	1950	93	900,000
1909	5	700,000	1951	51	4,200,000
1910	0	300,000	1952	54	1,200,000
1911	0	130,000	1953	40	520,000
1912	2	1,000,000	1954	55	460,000
1913	527	2,200,000	1955	302	4,000,000
1914	180	260,000	1956	42	280,000
1915	49	200,000	1957	82	1,400,000
1916	118	270,000	1958	47	900,000
1917	80	250,000	1959	25	580,000
1918	0	65,000	1960	32	390,000
1919	2	21,000	1961	52	580,000
1920	42	150,000	1962	19	300,000
1921	143	220,000	1963	39	590,000
1922	215	440,000	1964	100	2,300,000
1923	42	260,000	1965	119	3,000,000
1924	27	290,000	1966	31	420,000
1925	36	70,000	1967	34	1,700,000
1926	16	180,000	1968	31	1,300,000
1927	423	2,800,000	1969	297	3,100,000
1928	15	360,000	1970	135	800,000
1929	89	490,000	1971	74	900,000
1930	14	150,000	1972	554	10,600,000
1931	0	29,000	1973	148	4,100,000
1932	11	60,000	1974	121	1,400,000
1933	33	400,000	1975	107	3,000,000
1934	88	100,000	1976	193	5,800,000
1935	236	1,300,000	1977	210	2,300,000
1936	142	2,500,000	1978	143	1,200,000
1937	142	3,500,000	1979	121	5,900,000
1938	180	700,000	1980	82	2,000,000
1939	83	140,000	1981	84	1,400,000
1940	60	380,000	1982	155	3,700,000
1941	47	330,000	1983	204	3,800,000
1942	68	600,000	1984	126	4,800,000
1943	107	1,400,000	1985	304	3,000,000
1944	33	700,000			

period, 45% resulted from an increase in property values, 25% from an increase in the annual magnitude of flooding, and 30% from increased use of the flood plains. Thus although the investment in flood control has increased rapidly during the past decades, annual damages have also increased.

Annual flood damages for the United States were tabulated by the U.S. Water Resources Council until 1980 (20), but since then, they have been tabulated by the U.S. National Weather Service. The U.S. Army Corps of Engineers tabulates annual flood damages prevented by Corps projects.

Flood-control work began in the United States in 1718, soon after the site for the city of New Orleans was selected. Settlers were ordered to build levees to protect themselves and their neighbors from future floods. Until 1917, individuals and local government provided all funds for flood control in the United States, although the federal government did provide some funds for construction of levees along the Mississippi River in the interest of navigation. In 1917, an act of Congress authorized the construction of flood-control levees with federal funds.

Record floods occur as a result of either rainfall or rainfall and simultaneous snowmelt. In general, rainfall produces the highest peak flows, and the combination of rainfall and snowmelt produces the largest flood volumes. The largest flood in the United States, for which official records exist, occurred on the Mississippi River on May 1, 1927, and peaked at 2,270,400 ft^3/s (64,290 m^3/s) at Vicksburg, Mississippi (22). At that point, the Mississippi River has a drainage area of 1,142,100 mi^2 (2,958,000 km^2). Thus that record flood, which had an estimated recurrence interval of more than 100 yr, represented a peak flow per unit area of 1.99 ft^3/s/mi^2 (0.022 m^3/s/km^2). By contrast, Kawaikoi Creek at Waimea, Hawaii, with a drainage area of only 4.2 mi^2 (10.9 km^2) experienced a flood of 11,264 ft^3/s (319 m^3/s) on January 13, 1967, a peak flow per unit area of 2682 ft^3/s/mi^2 (29.3 m^3/s/km^2). The Kawaikoi Creek flood was also judged to be approximately a 100-yr event. Thus difference in flood characteristics is pronounced between large and small drainage areas. The primary difference is that a single rainstorm cannot uniformly cover a large drainage basin, whereas a small drainage basin may be totally subject to the peak intensity of the storm.

By contrast, the flood volumes produced by frontal storms that cover a large drainage area are likewise very large, and the large flow rates sometimes continue for many days. Floods produced by intense storms that cover small drainage areas peak quickly and recede quickly giving rise to large peak flow rates but relatively small runoff volumes.

The Mississippi River and Kawaikoi Creek floods resulted from severe rain storms. Snowmelt floods can also be extreme events. For example, the Judith River in Montana, a drainage area of 33 mi^2 (85.5 km^2), experiences regular snowmelt floods, as do almost all streams in northern climates. On June 6, 1927, a flood of 1116 ft^3/s (31.6 m^3/s) occurred, producing a peak discharge per unit area of 33.8 ft^3/s/mi^2 (0.37 m^3/s/km^2).

Table 9-2, pages 498–499, lists many record floods along with some of the drainage basin characteristics of these floods. Figure 9-1 shows a plot of peak

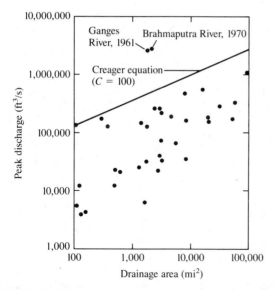

Figure 9-1   Peak discharge as a function of drainage area

flow rate for each of these floods as a function of drainage area and demonstrates that peak flow rates do not correlate with drainage area. A detailed study of several large historical floods from many rivers of the world was done by Creager (6). In his study, Creager developed envelope curves for the maximum floods of record in terms of the equation

$$Q_p = 46CA^{(0.894A^{-0.048})} \tag{9-1}$$

where Q_p = peak discharge in ft^3/s
A = drainage area in mi^2
C = an empirical coefficient dependent on drainage area characteristics
with a maximum value of approximately 100 for many areas

Figure 9-1 shows that Creager's equation would predict maximum flow rates above most of those in Table 9-2 but that two floods in Bangladesh substantially exceeded the envelope value. Although the floods used by Creager and those plotted in Fig. 9-1 are recorded maximum floods, most authorities now believe the record maximum flood is a poor indicator of the probable maximum flood.

Figure 9-2, page 500, compares probable maximum floods estimated for the Ohio-Tennessee region with peak flow rates calculated using Creager's equation (Eq. 9-1) and provides further evidence that although some recorded floods seem very large, they should always be considered carefully before they are assumed to be representative of the largest possible flood.

Table 9-2 Floods of Record from Various Drainage Basins (22)

Country	River	Date	Q Peak		Drainage Area		Mean Basin Elevation		Type
			(ft³/s)	(m³/s)	(mi²)	(km²)	(ft)	(m)	
Bangladesh	Brahmaputra	07/28/70	2,705,100	76,600	1,903	4,930	109	33.3	R*
Bangladesh	Ganges	09/01/61	2,585,030	73,200	1,842	4,770	123	37.5	R
Bulgaria	Iskar	06/29/57	33,200	940	3,232	8,370	2,315	706	R
Canada	Slocan	06/07/61	25,300	719	1,266	3,280	5,051	1,540	S†
Canada	Chilliwack	06/20/72	4,100	116	131	339	4,100	1,250	S
Canada	Chilco	06/20/50	16,920	479	3,232	8,370	5,346	1,630	S
Canada	Rocky	12/27/53	5,690	161	110	285	268	81.6	R
Canada	Churchill	06/27/57	240,850	6,820	28,840	74,700	1,702	519	S
Congo	Congo	11/17/61	2,683,920	76,000	1,343,600	3,480,000	656	200	R
Congo	Sangha	11/06/60	165,270	4,680	61,000	158,000	1,968	600	R
Czechoslovakia	Morava	12/03/41	22,110	626	2,706	7,010	1,614	492	S
Czechoslovakia	Becva	07/27/39	22,310	660	494	1,280	1,768	539	R
Finland	Kemijoki	05/01/73	170,220	4,820	19,610	50,800	951	290	S
France	Rhone	05/31/56	158,920	4,500	7,840	20,300	—	—	R
France	Durance	10/25/82	180,100	5,100	4,590	11,900	—	—	R
France	Seine	02/24/58	2,170	61.6	564	1,460	1,017	310	R
Gabon	Ogooue	11/18/61	480,280	13,600	7,880	204,000	1,476	450	R
Germany	Rhein	12/31/82	160,680	4,550	19,420	50,300	4,592	1,400	S
Ghana	Pra	07/17/57	35,900	1,020	8,030	20,800	1,312	400	R
Hungary	Danube	06/16/65	325,600	9,220	50,580	131,000	2,624	800	R
Hungary	Zagyva	06/12/65	6,390	181	1,625	4,210	649	198	R
India	Krishna	10/07/03	1,059,440	30,000	99,230	257,000	2,624	800	R
Italy	Magra	10/15/60	122,900	3,480	362	939	2,007	612	R
Italy	Alli	12/19/30	639	18.1	18	46	3,542	1,080	R
Japan	Mogami	08/29/67	137,730	3,900	1,363	3,530	1,692	516	R

Japan	Toyo	08/05/69	167,900	4,770	279	724	1,099	335	R
Maylasia	Perak	01/06/67	222,480	6,300	3,000	7,770	2,745	837	R
Morocco	Ouergha	12/18/63	248,260	7,030	2,390	6,190	2,460	750	R
Norway	Engera	05/07/34	4,130	117	152	394	2,722	830	S
Poland	Dunajec	06/30/58	123,600	3,300	1,676	4,340	2,620	799	R
Sweden	Baljanea	04/20/70	1,300	37	92	239	295	90	S
Sweden	Baljanea	11/23/63	1,510	43	92	239	295	90	R
USSR	Dnieper	05/01/08	64,270	1,820	5,444	14,100	738	225	S
USSR	Desna	04/20/31	281,600	8,000	31,429	81,400	656	200	S
United Kingdom	Don	05/24/32	12,250	347	486	1,260	564	172	R
United Kingdom	Trent	03/19/47	39,200	1,110	2,892	7,490	456	139	R
United Kingdom	Cannons Brook	07/01/58	500	14.2	8.3	21.4	246	75	R
United States	Ute Creek	05/15/41	629	17.8	32	83	—	—	S
United States	Arkansas	05/27/43	536,780	15,200	157,900	409,000	—	—	R
United States	Mississippi	05/01/27	2,277,800	64,500	1,142,900	2,960,000	—	—	R
United States	Kawaikoi	01/13/67	11,300	320	4.2	11	—	—	R
United States	Quaboag	08/19/55	12,740	362	150	390	—	—	
United States	Eel	12/23/64	752,200	21,300	3,130	8,100	—	—	R

* R = flood created by rainfall.
† S = flood created by rain and snowmelt.

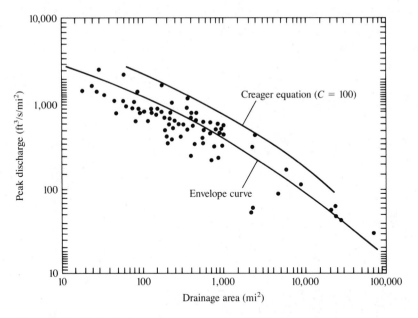

Figure 9-2 Peak discharge for computed probable maximum floods for rivers and streams in the region of Ohio, Tennessee, southern Pennsylvania, West Virginia, Kentucky, Alabama and Virginia (19)

9-2   Control of Flooding

As we mentioned, flood damages are created when river flows are large enough to cause flooding of those areas that are less often covered by water than the *main channel*. Many municipal and rural developments are located within the *flood plain*. As Fig. 9-3 shows, a river channel can be divided into zones. That part of the channel referred to as the *main channel* has been formed by frequently occurring flow rates. The bankfull capacity of the river is approximately the mean annual flood — that flood whose peak flow is exceeded approximately one out of two years. The Gumbel distribution (extreme value type

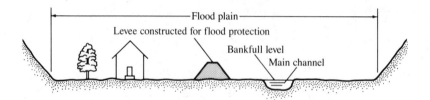

Figure 9-3 Illustration of definitions of flood plain, main channel and levee

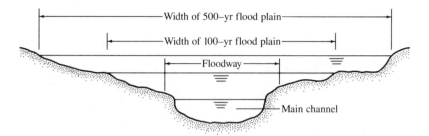

Figure 9-4   River channel and flood plain limits

I) a statistical distribution frequently used for maximum floods gives a recurrence interval of 2.33 yr for the mean annual flood (10). When a flood with a peak flow larger than the mean annual flood occurs, part of the overbank area is flooded. Thus the part of the river valley that is flooded (outside the main channel) is the flood plain, and its extent depends on the magnitude of flood being considered. Thus a 500-yr flood involves greater depths and a larger peak flow rate than a 100-yr flood and covers a greater width of the valley. Figure 9-4 shows a 500-yr and a 100-yr flood plain. The *floodway* is defined by the Federal Agency Management Agency as that part of the stream channel that could contain the 100-yr peak flow with not more than a 1.0-ft increase in depth above that which would occur if the entire cross section carried the 100-yr peak flow (7).

The general goal of flood-control engineering is to prevent or limit flood damages by controlling or managing the flood. In general, a flood can be controlled in only two ways: The peak flow rate of the flood must be reduced, or the capacity of the flood channel must be made large enough to prevent or limit overbank flooding.

The peak flow of a flood can be reduced by:

Storage of at least part of the flood water in an upstream reservoir or reservoirs

Land management upstream to increase infiltration, interception, and detention losses and, thus, both delay and reduce the rate and volume of runoff

Increasing channel storage through diversion to a bypass channel

Diverting flood flows into another river basin

The Sacramento River in California provides an excellent example of the effect of flood-control reservoirs in reducing flooding in the lower portions of the river. Table 9-3 shows the statistics on flood-control storage for several reservoirs in California, of which four are on the Sacramento River or its tributaries. These four reservoirs have succeeded in significantly reducing the peak flood flows occurring along the inhabited stretches of the river. During the January 1974 flood, the operation of Shasta Reservoir alone reduced the peak flood at Sacramento (160 mi downstream) from 330,000 to 128,000 ft^3/s. Figure 9-5

Table 9-3 Some Dams in the State of California with Flood-Control Capacity (4)*

(1) Reservoir	(2) Stream	(3) Completion Date	(4) Operating Agency	(5) Purposes	Storage Capacity		
					Total (acre-feet)	Flood-control (acre-feet)	(%)
Shasta	Sacramento R.	1943	USBR	F,N,I,P,Q	4,500,000	1,300,000	29
Pine Flat	Kings R.	1954	C of E	F,I,P	1,000,000	1,000,000	100
Oroville	Feather R.	1968	DWR	F,I,M,P	3,500,000	750,000	21
Isabella	Kern R.	1953	C of E	F,I	570,000	570,000	100
Folsom	American R.	1956	USBR	F,I,P	1,000,000	400,000	40
New Exchequer	Merced R.	1966	MID	F,I,P	1,000,000	400,000	40
Friant	San Joaquin R.	1941	USBR	F,I	520,000	390,000	75
New Don Pedro	Tuolumne R.	1971	M & TID	F,I,M,P	2,030,000	340,000	17
Camanche	Mokelumne R.	1963	EBMUD	F,M,P	432,000	200,000	46
New Hogan	Calaveras R.	1963	C of E	F,I	325,000	165,000	51
Black Butte	Stony Cr.	1963	C of E	F,I	160,000	150,000	94
Terminus	Kaweah R.	1962	C of E	F,I	150,000	150,000	100
Success	Tule R.	1961	C of E	F,I	80,000	80,000	100
Total					15,267,000	5,895,000	39

* Symbols: USBR, Bureau of Reclamation
C of E, Corps of Engineers
DWR, California Department of Water Resources
MID, Merced Irrigation District

M & TID, Modesto and Turlock Irrigation Districts
EB MUD, East Bay Municipal Utility District
F, Flood control
N, Navigation

I, Irrigation
P, Power
Q, Quality control
M, Municipal supply

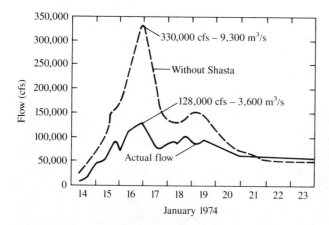

Figure 9-5   Flood hydrograph for the Sacramento River at Sacramento for the 1974 flood showing what the hydrograph would have been without Shasta Dam (4)

shows the actual flood hydrograph at Sacramento and the one that would have occurred without Shasta Dam. Figure 9-6 shows the Sacramento Weir, which diverts flood waters from the Sacramento River to the Yolo bypass — one of two flood bypass channels that can be used to bypass a portion of large floods to the west of Sacramento, California, and thus decrease the flooding depth at Sacramento.

Upstream improvement of land-use management within the United States has been the province of the Soil Conservation Service (SCS) of the U.S. Department of Agriculture since 1935, when the U.S. Congress enacted a policy of preventing soil erosion for the preservation of natural resources, control of floods, and related purposes (16). The SCS has continuously pursued the design and construction of small reservoirs and the implementation of sound land-management practices for preventing floods and conserving soil resources. By 1985, the SCS had constructed more than 23,000 reservoirs throughout the United States, providing some 7,031,700 acre-ft of storage.

Several famous flood-control projects exist within the United States. The Miami Conservancy District was created in Ohio to control flooding by the Miami River after a disastrous flood in March 1913 caused 360 deaths and more than $100 million (1913 dollars) in property damage (2). Arthur E. Morgan was chief engineer of the Conservancy District, and under his direction, many significant advances were made both in the engineering design of flood-control measures and in the hydraulic and hydrologic methodology required to analyze rainfall and floods for design (12, 15, 21).

To develop protection against flooding, Morgan proposed constructing a system of five reservoirs together with substantial channel improvements. The

Figure 9-6   Flood waters are passing through 48
gates that have been opened to relieve flooding in
Sacramento by allowing water to flow from the
Sacramento River in the foreground to the Yolo Bypass.
(Courtesy of the California Department of Water
Resources) (8)

reservoirs were formed by dams and acted to detain floods. During the 1913
flood, a total of 1,415,000 acre-ft had fallen as precipitation on the river basin.
The proposed detention reservoirs provided a storage capacity of 840,000 acre-ft,
and the peak outflow from the basins during a storm as severe as that of 1913
was designed to be not more than the bankfull capacity of the river downstream.
Figure 9-7 illustrates the typical form of an uncontrolled outlet from a detention
reservoir as used by Morgan.

The Tennessee Valley Authority (TVA) was created by the U.S. Congress
in 1933 to construct dams and reservoirs to promote navigation, control floods,
and generate electricity. The drainage basin of the Tennessee Valley covers
40,900 sq mi. Before TVA, the flow at the mouth of the Tennessee River had
ranged from a low of 4500 ft^3/s (127.8 m^3/s) to a high of 471,000 ft^3/s
(13,400 m^3/s) during the flood of 1897. Thirty-six major dams were constructed
by the TVA on the Tennessee River, and 14 were constructed on adjoining
rivers. Control of the system provides 12,000,000 acre-ft of reserved storage for
flood regulation on January 1 and 9,000,000 acre-ft on April 1 of each year.

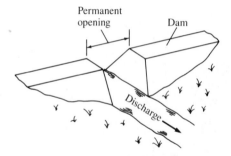

Figure 9-7 Uncontrolled outlet from a detention
basin or reservoir

This combination of available flood storage volume has provided substantial flood regulation and has increased the river low flow to 37,000 ft^3/s (1050 m^3/s). From its inception to 1983, the TVA system has prevented flood damages estimated at more than $2 billion (3).

On the Columbia River in the Pacific Northwest, a system of large dams has been developed by the United States and Canada (9). Figure 9-8 shows the location of the major single- and multiple-purpose projects that have been developed since construction of Bonneville and Grand Coulee dams was begun in 1933. A total of 210 projects (not all are shown in Fig. 9-8) provides an active storage of 62,000,000 acre-ft. The system provides benefits for navigation, hydroelectric generation, flood control, irrigation, and recreation. Figure 9-9 shows the reduction in peak flow of the Columbia River at the Dalles that has been achieved since 1948. Annual benefits due to flood reduction have been estimated at $507 million (9).

9-3 Flood Plain Management

The construction of reservoirs to store floodwaters or the improvement of river channels to lower flood levels has tended to increase development of the protected flood plains. As a result, when the design flood for the control measures is exceeded, damages are frequently more than would have occurred for the same flood before construction of the control measure. Laws have therefore been enacted to regulate the use of flood plains. This approach was given great impetus in the United States by the Flood Disaster Protection Act of 1973, the function of which was to keep people away from floods rather than floods away from people.

In the practice of flood-plain management, the extent of the flood plain for a particular design flood (usually the 100-yr or 500-yr floods in the United States) is defined. So defined, the flood plain is that area that would be inundated by the design flood if it should occur. The flood plain is frequently divided

Figure 9-8   Reservoirs in the Columbia River Basin (9)

into the *floodway*, where velocities are large enough to cause damage during flooding, and the peripheral area, which is primarily an area of overbank flood storage that is inundated during flooding but experience only small velocities. The *floodway* within communities that have adopted flood-plain regulations is usually subject to strong building restrictions; lesser restrictions apply to the peripheral area. In general, adopting a flood-plain management program is re- quired before federally subsidized flood insurance can be made available to property owners.

9-4   Delineation of the Flood Plain

Hydraulic analysis of flood-control structures and channels in the flood plain involves the use of principles and tools we presented in Chapter 4. The

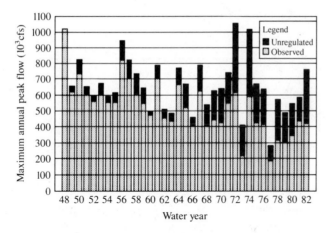

Figure 9-9 Comparison of historial observed flows on the Columbia at the Dalles with what the flows would have been without the reservoirs in the Columbia Basin (9)

flood plain defined in the United States is usually that for the 100-yr flood. The calculation of the 100-yr peak flow or the 100-yr flood hydrograph is performed using methods we presented in Chapter 2. If the length of the channel over which the flood plain is to be defined is relatively short, the methods of gradually varied flow (Chapter 4) can be used. These methods assume the flow is steady at the peak flow rate associated with the design flood. Cross sections must be surveyed along the river at small enough intervals to adequately define the valley geometry. Then values of the resistance coefficient must be estimated for both the main channel and the overbank areas. Table 4-1 and Figs. 4-2, 4-3, and 4-4 (pages 169–170) provide some typical values of Manning's n to be used in the computational process. Table 9-4 on the next page provides values of Manning's n that are more typical for overbank areas during flood flows. The overbank areas, because of typical growth of trees, grass, weeds, and other obstructions tend to exhibit greater resistance to flow than the main channel, with the result that velocities in the main channel (floodway) are always larger than those in the overbank (peripheral) area.

To delineate the extent of overbank flooding, it is necessary to compute the water surface profile in the channel, which involves the use of the energy equation (Eq. 4-28, page 199, or, the more compact form, Eq. 4-33, page 203). For flood flow in channels with overbank flow, it is convenient to use the kinetic-energy correction factor, α. Its use here differs from that in Chapter 5 in that we consider only the difference in mean velocities in the various parts of the channel, we do not consider the vertical variation of velocity. The coefficient is usually significantly larger than unity because the average velocity

Table 9-4 Values of Manning's n to Be Used for Overbank Areas Along Streams or Rivers

Channel Description	Average Value of n
Grassland	
Short grass	0.030
Tall grass	0.035
Cultivated ground	
Bare ground	0.030
Mature row crops	0.035
Mature field crops	0.040
Brushy areas	
Dense weeds and sparse brush	0.050
Brush-covered with some trees (winter)	0.050
Brush-covered with some trees (summer)	0.060
Dense brush (winter)	0.070
Dense brush (summer)	0.100
Forested	
Densely covered with willows (summer)	0.150
Cleared land with stumps; no new growth	0.040
Cleared land with stumps; dense new growth	0.060
Dense stands of large trees; flood stage below branches	0.100
Dense stands of large trees; flood stage reaching branches	0.120

in the overbank area is generally much smaller than the average velocity in the main channel. Thus Eq. (4-28) with α included is written between two sections a distance Δx apart as

$$y_a + \frac{\alpha_a V_a^2}{2g} + S_0 \Delta x = y_b + \frac{\alpha_b V_b^2}{2g} + S_f \Delta x \qquad (9\text{-}2)$$

or, in the form of Eq. (4-33),

$$E_a - E_b = (S_f - S_0) \Delta x \qquad (9\text{-}3)$$

In Eq. (9-2), y_a and y_b are depths at sections a and b, respectively, and can be in either the main channel or the overbank channel. To evaluate α, the definition given by Eq. (5-2), page 242, is applied in finite increment form:

$$\alpha = \frac{1}{V^3 A} \sum_{i=1}^{N} V_i^3 A_i \qquad (9\text{-}4)$$

where the subscript i indicates the subarea of which V_i and A_i are the average velocity and cross-sectional area, respectively, and N indicates the number of subareas that the section is divided into. V and A are, respectively, the average

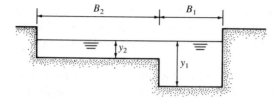

Figure 9-10 Definition sketch for a simple channel and overbank area

velocity and the total area of the cross section. Because $V = Q/A$, Eq. (9-4) can be expressed as

$$\alpha = \frac{A^2}{Q^3} \sum_{i=1}^{N} \left(\frac{Q_i^3}{A_i^2} \right) \tag{9-5}$$

where Q is the total discharge, and Q_i is the discharge in subsection i. For the case of a compound channel such as shown in Fig. 9-10, the free surface must be level across each section if the flow is to be one dimensional as assumed. Thus the friction slope between any two sections of a reach of channel will be the same for the main channel and the overbank channel.

The total discharge in the channel is the sum of the discharges in each part, or

$$Q = \sum_{i=1}^{N} V_i A_i \tag{9-6}$$

In calculating coordinates of the free surface profile in an open channel, Eq. (9-2) must be solved. In the solution, the energy slope S_f can be computed using either the Darcy-Weisbach relationship

$$S_f = \left(\frac{fQ^2}{8gRA^2} \right) \tag{9-7}$$

or the Manning relationship

$$S_f = \left(\frac{n^2 Q^2}{2.22 A^2 R^{4/3}} \right) \tag{9-8}$$

The Manning equation is used more frequently because the bottom roughness, shape, height, and texture vary greatly along a natural channel, and the greater accuracy afforded by using the resistance coefficient f is usually not warranted. The solution of Eq. (9-2) can be performed using standard numerical methods (14). The flow rate and depth in each subsection must be known at one cross section of the channel as well as the channel profile and cross sections of the stream throughout the length of interest. Channel roughness must be estimated for both the overbank and main channel between each set of known cross sections.

To proportion the flow properly between the subsections at each cross section, the relationship for resistance to flow in the channel must be considered. Equation (9-6) can be rewritten as

$$Q = Q_1 + Q_2 \tag{9-9}$$

where the subscripts 1 and 2 refer to subareas of the cross section. Moreover, the slope of the energy line in the main channel must equal that for the overbank area:

$$S_f = S_{f_1} = S_{f_2} \tag{9-10}$$

The conveyance for an open channel, K, is defined as $Q/\sqrt{S_f}$; thus

$$K = \frac{Q}{\sqrt{S_f}} = \frac{1.5}{n} AR^{2/3} \tag{9-11}$$

which can be conveniently used in the following development.

Using Eqs. (9-9) and (9-11), we can write

$$Q = K_1 \sqrt{S_{f_1}} + K_2 \sqrt{S_{f_2}} \tag{9-12}$$

or $$Q = (K_1 + K_2)\sqrt{S_f} \tag{9-13}$$

Equation (9-13) can be rearranged and generalized as

$$S_f = \frac{Q^2}{\left(\displaystyle\sum_{i=1}^{N} K_i\right)^2} \tag{9-14}$$

where again the subscript i refers to a subarea of the cross section, and N is the total number of subareas.

It is helpful to evaluate the energy correction coefficient in terms of the hydraulic conveyance and the areas of the subareas of the compound section. Using Eqs. (9-10), and (9-12), Eq. (9-4) can be rewritten as

$$\alpha = \frac{\left(\displaystyle\sum_{i=1}^{N} A_i\right)^2}{\left(\displaystyle\sum_{i=1}^{N} K_i\right)^3} \sum_{i=1}^{N} \left(\frac{K_i^{\,3}}{A_i^{\,2}}\right) \tag{9-15}$$

where again the subscript i refers to the subarea i of the compound cross section. Example 9-1 illustrates one method by which Eq. (9-2) can be solved.

EXAMPLE 9-1   The river having the channel whose cross section is shown in the accompanying figure is subject to a flood of 10,000 ft^3/s. Find the velocity in the overbank area and in the main channel at stations $0 + 00$ and $6 + 00$ if the depth shown occurs at station $0 + 00$. Manning's n is 0.05 in the overbank areas and 0.03 in the main channel. The channel slope is 0.00167.

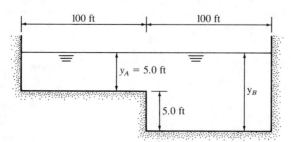

SOLUTION   To solve the problem, we use the total-head equation and the standard-step approach described in Chapter 4. The total-head equation is written between sections $0 + 00$ and $6 + 00$ as

$$y_0 + \alpha_0 \frac{V_0{}^2}{2g} + S_0 \Delta x = y_6 + \alpha_6 \frac{V_6{}^2}{2g} + S_f \Delta x \tag{9-16}$$

or
$$E_0 + S_0 \Delta x = E_6 + S_f \Delta x \tag{9-17}$$

Since the bottom slope is 0.00167, the difference in elevation between station $0 + 00$ and $6 + 00$ is

$$S_0 \Delta x = S_0 \cdot (x_6 - x_0) = 600(0.00167) = 1.0 \text{ ft} \tag{9-18}$$

The terms necessary to solve Eq. (9-17) are first computed for section $0 + 00$ in the top lines of the accompanying table, pages 512–513. As given and noted on columns 2 and 3, the depth at section $0 + 00$ is 5 ft and 10 ft in the subareas of the channel designated A and B, respectively. Thus under column 4, the area of the cross section is computed as 500 ft^2 and 1000 ft^2 for subareas A and B, respectively. The total area is then 1500 ft^2. The wetted perimeters of the subareas are $100 + 5 = 105$ ft for A and $100 + 5 + 10 = 115$ for B. Both are recorded in column 5. Because we will use Eq. (9-11) to compute conveyance, the hydraulic radius of each subarea has been computed as A/P, and the values 4.76 ft and 8.70 ft were obtained for A and B, respectively, and are listed in column 6. In column 7, the hydraulic radius of column 6 has been raised to the $\frac{2}{3}$ power. Column 8 shows the given values of Manning's n for each subarea.

(1) Section	(2) Sub Area	(3) y (ft)	(4) A (ft²)	(5) P (ft)	(6) R (ft)	(7) $R^{2/3}$	(8) n	(9) K	(10) K^3/A^2
0 + 00	A	5	500	105	4.76	2.83	0.05	42450	3.060×10^8
	B	10	1000	115	8.70	4.23	0.03	211500	94.609×10^8
			1500					253950	97.669×10^8
6 + 00	A	5.5	550	105.5	5.21	3.005	0.05	49582	4.029×10^8
	B	10.5	1050	115.5	9.09	4.356	0.03	228690	108.483×10^8
			1600					278272	112.512×10^8
6 + 00	A	5.7	570	105.7	5.39	3.08	0.05	52668	4.497×10^8
	B	10.7	1070	115.7	9.24	4.40	0.03	235400	113.934×10^8
			1640					288068	118.431×10^8
6 + 00	A	5.2	520	105.2	4.94	2.90	0.05	45240	3.424×10^8
	B	10.2	1020	115.2	8.85	4.28	0.03	218280	99.964×10^8
			1540					263520	103.388×10^8
6 + 00	A	5.1	510	105.1	4.85	2.87	0.05	43911	3.255×10^8
	B	10.1	1010	115.1	8.77	4.25	0.03	214625	96.917×10^8
			1520					258536	100.172×10^8

$Q = 10,000$ cfs and $S_0 = 0.00167$

The hydraulic conveyance K has been calculated for each subarea using Eq. (9-11) and has been recorded in column 9. Column 10 shows the ratios of the cube of the conveyance and the square of the area used in Eq. (9-15) to calculate the kinetic energy coefficient α. The sum of column 10 is multiplied by $(1500)^2$ from column 4 and divided by $(253,950)^3$ from column 9, as required by Eq. (9-15). The mean velocity for the total cross section V has been computed as Q/A by dividing the total discharge (10,000 cfs) by 1500 (the total area for the cross section) as totaled in column 4. The velocity head has been computed as $\alpha V^2/2g$ in column 13. The value of E_0, the specific energy at 0 + 00, has been computed as 10.93 in column 14 by adding y_B from column 2 to the velocity head from column 13. The slope of the energy gradient S_f listed in column 16 has been computed by dividing the total discharge (10,000 cfs) by the 253,950 from column 9 and squaring the result, as required by Eq. (9-14).

First iteration:

The next step is to calculate the depth of flow at section 6 + 00. To do this, it is necessary to iterate on a depth at section 6 + 00. An initial depth is assumed, as is shown in column 3. The remainder of this iteration determines whether the assumption is correct. As shown, a depth $y_6 = 10.5$ was assumed, which gives $y_A = 5.5$, as noted in column 3. Columns 4 through 15 are then calculated as was done for section 0 + 00. In column 17, the average energy

(11) α	(12) V (ft/s)	(13) $\alpha \dfrac{V^2}{2g}$ (ft)	(14) E (ft)	(15) S_f	(16) Avg S_f	(17) $S_f \Delta x$	(18) $(E_0 + S_0 \Delta x) - (E_6 + S_f \Delta x)$
1.342	6.67	0.93	10.93	0.00155			
1.337	6.25	0.81	11.31	0.00129	0.00142	0.85	-0.23
1.332	6.10	0.77	11.47	0.00120	0.00138	0.83	-0.37
1.340	6.49	0.88	11.08	0.00144	0.00150	0.90	-0.05
1.339	6.58	0.90	11.00	0.00150	0.00152	0.91	$+0.01$

gradient S_f has then been calculated as $\frac{1}{2}$ the sum of 0.0016 (the value for 0 + 00 in column 16) and 0.00129 (the value calculated for section 6 + 00 and shown in column 16). The total head loss between 0 + 00 and 6 + 00 has been estimated as 600 $S_f = 0.87$ and is shown in column 17. The value of the left-hand side of Eq. (9-16) minus the value of the right-hand side of Eq. (9-16) is shown in column 18. If the value in column 18 is zero, Eq. (9-16) has been satisfied, and the assumed depth is correct.

In order to obtain -0.23 ft in column 18 the value of $E_6 + S_f \Delta x (11.31 + 0.87)$ is subtracted from $E_0 + S_0 \Delta x (10.93 + 1.00)$. The result is not zero, indicating the assumed depth at 6 + 00 is incorrect. Therefore, a second iteration, with a new assumed depth, must be made.

Second iteration:
For the second iteration a depth $y_B = 10.7$ has been assumed and listed in column 3. Values for all 18 columns have been calculated as before, and again the computed values of $E_6 + S_f \Delta x (11.47 + 0.83)$ are subtracted from $E_0 + S_0 \Delta x (10.93 + 1.00)$, yielding a difference greater than that for the first iteration. The assumed depth is obviously still not correct; in fact, our correction after the first iteration was in the wrong direction. Our assumed depth should be slightly less than that for the first iteration. A third iteration must be performed.

Third iteration:

For the third iteration, a depth $y_B = 10.2$ has been assumed and is listed in column 3. Values for all columns have been calculated. The value of $(E_0 + S_0 \Delta x) - (E_6 + S_f \Delta x)$ is $11.93 - 11.98 = -0.05$, a better agreement than in the previous two iterations. However, since the assumed depth is still not correct, a fourth iteration is required.

Fourth iteration:

A depth of $y_B = 10.1$ was assumed and values for all columns calculated. The agreement between $E_0 + S_0 \Delta x$ and $E_6 + S_f \Delta x$ is now satisfactory. Thus the correct depth for section 6 + 00 is $y_B = 10.1$ ft. The velocity in the overbank areas and the main channel can now be computed in Eq. (9-7), since S_f must be the same for both subareas of the channel. Thus

$$V = \frac{1.5}{n} R^{2/3} S_f^{1/2} \tag{9-19}$$

For section 6 + 00,

$$V_A = \frac{1.5}{0.05} (2.87)(0.00150)^{1/2} = 3.33 \text{ ft/s}$$

$$V_B = \frac{1.5}{0.03} (4.25)(0.00150)^{1/2} = 8.23 \text{ ft/s}$$

(check Q) $Q = 3.33(510) + 8.23(1020) = 10090$

For section 0 + 00,

$$V_A = \frac{1.5}{0.05} (2.83)(0.0055)^{1/2} = 3.34 \text{ ft/s}$$

$$V_B = \frac{1.5}{0.03} (4.23)(0.00155)^{1/2} = 8.32 \text{ ft/s}$$

(check Q) $Q = 3.34(500) + 8.32(1000) = 9990$

Thus the velocities in the overbank area are less than half the velocity in the main channel. In each case, we checked to see if the indicated discharge equaled the 10,000 ft^3/s. The slight error that occurred for both 0 + 00 and 6 + 00 is due to the lack of precision in calculating the various column values used in calculating S_f. If desirable, further iterations could be performed using depth values with a precision to two decimal places. The calculated total Q will be in closer agreement with the known flow rate. ∎

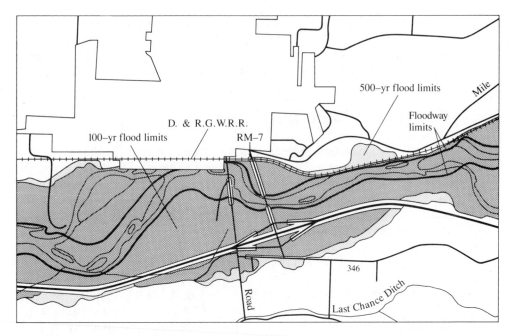

Figure 9-11   Plan of flood plain for Rifle, Colorado
on the Colorado River showing floodway, 100-year,
and 500-year flood limits (5)

Example 9-1 illustrates the method by which the depth of flooding and the
flow velocities are calculated for a given flood discharge. Once the depth has
been calculated at each section, the location of the limits of flooding can be
established on a contour map of the area. Figure 9-11 illustrates the plan view
of a flood plain as determined in an actual flood study.

In the process of flood routing through a river channel, the values of absolute roughness of the channel boundaries are not always apparent. Where
trees, brush, fences, and small buildings exist, there is no accurate method by
which resistance values can be determined from field observations. In these
cases, it may be possible to obtain water surface elevations observed during a
historical flood, or elevations may be estimated from debris lines formed during
the flood and observed later. The peak discharge which occurred during the
flood must be known and can frequently be obtained from gauged records on
the river. Using this known discharge, Eq. (9-2), page 508, is solved for a set of
assumed resistance coefficients. The process is somewhat difficult because the
value of the resistance coefficient (Manning's n or the Darcy-Weisbach f) must
be adjusted after each calculation of a free surface profile. If the calculated water
surface is below the one observed during the actual flood, the resistance coefficient must be increased for the next iteration. *HEC-2, Water Surface Profiles*

is a public domain computer program commonly used for this purpose that was developed by the U.S. Army Corps of Engineers (11). This program contains a subroutine that can be used directly to solve Eq. (9-2) for a known discharge and can also determine applicable values of n from historical flood depths.

9-5   Flood Plain Encroachment

As we mentioned, the *floodway* is generally considered that part of the river channel for which flood velocities are significantly higher than for the adjoining overbank areas. The definition of floodway (see Sec. 9-2) implies that encroachment on (or narrowing of) the channel could be performed until the water surface elevation during a 100-yr flood would not be increased more than 1.0 ft above the depth which would have occurred for the original channel. Encroachment could occur because of earth fills in the flood plain, construction of buildings, or construction of levees to confine the flood flows.

EXAMPLE 9-2   The channel shown in Fig. A has a longitudinal slope of 0.001. If the 100-yr flood peak is 100,000 cfs, determine the minimum width of channel that would meet the criteria in the definition of floodway. Assume the overbank area is covered with long grass and the main channel is a straight gravel-bed river. Further assume the encroachment occurs over a long enough distance so that flow occurs at normal depth throughout the part of the channel that is of interest.

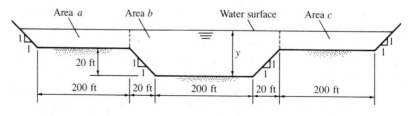

Figure A

SOLUTION   Since normal depth is to be assumed, the problem involves determining normal depth both before and after encroachment. First, we will determine normal depth y for conditions before encroachment. The channel is first divided into the three separate subareas a, b, and c, as shown in Fig. A. Using Eq. (9-6), we write

$$Q = V_a A_a + V_b A_b + V_c A_c$$

$$= \frac{1.5}{n_a} A_a R_a^{2/3} S_{f_a}^{1/2} + \frac{1.5}{n_b} A_b R_b^{2/3} S_{f_b}^{1/2} + \frac{1.5}{n_c} A_c R_c^{2/3} S_{f_c}^{1/2}$$

We know from Eq. (9-10) that for normal depth

$$S_{f_a} = S_{f_b} = S_{f_c} = 0.001$$

We obtain reasonable values of Manning's n from Tables 4-1 and 9-4. Table 4-1, page 169, indicates that $n_b = 0.025$, whereas Table 9-4, page 508, further indicates that $n_a = n_c = 0.035$. Thus

$$Q = 1.5(0.001)^{1/2}\left[\frac{A_a R_a^{2/3}}{0.035} + \frac{A_b R_b^{2/3}}{0.025} + \frac{A_c R_c^{2/3}}{0.035}\right]$$

or since $A_a = A_c$ and $R_a = R_c$,

$$Q = 1.355(2A_a R_a^{2/3} + 1.4A_b R_b^{2/3})$$

or $$Q = 1.90(1.429A_a R_a^{2/3} + A_b R_b^{2/3})$$

We may also write the following equations for area, wetted perimeter, and hydraulic radius at sections a and b:

$$A_a = \left(200 + \frac{y-20}{2}\right)(y-20) \qquad A_b = 220(20) + 240(y-20)$$

$$= \left(190 + \frac{y}{2}\right)(y-20) \qquad\qquad = 4400 + 240(y-20)$$

$$P_a = 200 + y(2)^{1/2} \qquad\qquad\qquad P_b = 200 + 2(800)^{1/2}$$

$$= 200 + 1.414y \qquad\qquad\qquad = 200 + 56.6 = 256.6$$

Thus

$$R_a = \frac{(190 + y/2)(y-20)}{200 + 1.414y} \qquad R_b = \frac{4400 + 240(y-20)}{256.6}$$

To solve these equations for Q, R_a, and R_b, we will assume values of y and then calculate the implied value of Q. This process of iteration continues until the calculated value of Q for the assumed value of y agrees with the given total flow rate. Table A on the next page gives the assumed values of y and the calculated flow rates. In the table, values of R_a and R_b are calculated from the assumed values of y. Q is then calculated from the resulting values of R_a and R_b. Each line is a separate iteration. In the first line, the calculated discharge is too small, so the assumed value of y is increased for the second iteration. Each of the other lines summarizes a new iteration, continuing until the calculated value of Q equals 99,846 cfs which differs from the given value by only 154 cfs which

Table A

y (ft)	A_a (ft²)	$R_a^{2/3}$	A_b (ft²)	$R_b^{2/3}$	Q (ft³/s)
20	0	0	4400	6.65	55,594
22	402	1.45	4880	7.12	67,600
24	808	2.28	5360	7.58	82,255
26	1218	2.98	5840	8.03	98,967
26.2	1259.2	3.04	5888	8.07	100,840
26.1	1238.6	3.01	5864	8.05	99,846

is a reasonable accuracy for this problem. Note that more precision in the assumed values of the depth will lead to closer agreement between the known and the calculated values of Q. Thus the normal depth before encroachment is 26.1 ft to the nearest 0.1 ft.

Now, it is necessary to determine the amount of encroachment (as shown in Fig. B) that will cause the depth to increase from 26.1 to 27.1 ft. We will use the same set of equations that we used in the previous step, except in this case y and Q are both known. We will assume the encroachment has vertical side walls, as Fig. B shows. The floodway width is indicated as B. Again denoting the left overbank area as a and the main channel as b, we can write

$$A_a = \frac{(B-240)}{2}(y-20) \qquad A_b = 6128$$

$$= (B-240)(3.55)$$

$$P_a = \frac{B-240}{2} + (y-20) \qquad P_b = 256.6$$

$$= \frac{B}{2} - 112.9$$

$$R_a = \frac{3.1(B-240)}{(B/2-112.9)} \qquad R_b = 23.79$$

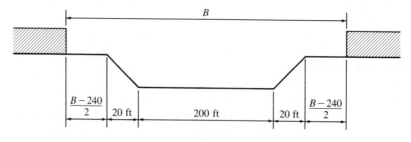

Figure B

Table B

B (ft)	A_b (ft²)	$R_b^{2/3}$	$A_b R_b^{2/3}$	Q (cfs)
240	0	0	50,491	95,932
260	71	2.58	50,491	96,430
400	568	3.49	50,491	101,315
350	391	3.41	50,491	99,553
360	426	3.43	50,491	99,899
365	444	3.44	50,491	100,079

Again, a trial-and-error solution is required, this time assuming values of B. Table B shows the steps of the calculation. The foregoing equations for R_a and R_b are used with each assumed value of B. The equation developed for Q in the first step is again used to calculate the value of Q once the R_a and R_b values have been determined. Each line of the table is again a separate iteration. Thus the width of the floodway would be slightly less than 365 ft. In our solution, we have assumed encroachment takes place equally on each side of the channel. In the second trial-and-error solution, the value of $A_b R_b^{2/3}$ is constant, since we know that the depth must be $y_b = 27.1$ ft. ■

9-6 Risk Within the Flood Plain

The concept of delineating the flood plain makes it possible to determine to a limited degree, the risk of damage due to flooding. That risk, coupled with the damages to be expected during a specific level of flooding, can be used to estimate the benefits of providing flood protection. A simple example will illustrate the analysis of risk. If the probability of a flood elevation exceeding a given level is p in any year, and if the damage that would be suffered as a result of a single flood of that level is D, the *probable annual damage* due to a flood of that magnitude would be

$$D_A = pD \tag{9-20}$$

This concept is overly simple, since floods having a long duration may create more damage than floods of shorter duration having the same peak stage. The concept of probable annual damages can be used to size flood-control measures for design. Suppose we consider a flood-control levee and consider what recurrence interval of flood should be used in the design of the levee. If the levee is designed high enough so that it will not be overtopped or damaged during a 100-yr flood, it will generally be free from significant damage by flooding due to lesser events, and homes or other facilities protected by the levee will not be damaged by smaller floods. However, to make the levee high enough to avoid all damage for all time, the probable maximum flood would need to be considered the design event. Building a levee to protect against that flood will be too expensive for consideration in most flood-control projects. The optimum design event to use is determined by an incremental economic analysis of the cost of flood protection and the resulting benefits. Example 9-3 illustrates the procedure.

EXAMPLE 9-3   It is desirable to construct a flood-control levee along a river channel in New Mexico. The expected useful life of the levee is assumed to be 100 yrs. Table A gives the annual cost (column 2) of designing and constructing the levee high enough to prevent flooding up to the flood stage. Column 1 of Table A lists the flood stage, or elevation of the water surface in the river for the associated peak flow. The expected magnitude of damage that would be caused by a single flood event of that magnitude without the levee is given in column 4. The average return period for a flood with the given stage is shown in column 5. Using the given data, determine the optimum height to which the levee should be constructed.

Table A

(1) Flood Stage (ft)	(2) Annual Project Costs ($)	(3) River Flow Rate (cfs)	(4) Expected Damage Cost if Flow Rate Occurs ($)	(5) Average Return Interval (yr)
2010	0	8,000	0	1.055
2012	6,300	12,200	25,000	2.257
2014	26,000	25,600	162,000	5.879
2016	66,000	65,000	885,000	14.472
2017	108,000	90,000	2,036,000	27.701
2018	155,000	138,400	3,187,000	56.818
2019	204,000	205,000	3,681,500	116.279
2020	256,000	293,500	4,176,000	250.000
2021	312,000	425,000	4,303,000	666.667
2022	246,046	603,700	4,430,000	3333.333

SOLUTION   Table B shows the calculations leading to the probable damages that could be expected if the levee were built high enough to protect against a flood having the stage given in column 1.

Columns 1 and 2 of Table B are the same as columns 1 and 3 in Table A. Thus in line 2 of Table B, the peak flow rate of 12,200 cfs would create a peak flood water surface elevation of 2012 and has an average return interval of 2.257 yr. Column 3 of Table B, shows that a flood of 12,200 cfs would occur, on the average, $100/2.257 = 44.31$ times in 100 yr. The number in column 4 of Table B shows the probable average number of times in 100 yr that a flood would occur within the given range in stage as shown in column 1. Thus the probable number of times that a flood would occur between stages 2010 ft and 2012 ft is $94.81 - 44.31 = 50.5$ (line 2). If the values in column 3 were divided by 100, the

Table B

	(1) Flood Stage (ft)	(2) River Flow Rate (cfs)	(3) Average Number of Times Flow is Exceeded in 100 yr	(4) Average Number of Times in 100-yr that Flow is in Range	(5) Average Damages in Range ($)	(6) Probable Damage Cost for 100 yr ($)
(1)	2010	8,000	94.81			
				50.5	12,500	631,250
(2)	2012	12,200	44.31			
				27.3	93,500	2,552,550
(3)	2014	25,600	17.01			
				10.1	523,500	5,287,350
(4)	2016	65,000	6.91			
				3.3	1,460,500	4,819,650
(5)	2017	90,000	3.61			
				1.85	2,611,500	4,831,275
(6)	2018	138,400	1.76			
				0.90	3,434,250	3,090,825
(7)	2019	205,000	0.86			
				0.46	3,928,750	1,807,225
(8)	2020	293,500	0.40			
				0.25	4,239,500	1,059,875
(9)	2021	425,000	0.15			
				0.12	4,366,500	523,980
(10)	2022	603,700	0.03			

result would be the average number of times per year that a flood could occur in that range. That concept is sometimes difficult to understand so a 100-yr period has been used here. The 100-yr period is arbitrary and does not imply that the expected useful life of the levee is 100 yr.

Column 5 of Table B lists the average magnitude of damage that would be expected for each occurrence of a flood with a stage between the given range. Table A shows that $0 and $25,000 in damages could be expected for floods having a stage of 2010 ft and 2012 ft, respectively if there were no levee. Thus an average damage per flood with a stage between 2010 ft and 2012 ft would be ($0 + $25,000)/2 = $12,500 as shown in line 2 of column 5. Column 6 shows the probable damages that should be expected in a 100-yr period from floods having stages between the given range. Thus the probable damages that should be expected during a 100-yr period is 50.5 × $12,500 = $631,250, as shown in line 2 of column 6.

The remaining lines of the table are completed similarly. The larger floods produce significantly greater damages for each flood. However, the probability of larger floods occurring is much smaller so that the probable damage to be expected during a 100-yr period becomes small for the large floods. Table B

Table C

(1) Level of Protection (ft)	(2) Annual Benefits ($)	(3) Incremental Benefits ($)	(4) Annual Project Costs ($)	(5) Incremental Costs ($)	(6) Benefits Minus Costs ($)
2010	0		0		
2012	6,312	6,312	6,300	6,300	12
2014	31,838	25,526	26,000	19,700	5,826
2016	84,711	52,873	66,000	40,000	18,873
2017	132,908	48,197	108,000	42,000	6,197
2018	181,221	48,313	155,000	47,000	1,313
2019	212,129	30,908	204,000	49,000	−18.992
2020	230,201	18,072	256,000	52,000	−33,919
2021	240,588	10,387	312,000	56,000	−45,613
2022	246,046	5,458	373,000	61,000	−54,542

shows that for this case, the probable maximum damages of $4,831,275 per 100 yr occurs for floods having stages between 2017 ft and 2018 ft. Note that $4,831,275 per 100 yr is equivalent to a probable annual damage of $48,313 per yr. Annual damages, rather than damages for 100 yr, are often used in economic analyses of flood protection systems.

The damage values calculated in Table B must be analyzed to determine the optimum height to which a levee should be constructed for flood-control benefits. According to the tabulated values, a levee constructed high enough so that it would not be overtopped (and strong enough to prevent structural failure) by floods having stages up to 2012 ft would prevent damages amounting to $631,250 in 100 yr. Thus the probable benefits to be realized by building the levee to that height would be $631,250 (line 2, column 6 of Table B) in 100 yr or $6312/yr. Column 2 of Table C gives the probable annual benefits that would be realized by building a levee to protect against floods having stages ranging up to the given elevation. These cumulative benefits are calculated by adding successive values from column 6 of Table B and dividing by 100 yr. Thus the annual benefits from constructing a levee to elevation 2014 ft (the levee would be slightly higher to allow for some settlement and freeboard) is ($631,250 + $2,552,550)/100 = $31,838/yr.

Column 4 of Table C (the same as column 2 of Table A) lists the cost of building levees of the given height. We furnish these values here without further explanation. However, they would be estimated by the engineer and would include the cost of right of way; engineering design; stripping the foundation; excavating, hauling, placing, and compacting the embankment; and riprap protection on the river side of the levee. The costs are given as annual costs, which would be determined by calculating the total engineering and construction cost of the levee, relating that to the equivalent annual investment cost, and then adding annual maintenance costs. The incremental costs shown in column 5 of

Table C are determined by subtracting adjoining values in column 4. The incremental costs increase nonlinearly with levee height because the bottom of the levee becomes wider as it becomes higher, necessitating much more embankment material for a 1-ft increase in height of a tall levee than for an equal increase of a short levee.

The following figure shows the incremental annual costs of construction and incremental annual benefits arising from constructing the levee as a function of the design flood for the levee. The figure shows that benefits increase faster than costs up to a levee height for elevation 2016 and increase slower than costs for higher design elevations.

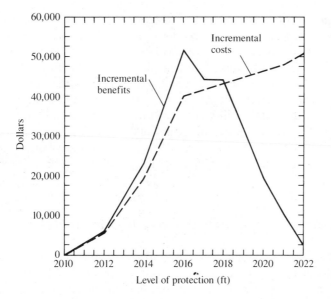

Column 6 of Table C shows that building a successively higher levee achieves increasing benefits that are more than the incremental costs up to protection from floods having stages to elevation 2018 ft. If the levee were increased in height to elevation 2019 ft, the increase in benefits realized would be less than the increase in cost. Thus it is not economically advantageous to build the levee to protect against floods having stages greater than 2018 ft. However, the benefit-cost ratio is greater than 1 up to protection from floods having stages of 2019 ft. Thus if one wished to build an economically feasible project with the maximum benefit, the levee would be constructed to protect from flood stages up to 2019 ft. However, from an optimum economic investment goal, the levee would be built to protect against stages only up to elevation 2018 ft.   ■

A similar analysis could be made for any flood control structure or channel improvement. However, such a procedure would have the inherent difficulty that each portion of the overall flood-control plan might end up being designed and constructed for a different design flood. Although there is nothing technically incorrect about such a procedure, the overall analysis required is prohibitive. Moreover, the damages estimated, based on current development conditions, will surely change in the future and render the currently estimated damages incorrect. This effect must be allowed for by increasing the current estimate expected flood damages.

Federal programs have adopted standard design recurrence intervals for the design of most flood protection facilities. The U.S. Army Corps of Engineers has responsibility for flood control on all navigable streams. Their practice has been to study flood-control projects for areas having a high degree of hazard using the "standard project flood," a flood that would be created by the most severe combination of meteorological and hydrologic conditions considered reasonably characteristic of the geographical region. In practice, the standard project flood is usually determined by transposing to the project area the most severe storm that has been observed in the region around the project and using hydrologic procedures for converting rainfall to runoff. Thus the standard project flood has an undefinable return interval. Its use has the advantage that many residents of the region will recall the storm and resulting flood and are able to relate to the value of protection from such a flood. In general, the U.S. Army Corps of Engineers has found that the standard project floods used for design have had peak discharges equal to 40 to 60% of the probable maximum flood (1).

9-7   Reservoir Operation for Flood Control

In Chapter 4, we briefly considered the routing of floods through reservoirs and channels. In each case, the routing process showed (see Example 4-11, page 220) that the peak rate of flow of the inflow hydrograph is reduced as a result of part of the flood volume being stored and released at a later time when the inflow flood has somewhat receded. This is shown in Fig. 9-5, page 503. For this reason dams and the reservoirs they create are frequently the most significant features of a flood-control plan. Both large and small reservoirs are used extensively to reduce the depth and extent of downstream flooding. In an urban setting, the reservoirs are used extensively to reduce the volume of runoff occurring quickly after the storm. The volume of runoff temporarily detained in the reservoir is released as rapidly as possible after the storm while maintaining downstream flows at less than that which would cause downstream flooding.

The reservoirs in urban settings are usually small detention ponds and are frequently built as a part of parks or recreational areas. In regional settings, large reservoirs, are common and are used to store flood waters from large drainage areas. The reservoirs created by large dams usually store water which can be released to meet multiple needs such as municipal and industrial, navigational, and the instream needs of aquatic life. Energy is usually generated by

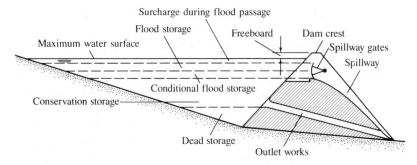

Figure 9-12   Allocation of flood-storage space in a multipurpose reservoir

hydroelectric plants as the flow is released. Because the water is usually stored during high flow periods, downstream flooding is also reduced. Table 9-3 lists several large reservoirs that have flood control as one of their benefits.

When flood-control storage is provided in a multi purpose project, the flood storage is the highest part of the reservoir storage. Figure 9-12 illustrates schematically the way storage is arranged. To be effective, flood-control storage must always be available when a flood occurs. Thus the flood-control storage is emptied as rapidly as possible after the flood has receded using release rates smaller than that which would produce stream levels higher than the downstream flood elevation. When the flood enters the reservoir, storage begins. The flood-control storage is allowed to fill, and outflow from the dam is controlled so that it does not exceed the discharge that would create flooding along the river downstream. If the flood inflow recedes rapidly enough, the spillway may not need to operate, and the outlet works (which usually have a large capacity for flood-control reservoirs) will be sufficient to pass the flood flow. However, after the passage of one flood, the flood-storage space must be emptied as rapidly as permissible so that the reserved flood-control storage will be available to store all or part of the next flood.

Depending on the climate in which the reservoir is located, the storage zone shown in Fig. 9-12 may be allocated differently. In a region like California, the rainfall season is almost totally confined to September through May. Flows due to snowmelt in the mountains may be large until the end of May. However, there is little probability that a significant flood can occur during the summer months. Thus during the summer months, the flood-storage zone is often allocated to conservation storage after the flood season. In regions such as the middle and eastern United States, where summer floods can be expected because of intense rainfall, the flood-storage zone in Fig. 9-12 will be reserved at all times of the year. Allocation of this storage is done by a "rule curve," which we discuss later.

A distinction should be made between the spillway design flood and the design flood for flood control. For a major dam, as we pointed out in Chapter

2, the spillway is frequently designed to pass the flood arising from inflow of the Probable Maximum Flood (PMF) to keep the dam safe from failure during such an event. In contrast, a flood-control project is designed to protect downstream areas against floods less than or equal to a certain magnitude. In Example 9-3, we illustrated economic considerations in arriving at a design flood and the necessary levee height for a levee project. A similar situation exists for flood-control storage. In the design of a dam, to provide flood protection, the total cost must include costs of the land for both the dam and reservoir, the dam itself, the spillway, the outlet works, roads, and any other parts of the structure associated with flood control. Thus if the stored flood water can be released through turbines which turn generators, then at least part of their cost could also be charged to flood control. Revenue generated by the sale of electricity will often greatly offset the cost of the project. Communication facilities and equipment to provide the hydrologic forecasts for control of the project are also part of the real costs.

The benefits from flood control are computed using the reduction in damages as a result of decreased downstream flood levels and reduced flood occurrence. The reductions in flood levels are computed by determining the depth of flood in the river channel that would occur downstream before construction of the dam for a flood of given return interval. Then the same flood is routed through the reservoir, and the flood stage arising from the peak routed outflow is computed. The reduction in damage as a result of the decreased level is then counted as a project benefit.

E X A M P L E 9 - 4   Table A gives the hydrograph for a 50-yr flood entering a reservoir. Route the flood though the reservoir, whose characteristics are also given in the table. Assume the routed outflow downstream must be limited to 8000 cfs if possible. What is the maximum elevation reached by the reservoir during the flood? If the channel characteristics at a particular location downstream are shown in the accompanying figure, calculate the reduction in flood level achieved by the reservoir. The top of the dam is at elevation 1135, the uncontrolled spillway crest is at elevation 1088, and the bottom of the flood-control pool is at elevation 1049.5. Figure 9-12, which illustrates the storage zones within the reservoir, should be used to identify the pertinent zones, and their respective elevations should be noted.

S O L U T I O N   At what level should the reservoir water surface be assumed to be at the beginning of the flood? Although it is possible that the reservoir will be lower, it is always assumed to be not higher than at the bottom of the flood-control pool when the inflow flood begins. Thus we will assume the reservoir is at elevation 1049.5 at time zero when the flood begins. Similarly, although it is not a factor in this problem, the reservoir level would be assumed to be at the top of the spillway crest at the beginning of a probable maximum flood. Equation (4-51), page 219, is used to perform the routing, and Table B, page 528, (which is similar to the table for Ex. 4-11, page 221) is formed to perform

Table A

Flood			Reservoir	
Time (hr)	Inflow (cfs)	Elevation (ft)	Storage (acre-ft)	$2S/\Delta t + O$ (acre-ft/hr)
0	2,000	600	975	1,148
4	5,000	700	53,900	27,611
8	10,000	800	272,800	137,061
12	25,000	900	723,000	362,161
16	33,000	1000	1,471,200	736,261
20	45,000	1049.5	1,969,500	985,411
24	50,000	1050	1,975,000	988,161
28	102,000	1052	1,997,000	999,161
32	72,000	1054	2,019,200	1,010,261
36	62,000	1056	2,041,500	1,021,411
40	60,000	1058	2,064,000	1,032,661
44	50,000	1060	2,086,600	1,043,961
48	45,000	1062	2,109,400	1,055,361
52	45,000	1064	2,132,300	1,066,811
56	41,000	1066	2,155,400	1,078,361
60	20,000	1068	2,178,700	1,090,011
64	15,000	1070	2,202,000	1,101,661
68	11,000	1072	2,225,500	1,113,461
72	9,000	1074	2,249,300	1,125,311
76	8,000	1076	2,273,100	1,137,211
80	6,000	1078	2,297,100	1,149,211
84	4,000	1080	2,321,300	1,161,311

the necessary calculations. We will use a 4-hr routing period in performing the calculations. The choice of this routing period is dictated by the base of the hydrograph and must be chosen small enough to define the hydrograph well. The last column of Table A is first calculated assuming the outflow O is 8000 cfs and $\Delta t = 4$ hr. Note that 8000 cfs is converted to 661 acre-ft/hr for use in the table.

The calculation steps for Table B are the same as those described in Example 4-11 with the exception of the outflow. For this example, the outflow was fixed at 8000 cfs (661 acre-ft/hr) after the first two time steps. For the initial step between 0 and 4 hr, the outflow was taken to be 2000 cfs (165 acre-ft/hr), or the same as the inflow. This assumption was made because the dam operator would not normally release more than the inflow up to the time when the inflow equaled the maximum permissible release of 8000 cfs. Similarly, between time 4 and 8 hr, the outflow was taken as equal to the inflow, or 5000 cfs (413 acre-ft/hr). To obtain the last column in Table B, the stage was interpolated from values of $(2S_1/\Delta t + O_1)$, as given in Table A.

Table B

Time (hr)	$I_1 + I_2$ (acre-ft/hr)	S_i (acre-ft)	O_i (acre-ft/hr)	$\frac{2S_i}{\Delta t} - O_i$ (acre-ft/hr)	$\frac{2S_i}{\Delta t} + O_i$ (acre-ft/hr)	Reservoir Elevation (ft)
0		1,969,500				1049.5
4	578		166	984,584	985,162	1049.4
8	1,240		413	984,336	985,576	1049.6
12	2,892		661	984,254	987,146	1049.8
16	4,793		661	985,824	990,617	1050.4
20	6,446		661	989,295	995,741	1051.4
24	7,851		661	994,419	1,002,270	1052.6
28	8,843		661	1,000,948	1,009,791	1053.9
32	14,380		661	1,008,469	1,022,849	1056.3
36	11,074		661	1,021,527	1,032,601	1058.0
40	10,083		661	1,031,279	1,041,362	1059.5
44	9,091		661	1,040,040	1,049,131	1060.9
48	7,851		661	1,047,807	1,055,660	1062.0
52	7,438		661	1,054,338	1,061,776	1063.1
56	7,107		661	1,060,454	1,067,561	1064.1
60	5,041		661	1,066,239	1,071,280	1064.8
64	2,892		661	1,069,958	1,072,850	1065.0
68	2,149		661	1,071,528	1,073,677	1065.2
72	1,653		661	1,072,355	1,074,008	1065.2
76	1,405		661	1,072,686	1,074,091	1065.3
80	1,157		661	1,072,769	1,073,926	1065.2
84	826		661	1,072,604	1,073,430	1065.1

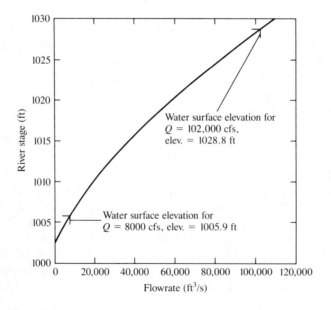

Table B shows that the maximum elevation reached by the reservoir was 1065.3 ft, or nearly 23 ft below the spillway crest. Thus it is possible to pass the given flood without exceeding an outflow of 8000 cfs from the reservoir. The 50-yr flood is thus reduced from a peak inflow of 102,000 cfs to a downstream peak of 8000 cfs as a result of flood-control storage provided by the reservoir. The figure shows that the water surface elevation that would have occurred for the flood of 102,000 cfs would have been 1028.8 ft, whereas for the reduced discharge of 8000 cfs, it is reduced to 1005.9 ft.   ■

When reservoirs are constructed for multiple purposes, of which flood control is only one, the space reserved for flood control can sometimes be variable depending on the season and known hydrologic conditions (conditional flood-control storage shown in Fig. 9-12). For example, many major reservoirs constructed by the federal or state governments in California have been constructed to provide conservation storage, flood control, and hydroelectric generation. Examples of these projects include Shasta Dam, built by the U.S. Bureau of Reclamation (18); New Melones Dam, built by the U.S. Army Corp of Engineers (17); and Oroville Dam, built by the California Department of Water Resources (5). All three reservoirs have drainage areas for which the upper reaches (above approximately 6000 feet) often contain significant volumes of water in the form of snow when the runoff season begins. Snow surveys made on established snow courses in the Sierra Nevada mountains provide an estimate of the total volume of water contained in the existing snowpack. Results of these snow surveys are normally published beginning in January, with new data being measured each month. The amount of water that will run off because of snowmelt is estimated, and the result is used to estimate the amount of runoff that could be stored without violating flood-control space. Figure 9-13 shows the amounts of flood storage, conservation storage, and surcharge provided within New Melones Reservoir and the way in which the reservoir is seasonally programmed for control. The graph shown constitutes a rule curve for reservoir operation. As

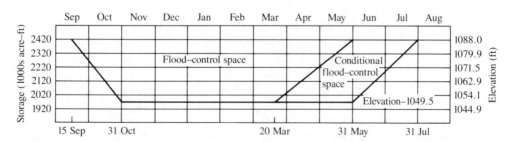

Figure 9-13   Rule curve for operation of flood-control storage space in the reservoir behind New Melones Dam on the Stanilaus River in California (17)

shown in Fig. 9-13, the bottom of the flood-control storage is at elevation 1049.5 ft, and 1,970,000 acre-ft of water can be stored in the reservoir below that elevation. The spillway crest is at elevation 1088.0 ft, and storage above that level is temporary flood surcharge, since the spillway is uncontrolled.

As the rule curve of Fig. 9-13 shows, no flood-control space needs to be retained between July 31 and September 15 of any year. Because a significant amount of rainfall is extremely rare in California between these dates. Beginning on September 15, the reservoir must be drawn down so that the reservoir water surface is down to elevation 1049.5 ft by October 31. Normally, this drawdown is accomplished through required releases of water for irrigation and municipal and industrial uses. If for some reason, those releases did not require all the water stored above elevation 1049.5, water would have to be released and wasted to evacuate the required flood-control space. For a multireservoir system, such as exists for tributaries to the Sacramento River in California's Central Valley, coordinated operation makes efficient overall operation possible, and unnecessary release of water is infrequent. When releases need to be made in one reservoir to empty required flood-control space, that water can usually be used to meet downstream demands that might otherwise have been met by releasing water from another reservoir in the system.

From October 31 to March 20, the entire flood-control space is reserved for flood control. Whenever water is stored in the flood-control space it must be released as rapidly as possible without exceeding 8000 cfs downstream. After March 20, the amount of flood-control space to be maintained is subject to the amount of runoff to be expected before July 31. Thus if on March 20, snow surveys indicated that the total runoff from snowmelt would be approximately 450,000 acre-ft, the lower curve must be followed, and the entire 450,000 acre-ft of flood-control space must be reserved until May 31, after which runoff can be stored until the reservoir is full on July 31.

If, however, the snow surveys indicate a potential snowmelt runoff of only 200,000 acre-ft, water may be stored in the reservoir as long as 200,000 acre-ft is reserved for the snowmelt runoff. Thus on March 20, the dam operator would begin to store water and could store up to 250,000 acre-ft (450,000 − 200,000) without making flood-control releases. The total storage in the reservoir could be as large as 2,220,000 acre-ft before flood-control releases were required.

Figure 9-14 shows the actual operation of the Shasta Reservoir during 1983–1984 (5). It also shows the flood-control space that must be allocated in Shasta Reservoir for each month of the runoff season as well as the actual storage. Some encroachment on flood-control storage occurred early in the year. On December 11, 1983, the peak inflow to Shasta is seen to have been approximately 65,000 cfs, but releases were held to only slightly more than 30,000 cfs until December 22, when storage was back to that required by the rule curve. On December 23, a second large inflow occurred, but it did not exceed 35,000 cfs. Between February 1 and February 10, the snowmelt forecasts indicated that flood-control storage could be reduced, and subsequently releases

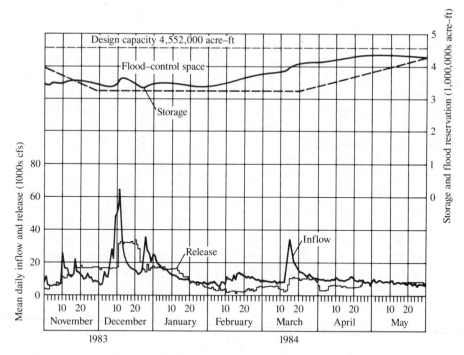

Figure 9-14   Operation of Shasta Dam flood-control space during the 1983-1984 flood on the Sacramento River (5)

were kept to less than inflow so that storage in the reservoir increased to approximately 4,300,000 acre-ft by April 23, 1984. Figure 9-14 is typical of the operation of multipurpose reservoirs subject to snowmelt inflow.

The operation of reservoirs for flood control, thus depends strongly on the hydrology of the drainage area upstream from the dam and the flooding characteristics of the river downstream. Detailed hydrologic studies need to be performed before a rule curve such as that shown in Fig. 9-13 can be constructed.

Historical flood hydrographs need to be assembled and carefully examined to develop a unit hydrograph for floods in the basin. The unit hydrograph is then used to develop hydrographs for each flood to be routed through the reservoir. By routing a range of floods through the reservoir and downstream without the reservoir, the resulting depths of flooding can be compared, as was done in Example 9-4. Damages prevented by the dam can then be evaluated. Damages prevented are considered benefits due to the provision of flood control and can be used to assess the economic feasibility of the flood-control aspects of the project.

9-8   Nonstructural Aspects of Flood Control

Many reservoirs and levees were built for flood-control purposes in the United States before 1960. However, as we mentioned, the annual flood damages have continued to increase. The chief reason for these increased damages is the increased development in the flood plain, which occurs because of the reduced risk of flooding resulting from the construction of levees or upstream flood-control reservoirs. The great cost of flood-control structures has led to an emphasis on nonstructural aspects of flood control. Nonstructural aspects can consist of channel improvement by straightening, cleaning, and increasing the cross-sectional area of the channel which increases the channel capacity. However, with these measures, laws must be instituted to control the development of the flood plains. In the United States, this control is called *flood-plain zoning*. As an incentive for communities to enact flood-plain zoning, the U.S. government makes low-cost, federally subsidized flood insurance available to property owners within the flood plain for communities where flood-plain zoning ordinances have been enacted. Delineation of the flood plain must still be performed as we described.

PROBLEMS

9-1   A river channel has the cross section shown in the figure. The channel slope is a uniform 0.02%. If Manning's n for the channel is 0.025, determine the flow depth for a probable maximum flood. The river is in the Ohio-Tennessee region of the United States and has a drainage area of 500 mi^2.

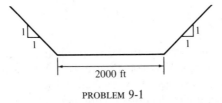

PROBLEM 9-1

9-2   A river with the cross section shown, experienced a 50,750 cfs flood and produced the depth shown. Determine Manning's n for the main channel and overbank area assuming they are equal. The channel has a uniform slope of 0.45%.

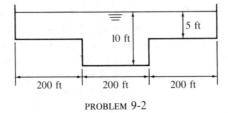

PROBLEM 9-2

9-3   If, in Prob. 9-2, Manning's n is known to be 0.020 in the main channel, determine the value of n for the overbank area. What is the flow rate in the main channel and in the overbank areas? What is the flow velocity in the main channel and in the overbank areas?

9-4   For the cross section shown, determine the width of the flood plain for a 100-yr summer flood of 100,000 cfs. Assume a uniform channel slope of 0.09%. Further assume the overbank area is brush-covered with some trees and the main channel is straight with a gravel bed.

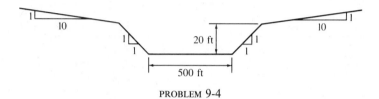

PROBLEM 9-4

9-5   The compound cross section shown has Manning's n values of 0.025 for the main channel and 0.070 for the overbank areas. The uniform channel slope is 0.004. Determine the depth and flow rate in the main channel, the left overbank area, and the right overbank area. The flood flow is 45,400 cfs.

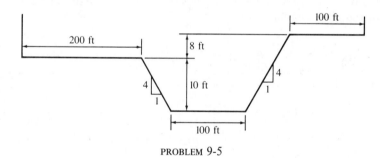

PROBLEM 9-5

9-6   What would be the depth of flow for the cross section shown for Prob. 9-5 if the flow rate were 85,000 cfs?

9-7   Levees are to be built on both sides of the river whose cross section is shown in the figure. Protection is required from the 50-yr flood, which has a magnitude of 172,000 cfs. The overbank areas are covered with tall

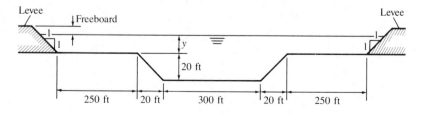

grass, and the main channel is covered with gravel beds and some boulders. How high must the levee be if it is to have 2 ft of freeboard? What velocity of flow will occur near the levee? The uniform slope of the channel is 0.15%.

9-8   The channel cross section shown has been constructed for flood protection. At what flow rate would the levees be overtopped. Assume the channel has a uniform slope of 0.008 and that Manning's n is 0.060 and 0.040 for the overbank and the main channel, respectively.

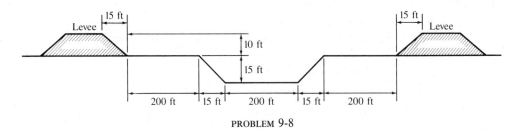

PROBLEM 9-8

9-9   The 100-yr peak flow for the river whose cross section is shown in the figure is 56,000 cfs. If in normal conditions, the river can carry the 100-yr flow with a water surface elevation of 185 ft, what is the width of the floodway? Assume Manning's n is 0.06 for the overbank areas and 0.035 for the main channel.

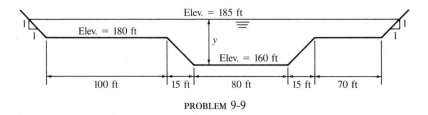

PROBLEM 9-9

9-10   Levees are constructed that contract a river as shown in the figure. A 50-yr flood of 50,000 cfs creates a depth of 30 ft as shown. Determine the flood water surface profile for a distance of 2000 ft upstream if the channel slope is 0.0015, assume $n = 0.030$.

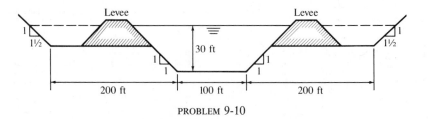

PROBLEM 9-10

9-11   For the cross section used in Prob. 9-2, the Manning's n values are known to be 0.050 for the overbank areas and 0.030 for the main channel. For a flow of 65,000 cfs, the depth in the main channel at a specific section is 14 ft. Calculate the depth of flow in the main channel that will exist at a similar section 4000 ft downstream. Assume the slope between the two sections is 0.005.

9-12   A local agency desires to construct a stream gauge on a nearby stream. The life expectancy of the recorder and other equipment is 10 yr. Floods equal to or smaller than 10,000 cfs will not damage the structure or equipment, but floods greater than 10,000 cfs will destroy the entire installation. Initial cost of equipment and installation is $12,000. Annual maintenance including supplies is $600. If the probability of exceeding 10,000 cfs is 1%, what is the total probable annual cost exclusive of the initial investment for the installation?

9-13   A pipeline is to be constructed across a major river by burying the pipeline beneath the riverbed. Burial depth will be sufficient to prevent the pipe from being damaged during a 25-yr flood. The original cost of constructing the crossing is $75,000. If a 50-yr flood will destroy the crossing and require totally new construction, and damages are linearly proportional to the return period between the 25- and 50-yr events, determine the probable annual cost during the desired 25-yr life of the structure. (Neglect capitalization of the initial cost.)

9-14   The figure gives a graph of return period versus peak flow rate at a location where a bridge is to be constructed. The bridge will be designed

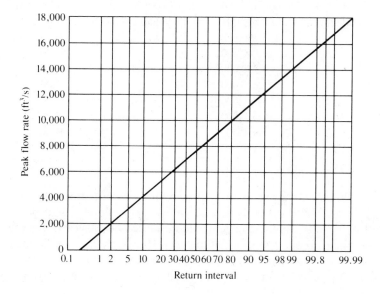

so that it sustains minimal damage during a 10-yr flood but is destroyed by a 50-yr flood. Assuming damage is linearly proportional to the difference between discharge and design discharge between the 10-yr and 50-yr floods and initial construction of the bridge costs $200,000, calculate the average annual damage over the 20-yr life of the bridge. (Neglect capitalization of the initial cost.)

9-15   The table lists flood stages and the flooding damages that have been found to occur during floods of the indicated stage. Column 3 presents the percentage chance that a flood can equal or exceed that stage. Column 4 presents the annualized cost of constructing a levee high enough to prevent all flooding by a flood of the indicated state. Is it economically feasible to construct a levy? If it is feasible, to what height should it be constructed for
a. The maximum benefit
b. The most economical solution

(1) Flood Stage with No Protection (ft)	(2) Damage for One Flood at Stage ($)	(3) Chance That Flood Could Be Exceeded (%)	(4) Annual Cost of Constructing a Levy to Prevent Flooding ($)
35	0	49.50	0
36	100,000	41.50	4,000
37	200,000	33.50	9,000
38	300,000	25.50	14,000
39	450,000	18.50	22,000
40	600,000	13.50	31,000
41	800,000	9.10	41,000
42	1,000,000	5.78	51,000
44	1,250,000	2.53	71,000
46	1,550,000	0.95	91,000
48	1,950,000	0.30	121,000
50	2,450,000	0.10	151,000
52	3,950,000	0.05	191,000
Above 52	6,150,000	0.0	—

9-16   A 36-in. diameter culvert is to be built across a highway and designed to pass a 10-yr flood without damage. A 100-yr flood would presumably destroy the structure. Damage would be caused by erosion of the roadway, flooding upstream, and scour of the downstream channel for floods greater than the 10-yr design flood. The table gives the probability of equaling or exceeding for each flood. The damages that would occur during a flood

Discharge (ft³/s)	Probability of Equaling or Exceeding (%)
30	10.0
35	8.0
40	6.0
45	5.0
50	4.0
55	3.5
60	2.5
65	2.0
70	1.5
75	1.2
100	1.0
150	0

can be computed as

$$\$D = 75 + 7(Q - Q_D)$$

where $\$D$ is the damage caused by one flood, Q is the flood discharge, and Q_D is the design discharge in ft³/s for the culvert. What is the probable annual flood damage that should be planned for in estimating maintenance for this culvert?

9-17 For the New Melones Reservoir, the rule curve for operation is given in Fig. 9-13, page 529. If a flood were to occur beginning October 31 with the reservoir at elevation 1049.5, would it be possible to prevent the reservoir level from exceeding 1088.0 without releasing more than 8000 ft³/s? The inflow hydrograph of the flood is given in the accompanying figure.

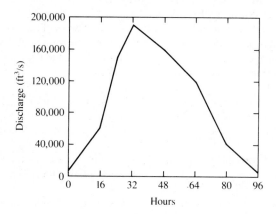

9-18   Use Fig. 9-13, page 529, to determine the maximum volume of inflow that could be reserved in New Melones Reservoir without exceeding an outflow of 8000 ft^3/s.

9-19   A dam is to be built solely for flood control. The annualized cost of the dam is shown in the table. The table also shows the return period for the flood volume that could be passed through the dam without exceeding flood stages downstream for each height of dam considered and the damages that would occur downstream if the dam were not there. Determine the indicated height of dam that should be constructed if the most economical solution is desired.

Height of Dam (ft)	Probability of Equaling or Exceeding Flood Volume That Could Be Passed Without Exceeding Downstream Flood Stage (%)	Annual Cost of Dam ($)	Damages Produced by the Flood ($)
100	1.0	50,000	6,000,000
150	0.5	70,000	15,000,000
200	0.2	140,000	76,000,000
250	0.1	400,000	160,000,000
300	0.05	1,500,000	400,000,000

REFERENCES

1. Beard, L.R. "Statistical Methods In Hydrology." U.S. Army Engineer District, Corps of Engineers, Sacramento, Calif. (January 1962).
2. Bock, C.A. "History of the Miami Flood Control Project." *Technical Reports Part II*, State of Ohio, The Miami Conservancy District, Dayton, Ohio, 1918.
3. Brown, B.W., and R.A. Shelton. "Fifty Years of Operation of the TVA Reservoir System." *Accomplishments and Impacts of Reservoirs*, Proceedings of the Symposium of the ASCE Water Resources Planning and Management Division (October 1983).
4. California Flood Management. "An Evaluation of Flood Damage Prevention Programs." *Bulletin 199*, State of California, Dept. of Water Resources, Sacramento, Calif. (September 1980).
5. "California High Water, 1983–84." *Bulletin 69–84*, State of California, Dept. of Water Resources, Sacramento, Calif. (March 1985).
6. Creager, W.P., and J.D. Justin. *Hydroelectric Handbook*. John Wiley & Sons, New York, 1950.
7. Federal Emergency Management Agency. *Flood Insurance Study, Guidelines and Specifications for Study Contractors*. Federal Insurance Administration, Washington, D.C. (September 1985).
8. "Flood! December 1964–January 1965." *Bulletin 161*, State of California, Dept. of Water Resources, Sacramento, Calif. (January 1965).

9. Green, G.G., and D.D. Speers. "Columbia Basin Reservoir Accomplishments." *Accomplishments and Impacts of Reservoirs*, Proceedings of the Symposium of the ASCE Water Resources Planning and Management Division (October 1983).

10. Haan, C.T. *Statistical Methods in Hydrology*. Iowa State University Press, Ames, Iowa, 1977.

11. *HEC-2, Water Surface Profiles, Users Manual*. U.S. Army Corps of Engineers, Hydrologic Engineering Center, Davis, Calif. (August 1979).

12. Houk, Ivan E. "Calculation of Flow in Open Channels." *Technical Reports Part IV*, State of Ohio, Miami Conservancy District, Dayton, Ohio, 1918.

13. Hoyt, W.G., and W.B. Langbein. *Floods*. Princeton University Press, Princeton, N.J., 1955.

14. McCracken, D.D., and W.S. Doun. *Numerical Methods and Fortran Programming*. John Wiley & Sons, New York, 1964.

15. Morgan, Arthur E. "The Miami Valley and the 1913 Flood." *Technical Reports Part I*, State of Ohio, The Miami Conservancy District, Dayton, Ohio, 1917.

16. Schneider, G.R. "History and Future of Flood Control." *Trans. ASCE*, CT (1953), pp. 1042–99.

17. U.S. Army Corps of Engineers. "New Melones Dam and Lake, Report on Reservoir Regulation for Flood Control." Appendix V to Master Manual of Reservoir Regulation, San Joaquin River Basin, Sacramento District, Sacramento, Calif. (January 1980).

18. U.S. Bureau of Reclamation. Project Data. Department of the Interior, Denver, Colo. 1981 (August 1977).

19. U.S. Nuclear Regulatory Commission. Regulatory Guide 1.59, rev. 2, "Design Basis Floods for Nuclear Power Plants." Rules and Regulations, title 10, chapter 1, Code of Federal Regulations, Washington D.C.

20. U.S. Water Resources Council. *Estimated Flood Damages*. Washington, D.C., 1977.

21. Woodward, Sherman M. "Hydraulics of the Miami Flood Control Project." *Technical Reports Part VII*, State of Ohio, Miami Conservancy District, Dayton, Ohio, 1920.

22. "World Catalogue of Very Large Floods." *Studies and Reports on Hydrology*. The UNESCO Press, Paris, France, 1976.

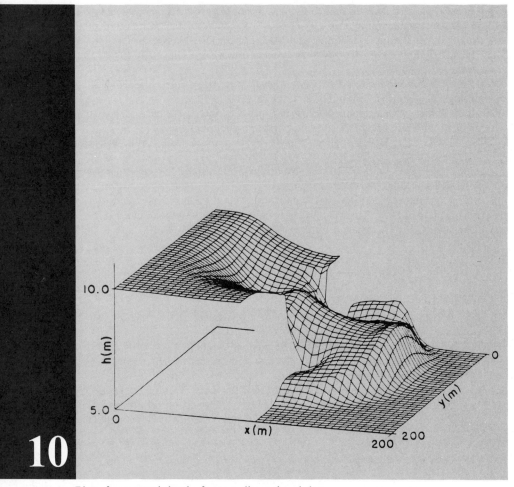

Plot of computed depths for two-dimensional dam
break flow (Courtesy of Robert J. Fennema)

Mathematical
Modeling of
Hydraulic
Systems

10

We discussed traditional methods of analysis in Chapters 1 through 9 and introduced physical models in Chapter 7. In this chapter, we will present a description of mathematical models and illustrate their use as a design tool. Most of the analysis procedures discussed in this and the following chapters are suitable for digital computer solutions.

We first outline the steps involved in the development of a mathematical model. Then we present a number of numerical analysis procedures. For illustration, we include the application of these procedures for the solution of several typical hydraulic problems.

10-1 Mathematical Models

A *mathematical model* is a representation of the behavior of a particular system in the form of mathematical equations. For specified parameters, the system response may be determined from this model. Though usually used as a design tool, these models may also be used for real-time control or operation of the system. For example, the mathematical model of a municipal pipe network may be used for its design, for investigating the effects of various modifications, and for optimal operation of pumps, reservoirs, and so forth. If certain components of a system fail, necessary remedial measures and operational strategies may be decided fairly quickly thus making the "let us do this to see what happens" approach unnecessary.

Usually, the following steps are involved in the development of a mathematical model:

1. Derivation of governing equations
2. Selection of solution procedures
3. Development and verification of computer codes

We briefly describe each of these steps in the following paragraphs.

Derivation of Governing Equations

The "physics" of the phenomenon has to be first understood to derive the governing equations. In this derivation assumptions are made so that the resulting equations are simple but adequately describe the phenomenon. If equations are too complex, they may not be easily amenable to a solution. For example, if the transverse and vertical components of the flow velocity are small, they may be neglected, thereby resulting in a one-dimensional model.

Governing equations are usually derived by applying the well-known laws of mechanics, such as conservation of mass, momentum, and energy. For simplification, some of the variables may be eliminated by combining different equations. By using reference variables, the governing equations may be non-dimensionalized.

Table 10-1 Typical Governing Equations in Hydraulic Engineering

---

Algebraic Equations

Single equation — energy equation between two channel sections:

$$z_1 + y_1 + \frac{V_1^2}{2g} = z_2 + y_2 + \frac{V_2^2}{2g} + h_f$$

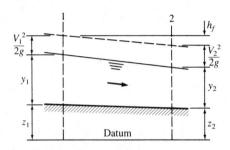

Two equations — energy and continuity equations at a channel junction:

$$y_1 + \frac{V_1^2}{2g} = \Delta z + y_2 + \frac{V_2^2}{2g} + h_L$$

$$A_1 V_1 = A_2 V_2$$

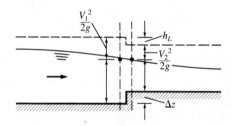

System of equations — Kirchhoff laws for a pipe network:

$$\sum_{i=1}^{n} h_{f_i}^{j} = 0$$

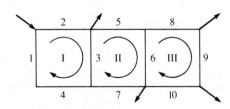

where $h_{f_i}^{j} = c_i Q_i |Q_i|$ = head loss in the ith pipe of jth loop. Similar equations may be written for the other loops.

Ordinary Differential Equations

Single equation — Equation for the variation of flow depth in gradually varied flow:

$$\frac{dy}{dx} = \frac{S_0 - S_f}{1 - (\mathrm{Fr})^2}$$

where Fr = Froude number, S_0 = slope of channel bottom, and S_f = slope of the energy grade line.

Two equations — water-level oscillations in a surge tank:

$$\frac{dz}{dt} = \frac{Q - Q_{\mathrm{tur}}}{A_s}$$

$$\frac{dQ}{dt} = \frac{gL}{A_t}(-z - cQ|Q|)$$

where Q_{tur} = turbine flow, and c = coefficient of head losses in the tunnel.

Partial Differential Equations

Hyperbolic — unsteady flow in pipes:

$$\frac{\partial H}{\partial t} + \frac{a^2}{gA}\frac{\partial Q}{\partial x} = 0$$

$$\frac{\partial Q}{\partial t} + gA\frac{\partial H}{\partial x} + \frac{fQ|Q|}{2DA} = 0$$

where H = piezometric head, Q = discharge, a = wave speed, A = pipe cross-sectional area, D = pipe diameter, and f = Darcy-Weisbach friction factor.

Parabolic — diffusion of a tracer:

$$\frac{\partial c}{\partial t} = D\frac{\partial^2 c}{\partial x^2}$$

where c = concentration of the tracer, and D = molecular diffusion coefficient.

Elliptic — stream function for irrotational flow:

$$\frac{\partial^2 \psi}{\partial x^2} + \frac{\partial^2 \psi}{\partial y^2} = 0$$

where ψ = stream function.

The governing equations may be algebraic, ordinary differential equations, or partial differential equations, depending on the phenomenon under consideration. Physical phenomena representing the rate of change of one or more variables with respect to other variables are described by differential equations. If the dependent variables are functions of only one independent variable, then the resulting equations are ordinary differential equations. However, if the dependent variables are functions of more than one independent variable, the resulting equations are partial differential equations.

Table 10-1 presents typical examples of different types of equations.

Selection of Solution Procedures

Because a closed-form solution of nonlinear equations is not usually available, numerical methods are used for their solution. Accuracy, efficiency, and simplicity are the main criteria used during the selection of a numerical method for a particular application.

Several different numerical methods are available for the solution of governing equations depending on their type. A list of these methods follows (1, 4):

Algebraic equations
> Linear equations: Gauss elimination, Gauss-Seidel methods
> Nonlinear equations: Newton-Raphson method, method of successive approximations, bisection method

Ordinary differential equations
> Single-step methods: Euler, Improved Euler, Runge-Kutta
> Multistep methods: Predictor-corrector, Adam-Bashforth-Moulton

Partial differential equations
> Method of characteristics, Finite-difference methods, Finite-element method, Boundary-integral method, Spectral method, Pseudo-Spectral method

Development and Verification of Computer Codes

A computer code may be divided into three parts: reading the input data, specifying the solution algorithm, and printing the computed results and preparing the necessary plots. The input data give the system parameters and may specify the layout for the program output.

Once a computer program has been developed and its computed results appear reasonable, it is necessary to verify their validity by comparing them with some reliable data. These data may be the exact solution for idealized and simplified cases, or they may be obtained from laboratory and field experiments. The experimental accuracy of the field or laboratory results should be considered while doing these comparisons. Moreover, the code should be verified over the complete range of system parameters.

10-2   Errors in Numerical Analysis*

To confidently use the computed results, it is usually necessary to have an estimate of the errors. Two types of errors are introduced into the numerical computations: truncation errors and roundoff errors. Superficially, these errors may appear small, but in repetitive calculations, they may grow and mask the actual results.

A *truncation error* is caused by the truncation of an infinite series after a finite number of terms. For example, sin x may be approximated by an infinite series. For actual calculations, however, we can include only a finite number of terms; for example, $\sin x = x - \dfrac{x^3}{3!} + \dfrac{x^5}{5!}$, in which terms higher than x^5 have been neglected. These neglected terms introduce the truncation error.

A *roundoff error* is introduced into the computations because irrational numbers are represented by a finite number of digits in the computer. For example, 4/3 will be represented as 1.333333 in machine computations involving seven significant digits.

10-3   Interpolations

It often is necessary to interpolate results from tabulated data. For example, the discharge through a spillway may be determined at a number of water levels by using hydraulic models. The flows at intermediate water levels may then be interpolated from the measured data. Depending on the curve selected to describe the relationship between the dependent and independent variables, these interpolations may be linear or higher order.

In this section we will discuss *parabolic interpolation*. Let us assume we have values of y at equal intervals of x, and we want to interpolate the value of y at an intermediate value of x such that $x_i < x < x_{i+1}$ (Fig. 10-1, page 546). Let us have new coordinate axes, x', y', with the origin at point B, having coordinates (x_i, y_i), and pass a parabola through points A, B, and C, having the coordinates (x_{i-1}, y_{i-1}), (x_i, y_i), and (x_{i+1}, y_{i+1}), respectively. The equation of this parabola in terms of the new coordinates, x' and y', will be

$$y' = ax' + bx'^2 \qquad\qquad (10\text{-}1)$$

---

* For more detailed information about numerical analysis, see Chapra (1) and McCracken (4).

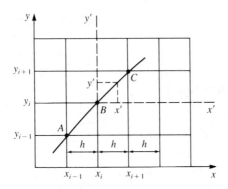

Figure 10-1   Parabolic interpolation

By substituting into this equation the coordinates of points A and C, (x_{i-1}, y_{i-1}) and (x_{i+1}, y_{i+1}), denoting the interval of x at which values of y are specified as h $(h = x_{i+1} - x_i = x_i - x_{i-1})$, and noting that $y' = y - y_i$, we obtain

$$y_{i-1} - y_i = -ah + bh^2 \tag{10-2}$$

$$y_{i+1} - y_i = ah + bh^2 \tag{10-3}$$

Simultaneous solution of these equations for a and b yields

$$a = \frac{1}{2} \frac{y_{i+1} - y_{i-1}}{h}$$

$$b = \frac{1}{2} \frac{y_{i+1} - 2y_i + y_{i-1}}{h^2} \tag{10-4}$$

By substituting into Eq. (10-1), these expressions for a and b and $r = (x - x_i)/h$, and writing x' and y' in terms of x and y, we obtain

$$y = y_i + \frac{1}{2} r[y_{i+1} - y_{i-1} + r(y_{i+1} + y_{i-1} - 2y_i)] \tag{10-5}$$

This equation may be used to interpolate values of y from the given coordinates of points A, B, and C. *Note*: This equation is valid only for equally spaced grid points.

Figure 10-2 lists a subroutine for parabolic interpolation using this expression.

E X A M P L E   1 0 - 1   The following table lists the flows over a spillway for different reservoir levels. Determine the flows at water-level elevations of 102 m, 114 m, 123 m, and 138 m.

Spillway Rating Curve	
Reservoir level (m)	Discharge (m³/s)
100	0.0
105	33.5
110	96.6
115	178.6
120	277.0
125	393.7
130	519.3
135	656.2
140	804.5

SOLUTION   The interval at which spillway flows are stored for different reservoir levels is 5 m. We will write a short program to read the following input data: reservoir elevation, corresponding flows, and the reservoir levels at which flows have to be interpolated. We will use subscripted arrays for these variables and a subroutine for parabolic interpolation. Input to the subroutine is through its argument list. For each reservoir level, the flows are interpolated in the subroutine from the stored data, then they are printed in the main program. Figure 10-2 lists the computer program, its input data, and program output.   ∎

Figure 10-2   Program for parabolic interpolation

Program Listing

```
C      PROGRAM FOR PARABOLIC INTERPOLATION
C
C      **** NOTATION  ****
C
C      DX = INTERVAL FOR STORING INDEPENDENT VARIABLE
C      N  = NUMBER OF DATA POINTS TO BE STORED
C      NI = NUMBER OF INTERMEDIATE VALUES OF X AT WHICH Y IS TO
C           BE INTERPOLATED
C      X  = ARRAY FOR INDEPENDENT VARIABLE
C      XI = ARRAY OF INTERMEDIATE VALUES OF X AT WHICH Y IS TO BE
C           INTERPOLATED
C      Y  = ARRAY OF DEPENDENT VARIABLE
C
       DIMENSION X(50),Y(50),XI(50)
       READ (5,*) N,DX,(X(I), I=1,N)
       READ (5,*) (Y(I), I=1,N)
       WRITE (6,20)
20     FORMAT(5X, 'X',8X,'Y')
       DO 30 I=1,N
       WRITE (6,25) X(I),Y(I)
25     FORMAT(F8.2,2X,F8.2)
30     CONTINUE
       READ(5,*) NI, (XI(I),I=1,NI)
       WRITE (6,40)
40     FORMAT(/5X,'INTERPOLATED VALUES'/)
       WRITE(6,42)
42     FORMAT(5X,'XI',10X,'YI')
       DO 50 I = 1,NI
       XINT=XI(I)
       CALL INT(XINT,X,Y,DX,YI)
       WRITE(6,45) XI(I),YI
45     FORMAT(2F10.2)
```

(continued)

Figure 10-2   (continued)

```
50    CONTINUE
      STOP
      END
      SUBROUTINE INT(XI,X,Y,DX,YI)
      DIMENSION X(50),Y(50)
      J=1+(XI-X(1))/DX
      R=(XI-X(J))/DX
      IF (J.EQ.1) R=R-1.
      IF (J.LT.2) J=2
      YI=Y(J)+0.5*R*(Y(J+1)-Y(J-1)+R*(Y(J+1)+Y(J-1)-2.*Y(J)))
      RETURN
      END
```

Input Data

```
9,5.,100.,105.,110.,115.,120.,125.,130.,135.,140.
0.0,33.5,96.6,178.6,277.,393.7,519.3,656.2,804.5
4,102.,114.,123.,138.
```

Program Output

```
      X          Y
   100.00        .00
   105.00      33.50
   110.00      96.60
   115.00     178.60
   120.00     277.00
   125.00     393.70
   130.00     519.30
   135.00     656.20
   140.00     804.50

      INTERPOLATED VALUES

      XI          YI
   102.00       9.85
   114.00     160.69
   123.00     344.82
   138.00     743.81
```

10-4 Nonlinear Algebraic Equations

Two commonly used methods for determining the roots of nonlinear algebraic equations are the bisection method and the Newton-Raphson method. Details of these methods are given in the following paragraphs.

Bisection Method

Assume that we have to determine the roots of the equation

$$F(x) = 0 \tag{10-6}$$

The x-coordinates of points where the graph of $F(x)$ versus x intersects the x-axis are the roots of Eq. (10-6). We will outline a step-by-step procedure for determining one of the roots; similarly, other roots may be determined one by one.

1. Estimate two values of x, namely, x_n and x_p, so that $F(x_n)$ is negative and $F(x_p)$ is positive (see Fig. 10-3).

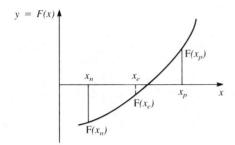

Figure 10-3 Bisection method

2. Compute $x_e = \dfrac{1}{2}(x_n + x_p)$.

3. Determine $F(x_e)$. If $|F(x_e)| < \epsilon$, where ϵ is specified tolerance, then x_e is the root of Eq. (10-6); otherwise, set $x_n = x_e$ if $F(x_e) < 0$, and set $x_p = x_e$ if $F(x_e) > 0$.

4. Repeat steps 2 and 3 until a solution is obtained.

Convergence to the solution in this method is usually assured although it is usually slower than that in the Newton-Raphson method, which we discuss in the next section.

In all iterative procedures, it is always a good idea to introduce a counter to avoid an unlimited number of iterations due to errors in logic, computation, programming, or other reasons. Iterations may be stopped if the number exceeds a specified number, say 50.

EXAMPLE 10-2 Use the bisection method to determine the normal depth in a trapezoidal channel carrying a flow of 110 m^3/s. The channel bottom slope is 0.0001, Manning's $n = 0.013$, channel bottom width is 20 m, and the side slopes of the channel are 2 horizontal to 1 vertical.

SOLUTION To determine the normal depth, we have to solve Manning's equation. In this case, we use the SI version:

$$Q = \frac{1}{n} AR^{2/3}S_0^{1/2} \qquad (10\text{-}7)$$

where Q is the rate of discharge, n is Manning's constant, A is the flow area, R is the hydraulic radius, and S_0 is the slope of the channel bottom. Since n, Q, and S_0 are specified, and both A and R are functions of flow depth y, we can write Eq. (10-7) as

$$F(y) = Q - \frac{1}{n} AR^{2/3}S_0^{1/2} = 0 \qquad (10\text{-}8)$$

The value of y that satisfies Eq. (10-8) is the normal depth. We may solve this equation by the bisection method as follows.

Let us select two values of y — 0.2 m and 20 m — so that $F(y)$ is positive for one of them and negative for the other. If this is not the case increase the higher value, or decrease the lower value, and then follow the preceding steps to solve Eq. (10-8). Figure 10-4 lists a computer program written for this purpose, its input data, and program output.   ■

Figure 10-4   Program for computing normal depth
using the bisection method

Program Listing

```
      C       COMPUTATION OF NORMAL DEPTH IN A TRAPEZOIDAL CHANNEL BY
      C         USING BISECTION METHOD
      C
      C       ********************NOTATION ************************
      C
      C       AN = MANNING'S N;
      C       BO = CHANNEL-BOTTOM WIDTH;
      C       Q = DISCHARGE;
      C       S = SLOPE OF CHANNEL SIDES, S:HORIZONTAL TO 1 VERTICAL;
      C       SO = CHANNEL BOTTOM SLOPE;
      C       Y = NORMAL DEPTH;
      C       YP, YN = DEPTH ESTIMATES SUCH THAT YP<Y<YN
      C
              AR(Y)=(BO+S*Y)*Y
              P(Y)=BO+2.*Y*SQRT(1.+S*S)
              READ(5,*) AN,Q,SO,S,BO,YP,YN
              WRITE(6,10)AN,Q,SO,S,BO
      10      FORMAT(5X,'N ='F5.3,3X,'Q =',F8.3,' M3/S',3X,'SO =',F6.4,
             1 3X,'S =',F4.2,3X,'BO =',F6.2,' M')
              WRITE(6,15) YP,YN
      15      FORMAT(/5X,'INITIAL ESTIMATED FLOW DEPTHS:',2X, 'YP =',
             1    F4.2,' M',3X,'YN =',F6.2,' M')
              K = 0
      20      Y=0.5*(YP+YN)
              K=K+1
              IF (K.GT.50) GO TO 60
              F=Q-(AR(Y)**1.667*SQRT(SO))/(AN*P(Y)**0.667)
              IF (F.LT.0.0) YN=Y
              IF (F.GT.0.0) YP=Y
              IF (ABS(YP-YN).LE.0.001) GO TO 40
              GO TO 20
      40      Y=0.5*(YP+YN)
              WRITE(6,50) Y
      50      FORMAT(/5X, 'NORMAL DEPTH =',F6.3,' M')
              GO TO 80
      60      WRITE(6,70)
      70      FORMAT(10X,'ITERATIONS FAILED')
      80      STOP
              END
```

Input Data

```
      0.013,110.,0.0001,2.,20.,0.2,20.
```

Program Output

```
      N = .013   Q = 110.000 M3/S   SO = .0001   S =2.00   BO = 20.00 M

      INITIAL ESTIMATED FLOW DEPTHS:  YP = .20 M   YN = 20.00 M

      NORMAL DEPTH = 3.069 M
```

Newton-Raphson Method

Assume $x = x_1$ is one of the roots of Eq. (10-6), that is,

$$F(x_1) = 0 \tag{10-9}$$

Let the initial estimate for this root be x_0. Expanding Eq. (10-9) in a Taylor series, and neglecting terms of second and higher order, we obtain

$$F(x_1) = F(x_0) + (x_1 - x_0)F'(x_0) = 0 \tag{10-10}$$

where $F'(x_0)$ denotes dF/dx evaluated at $x = x_0$. However, since the initial estimate, x_0, may not be very close to the root of Eq. (10-6), an iterative procedure is used to refine the solution. Let us call x_1 determined from Eq. (10-10) as x_n. Then, from Eq. (10-10), it follows that

$$x_n = x_0 - \frac{F(x_0)}{F'(x_0)} \tag{10-11}$$

If $|(x_n - x_0)| < \epsilon$, where ϵ is the specified tolerance, then x_n is the root of Eq. (10-6). Otherwise, set $x_0 = x_n$, and repeat the above procedure until a solution is obtained.

Figure 10-5 shows a geometrical representation of the convergence of the iterative process.

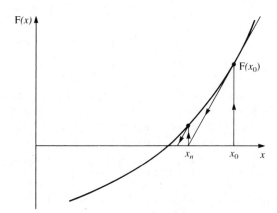

Figure 10-5   Newton-Raphson method

The disadvantages of the Newton-Raphson method are

1. It is necessary to determine the first derivative of the function F, and an expression for this may not be easily available. The value of the derivative

in such cases may be obtained by using numerical methods, such as secant method.

2. The procedure fails if $dF/dx = 0$.
3. The iterations may diverge if $F''(x)$ becomes too large.
4. It is difficult to apply if a table of values is given instead of a mathematical expression.

The advantages of the method are

1. It can be easily extended to a system of two or more equations.
2. It converges fast, thereby reducing the required number of iterations.

EXAMPLE 10-3   Solve Example 10-2 using the Newton-Raphson method.

SOLUTION   To use the Newton-Raphson method, we need the derivative of function F. This may be obtained by differentiating Eq. (10-8) with respect to y as follows:

$$\frac{dF}{dy} = \frac{d}{dy}\left[Q - \frac{1}{n}AR^{2/3}S_0^{1/2}\right] \tag{10-12}$$

Since we are determining the normal depth for specified values of Q, n, and S_0, these are constants for differentiation purposes. In addition, $R = A/P$. Hence we may write this equation as

$$\frac{dF}{dy} = -\frac{S_0^{1/2}}{n}\frac{d}{dy}(A^{5/3}P^{-2/3}) \tag{10-13}$$

$$= -\frac{S_0^{1/2}}{n}\left[-\frac{2}{3}A^{5/3}P^{-5/3}\frac{dP}{dy} + \frac{5}{3}P^{-2/3}A^{2/3}\frac{dA}{dy}\right] \tag{10-14}$$

For a trapezoidal channel, $dP/dy = 2\sqrt{1 + s^2}$, where s is the channel side slope (s horizontal to 1 vertical). If the top water surface width, B, for a channel section is continuous with change in flow depth, then the change in flow area, ΔA, for a small change in the flow depth, Δy, is $B\,\Delta y$, in which we have neglected the higher-order terms. (These terms correspond to the area of the small triangles, shown shaded in Fig. 10-6.) In the limit, as $\Delta y \to 0$, we can write $dA = B\,dy$, or $\frac{dA}{dy} = B$. By substituting these expressions for dP/dy and dA/dy into Eq. (10-14), and by simplifying the resulting equation, we obtain

$$\frac{dF}{dy} = \frac{S_0^{1/2}}{3n}[4R^{5/3}\sqrt{1 + s^2} - 5BR^{2/3}] \tag{10-15}$$

A computer program was developed using this relationship for $F'(y)$. The initial estimate for the normal depth was specified as 1 m, and the tolerance

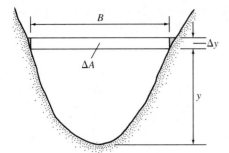

Figure 10-6   Definition sketch

for the convergence of the iterations as 0.001 m. Figure 10-7 lists the computer program, its input data, and program output.   ∎

Figure 10-7   Program for computing normal depth using Newton-Raphson method

Program Listing

```
C       COMPUTATION OF NORMAL DEPTH IN A TRAPEZOIDAL CHANNEL
C         USING NEWTON-RAPHSON METHOD
C
C       ******************** NOTATION ******************
C
C       AN= MANNING'S N;
C       BO = CHANNEL-BOTTOM WIDTH;
C       Q = DISCHARGE;
C       S = SLOPE OF CHANNEL SIDES, S: HORIZONTAL TO 1 VERTICAL;
C       SO = CHANNEL-BOTTOM SLOPE;
C       Y = NORMAL DEPTH;
C       YI = INITIAL ESTIMATE FOR NORMAL DEPTH.
C
        AR(Y)= Y*(BO+S*Y)
        P(Y) = BO+2.*Y*SQRT(1.+S*S)
        READ(5,*) AN,Q,BO,SO,S,YI
        WRITE(6,20) AN,Q,BO,SO,S,YI
20      FORMAT(5X,'N =',F5.3,3X,'Q =',F7.3,' M3/S',3X,'BO =',
     1     F5.2,' M',3X,'SO =',F6.4,3X,'S =',F4.2,3X,'YI =',F5.2)
        K=0
30      K=K+1
        IF (K.GT.50) GO TO 80
        R=AR(YI)/P(YI)
        B = BO+2.*S*YI
        F=Q-(AR(YI)*R**0.667*SQRT(SO))/AN
        FD=(SQRT(SO)/(3.*AN))*(4.*SQRT(1.+S*S)*R**1.667
     1        - 5.*B*R**0.667)
        DY=F/FD
        IF (ABS(DY).LE.0.001) GO TO 60
        YI=YI-DY
        GO TO 30
60      WRITE(6,70) YI
70      FORMAT(5X,'NORMAL DEPTH =',F6.2,' M')
        GO TO 90
80      WRITE(6,85)
85      FORMAT(10X,'ITERATIONS FAILED')
90      STOP
        END
```

Input Data

```
0.013,110.,20.,0.0001,2.,1.
```

Program Output

```
        N = .013   Q =110.000 M3/S   BO =20.00 M   SO = .0001
S =2.00   YI = 1.00
        NORMAL DEPTH =   3.07 M
```

Let us discuss how we may extend this procedure (2) for the following system of n equations in n unknowns, $x_1, x_2, \ldots, x_n$:

$$
\left.
\begin{aligned}
F_1(x_1, x_2, \ldots, x_n) &= 0 \\
F_2(x_1, x_2, \ldots, x_n) &= 0 \\
&\cdots\cdots\cdots\cdots\cdots \\
&\cdots\cdots\cdots\cdots\cdots \\
F_n(x_1, x_2, \ldots, x_n) &= 0
\end{aligned}
\right\}
\tag{10-16}
$$

Let the initial estimate for the roots be $x_1^{(0)}, x_2^{(0)}, \ldots, x_n^{(0)}$, where the superscript indicates the number of the iteration — 0 for initial estimate, 1 for values obtained after one iteration, and so on. Similar to the single equation, Eq. (10-6), we may expand and rearrange Eq. (10-16) as

$$
\begin{bmatrix}
\dfrac{\partial F_1}{\partial x_1} & \dfrac{\partial F_1}{\partial x_2} & \cdots & \dfrac{\partial F_1}{\partial x_j} & \cdots & \dfrac{\partial F_1}{\partial x_n} \\[2mm]
\dfrac{\partial F_2}{\partial x_1} & \dfrac{\partial F_2}{\partial x_2} & \cdots & \dfrac{\partial F_2}{\partial x_j} & \cdots & \dfrac{\partial F_2}{\partial x_n} \\[2mm]
\cdots & \cdots & \cdots & \cdots & \cdots & \cdots \\
\cdots & \cdots & \cdots & \cdots & \cdots & \cdots \\
\dfrac{\partial F_i}{\partial x_1} & \dfrac{\partial F_i}{\partial x_2} & \cdots & \dfrac{\partial F_i}{\partial x_j} & \cdots & \dfrac{\partial F_i}{\partial x_n} \\[2mm]
\cdots & \cdots & \cdots & \cdots & \cdots & \cdots \\
\cdots & \cdots & \cdots & \cdots & \cdots & \cdots \\
\dfrac{\partial F_n}{\partial x_1} & \dfrac{\partial F_n}{\partial x_2} & \cdots & \dfrac{\partial F_n}{\partial x_j} & \cdots & \dfrac{\partial F_n}{\partial x_n}
\end{bmatrix}^{(0)}
\begin{Bmatrix}
\Delta x_1 \\
\Delta x_2 \\
\vdots \\
\Delta x_i \\
\vdots \\
\Delta x_n
\end{Bmatrix}
= -
\begin{Bmatrix}
F_1 \\
F_2 \\
\vdots \\
F_i \\
\vdots \\
F_n
\end{Bmatrix}^{(0)}
\tag{10-17}
$$

where the functions and their partial derivatives are evaluated at $x_1^{(0)}, x_2^{(0)}, \ldots, x_n^{(0)}$. Then after solving for the Δx_i's, a better estimate for the solution is

$$
\left.
\begin{aligned}
x_1^{(1)} &= x_1^{(0)} + \Delta x_1 \\
x_2^{(1)} &= x_2^{(0)} + \Delta x_2 \\
&\cdots\cdots\cdots\cdots \\
&\cdots\cdots\cdots\cdots \\
x_n^{(1)} &= x_n^{(0)} + \Delta x_n
\end{aligned}
\right\}
\tag{10-18}
$$

If $(|\Delta x_1| + |\Delta x_2| + \cdots + |\Delta x_n|) < \epsilon$, where ϵ is the specified tolerance, then $x_1^{(1)}, x_2^{(1)}, \ldots, x_n^{(1)}$ are the roots of Eq. (10-16); otherwise, set

$$\left. \begin{aligned} x_1^{(0)} &= x_1^{(1)} \\ x_2^{(0)} &= x_2^{(1)} \\ &\cdots\cdots\cdots \\ &\cdots\cdots\cdots \\ x_n^{(0)} &= x_n^{(1)} \end{aligned} \right\} \tag{10-19}$$

Repeat the above procedure until a solution is obtained.

10-5 Quadrature

We want to numerically determine the value of the integral

$$I = \int_a^b f(x)\, dx \tag{10-20}$$

Evaluating this integral is equivalent to determining the area, I, under the curve $y = f(x)$ and the x axis between $x = a$ and $x = b$ (Fig. 10-8). To evaluate this integral numerically, we divide the interval between $x = a$ and $x = b$ into a number of subintervals, approximately compute the area for each subinterval, and then sum these subareas to determine the total area. Let us divide the interval $x = a$ to $x = b$ into N equal subintervals of length, h:

$$h = \frac{b - a}{N} \tag{10-21}$$

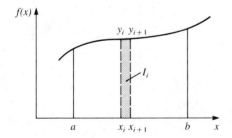

Figure 10-8 Numerical evaluation of integrals

If we consider each subarea as a trapezoid, the area of the trapezoid between $x = x_i$ and $x = x_{i+1}$ is

$$I_i = 0.5h(y_i + y_{i+1}) \tag{10-22}$$

Adding these N subareas, we obtain

$$I = I_h = \sum I_i = \frac{h}{2}(y_1 + 2y_2 + \cdots + 2y_N + y_{N+1}) \tag{10-23}$$

This is referred to as the *trapezoidal rule*, in which the truncation error, $e_t < (h^2/12)(b - a)$, and the roundoff error, e_r, is proportional to $1/h$. Theoretically, $I_h \to I$ as $h \to 0$. In reality, however, this is not the case, since the roundoff error becomes dominant as h becomes smaller and smaller.

If N is an even number, we may use *Simpson's rule* to evaluate the integral, I_h:

$$I_h = \frac{h}{3}(y_1 + 4y_2 + 2y_3 + 4y_4 + \cdots + 2y_{N-1} + 4y_N + y_{N+1}) \tag{10-24}$$

In this rule, the truncation error is proportional to h^4, and the roundoff error is proportional to $1/h$.

10-6 Linear Algebraic Equations

The Gauss elimination technique may be used to solve a system of independent linear equations (4). Assume that we have n equations in n unknowns $x_1, x_2, \ldots, x_n$ and that we have rearranged the equations, if necessary, so that $a_{11} \neq 0$.

$$\left. \begin{aligned} a_{11}x_1 + a_{12}x_2 + \cdots + a_{1j}x_j + \cdots + a_{1n}x_n &= b_1 \\ a_{21}x_1 + a_{22}x_2 + \cdots + a_{2j}x_j + \cdots + a_{2n}x_n &= b_2 \\ \cdots\cdots\cdots\cdots\cdots\cdots\cdots\cdots\cdots\cdots\cdots \\ a_{i1}x_1 + a_{i2}x_2 + \cdots + a_{ij}x_j + \cdots + a_{in}x_n &= b_i \\ \cdots\cdots\cdots\cdots\cdots\cdots\cdots\cdots\cdots\cdots\cdots \\ \cdots\cdots\cdots\cdots\cdots\cdots\cdots\cdots\cdots\cdots\cdots \\ a_{n1}x_1 + a_{n2}x_2 + \cdots + a_{nj}x_j + \cdots + a_{nn}x_n &= b_n \end{aligned} \right\} \tag{10-25}$$

Let us define $n - 1$ multipliers

$$m_i = \frac{a_{i1}}{a_{11}} \qquad i = 2, 3, \ldots, n \tag{10-26}$$

Subtract m_i times the first equation from the ith equation, and let

$$a'_{ij} = a_{ij} - m_i a_{1j} \qquad i = 2, 3, \ldots, n \quad \text{and} \quad j = 1, 2, \ldots, n$$
$$b'_i = b_i - m_i b_1 \tag{10-27}$$

Then, the resulting equations can be written as

$$
\left.
\begin{aligned}
a_{11}x_1 + a_{12}x_2 + \cdots + a_{1j}x_j + \cdots + a_{1n}x_n &= b_1 \\
0 \quad + a'_{22}x_2 + \cdots + a'_{2j}x_j + \cdots + a'_{2n}x_n &= b'_2 \\
\cdots\cdots\cdots\cdots\cdots\cdots\cdots\cdots\cdots\cdots\cdots\cdots\cdots \\
\cdots\cdots\cdots\cdots\cdots\cdots\cdots\cdots\cdots\cdots\cdots\cdots\cdots \\
0 \quad + a'_{n2}x_2 + \cdots + a'_{nj}x_j + \cdots + a'_{nn}x_n &= b'_n
\end{aligned}
\right\} \tag{10-28}
$$

If we repeat this procedure, then after the kth time, we have eliminated x_k by defining multipliers

$$m_i^{(k-1)} = \frac{a_{ik}^{(k-1)}}{a_{kk}^{(k-1)}} \qquad i = k+1, \ldots, n \tag{10-29}$$

where $a_{kk}^{(k-1)} \neq 0$. And

$$a_{ij}^{(k)} = a_{ij}^{(k-1)} - m_i^{(k-1)} a_{kj}^{(k-1)}$$
$$b_i^{(k)} = b_i^{(k-1)} - m_i^{(k-1)} b_k^{(k-1)}$$
$$i = k+1, \ldots, n \qquad j = k, \ldots, n; \quad \text{and} \quad k = 1, 2, \ldots, n-1 \tag{10-30}$$

When $k = n - 1$, we have eliminated all unknowns x_1 to x_{n-1} from the last equation, and the resulting equations are

$$
\left.
\begin{aligned}
a_{11}x_1 + a_{12}x_2 + \cdots + a_{1j}x_j + \cdots + a_{1n}x_n &= b_1 \\
a'_{22}x_2 + \cdots + a'_{2j}x_j + \cdots + a'_{2n}x_n &= b'_2 \\
\cdots\cdots\cdots\cdots\cdots\cdots\cdots\cdots\cdots\cdots\cdots \\
a_{jj}^{(j-1)}x_j + \cdots + a_{jn}^{(j-1)}x_n &= b_j^{(j-1)} \\
\cdots\cdots\cdots\cdots\cdots\cdots\cdots\cdots\cdots\cdots\cdots \\
a_{nn}^{(n-1)}x_n &= b_n^{(n-1)}
\end{aligned}
\right\} \tag{10-31}
$$

Now, we can determine the value of x_n from the last equation. Then substituting this computed value of x_n into next to the last equation, we can determine x_{n-1} and proceed similarly to determine the remaining unknowns:

$$x_n = \frac{b_n^{(n-1)}}{a_{nn}^{(n-1)}}$$

$$x_{n-1} = \frac{b_{n-1}^{(n-2)} - a_{n-1,n}^{(n-2)}x_n}{a_{n-1,n-1}^{(n-2)}}$$

$$\cdots\cdots\cdots\cdots\cdots\cdots\cdots\cdots\cdots$$

$$x_j = \frac{b_j^{(j-1)} - a_{jn}^{(j-1)}x_n - \cdots - a_{j,j+1}^{(j-1)}x_{j+1}}{a_{jj}^{(j-1)}} \qquad j = n-2, \ldots, 1$$

$$(10\text{-}32)$$

10-7 Ordinary Differential Equations

Let us consider a first-order differential equation

$$\frac{dy}{dx} = f(x, y) \tag{10-33}$$

which represents the rate of change of y with respect to x and is equal to function $f(x, y)$. A closed-form solution, that is, a direct relationship between x and y, may not be available if $f(x, y)$ is nonlinear. Therefore, numerical methods are used. These methods may be divided into two categories — single-step and multistep — with each type having several different methods. For simplicity and ease of understanding, we will discuss the use of these methods for the numerical integration of single, first-order equations. Higher-order equations may be reduced to a system of first-order equations; then these methods may be applied similarly.

Euler Method

Assume that we know the value of y at $x = x_i$ and we want to determine y at $x = x_{i+1}$. The known value may be the initial condition or may have been computed during the previous computations. From Eq. (10-33), it follows that we can determine the rate of change of y at $x = x_i$:

$$y_i' = f(x_i, y_i) \tag{10-34}$$

where the subscript i refers to the quantities at distance x_i, and a prime, $'$, indicates the derivative with respect to x. Assuming that this rate of change remains constant in the interval $x = x_i$ to $x = x_{i+1}$, we can write

$$y_{i+1} = y_i + y_i'(x_{i+1} - x_i)$$

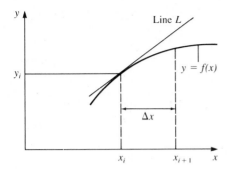

Figure 10-9   Euler method

Substitution of Eq. (10-34) into this equation yields

$$y_{i+1} = y_i + f(x_i, y_i)\Delta x \tag{10-35}$$

where $\Delta x = (x_{i+1} - x_i)$

Figure 10-9 shows a geometrical representation of this method. Equation (10-35) is the equation of a straight line (line L in Fig. 10-9) passing through the point (x_i, y_i), and y_i' is the slope of the tangent to the curve at (x_i, y_i). If the rate of change of y remained constant in the interval x_i to x_{i+1}, then Eq. (10-35) will give exact results. However, the rate is not usually constant; thus, a small error is introduced at each step. By expanding y_{i+1} into a Taylor's series and comparing it with Eq. (10-35), we can prove that we are including terms up to only the first power of x. Therefore, Euler's method is referred to as first-order accurate. This method although simple to program, may become unstable for large values of Δx since truncation or roundoff errors are amplified as the value of x increases.

In the Euler method, we used the slope of the tangent at only one point. By using the slope at more than one point, we can improve the accuracy of the results, as we discuss in the following paragraphs.

Improved Euler Method

In this method, the average of the slope of the tangent to the solution curve at points (x_i, y_i) and $(x_{i+1}, y_i + y_i'\Delta x)$ is used (Fig. 10-10, page 560):

$$y_{i+1} = y_i + 0.5[f(x_i, y_i) + f(x_{i+1}, y_i + y_i'\Delta x)]\Delta x \tag{10-36}$$

EXAMPLE 10-4   A trapezoidal channel having a bottom width of 20 m, side slopes of 2 horizontal to 1 vertical, and a bottom slope of 0.0001 carries a flow of 110 m³/s. A bridge is planned on this channel that will raise the flow depth at

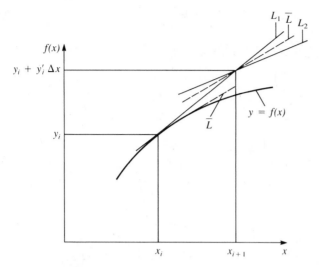

Figure 10-10   Improved Euler method

the bridge to 5 m. Compute the backwater profile in the channel. Assume the Manning n is 0.013.

SOLUTION   Modern methods for computing the water surface profiles may be divided into two categories: methods based on the solution of the energy equation between different channel sections and methods based on the solution of the differential equation describing the rate of change of depth with distance. We will follow the second approach in this example.

The following differential equation describes the rate of change of flow depth with distance for gradually varied flows in a prismatic channel (3):

$$\frac{dy}{dx} = \frac{S_0 - S_f}{1 - \dfrac{BQ^2}{gA^3}} \tag{10-37}$$

where y is flow depth at distance x, S_0 is slope of the channel bottom, S_f is slope of the energy-grade line, B is top water surface width, Q is discharge, A is flow area, and g is acceleration due to gravity. Eq. (10-37) is the same as Eq. (4-32) but is in a slightly different form.

We will use the improved Euler method to numerically integrate Eq. (10-37). The initial condition is the flow depth at the bridge, $y = 5.0$ m at $x = 0$. Since the computation of the water levels proceed in the upstream direction, Δx is negative.

Figure 10-11 lists a computer program using the improved Euler method, its input data, and program output.   ∎

Figure 10-11   Computer program for computing
water-surface using improved Euler method

Program Listing

```
C     COMPUTATION OF WATER-SURFACE PROFILE BY USING
C        IMPROVED EULER METHOD
C
C     *************** NOTATION ******************
C     A=FLOW AREA;
C     B=TOP WATER-SURFACE WIDTH;
C     BO=CHANNEL-BOTTOM WIDTH;
C     P = WETTED PERIMETER;
C     Q = DISCHARGE;
C     MN = MANNING'S N;
C     S = CHANNEL-SIDE SLOPE, S HORIZONTAL : 1 VERTICAL;
C     SO = CHANNEL-BOTTOM SLOPE;
C     X = DISTANCE ALONG CHANNEL BOTTOM, POSITIVE IN THE DOWNSTREAM
C             DIRECTION;
C     Y = FLOW DEPTH
C     YD = DEPTH AT DOWNSTREAM END.
C
      REAL MN
      DIMENSION X(100)
      AR(Y)=Y*(BO+S*Y)
      WP(Y)=BO+2.*Y*SQRT(1.+S*S)
      READ (5,*) BO,S,SO,MN,Q,YD
      READ (5,*) N,(X(I),I=1,N)
      WRITE(6,10) BO,S,SO,MN,Q,YD
10    FORMAT(2X,'B =',F5.1,' M',2X,'S =',F4.1,2X,'SO =',F6.4,
     1    2X,'N =',F5.3,2X,'Q =',F7.3,' M3/S',2X,'YD =',F5.3,' M')
      Q2=Q*Q
      QN2=(MN*Q)**2
      Y=YD
      WRITE(6,15)
15    FORMAT(6X,'X',10X,'Y')
      WRITE(6,20) X(1),Y
      DO 30 I = 2,N
      DX=X(I)-X(I-1)
      A=AR(Y)
      P=WP(Y)
      R=A/P
      SF1=QN2/(A*A*R**1.333)
      B=BO+2.*S*Y
      DY1=(SO-SF1)/(1-(B*Q2)/(9.81*A**3))
      Y2=Y+DY1*DX
      A=AR(Y2)
      P=WP(Y2)
      R=A/P
      SF2=QN2/(A*A*R**1.333)
      B=BO+2.*S*Y2
      DY2=(SO-SF2)/(1.-(B*Q2)/(9.81*A**3))
      Y= Y+0.5*(DY1+DY2)*DX
      WRITE(6,20) X(I),Y
20    FORMAT(F10.1,F10.3)
30    CONTINUE
      STOP
      END
```

Input Data

```
20.,2.,0.0001,0.013,110.,5.
10,0.0,-50.,-100.,-200.,-300.,-400.,-500.,-750.,-1000.,-1500.
```

Program Output

```
B = 20.0 M  S = 2.0   SO = .0001  N = .013  Q =110.000 M3/S  YD =5.000 M
    X          Y
      .0      5.000
   -50.0      4.996
  -100.0      4.992
  -200.0      4.983
  -300.0      4.975
  -400.0      4.966
  -500.0      4.958
  -750.0      4.937
 -1000.0      4.916
 -1500.0      4.875
```

Fourth-order Runge-Kutta Method

In this method, the slope of the curve is determined from the following equations:

$$k_1 = f(x_i, y_i)$$

$$k_2 = f\left(x_i + \frac{1}{2}\Delta x, y_i + \frac{1}{2}k_1\Delta x\right)$$

$$k_3 = f\left(x_i + \frac{1}{2}\Delta x, y_i + \frac{1}{2}k_2\Delta x\right) \qquad (10\text{-}38)$$

$$k_4 = f(x_i + \Delta x, y_i + k_3\Delta x)$$

Then,

$$y_{i+1} = y_i + \frac{\Delta x}{6}(k_1 + 2k_2 + 2k_3 + k_4) \qquad (10\text{-}39)$$

Predictor-corrector Method

In the single-step methods discussed previously, we used the known information at point x_i, and to improve accuracy, we used the value of the function at x_i, $x_i + \Delta x$, $x_i + 1/2\,\Delta x$. In the predictor-corrector method, we do not compute the function at several points but rather predict the dependent variable first by using values computed during the previous steps, "correct" this predicted value, and then "recorrect" the corrected value if necessary. This iterative procedure is continued until a solution of a desired accuracy is obtained.

Several predictor-corrector methods are available. However, to conserve space, we will discuss only one of them.

In the predictor part, let us use the Euler method to predict the value of y_{i+1}:

$$y_{i+1}^{(0)} = y_i + f(x_i, y_i)\Delta x \qquad (10\text{-}40)$$

where the superscript indicates the number of the iteration (zero iteration is the predicted value). Then, we may correct it by using the following equation:

$$y_{i+1}^{(1)} = y_i + \frac{1}{2}\left[f(x_i, y_i) + f(x_{i+1}, y_{i+1}^{(0)})\right]\Delta x \qquad (10\text{-}41)$$

Now, we may recorrect $y_{i+1}^{(1)}$ to obtain a better value:

$$y_{i+1}^{(2)} = y_i + \frac{1}{2}\left[f(x_i, y_i) + f(x_{i+1}, y_{i+1}^{(1)})\right]\Delta x \qquad (10\text{-}42)$$

Thus, the jth iteration is

$$y_{i+1}^{(j)} = y_i + \frac{1}{2} \left[f(x_i, y_i) + f(x_{i+1}, y_{i+1}^{(j-1)}) \right] \Delta x \tag{10-43}$$

We continue the iterations until $\left| y_{i+1}^{(j)} - y_{i+1}^{(j-1)} \right|$ is less than a specified toler-
ance. Then, we increment the value of x and repeat the above procedure.

EXAMPLE 10-5 Route the inflow hydrograph listed in Table A through a
detention reservoir 420×420 ft having side slopes of 2 horizontal to 1 vertical.
The rating curve for the outflow facilities is given in Table B. The reservoir level
at time $t = 0$ is at $z = 0$.

Table A

Inflow Hydrograph

Time (min)	Inflow (cfs)
0	0
15	5
30	16
45	39
60	104
75	322
90	555
105	722
120	626
135	481
150	318
165	162

Table B

Outflow Rating Curve

Stage (ft)	Outflow (cfs)
0	0
1	20
2	48
3	100
4	160
5	210
6	260
7	300
8	330
9	360
10	380

SOLUTION Assuming that the reservoir water surface remains level, the
following equation can be written for the variation of water level in the reservoir
with time:

$$\frac{dz}{dt} = \frac{Q_{in} - Q_{out}}{A} \tag{10-44}$$

where A is horizontal surface area of the reservoir, z is elevation of the water
surface in the reservoir above a specified datum, t is time, Q_{in} is inflow into the
reservoir, and Q_{out} is outflow from the reservoir.

Figure 10-12 lists a computer program using the predictor-corrector method
to solve Eq. (10-44). A routing interval of 15 min was used. Linear interpolations
are used to determine the outflow at intermediate levels from the stored data.
Since the variation of the reservoir water level is very gradual, no iterations are
used to refine the solution. ∎

Figure 10-12   Program for reservoir routing using predictor-corrector method

Program Listing

```
C       RESERVOIR ROUTING USING PREDICTOR-CORRECTOR METHOD
C
C       ************** NOTATION ****************
C
C       AR(I) = SURFACE AREA OF RESERVOIR AT LEVEL ELAR(I);
C       DT = ROUTING INTERVAL;
C       NAR = NUMBER OF POINTS ON THE STAGE VS. AREA CURVE;
C       NQO = NUMBER OF POINTS ON THE STAGE VS. OUTFLOW CURVE;
C       NT = NUMBER OF POINTS ON THE INFLOW VS TIME CURVE;
C       QIN(I) = INFLOW AT TIME(I);
C       QO(I) = OUTFLOW AT WATER LEVEL ELQO(I);
C       QO1 = OUTFLOW AT TIME T = 0;
C       Z = RESERVOIR LEVEL ABOVE DATUM;
C       Z1 = RESERVOIR LEVEL AT TIME T = 0;
C       TSTOP = TIME UPTO WHICH ROUTING IS TO BE COMPUTED.
C
        DIMENSION AR(50),ELAR(50),ELQO(50),QO(50),TIME(50),QIN(50)
C
C       INITIAL CONDITIONS
C
        READ(5,*) Z1,QO1,DT,TSTOP
        WRITE(6,20) Z1,QO1,DT,TSTOP
20      FORMAT(//5X,'INITIAL RESERVOIR WATER LEVEL =',F7.2,' M'/
       1 5X,'INITIAL OUTFLOW =',F5.1,' M3/S'/
       2 5X,'ROUTING INTERVAL =',F6.1,' S'/
       3 5X,'TIME UPTO WHICH ROUTING IS TO BE DONE =',F8.1,' S'/)
C
C       STAGE VS. RESERVOIR-SURFACE AREA CURVE
C
        READ (5,*) NAR,(ELAR(I),AR(I),I=1,NAR)
        WRITE(6,30)
30      FORMAT(15X,'STAGE',5X,'RESERVOIR SURFACE AREA'/
       1  16X,'(M)',13X,'(SQ. M)')
        WRITE(6,40) (ELAR(I),AR(I),I=1,NAR)
40      FORMAT(9X,F10.1,10X,F10.1)
C
C       STAGE-OUTFLOW CURVE
C
        READ (5,*) NQO, (ELQO(I),QO(I),I=1,NQO)
        WRITE(6,60)
60      FORMAT(/15X,'STAGE',9X,' OUTFLOW'/
       1  16X,'(M)',12X,'(M3/S)')
        WRITE(6,70) (ELQO(I),QO(I),I=1,NQO)
70      FORMAT(9X,F10.1,8X,F10.1)
C
C       INFLOW HYDROGRAPH
C
        READ(5,*) NT, (TIME(I),QIN(I), I=1,NT)
        WRITE(6,90)
90      FORMAT(/14X,'TIME',12X,'INFLOW'/
       1  14X,'(S)',13X,'(M3/S)')
        WRITE(6,100) (TIME(I),QIN(I),I=1,NT)
100     FORMAT(8X,F10.1,8X,F10.2)
        T = 0.
        WRITE(6,125)
125     FORMAT(//13X,'TIME',11X,'INFLOW',4X,'RESERVOIR LEVEL',2X,
       1    'OUTFLOW'/13X,'(S)',12X,'(M3/S)',9X,'(M)',11X,'M3/S)')
        WRITE(6,240) T,QIN(1),Z1,QO1
C
C         PREDICTOR PART
C
135     T = T+DT
        DO 140 I=1,NT
        IF (T.LT.TIME(I)) GO TO 150
140     CONTINUE
150     I1=I-1
        QIP=QIN(I1)+(T-TIME(I1))/(TIME(I)-TIME(I1))*(QIN(I)-QIN(I1))
        DO 160 I=1,NQO
        IF (Z1.LT.ELQO(I)) GO TO 170
160     CONTINUE
170     I1=I-1
        QOP=QO(I1)+(Z1-ELQO(I1))/(ELQO(I)-ELQO(I1))*(QO(I)-QO(I1))
```

```
            DO 180 I=1,NAR
            IF (Z1.LT.ELAR(I)) GO TO 190
180         CONTINUE
190         I1=I-1
            ARP=AR(I1)+(Z1-ELAR(I1))/(ELAR(I)-ELAR(I1))*(AR(I)-AR(I1))
            DZP=(QIP-QOP)/ARP
            ZP=Z1+DZP*DT
C
C           CORRECTOR PART
C
            DO 200 I=1,NQO
            IF (ZP.LT.ELQO(I)) GO TO 210
200         CONTINUE
210         I1=I-1
            QOC=QO(I1)+(ZP-ELQO(I1))/(ELQO(I)-ELQO(I1))*(QO(I)-QO(I1))
            DO 220 I=1,NAR
            IF (ZP.LT.ELAR(I)) GO TO 230
220         CONTINUE
230         I1=I-1
            ARC=AR(I1)+(ZP-ELAR(I1))/(ELAR(I)-ELAR(I1))*(AR(I)-AR(I1))
            DZC=(QIP-QOC)/ARC
            Z2=Z1+0.5*DT*(DZC+DZP)
            DO 232 I=1,NQO
            IF (Z2.LT.ELQO(I)) GO TO 235
232         CONTINUE
235         I1=I-1
            Q2=QO(I1)+(Z2-ELQO(I1))/(ELQO(I)-ELQO(I1))*(QO(I)-QO(I1))
            WRITE(6,240) T,QIP,Z2,Q2
240         FORMAT(9X,F10.2,5X,F10.2,2X,F10.2,5X,F10.2)
            IF (T.GT.TSTOP) GO TO 250
            Z1=Z2
            GO TO 135
250         STOP
            END
```

Input Data

```
0.,0.,900.,10800.
11,0.,176400.,1.,179800.,2.,183200.,3.,186600.,4.,190100.,5.,193600.
6.,197100.,7.,200700.,8.,204300.,9.,207900.,10.,211600.
11,0.,0.,1.,20.,2.,48.,3.,100.,4.,160.,5.,210.,6.,260.,7.,300.,8.,330.,
9.,360.,10.,380.
12,0.,0.,900.,5.,1800.,16.,2700.,39.,3600.,104.,4500.,322.,5400.,555.,
6300.,722.,7200.,626.,8100.,481.,9000.,318.,9900.,162.
```

Program Output

```
INITIAL RESERVOIR WATER LEVEL =    .00 M
INITIAL OUTFLOW =    .0 M3/S
ROUTING INTERVAL = 900.0 S
TIME UPTO WHICH ROUTING IS TO BE DONE = 10800.0 S
```

STAGE (M)	RESERVOIR SURFACE AREA (SQ. M)
.0	176400.0
1.0	179800.0
2.0	183200.0
3.0	186600.0
4.0	190100.0
5.0	193600.0
6.0	197100.0
7.0	200700.0
8.0	204300.0
9.0	207900.0
10.0	211600.0

STAGE (M)	OUTFLOW (M3/S)
.0	.0
1.0	20.0
2.0	48.0

(continued)

Figure 10-12   (continued)

Program Output

3.0	100.0
4.0	160.0
5.0	210.0
6.0	260.0
7.0	300.0
8.0	330.0
9.0	360.0
10.0	380.0

TIME (S)	INFLOW (M3/S)
.0	.00
900.0	5.00
1800.0	16.00
2700.0	39.00
3600.0	104.00
4500.0	322.00
5400.0	555.00
6300.0	722.00
7200.0	626.00
8100.0	481.00
9000.0	318.00
9900.0	162.00

TIME (S)	INFLOW (M3/S)	RESERVOIR LEVEL (M)	OUTFLOW M3/S)
.00	.00	.00	.00
900.00	5.00	.02	.48
1800.00	16.00	.10	1.98
2700.00	39.00	.28	5.56
3600.00	104.00	.75	15.00
4500.00	322.00	2.15	56.01
5400.00	555.00	4.23	171.55
6300.00	722.00	6.49	279.73
7200.00	626.00	7.92	327.72
8100.00	481.00	8.55	346.56
9000.00	318.00	8.44	343.06
9900.00	162.00	7.69	320.73
10800.00	176.73	7.09	302.77
11700.00	191.45	6.63	285.39

10-8 Finite-Difference Approximations

To numerically integrate the ordinary or partial differential equations, we replace the derivative terms by finite-difference approximations and then solve the resulting algebraic equations (1, 4). To facilitate understanding, let us consider the Taylor series expansion of function $f(x)$ about some known point x_0:

$$f(x_0 + \Delta x) = f(x_0) + \Delta x f'(x_0) + \frac{(\Delta x)^2}{2!} f''(x_0)$$

$$+ \frac{(\Delta x)^3}{3!} f'''(x_0) + O[(\Delta x)^4] \tag{10-45}$$

$$f(x_0 - \Delta x) = f(x_0) - \Delta x f'(x_0) + \frac{(\Delta x)^2}{2!} f''(x_0)$$

$$- \frac{(\Delta x)^3}{3!} f'''(x_0) + O[(\Delta x)^4] \tag{10-46}$$

where $O[(\Delta x)^4]$ denotes terms containing fourth and higher power of Δx. Rearranging and dividing throughout by Δx, we can write Eqs. (10-45) and (10-46) as

$$f'(x_0) = \frac{f(x_0 + \Delta x) - f(x_0)}{\Delta x} + O(\Delta x) \tag{10-47}$$

$$f'(x_0) = \frac{f(x_0) - f(x_0 - \Delta x)}{\Delta x} + O(\Delta x) \tag{10-48}$$

If we neglect the leading error terms, $O(\Delta x)$, in these equations, we obtain the following expressions for the finite-difference approximations:

$$f'(x_0) = \frac{f(x_0 + \Delta x) - f(x_0)}{\Delta x} \tag{10-49}$$

$$f'(x_0) = \frac{f(x_0) - f(x_0 - \Delta x)}{\Delta x} \tag{10-50}$$

Eq. (10-49) is called *forward finite difference*, and Eq. (10-50) is called *backward finite difference*. Both are first-order accurate, that is, the leading error in both is of the first order. Referring to Fig. 10-13, we can say that we replace the slope of the tangent to the curve at $x = x_0$ by the slope of the chord line PB in the forward finite difference and by the slope of the chord line AP in the backward finite difference.

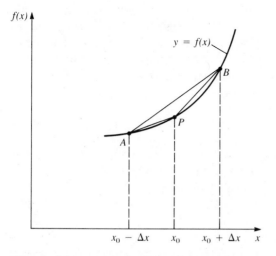

Figure 10-13   Finite-difference approximation

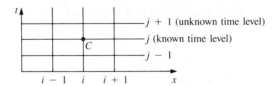

Figure 10-14   Computational grid

Subtracting Eq. (10-46) from Eq. (10-45), dividing throughout by Δx and neglecting higher-order terms, we obtain

$$f'(x_0) = \frac{f(x_0 + \Delta x) - f(x_0 - \Delta x)}{2\,\Delta x} \tag{10-51}$$

This is referred to as the *central finite difference*, and it is second-order accurate. Again, referring to Fig. 10-13, we are replacing the slope of the tangent to the curve at $x = x_0$ by the slope of the chord line AB in the central finite difference.

Let us now consider a function f that is a function of independent variables x and t. Assume that we have divided the x-t plane into a grid having spatial spacing of Δx and timewise spacing of Δt, as shown in Fig. 10-14. For brevity, we will designate the function f at point C as follows:

$$f_C = f(x_0, t_0) = f(i\,\Delta x, j\,\Delta t) = f_i^{\,j} \tag{10-52}$$

Let us say that we know the values of the dependent variables at j time level* (called the known time level) and we want to determine their values at the $j + 1$ time level (referred to as the unknown time level). To replace the spatial derivatives, that is, with respect to x, we have the choice of not only using the forward, backward, or central finite differences but also whether to use these differences at the known or unknown time levels. Spatial derivatives written at the known time level are *explicit finite differences*, and those at the unknown time level are *implicit finite differences*.[†]

PROBLEMS

10-1   For flood routing through a reservoir, we have to compute the reservoir area at intermediate levels from data stored at equal intervals. Write a computer program to parabolically interpolate the areas from the specified data.

---

[*] For brevity, we replace $j\,\Delta t$ time as j time level and $i\,\Delta x$ as the ith grid.
[†] We discuss the application of these methods for the analysis of unsteady flows in open channels in Chapter 12.

10-2   Modify the computer program in Fig. 10-12, page 564, to use parabolic interpolation instead of linear interpolation for the stage versus outflow curve. Compare the computed results for these two types of interpolations. In addition, study the effects of changing the routing interval on the computed outflows.

10-3   Derive expressions for corrections, Δx and Δy, in the Newton-Raphson method for the following nonlinear equations:

$$F(x, y) = 0$$

$$G(x, y) = 0$$

10-4   Write a computer program to determine the critical depth in a trapezoidal channel by using the bisection method. Use the data in Example 10-2, pages 549–550.

10-5   Use the Newton-Raphson method to determine the critical depth in a trapezoidal channel, and compare the number of iterations required with that in the bisection method.

10-6   To compute the water surface profile in a channel, we solve the energy equation between two consecutive channel sections. Instead of considering two sections at a time, we can write these equations for the entire channel length and then solve all equations simultaneously using the Newton-Raphson method. The flow depth for the given discharge will be known at the downstream end for subcritical flows and at the upstream end for super-critical flows. Write a computer program to compute the water surface profile in the channel of Example 10-4, pages 559–560, and compare the results with those obtained by numerically integrating the differential equation, Eq. (10-37).

10-7   Using the trapezoidal rule, numerically evaluate the integral $\int_0^\pi \sin x \, dx$ by dividing the interval 0 to π into 1 to 2000 subintervals; compute the error in each case (exact value of the integral is 2); and plot a curve between the number of subintervals and the error. Why does the error first decrease and then increase as the number of subintervals increases from 1 to 2000?

10-8   The following two equations relate the flow variables at a channel transition:

$$A_1 V_1 = A_2 V_2$$

$$z_1 + y_1 + \frac{V_1^{\,2}}{2g} = z_2 + y_2 + \frac{V_2^{\,2}}{2g} + k \frac{\left| V_1^{\,2} - V_2^{\,2} \right|}{2g}$$

where A is flow area, V is flow velocity, z is height of channel bottom above datum, y is flow depth, k is coefficient of head losses at the transition, and subscripts 1 and 2 denote variables on the upstream and downstream side of the transition.

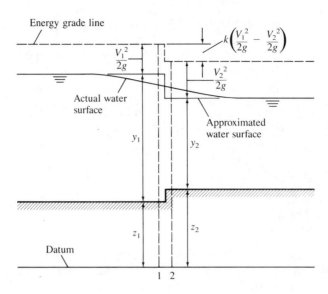

For given values of flow depth and flow velocity upstream of the transition, write a computer program to determine the flow depth and flow velocity downstream of the transition by using the Newton-Raphson method. Use the expressions developed in Prob. 10-3.

Use this program to determine y_2 and V_2 in a 10-m wide rectangular channel having a step rise of 0.1 m in the channel bottom. Flow velocity and flow depth upstream of the transition are 3 m/s and 2 m, respectively.

10-9   Use the computer programs of Figs. 10-4, page 550, and 10-7, page 553, to solve Probs. 4-6 and 4-9, page 226.

10-10   Write a computer program to determine the alternate depth in a channel, and use this program to solve Prob. 4-20, page 227.

10-11   Solve Prob. 4-27, page 228, by using the computer programs of Probs. 10-4 and 10-5.

10-12   Solve Probs. 4-30 and 4-31, page 229, by using the computer program of Prob. 10-7.

10-13   Route the flood hydrograph of Example 10-5, page 563, through the reservoir using the Improved Euler method. Compare the results with those obtained by using the predictor-corrector method.

10-14   Investigate the effects of routing interval on the peak of outflow hydrograph by solving Example 10-5, page 563. Use the Euler method for solving the governing equation.

10-15 Write a computer program to compute the backwater profile of Prob. 4-50, page 233, using the Euler method and the fourth-order Runge-Kutta method. Compare your results with those obtained by using the methods discussed in Chapter 4.

REFERENCES

1. Chapra, S.C., and R.P. Canale. *Numerical Methods for Engineers with Personal Computer Applications*. McGraw-Hill, New York, 1985.
2. Chaudhry, M.H. *Applied Hydraulic Transients*, 2d ed. Van Nostrand Reinhold, New York, 1987.
3. Chow, V.T. *Open Channel Hydraulics*. McGraw-Hill, New York, 1958.
4. McCracken, D.D., and W.S. Dorn. *Numerical Methods and FORTRAN Programming*. John Wiley & Sons, New York, 1964.

11

Oigawa Power Plant: Collapsed penstock due to subatmospheric transient pressure (1) ("Water-Hammer Damage to Oigawa Power Station," by Bonin from JOURNAL OF ENGINEERING FOR POWER, April 1960. Courtesy of The American Society of Mechanical Engineers.)

Unsteady Closed-Conduit Flows

Spectacular accidents have occurred because of transient-state pressures (intermediate flow while changing from one steady state to another is called transient flow) exceeding the design pressure of a conduit (1, 2, 4, 7, 9). These accidents, which are due to design or operating errors or equipment malfunction, have resulted in loss of life and money. Photos of damage due to transient-state pressures are shown in Fig. 11-1.

In this chapter, we discuss what transient flows are, how they are produced, how they are analyzed, and the different methods available for their control. Several commonly used terms are first defined. Expressions are derived for the wave speed and for the pressure rise in a closed conduit caused by an instantaneous change in flow velocity. Equations describing the unsteady flows are developed and the method of characteristics for their numerical integration is then presented. Transients caused by pumps are then discussed. The chapter concludes with a discussion of various methods for keeping the transient conditions within prescribed limits.

11-1 Definitions

As we discussed previously, a flow is *steady* if the flow velocity at a given location does not vary with time. If the flow velocity at a point does vary with time, the flow is *unsteady*.

Figure 11-1 Damage caused by transient-state pressures: (left) Burst pipe due to high transient pressures (1) ("Water-Hammer Damage to Oigawa Power Station," by Bonin from JOURNAL OF ENGINEERING FOR POWER, April 1960. Courtesy of The American Society of Mechanical Engineers.) (right) Failed pump casing due to high transient pressures. (Courtesy of A. B. Almeida)

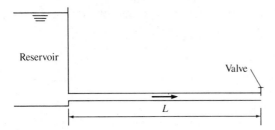

Figure 11-2   Piping system

When the flow conditions are changed from one steady state to another, the intermediate-stage flow is referred to as *transient flow* (2, 11). In the past, the terms *water hammer* and *oil hammer* were used for transient flows; at present the terms *hydraulic transients* and *fluid transients* are more commonly used.

The following discussion of different flow conditions in a piping system will help you understand the preceding definitions.

Let us consider a pipeline of length L in which water is flowing from a constant-level, upstream reservoir to a valve located at the downstream end, as shown in Fig. 11-2. Assume that the valve is instantaneously closed at time $t = t_0$ from the full-open position to the half-open position. This will reduce the flow velocity through the valve, thereby increasing the pressure at the valve. The increased pressure will produce a pressure wave that will travel back and forth in the pipeline until it is dissipated because of friction and the flow conditions have become steady again. This time, when the flow conditions have become steady again, we will call t_f.

Based on the preceding definitions, we may classify these flow regimes into the following categories:

1.   Steady flow for $t < t_0$
2.   Transient flow for $t_0 \leqslant t < t_f$
3.   Steady flow again for $t \geqslant t_f$

Transient-state pressures are sometimes reduced to the vapor pressure of a liquid that results in separating the liquid column at that section; this is referred to as *liquid-column separation*. If the flow conditions are repeated after a fixed time interval, the flow is called *periodic flow*, and the time interval at which the conditions are repeated is called *period*.

The analysis of transient-state conditions in closed conduits may be classified into two categories: lumped-system approach and distributed-system approach. In the *lumped-system approach*, the conduit walls are assumed rigid, and the liquid in the conduit is assumed incompressible, that is, it behaves like a rigid mass so that the flow velocity at any given instant of time is the same from one end of the conduit to the other. In other words, the flow variables are functions of time only. Therefore, ordinary differential equations describe the system behavior. The flow velocity in each conduit may be considered individually in a multiconduit system. In the *distributed-system approach*, the liquid is assumed to be slightly compressible. Therefore, the flow velocity may vary along the length of the conduit in addition to the variation in time. That is, the flow variables are now functions of not only time but also of distance. Partial differential equations therefore describe the system behavior. If the rate of change of flow velocity is slow, a lumped-system approach yields acceptable results; for rapid changes, however, a distributed-system approach must be used. The distributed-system approach is somewhat more complex than the lumped-system approach. In Sec. 11-2, we apply the lumped-system approach to derive an expression for the time required to establish flow in a conduit, and in Sec. 11-13 we apply this approach for the analysis of water level oscillations in a surge tank. In the remainder of this chapter, we deal with the analyses based on the distributed-system approach.

11-2 Time for Flow Establishment in a Pipe

Let us consider the piping system shown in Fig. 11-3, in which the valve is fully opened from the fully closed position at time $t = 0$. As a result, the pressure force acting on the liquid in the pipe is greater at the reservoir end than that at the valve end. Therefore, liquid in the pipeline begins to accelerate. The flow velocity will keep on increasing until the unbalanced pressure force is equal to the frictional resistance in the pipe. We are interested in determining the time when the flow is fully established, that is, when the flow becomes steady. To do this, we will apply the momentum equation to the control volume shown in Fig. 11-3.

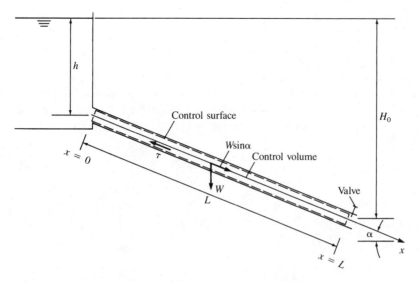

Figure 11-3 Control volume for flow establishment

Assume that the control volume is fixed and does not change shape. Let the x axis be along the pipe axis, and let the positive flow direction be from the reservoir to the valve. The one-dimensional form of momentum equation for a control volume that is fixed in space and does not change shape may be written as (6)

$$\sum F = \frac{d}{dt} \int_{x_1}^{x_2} \rho V A \, dx + (\rho A V^2)_{\text{out}} - (\rho A V^2)_{\text{in}} \tag{11-1}$$

where $\sum F$ is the sum of all forces acting on the system in the x direction, V is the flow velocity in the x direction, ρ is the mass density of the fluid, A is the flow area, and the subscripts in and out refer to the inflow and outflow quantities from the control volume.

If the liquid is assumed incompressible and the pipe is rigid, then at any instant, the velocity along the pipe length from $x = 0$ to $x = L$ will be the same. Since the flow velocity, flow area, and mass density of the liquid are assumed the same along the pipe length $(\rho A V^2)_{\text{in}} = (\rho A V^2)_{\text{out}}$. The term on the left-hand side is the sum of all the forces acting in the x direction on the system within the control volume. Substituting expressions for these forces and for the first term on the right-hand side, and noting that the end sections of the control volume are fixed, we obtain

$$pA + \gamma AL \sin \alpha - \tau_0 \pi DL = \frac{d}{dt} (V \rho AL) \tag{11-2}$$

where $p = \gamma \left(h - \dfrac{V^2}{2g} \right)$ (if the entrance losses are neglected)

$\alpha = $ pipe slope
$D = $ pipe diameter
$L = $ pipe length
$\gamma = $ specific weight of fluid
$\tau_0 = $ shear stress at the pipe wall

Let us replace the frictional force $\tau_0 \pi DL$ by its equivalent, $\gamma h_f A$, and substitute $\rho = \gamma/g$, $H_0 = h + L \sin \alpha$, for h_f from the Darcy-Weisbach friction formula, $h_f = (fL/D)V^2/(2g)$. In this expression, f is the friction factor and g is acceleration due to gravity. Simplifying the resulting equation yields

$$H_0 - \frac{fL}{D} \cdot \frac{V^2}{2g} - \frac{V^2}{2g} = \frac{L}{g} \cdot \frac{dV}{dt} \tag{11-3}$$

When the flow is fully established, $dV/dt = 0$. Hence it follows from Eq. (11-3) that the final flow velocity, V_0, will be such that $H_0 = [1 + (fL/D)]V_0^2/(2g)$. By using this relationship, we can write Eq. (11-3) as

$$dt = \frac{2LD}{D + fL} \cdot \frac{dV}{V_0^2 - V^2}$$

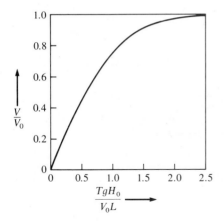

Figure 11-4 Velocity-time relation for flow establishment

Integrating both sides, noting that $V = 0$ at $t = 0$, and designating the time for flow establishment as T, we obtain

$$T = \frac{LV_0}{2gH_0} \ln \left[\frac{1 + \dfrac{V}{V_0}}{1 - \dfrac{V}{V_0}} \right] \tag{11-4}$$

A nondimensional plot of this equation is shown in Fig. 11-4. It is clear from this figure and from Eq. (11-4) that V tends to V_0 asymptotically. That is, flow velocity in the pipe does not become equal to V_0 in a finite interval of time. However, for practical purposes, the flow may be assumed to be fully established when $V = 0.99V_0$. Moreover, it can be seen that the flow velocity increases very rapidly at the beginning, but then the rate of increase decreases as the time progresses. This is because the frictional resistance is small when the flow velocity is small, and therefore, the motive force is high. When the flow velocity increases, the frictional resistance is increased thereby decreasing the motive force and reducing the rate of acceleration of the liquid in the pipe.

11-3 Pressure Change Produced by a Velocity Change

In the previous section, we derived an expression for the time required to establish flow in a pipe assuming the fluid to be incompressible. However, if the flow changes are rapid, the fluid compressibility has to be taken into consideration. As a result, the flow changes are not experienced instantaneously throughout the system; rather pressure waves move back and forth in the piping

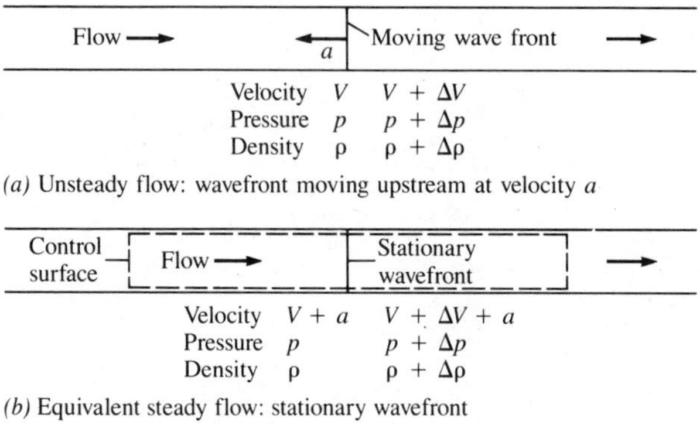

(a) Unsteady flow: wavefront moving upstream at velocity a

(b) Equivalent steady flow: stationary wavefront

Figure 11-5   Definition sketch

system. In this section, we will derive an expression for the change in pressure produced by an instantaneous change in the flow velocity by assuming the pipe walls to be rigid and the liquid to be slightly compressible. The walls are rigid if the pipe diameter does not increase or decrease with a change in the pressure and the liquid is slightly compressible if the mass density of the liquid changes due to a change in pressure, although this change is very small.

Let us consider a pipeline (Fig. 11-5) in which the flow velocity at the downstream end is changed from V to $V + \Delta V$, thereby changing the pressure from p to $p + \Delta p$. This change in pressure will produce a pressure wave that will propagate in the upstream direction (upstream and downstream are with respect to undisturbed initial flow). The pressure on the upstream side of this wave will be p, whereas the pressure on the downstream side of this wave will be $p + \Delta p$. Let us denote the speed of this wave by a, and let us consider the downstream flow direction positive.

We can transform the unsteady-flow situation of Fig. 11-5a to a steady-flow situation by letting the velocity reference system move with the pressure wave. Then, the flow velocities will be as shown in Fig. 11-5b. We will use the momentum equation (Eq. 11-1) with control volume approach to solve for Δp. We let the control surface move with the wave front so that we have steady flow with respect to the moving coordinate system.

Because we have steady flow, the first term on the right-hand side of the momentum equation (Eq. 11-1) is zero. Referring to Fig. 11-5b and introducing the forces and velocities into Eq. (11-1) yield

$$pA - (p + \Delta p)A = (V + a + \Delta V)(\rho + \Delta \rho)(V + a + \Delta V)A$$
$$- (V + a)\rho(V + a)A \quad (11\text{-}5)$$

By simplifying and discarding terms of higher order, this equation becomes

$$-\Delta p = 2\rho V \, \Delta V + 2\rho \, \Delta Va + \Delta \rho(V^2 + 2Va + a^2) \quad (11\text{-}6)$$

The general form of the equation for conservation of mass for one-dimensional flows may be written as

$$0 = \frac{d}{dt} \int_{x_1}^{x_2} \rho A \, dx + (\rho VA)_{out} - (\rho VA)_{in} \qquad (11\text{-}7)$$

As we discussed previously, we have steady flow. Therefore, the first term on the right-hand side of Eq. (11-7) is zero. Referring to Fig. 11-5b and introducing the velocities into Eq. (11-7), we obtain

$$0 = (\rho + \Delta\rho)(V + a + \Delta V)A - \rho(V + a)A \qquad (11\text{-}8)$$

Simplifying this equation, we have

$$\Delta\rho = -\frac{\rho \, \Delta V}{V + a} \qquad (11\text{-}9)$$

In most of the real-life situations, $V \ll a$ for example, V is usually less than 20 m/s, and a is usually in the range of 1000 to 1400 m/s (2, 4, 11). Hence we may approximate $(V + a)$ as a, and Eq. (11-9) may be written as

$$\Delta\rho = -\frac{\rho \, \Delta V}{a} \qquad (11\text{-}10)$$

Now, by substituting Eq. (11-10) into Eq. (11-6), discarding terms of higher order, and simplifying, we obtain

$$\Delta p = -\rho \, \Delta V a \qquad (11\text{-}11)$$

or, since $\Delta p = \gamma \, \Delta H = \rho g \, \Delta H$, we can write Eq. (11-11) as

$$\Delta H = -\frac{a}{g} \Delta V \qquad (11\text{-}12)$$

In other words, the change in pressure head due to an instantaneous change in flow velocity is approximately 100 times the change in the flow velocity. No wonder very high transient pressures occurred that caused the failures shown in Fig. 11-1.

Equation (11-12) gives the pressure head change caused by an instantaneous velocity change at the downstream end of a pipe. By doing a similar derivation, it can be shown that the pressure change caused by an instantaneous change in the flow velocity at the upstream end of a pipeline and the wave moving in the downstream direction is given by the equation

$$\Delta H = \frac{a}{g} \Delta V \qquad (11\text{-}13)$$

In this case, there is no negative sign; in other words, the pressure rises for an increase in the flow velocity, and it reduces by a decrease in the flow velocity.

11-4 Wave Speed

From Eq. (11-10), it follows that

$$\Delta V = -\frac{\Delta \rho}{\rho} a \tag{11-14}$$

Now, the bulk modulus of elasticity of a liquid may be defined (6) as

$$K = \frac{\Delta p}{\dfrac{\Delta \rho}{\rho}} = \frac{\rho g \, \Delta H}{\dfrac{\Delta \rho}{\rho}} \tag{11-15}$$

Hence it follows from Eqs. (11-14) and (11-15) that

$$a = -\frac{K \, \Delta V}{\rho g \, \Delta H} \tag{11-16}$$

By substituting the expression for $\Delta V / \Delta H$ from Eq. (11-12) into Eq. (11-16), and simplifying the resulting equation, we obtain

$$a = \sqrt{\frac{K}{\rho}} \tag{11-17}$$

If we had assumed the conduit walls to be slightly deformable instead of rigid, then as shown in Sec. 11-6, Eq. (11-17) would be modified to

$$a = \sqrt{\frac{\dfrac{K}{\rho}}{1 + \left(\dfrac{KD}{eE}\right)}} \tag{11-18}$$

where D is the inside diameter of the conduit, e is the wall thickness, and E is the modulus of elasticity of the conduit-wall material (2, 11).

EXAMPLE 11-1 Determine the rise in pressure head in a 1-m diameter pipeline if a valve is instantaneously closed at the downstream end. The initial steady-state flow velocity in the pipeline is 1.5 m/s. The pipeline is carrying water. The bulk modulus of elasticity and the mass density for water are 2.19 GPa and 999 kg/m³. Assume the pipe walls are rigid.

SOLUTION   To compute the increase in pressure head, we have to first determine the value for the wave speed, a. This value may be determined from Eq. (11-17):

$$a = \sqrt{\frac{K}{\rho}}$$

$$= \sqrt{\frac{2.19 \times 10^9}{999}}$$

$$= 1480.6 \text{ m/s}$$

Now, we determine the instantaneous change in the flow velocity ΔV:

$$\Delta V = 0 - 1.5$$

$$= -1.5 \text{ m/s}$$

By substituting the values for ΔV, a, and $g = 9.81 \text{ m/s}^2$ into Eq. (11-12), we obtain

$$\Delta H = -\frac{1480.6}{9.81}(-1.5)$$

$$= 226.4 \text{ m}$$

The positive sign indicates that the pressure head increases.   ■

11-5 Pressure Wave Propagation and Reflections

Once a pressure wave is produced in a pipeline, it propagates back and forth in the pipeline until it is dissipated because of friction. This wave is reflected and transmitted at different boundaries. To illustrate the propagation and reflection of pressure waves in a pipeline, let us consider the piping system shown in Fig. 11-6a on the next page. Let the pipe length be L, wave speed be a, and initial piezometric head and flow velocity be H_0 and V_0. Let us assume the system is frictionless, the pipe walls are elastic, and the liquid inside the pipeline is slightly compressible. The transient-state conditions are produced by instantaneously closing a downstream valve at time $t = 0$.

Figure 11-6 shows the sequence of events following the valve closure. These events may be divided into the following four parts:

1. (Fig. 11-6a) Pressure wave propagation toward the reservoir ($0 < t < L/a$): The pressure wave will reach the upstream reservoir in L/a seconds. For any time less than L/a, the wave will be between the valve and the reservoir. On the reservoir side of the wave, flow will be undisturbed, flow velocity will be V_0, the piezometric head will be H_0 (assuming $V_0^2/2g$ is negligible

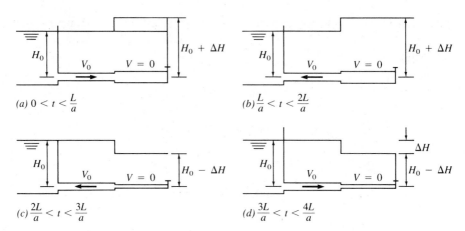

$(a)\ 0 < t < \dfrac{L}{a}$

$(b)\ \dfrac{L}{a} < t < \dfrac{2L}{a}$

$(c)\ \dfrac{2L}{a} < t < \dfrac{3L}{a}$

$(d)\ \dfrac{3L}{a} < t < \dfrac{4L}{a}$

Figure 11-6   Sequence of events following valve closure

compared to H_0), and the pipe diameter will be the same as during the initial steady-state conditions. However, on the valve side of the wave front, that is, behind the wave, the flow velocity will be zero, piezometric head will be $H_0 + \Delta H$, and the pipe diameter will be increased because of inside pressure being higher than the initial steady-state value.

2. (Fig. 11-6b) Pressure wave reflection at the reservoir and propagation toward the valve ($L/a < t < 2L/a$):

    At time $t = L/a$, the pressure wave will reach the upstream reservoir, and the piezometric head throughout the pipe length will be $H_0 + \Delta H$. However, the head on the upstream side of the pipe entrance will be H_0, since the reservoir level is assumed to remain constant. Physically, it is not possible to have the pressure head on one side of a fluid section H_0 and, on the other side, $H_0 + \Delta H$ and be in stable equilibrium. Thus, the fluid will begin to flow toward the reservoir with velocity V_0, and the head will drop to H_0. A pressure wave will therefore now propagate toward the valve. In front of this wave (on the valve side), the flow velocity will be zero, pressure head will be $H_0 + \Delta H$, and the pipe will be expanded. Behind the wave (on the reservoir side), however, the flow velocity will be $-V_0$, that is, toward the reservoir, the pressure head will be H_0, and the pipe diameter will be the same as that during the steady state.

3. (Fig. 11-6c) Pressure wave reflection at the valve and propagation toward the reservoir ($2L/a < t < 3L/a$):

    The wave reflected from the reservoir will reach the valve at $t = 2L/a$. Since the valve is completely closed, it is not possible to maintain a flow velocity of $-V_0$ at the valve. Therefore, the flow velocity instantaneously becomes zero (that is, $\Delta V = V_0$), and the pressure head drops to $H_0 - \Delta H$. This can be seen by substituting $\Delta V = V_0$ into Eq. (11-12). Now, this negative pressure wave propagates toward the reservoir. On the front side of the wave, the head is H_0, the flow velocity is $-V_0$, and the pipe diameter is the same as that during the initial steady-state conditions; behind the wave, however, the pressure head is $H_0 - \Delta H$, flow velocity is zero, and the pipe diameter is reduced.

4. (Fig. 11-6d) Pressure wave reflection at the reservoir and propagation toward the valve ($3L/a < t < 4L/a$):

As the negative wave reaches the upstream reservoir, we have an unstable situation again; that is, the pressure head on the reservoir side of the entrance is H_0, and the pressure head on the valve side is $H_0 - \Delta H$. Therefore, the negative pressure wave is now reflected as a positive pressure wave. On the valve side of this wave, the pressure head is $H_0 - \Delta H$, the flow velocity is zero, and the pipe diameter is reduced. On the reservoir side of the wave, however, the pressure head is H_0, the flow velocity is V_0, and the pipe diameter is the same as that during the initial steady-state conditions. As this wave reaches the valve at $t = 4L/a$, we have the same conditions as at $t = 0$ except that the valve is now closed. Therefore, the above sequence of events starts all over again.

Since, we are assuming a frictionless system, the pressure wave travels back and forth in the pipeline indefinitely with the same flow conditions being repeated every $4L/a$ seconds. The time interval, $4L/a$, after which conditions are repeated, is referred to as the theoretical period of the pipeline (2, 11). Figure 11-7 shows the variation of pressure head at the valve with respect to time. The variation of pressure head at other locations may similarly be plotted (see Prob. 11-2, page 606).

11-6   Governing Equations

To analyze the transient-state conditions in a pipeline, we need the equations describing these flows. In this section, we derive the equations by making the following assumptions: The fluid is slightly compressible, the walls of the conduit are linearly elastic and are slightly deformable, and the head losses during the transient state may be computed by using the steady-state formula.

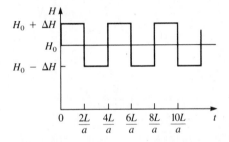

Figure 11-7   Variation of pressure at valve

Continuity Equation

To derive the continuity equation, we will apply the law of conservation of mass to the control volume shown in Fig. 11-8. Let the velocity (with respect to the coordinate axes) of sections 1 and 2, because of the contraction or expansion of the control volume, be W_1 and W_2, respectively. The distance x, flow velocity V, and discharge Q will be considered positive in the downstream direction.

Hence applying Eq. (11-7), page 579, to the control volume shown in Fig. 11-8, and using relative flow velocity at sections 1 and 2 to allow for velocity of these two sections, we obtain

$$\frac{d}{dt}\int_{x_1}^{x_2} \rho A \, dx + \rho_2 A_2(V_2 - W_2) - \rho_1 A_1(V_1 - W_1) = 0 \tag{11-19}$$

Applying Leibnitz's rule* to the first term on the left-hand side of this equation, and noting that $dx_2/dt = W_2$ and $dx_1/dt = W_1$, this equation simplifies to

$$\int_{x_1}^{x_2} \frac{\partial}{\partial t}(\rho A) \, dx + (\rho AV)_2 - (\rho AV)_1 = 0 \tag{11-20}$$

By using the mean value theorem,† dividing throughout by $\Delta x = x_2 - x_1$ and letting Δx approach zero, Eq. (11-20) may be written as

$$\frac{\partial}{\partial t}(\rho A) + \frac{\partial}{\partial x}(\rho AV) = 0 \tag{11-21}$$

Expanding the terms in the parentheses, rearranging various terms, using expressions for total derivatives, and dividing throughout by ρA, we obtain

$$\frac{1}{\rho}\frac{d\rho}{dt} + \frac{1}{A}\frac{dA}{dt} + \frac{\partial V}{\partial x} = 0 \tag{11-22}$$

Let us express the derivatives of ρ and A in terms of commonly used variables p and V as follows.

---

* According to this rule (10),

$$\frac{d}{dt}\int_{f_1(t)}^{f_2(t)} F(x, t) \, dx = \int_{f_1(t)}^{f_2(t)} \frac{\partial}{\partial t} F(x, t) \, dx + F(f_2(t), t)\frac{df_2}{dt} - F(f_1(t), t)\frac{df_1}{dt}$$

provided f_1 and f_2 are differentiable functions of t, and $F(x, t)$ and $\partial F/\partial t$ are continuous in x and t.
† According to this theorem, $\int F(x)\,dx = (x_2 - x_1)F(\xi)$, where $x_1 < \xi < x_2$.

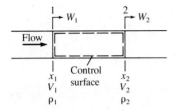

Figure 11-8   Notation for continuity equation

We define the bulk modulus of elasticity, K, of a fluid Eq. (11-15) as

$$K = \frac{dp}{\dfrac{d\rho}{\rho}}$$

We can write this equation as

$$\frac{d\rho}{dt} = \frac{\rho}{K}\frac{dp}{dt} \tag{11-23}$$

Now, area of the conduit, $A = \pi R^2$, where R is the radius of the conduit. Hence $dA/dt = 2\pi R \, dR/dt$. In terms of strain, ϵ, this may be written as (8)

$$\frac{dA}{dt} = 2A\frac{d\epsilon}{dt}$$

or

$$\frac{1}{A}\frac{dA}{dt} = 2\frac{d\epsilon}{dt} \tag{11-24}$$

To simplify the derivation, let us assume the conduit has expansion joints throughout its length so that the axial stress will be zero. Now, the hoop stress (8) in a thin-walled conduit having inside pressure p is given by the expression

$$\sigma_2 = \frac{pD}{2e} \tag{11-25}$$

where e is the thickness of the conduit walls. By taking the time derivative of Eq. (11-25), noting that dD/dt is small and therefore may be neglected, and using the relationship between stress and strain ($\epsilon = \sigma_2/E$), we obtain

$$\frac{d\epsilon}{dt} = \frac{D}{2eE}\frac{dp}{dt} \tag{11-26}$$

It follows from Eqs. (11-24) and (11-26) that

$$\frac{1}{A}\frac{dA}{dt} = \frac{D}{eE}\frac{dp}{dt} \tag{11-27}$$

Substituting Eqs. (11-23) and (11-27) into Eq. (11-22) and simplifying the resulting equation yields

$$\frac{\partial V}{\partial x} + \frac{1}{K}\left[1 + \frac{1}{eE/DK}\right]\frac{dp}{dt} = 0 \tag{11-28}$$

Let us define $a^2 = \dfrac{\dfrac{K}{\rho}}{1 + \dfrac{DK}{eE}}$ (11-29)

As we will see in the next section, a is the wave speed with which pressure waves travel back and forth.

Substituting Eq. (11-29) and the expression for the total derivative into Eq. (11-28) yields

$$\frac{\partial p}{\partial t} + V\frac{\partial p}{\partial x} + \rho a^2 \frac{\partial V}{\partial x} = 0 \tag{11-30}$$

This equation is called the *continuity equation*.

Momentum Equation

For an expanding or contracting control volume (see Fig. 11-9, page 588), Eq. (11-1) is modified to

$$\frac{d}{dt}\int_{x_1}^{x_2} AV\rho\,dx + [\rho A(V-W)V]_2 - [\rho A(V-W)V]_1 = \sum F \tag{11-31}$$

Applying Leibnitz's rule to the first term on the left-hand side of this equation and noting that $dx_1/dt = W_1$ and $dx_2/dt = W_2$, we obtain

$$\int_{x_1}^{x_2} \frac{\partial}{\partial t}(\rho AV)\,dx + (\rho AV)_2 W_2 - (\rho AV)_1 W_1 + [\rho A(V-W)V]_2$$
$$- [\rho A(V-W)]_1 = \sum F \tag{11-32}$$

Applying the mean-value theorem to the first term of this equation and dividing throughout by Δx yield

$$\frac{\partial}{\partial t}(\rho AV) + \frac{(\rho AV^2)_2 - (\rho AV^2)_1}{\Delta x} = \sum \frac{F}{\Delta x} \tag{11-33}$$

The following forces are acting on the system in the control volume (pipe is assumed horizontal):

Pressure force at section 1, $F_{p_1} = p_1 A$ $\qquad$ (11-34a)

where p is pressure intensity, and the subscript, 1, refers to the section.

Pressure force at section 2, $F_{p_2} = p_2 A$ $\qquad$ (11-34b)

If the Darcy-Weisbach friction formula is used for computing the losses due to friction, the shear stress between the fluid and the conduit walls, $\tau_0 = \rho f V|V|/8$, where f is the Darcy-Weisbach friction factor, and V^2 is written as $V|V|$ to automatically account for reverse flows. Therefore,

Shear force, $F_s = \tau_0 \pi D \, \Delta x$

$$= \rho \frac{fV|V|}{8} \pi D \, \Delta x \tag{11-34c}$$

Substituting these expressions for the various forces yields

$$\sum F = p_1 A - p_2 A - \rho \frac{fV|V|}{8} \pi D \, \Delta x \tag{11-35}$$

Substituting Eq. (11-35) into Eq. (11-33) and letting Δx approach zero in the limit yield

$$\frac{\partial}{\partial t}(\rho AV) + \frac{\partial}{\partial x}(\rho AV^2) + A\frac{\partial p}{\partial x} + \rho \frac{fV|V|}{8} \pi D = 0 \tag{11-36}$$

Expanding the terms in parentheses and rearranging the terms of the resulting equation gives

$$V\left[\frac{\partial}{\partial t}(\rho A) + \frac{\partial}{\partial x}(\rho AV)\right] + \rho A\frac{\partial V}{\partial t} + \rho AV\frac{\partial V}{\partial x} + A\frac{\partial p}{\partial x} + \rho Af\frac{V|V|}{2D} = 0 \tag{11-37}$$

According to the continuity equation (Eq. 11-21, page 584), the sum of the two terms in the brackets is zero. Hence dropping these terms and dividing the resulting equation by ρA, we obtain

$$\frac{\partial V}{\partial t} + V\frac{\partial V}{\partial x} + \frac{1}{\rho}\frac{\partial p}{\partial x} + \frac{fV|V|}{2D} = 0 \tag{11-38}$$

This equation is called the *momentum equation*.

Simplified Equations

In most engineering applications, the terms $V\dfrac{\partial p}{\partial x}$ and $V\dfrac{\partial V}{\partial x}$ of the governing equations (Eqs. 11-30 and 11-38) are very small as compared to the other terms. Therefore, these terms may be neglected. In hydraulic engineering applications, the pressure in the pipeline is expressed in terms of the piezometric head, H, above a specified datum, and the discharge, Q, is used as the second variable instead of flow velocity V. Now, the pressure intensity $p = \rho g H$. If we assume the fluid is slightly compressible and the conduit walls are slightly deformable, we may neglect the variation of ρ and flow area A caused by variation of the inside pressure. Hence we can write, $\dfrac{\partial p}{\partial t} = \rho g \dfrac{\partial H}{\partial t}$, and $\dfrac{\partial p}{\partial x} = \rho g \dfrac{\partial H}{\partial x}$. Substituting these relationships and $Q = VA$ into Eqs. (11-30) and (11-38), and neglecting $V\dfrac{\partial V}{\partial x}$ and $\rho g V\dfrac{\partial H}{\partial x}$, we obtain

$$\frac{\partial H}{\partial t} + \frac{a^2}{gA}\frac{\partial Q}{\partial x} = 0 \tag{11-39}$$

$$\frac{\partial Q}{\partial t} + gA\frac{\partial H}{\partial x} + RQ|Q| = 0 \tag{11-40}$$

where $R = f/(2DA)$.

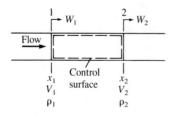

Figure 11-9 Notation for momentum equation

Steady-state equations corresponding to these equations may be obtained by substituting $\partial H/\partial t = 0$ and $\partial Q/\partial t = 0$. Hence it follows from Eq. (11-39) that $dQ/dx = 0$, that is, Q is constant along the pipe length. Similarly, by substituting $\partial Q/\partial t = 0$ into Eq. (11-40), and writing it in finite-difference form, we obtain $\Delta H = f\,\Delta x Q^2/(2gDA^2)$. This is the same as the Darcy-Weisbach friction formula.

11-7 Solution of Momentum and Continuity Equations

The continuity and momentum equations (Eqs. 11-39 and 11-40) are a set of partial differential equations; in other words, the piezometric head and discharge are functions of time t and distance x. The time and distance are referred to as the independent variables, whereas the head and discharge are referred to as the dependent variables. In transient-flow computations, we are interested in determining the variation of H and Q with respect to both x and t.

Assume the piping system is frictionless, that is, $f = 0$. Let us also eliminate Q from the continuity and momentum equations as follows: Multiply Eq. (11-40) by $a^2/(gA)$, substitute $R = 0$, differentiate it with respect to x, and then subtract it from an equation obtained by differentiating Eq. (11-39) with respect to t. Then,

$$\frac{\partial^2 H}{\partial t^2} = a^2 \frac{\partial^2 H}{\partial x^2} \tag{11-41}$$

This is the well-known wave equation describing propagation of pressure waves at wave velocity a. Most of the computational procedures for water hammer analysis — graphical, arithmetical integration, or closed-form solutions in one form or another — before the availability of digital computer, used solutions of this equation. Various innovations were devised to include the head losses in the analysis. With the availability of high-speed digital computers, however, it is now possible to obtain a numerical solution of the governing equations with the nonlinear head-loss term included. We present such a method in the next section. This method, called the method of characteristics, was introduced in 1789 by Monge (5) for a graphical solution of partial differential equations. Gray (3) was the first to use it in 1956 for the analysis of water hammer; however, several of Streeter's publications (11) made it popular for such analyses.

11-8 Method of Characteristics

Multiplying Eq. (11-39) by an unknown multiplier, λ, and adding the resulting equation to Eq. (11-40), we obtain the following equation.

$$\left[\frac{\partial Q}{\partial t} + \frac{\lambda a^2}{gA} \frac{\partial Q}{\partial x} \right] + \lambda \left[\frac{\partial H}{\partial t} + \frac{gA}{\lambda} \frac{\partial H}{\partial x} \right] + RQ|Q| = 0 \tag{11-42}$$

As we discussed above, both Q and H are functions of x and t, that is, $Q = Q(x, t)$, and $H = H(x, t)$. Therefore, the total derivatives may be written as

$$\frac{dH}{dt} = \frac{\partial H}{\partial t} + \frac{\partial H}{\partial x} \frac{dx}{dt} \tag{11-43a}$$

$$\frac{dQ}{dt} = \frac{\partial Q}{\partial t} + \frac{\partial Q}{\partial x} \frac{dx}{dt} \tag{11-43b}$$

Now, let us compare Eqs. (11-43a) and (11-43b) with the expressions in the brackets of Eq. (11-42). If we select the unknown multiplier, λ, so that

$$\frac{\lambda a^2}{gA} = \frac{dx}{dt} = \frac{gA}{\lambda} \tag{11-44}$$

then the right-hand side of Eq. (11-43b) is the same as the first expression in the brackets on the left-hand side of Eq. (11-42), and the right-hand side of Eq. (11-43a) is the same as the second expression in the brackets on the left-hand side of Eq. (11-42). It follows from Eq. (11-44) that

$$\lambda = \pm \frac{gA}{a} \tag{11-45}$$

and $\quad \dfrac{dx}{dt} = \pm a \tag{11-46}$

By specifying λ as given by Eq. (11-45), we may write Eq. (11-42) as follows.

If $\quad \dfrac{dx}{dt} = a \tag{11-47}$

then $\quad \dfrac{dQ}{dt} + \dfrac{gA}{a} \dfrac{dH}{dt} + RQ|Q| = 0 \tag{11-48}$

And, if $\quad \dfrac{dx}{dt} = -a \tag{11-49}$

then $\quad \dfrac{dQ}{dt} - \dfrac{gA}{a} \dfrac{dH}{dt} + RQ|Q| = 0 \tag{11-50}$

We now have two ordinary differential equations, Eqs. (11-48) and (11-50) in H and Q instead of the partial differential equations, since we have eliminated the independent variable, x. However, in obtaining this simplification, we have paid a price. Equations (13-39) and (11-40) were valid everywhere in the x-t plane; this is not the case with Eqs. (11-48) and (11-50). Equation (11-48) is valid

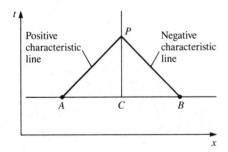

Figure 11-10   Characteristic lines

only if Eq. (11-47) is satisfied, and Eq. (11-50) is valid only if Eq. (11-49) is satisfied. In the x-t plane, Eqs. (11-47) and (11-49) describe two straight lines AP and BP, respectively, as shown in Fig. 11-10. These two lines are referred to as the characteristic lines; line AP is called the positive characteristic, and line BP is called the negative characteristic. Moreover, during the transformation from the partial differential equations to the ordinary differential equations, we have not made any approximation at all; that is, Eqs. (11-48) and (11-50) are as valid as the original governing equations (Eqs. 11-39 and 11-40) except that the former are valid only along the characteristic lines.

We will now discuss how to solve Eqs. (11-48) and (11-50). Let us multiply these equations by dt and integrate along the characteristic lines AP and BP. This procedure yields:

Along the positive characteristic line AP:

$$\int_A^P dQ + \frac{gA}{a} \int_A^P dH + R \int_A^P Q|Q|\, dt = 0 \qquad (11\text{-}51)$$

Along the negative characteristic line BP:

$$\int_B^P dQ - \frac{gA}{a} \int_B^P dH + R \int_B^P Q|Q|\, dt = 0 \qquad (11\text{-}52)$$

The first two integrals on the left-hand sides of these equations can be exactly evaluated. However, this is not the case with the friction-loss term, since we do not a priori know the variation of either Q or H along the characteristic lines. Therefore, we have to make some approximations to evaluate the integral of the friction-loss term. Several procedures have been proposed in the literature for this purpose. The simplest of these is to use the value of Q at point A for the positive characteristic line, and to use the value of Q at point B for the negative characteristic line. Then, Eqs. (11-51) and (11-52) are simplified to

$$Q_P - Q_A + \frac{gA}{a}(H_P - H_A) + RQ_A|Q_A|\,\Delta t = 0 \qquad (11\text{-}53)$$

and $\quad Q_P - Q_B - \dfrac{gA}{a}(H_P - H_B) + RQ_B|Q_B|\Delta t = 0$ $\qquad$ (11-54)

where the subscripts A, B, and P refer to the variables corresponding to points in the x-t plane (Fig. 11-10). These two equations may be written as

$$Q_P = C_p - C_a H_P \qquad (11\text{-}55)$$

and $\quad Q_P = C_n + C_a H_P$ $\qquad$ (11-56)

where $\quad C_p = Q_A + C_a H_A - RQ_A|Q_A|\Delta t$ $\qquad$ (11-57)

$$C_n = Q_B - C_a H_B - RQ_B|Q_B|\Delta t \qquad (11\text{-}58)$$

$$C_a = \dfrac{gA}{a} \qquad (11\text{-}59)$$

Assume that we know the values of H and Q at points A and B and that we want to determine their values at point P (Fig. 11-10). They may be determined from Eqs. (11-55) and (11-56). The following discussion for the analysis of transient-state conditions in a single pipeline (Fig. 11-2) should help you to understand the computational procedure.

The pipeline is divided into a number of reaches. The ends of a reach are called *sections, nodes,* or *grid points.* The nodes at the upstream end and at the downstream end of a pipe are called *boundary nodes,* and the remaining nodes are called the *interior nodes.* To start the calculations, the piezometric head and discharge at $t = t_0$ are determined at the computational nodes. These are called the *initial conditions.* Then, by using Eqs. (11-55) and (11-56), the conditions at the interior nodes at time $t_0 + \Delta t$ are computed. At the boundaries, however, we have only one equation: Eq. (11-55) at the downstream end and Eq. (11-56) at the upstream end. To determine the second unknown from these equations at the boundary nodes, we need another equation. This additional equation is provided by the condition imposed by the boundary. By solving this equation simultaneously with the positive or negative characteristic equations, we develop the boundary conditions, which are then used to determine the transient conditions at the boundaries. To illustrate this procedure, we will develop in the following section boundary conditions for a constant-level upstream reservoir, for a downstream reservoir, for a dead end, for an opening or closing valve, and for a series junction.

11-9 Boundary Conditions

As we mentioned in Sec. 11-8 we may develop the boundary conditions by solving the positive or negative characteristic equations simultaneously with the condition imposed by the boundary. This condition may be in the form of

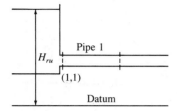

Figure 11-11 Upstream reservoir

specifying head, discharge, or a relationship between the head and discharge. For example, head is constant in the case of a constant-level reservoir, flow is always zero at a dead end, and the flow through an orifice is related to the head loss through the orifice. The following simple examples should clarify the development of the boundary conditions. In these derivations, we will use two subscripts to denote variables at different nodes: The first subscript will denote the number of the pipe and the second subscript will refer to the number of the node on that pipe. If a pipe is divided into n reaches, and the first node is numbered as 1, the last node will be $n + 1$.

Constant-Level Upstream Reservoir

In this case (Fig. 11-11), we are assuming that the water surface in the reservoir or tank remains at the same level independent of the flow conditions in the pipeline. This will be true if the reservoir volume is large. Hence if we refer to the pipe at the upstream end of the pipeline as 1, we may write that

$$H_{P1,1} = H_{ru} \tag{11-60}$$

where H_{ru} is the elevation of the water level in the reservoir above the datum.

Now, at the upstream end, we have the negative characteristic equation. Hence substituting Eq. (11-60) into Eq. (11-56), we obtain

$$Q_{P1,1} = C_n + C_a H_{ru} \tag{11-61}$$

Thus, we determine the head at an upstream reservoir from Eq. (11-60) and the discharge from Eq. (11-61).

Constant-Level Downstream Reservoir

In this case (Fig. 11-12), the head at the last node of pipe i will always be equal to the height of the water level in the tank above the datum, H_{rd}:

$$H_{Pi,n+1} = H_{rd} \tag{11-62}$$

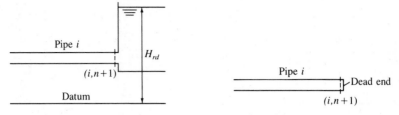

Figure 11-12   Downstream reservoir   **Figure 11-13**   Dead end

At the downstream end, we have the positive characteristic equation linking the boundary node to the rest of the pipeline. Substituting Eq. (11-62) into the positive characteristic equation, Eq. (11-55), we obtain

$$Q_{Pi,n+1} = C_p - C_a H_{rd} \tag{11-63}$$

Dead End

At a dead end located at the end of pipe i, see Fig. 11-13 at right above, the discharge is always zero:

$$Q_{Pi,n+1} = 0 \tag{11-64}$$

At the last node of pipe i, we have the positive characteristic equation. Hence it follows from Eqs. (11-55) and (11-64) that

$$H_{Pi,n+1} = \frac{C_p}{C_a} \tag{11-65}$$

Downstream Valve

In the previous three boundaries, either the head or discharge was specified. However, for a valve, we specify a relationship between the head (see Fig. 11-14) losses through the valve and the discharge. Denoting the steady-state values by subscript 0, the discharge through a valve is given by the following equation:

$$Q_0 = C_d A_{vo} \sqrt{2gH_0} \tag{11-66}$$

where C_d is the coefficient of discharge, A_{vo} is the area of the valve opening, and H_0 is the drop in head for a discharge of Q_0. By assuming that a similar relationship is valid for the transient-state conditions, we may write

$$Q_{Pi,n+1} = (C_d A_v)_P \sqrt{2gH_{Pi,n+1}} \tag{11-67}$$

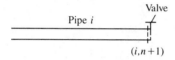

Figure 11-14   Downstream valve

where subscript P denotes values of Q and H at the end of a computational time interval.

Dividing Eq. (11-67) by Eq. (11-66), and squaring both sides, we obtain

$$Q^2_{Pi,n+1} = (Q_0\tau)^2 \frac{H_{Pi,n+1}}{H_0} \tag{11-68}$$

where the effective valve opening is $\tau = (C_dA_v)_P/(C_dA_v)_0$. For the last section on pipe i, we have the positive characteristic equation. Eliminating $H_{Pi,n+1}$ from Eq. (11-68) and the positive characteristic equation (Eq. 11-55), and simplifying the resulting equation, we obtain

$$Q^2_{Pi,n+1} + C_vQ_{Pi,n+1} - C_pC_v = 0 \tag{11-69}$$

where $C_v = (\tau Q_0)^2/(C_aH_0)$. Solving for $Q_{Pi,n+1}$ and neglecting the negative sign with the radical term, we obtain

$$Q_{Pi,n+1} = 0.5(-C_v + \sqrt{C_v^2 + 4C_pC_v}) \tag{11-70}$$

Now, $H_{Pi,n+1}$ may be determined from Eq. (11-55).

To compute the transient-state conditions caused by an opening or closing valve, the variation of τ with respect to time is needed. This relationship may be specified either by describing the variation by an expression or by giving the values of τ at discrete times. At intermediate times, the value of τ may be interpolated from the tabulated values. For example, if the values of τ are stored at an interval of 1 s, the value of τ at $t = 4.3$ s may be interpolated from the specified τ values at 4 and 5 s.

Series Junction

A boundary where two pipes having different diameters, wall materials, or friction factors are connected is referred to as a series junction. There are two nodes at a series junction, as shown in Fig. 11-15. Hence, there are four unknowns—head and discharge for each node—and we need four equations for a unique solution. These equations are as follows.

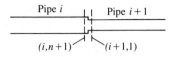

Figure 11-15   Series junction

Positive characteristic equation:

$$Q_{Pi,n+1} = C_p - C_{ai}H_{Pi,n+1}$$ (11-71)

Negative characteristic equation:

$$Q_{Pi+1,1} = C_n + C_{ai+1}H_{Pi+1,1}$$ (11-72)

Continuity equation:

$$Q_{Pi,n+1} = Q_{Pi+1,1}$$ (11-73)

Energy equation:

$$H_{Pi,n+1} = H_{Pi+1,1}$$ (11-74)

In Eq. (11-74) we have neglected the head losses at the junction and the difference in the velocity heads in pipes i and $i + 1$.

Eliminating $Q_{Pi,n+1}$, $Q_{Pi+1,1}$, and $H_{Pi+1,1}$ from Eqs. (11-71–11-74), we obtain

$$H_{Pi,n+1} = \frac{C_p - C_n}{C_{ai} + C_{ai+1}}$$ (11-75)

The remaining variables, $H_{Pi+1,1}$, $Q_{Pi,n+1}$, and $Q_{Pi+1,1}$ may now be determined from Eqs. (11-71–11-74).

EXAMPLE 11-2   Develop the boundary conditions for a centrifugal pump operating at constant speed. Assume the pipe between the upstream reservoir and the suction flange of the pump is short.

SOLUTION   The head-discharge relationship for a centrifugal pump operating at constant speed may be approximated as

$$H_P = H_{sh} - kQ_P^2$$ (11-76)

where H_{sh} is the shutoff head, which is the head developed by the pump when there is no discharge, and k is a constant.

For the node on the discharge flange of the pump, we have the negative characteristic equation (Eq. 11-56). Eliminating H_P from Eqs. (11-56) and (11-76) yields

$$C_a k Q_P{}^2 + Q_P - (C_n + C_a H_{sh}) = 0 \tag{11-77}$$

Solving this equation and neglecting the negative sign with the radical term, we obtain

$$Q_P = \frac{-1 + \sqrt{1 + 4\, C_a k (C_n + C_a H_{sh})}}{2C_a k} \tag{11-78}$$

Now, H_P may be determined from Eq. (11-76).　　　　　■

11-10 Computational Procedure

To compute the transient-state conditions in a pipeline, each pipe is divided into a number of reaches. For short pipes, only one reach may be used. It is necessary that the same computational time interval be used for all pipes so that conditions at the junction may be computed without interpolation or extrapolation. For the computational procedure presented in the previous sections to be stable, the computational time interval and reach length in each pipe must satisfy the following stability condition, commonly referred to as Courant's condition:

$$\Delta x \geqslant a\,\Delta t \tag{11-79}$$

While selecting the computational time interval and the number of reaches in which a pipe is subdivided, it is necessary to satisfy Eq. (11-79). However, better results are obtained if $\Delta x = a\,\Delta t$. For this purpose, the wave speed may be slightly adjusted, if necessary, so that $\Delta x = a\,\Delta t$ for each pipe in the system.

Now, the initial conditions (the piezometric head and discharge) at time t_0, at all the grid points are computed. From these conditions, the head and discharge at time $= t_0 + \Delta t$ at the interior points are computed by using Eqs. (11-55) and (11-56). The boundary conditions are used to compute the flow conditions at the boundaries. Thus, we now know the head and discharge at time $t_0 + \Delta t$. By repeating the above procedure, the head and discharge may be computed for any length of time.

If the computational time interval is short, the computed conditions may be printed after a number of time steps.

Figure 11-16, pages 598–601, lists a FORTRAN computer program for the analysis of transient-state conditions in a pipe generated by opening or closing a downstream valve. The valve is discharging into atmosphere, and there is a constant-level reservoir at the upstream end. An equation is used to specify the variation of τ with time, and the friction losses are computed using the Darcy-Weisbach formula. The input data and the output of the program are given at the end of the program.

Figure 11-17, page 602, shows a plot of the variation of pressure head at the valve with respect to time. To show the difference between the transient pressures produced by instantaneous and gradual valve closure, transient pressures generated by instantaneous valve closure are shown in this figure by a broken line.

Figure 11-16   Computer program for analysis of transients in a pipeline caused by opening or closing a valve

Program Listing

```
C
C     ANALYSIS OF TRANSIENTS IN A PIPELINE CAUSED BY OPENING OR
C     CLOSING OF A DOWNSTREAM VALVE
C
C     *************** NOTATION ********************
C     A = WAVE SPEED;
C     AR = PIPE CROSS-SECTIONAL AREA;
C     D = PIPE DIAMETER;
C     DT = COMPUTATIONAL TIME INTERVAL;
C     F = DARCY-WEISBACH FRICTION FACTOR;
C     H = PIEZOMETRIC HEAD AT BEGINNING OF TIME INTERVAL;
C     HP = PIEZOMETRIC HEAD AT END OF TIME INTERVAL;
C     HRES = RESERVOIR LEVEL ABOVE DATUM;
C     HS = VALVE HEAD LOSS FOR FLOW OF QS;
C     IPRINT = TIME INTERVALS AFTER WHICH CONDITIONS ARE TO BE
C     PRINTED;
C     L = PIPE LENGTH;
C     N = NUMBER OF REACHES INTO WHICH PIPE IS SUB-DIVIDED;
C     Q = DISCHARGE AT THE BEGINNING OF TIME INTERVAL;
C     QO = STEADY-STATE DISCHARGE;
C     QP = DISCHARGE AT END OF TIME INTERVAL;
C     QS = VALVE DISCHARGE;
C     TAU = RELATIVE VALVE OPENING;
C     TAUF = FINAL VALVE OPENING;
C     TAUO = INITIAL VALVE OPENING;
C     TLAST = TIME UPTO WHICH CONDITIONS ARE TO BE COMPUTED;
C     TV = VALVE OPENING OR CLOSING TIME.
C
      REAL L
      DIMENSION H(100),Q(100),HP(100),QP(100)
C
C     READING AND WRITING OF INPUT DATA
C
      READ(5,*) N, IPRINT,QO,HRES,TLAST
      WRITE(6,20) N,IPRINT,QO,HRES,TLAST
20    FORMAT(3X,'N = ',I2,2X,'IPRINT =',I2,2X,'QO =',F6.3,' M3/S',
     1      2X,'HRES =',F7.2, ' M',2X,'TLAST =',F6.1,' S'/)
      READ (5,*) L,D,A,F
      WRITE (6,40) L,D,A,F
40    FORMAT(3X,'L = ',F7.1, ' M',2X,'D =',F5.2,' M',2X,'A =',
     1      F7.1,' M/S'/2X,'F =',F6.3/)
      READ(5,*) TV,TAUO,TAUF,HS,QS
      WRITE(6,60) TV,TAUO,TAUF,HS,QS
60    FORMAT(3X,'TV =',F5.2,2X,'TAUO =',F6.3,2X,'TAUF =',F5.3,2X,
     1      'HS =',F7.2,' M',2X,'QS =',F6.3,' M3/S'/)
C
C     COMPUTATION OF PIPE CONSTANTS
C
      AR = .7854*D*D
      CA=9.81*AR/A
      DT=L/(N*A)
      CF=F*DT/(2.*D*AR)
      F=F*L/(19.62*D*N*AR*AR)
C
```

```
C       STEADY-STATE CONDITIONS
C
        H(1)=HRES
        NN = N+1
        DH=F*QO*QO
        DO 80 I=1,NN
        H(I)=HRES-(I-1)*DH
        Q(I)=QO
80      CONTINUE
        K=0
        TAU=TAUO
        T=0.
100     WRITE(6,110) T,TAU
110     FORMAT(/3X,'T =',F6.2,' S',2X,'TAU =',F5.3)
        WRITE(6,120) (H(I),I=1,NN)
120     FORMAT(5X,'H =',15F8.2)
        WRITE(6,130) (Q(I),I=1,NN)
        K=0
130     FORMAT(5X,'Q =',15F8.3)
150     T=T+DT
        K=K+1
        IF (T.GT.TLAST) STOP
C
C       UPSTREAM RESERVOIR
C
        HP(1)=HRES
        CN=Q(2)-H(2)*CA-CF*Q(2)*ABS(Q(2))
        QP(1)=CN+CA*HRES
C
C       INTERIOR POINTS
C
        DO 160 J=2,N
        CN=Q(J+1)-CA*H(J+1)-CF*Q(J+1)*ABS(Q(J+1))
        CP=Q(J-1)+CA*H(J-1)-CF*Q(J-1)*ABS(Q(J-1))
        QP(J)=0.5*(CP+CN)
        HP(J)=(CP-QP(J))/CA
160     CONTINUE
C
C       DOWNSTREAM VALVE
C
        CP=Q(N)+CA*H(N)-CF*Q(N)*ABS(Q(N))
        IF (T.GE.TV) GO TO 180
        TAU=TAUO+T*(TAUF-TAUO)/TV
        GO TO 190
180     TAU=TAUF
190     IF (TAU.LE.0.0) GO TO 200
        CV=(QS*TAU)**2/(HS*CA)
        QP(NN)=0.5*(-CV+SQRT(CV*CV+4.*CP*CV))
        HP(NN)=(CP-QP(NN))/CA
        GO TO 210
200     QP(NN)=0.
        HP(NN)=CP/CA
C       STORING VARIABLES FOR NEXT TIME STEP

210     DO 230 J=1,NN
        Q(J)=QP(J)
        H(J)=HP(J)
230     CONTINUE
        IF (K.EQ.IPRINT) GO TO 100
        GO TO 150
300     STOP
        END
```

Input Data

```
8,4,1.,200.99,16.
1000.,1.,1000.,0.012
4.0,1.,0.,200.,1.
```

Program Output

```
N =  8  IPRINT = 4  QO = 1.000 M3/S  HRES = 200.99 M  TLAST = 16.0 S

L =  1000.0 M  D = 1.00 M  A = 1000.0 M/S
F =  .012
```

(continued)

Figure 11-16
Program Output (continued)

```
TV = 4.00   TAUO = 1.000   TAUF = .000   HS = 200.00 M   QS = 1.000 M3/S


T =    .00 S   TAU =1.000
  H = 200.99 200.87 200.74 200.62 200.49 200.37 200.25 200.12 200.00
  Q =  1.000  1.000  1.000  1.000  1.000  1.000  1.000  1.000  1.000

T =   .50 S   TAU = .875
  H = 200.99 200.87 200.74 200.62 200.49 203.45 206.47 209.55 212.69
  Q =  1.000  1.000  1.000  1.000  1.000   .976   .952   .927   .902

T =  1.00 S   TAU = .750
  H = 200.99 203.93 206.94 210.01 213.14 216.33 219.59 222.92 226.31
  Q =  1.000   .976   .952   .928   .903   .877   .851   .825   .798

T =  1.50 S   TAU = .625
  H = 200.99 207.42 213.84 220.28 226.72 230.17 233.68 237.27 240.93
  Q =   .806   .806   .804   .802   .798   .771   .743   .715   .686

T =  2.00 S   TAU = .500
  H = 200.99 207.92 214.85 221.79 228.73 235.68 242.64 249.61 256.59
  Q =   .598   .598   .596   .594   .590   .586   .580   .574   .566

T =  2.50 S   TAU = .375
  H = 200.99 208.45 215.92 223.38 230.86 236.01 241.05 245.95 250.72
  Q =   .376   .375   .373   .371   .367   .380   .393   .407   .420

T =  3.00 S   TAU = .250
  H = 200.99 206.69 212.26 217.69 222.98 228.14 233.14 238.00 242.71
  Q =   .137   .154   .171   .188   .206   .223   .240   .258   .275

T =  3.50 S   TAU = .125
  H = 200.99 203.98 206.96 209.92 212.85 217.94 222.87 227.62 232.19
  Q =   .036   .037   .038   .041   .045   .067   .089   .112   .135

T =  4.00 S   TAU = .000
  H = 200.99 203.32 205.64 207.93 210.20 212.42 214.59 216.69 218.72
  Q =  -.046  -.045  -.043  -.040  -.035  -.028  -.020  -.011   .000

T =  4.50 S   TAU = .000
  H = 200.99 202.49 203.98 205.44 206.86 206.35 205.98 205.76 205.69
  Q =  -.106  -.105  -.102  -.098  -.091  -.069  -.046  -.023   .000

T =  5.00 S   TAU = .000
  H = 200.99 199.57 198.35 197.32 196.48 195.83 195.37 195.09 195.00
  Q =  -.137  -.121  -.105  -.088  -.071  -.053  -.036  -.018   .000

T =  5.50 S   TAU = .000
  H = 200.99 198.00 195.03 192.07 189.14 188.32 187.74 187.39 187.28
  Q =  -.036  -.037  -.038  -.041  -.045  -.034  -.023  -.012   .000

T =  6.00 S   TAU = .000
  H = 200.99 198.66 196.35 194.05 191.79 189.57 187.40 185.30 183.27
  Q =   .046   .045   .043   .040   .035   .028   .020   .011   .000

T =  6.50 S   TAU = .000
  H = 200.99 199.49 198.00 196.54 195.12 195.64 196.00 196.22 196.30
  Q =   .106   .105   .102   .098   .091   .069   .046   .023   .000

T =  7.00 S   TAU = .000
  H = 200.99 202.41 203.63 204.66 205.50 206.14 206.61 206.88 206.98
  Q =   .136   .121   .105   .088   .071   .053   .036   .018   .000

T =  7.50 S   TAU = .000
  H = 200.99 203.97 206.95 209.91 212.84 213.65 214.23 214.58 214.69
  Q =   .036   .037   .038   .041   .045   .034   .023   .012   .000

T =  8.00 S   TAU = .000
  H = 200.99 203.32 205.63 207.92 210.19 212.41 214.57 216.67 218.70
  Q =  -.046  -.045  -.043  -.040  -.035  -.028  -.020  -.011   .000

T =  8.50 S   TAU = .000
  H = 200.99 202.49 203.97 205.43 206.85 206.34 205.97 205.75 205.68
  Q =  -.106  -.105  -.102  -.098  -.091  -.069  -.046  -.023   .000
```

```
T =  9.00 S  TAU = .000
H = 200.99 199.57 198.35 197.32 196.49 195.84 195.38 195.10 195.01
Q =   -.136  -.121  -.105  -.088  -.071  -.053  -.036  -.018   .000

T =  9.50 S  TAU = .000
H = 200.99 198.01 195.04 192.08 189.15 188.34 187.76 187.41 187.30
Q =   -.036  -.037  -.038  -.041  -.045  -.034  -.023  -.012   .000

T = 10.00 S  TAU = .000
H = 200.99 198.67 196.35 194.06 191.80 189.58 187.42 185.32 183.30
Q =    .046   .045   .043   .040   .035   .028   .020   .011   .000

T = 10.50 S  TAU = .000
H = 200.99 199.49 198.01 196.55 195.13 195.65 196.01 196.23 196.30
Q =    .105   .105   .102   .097   .091   .069   .046   .023   .000

T = 11.00 S  TAU = .000
H = 200.99 202.41 203.63 204.65 205.49 206.14 206.60 206.88 206.97
Q =    .136   .121   .105   .088   .071   .053   .036   .018   .000

T = 11.50 S  TAU = .000
H = 200.99 203.97 206.94 209.89 212.82 213.63 214.21 214.56 214.67
Q =    .036   .037   .038   .041   .045   .034   .023   .012   .000

T = 12.00 S  TAU = .000
H = 200.99 203.31 205.62 207.92 210.17 212.39 214.55 216.65 218.67
Q =   -.046  -.045  -.043  -.040  -.035  -.028  -.020  -.011   .000

T = 12.50 S  TAU = .000
H = 200.99 202.49 203.97 205.42 206.84 206.33 205.97 205.75 205.67
Q =   -.105  -.105  -.102  -.097  -.091  -.069  -.046  -.023   .000

T = 13.00 S  TAU = .000
H = 200.99 199.58 198.36 197.33 196.49 195.85 195.38 195.11 195.02
Q =   -.136  -.121  -.104  -.088  -.071  -.053  -.036  -.018   .000

T = 13.50 S  TAU = .000
H = 200.99 198.01 195.04 192.09 189.17 188.36 187.78 187.43 187.32
Q =   -.036  -.037  -.038  -.041  -.045  -.034  -.023  -.012   .000

T = 14.00 S  TAU = .000
H = 200.99 198.67 196.36 194.07 191.81 189.60 187.44 185.34 183.32
Q =    .046   .045   .043   .040   .035   .028   .020   .011   .000

T = 14.50 S  TAU = .000
H = 200.99 199.50 198.01 196.56 195.14 195.65 196.02 196.24 196.31
Q =    .105   .104   .102   .097   .091   .069   .046   .023   .000

T = 15.00 S  TAU = .000
H = 200.99 202.40 203.62 204.65 205.48 206.13 206.59 206.87 206.96
Q =    .136   .121   .104   .088   .071   .053   .036   .018   .000

T = 15.50 S  TAU = .000
H = 200.99 203.97 206.93 209.88 212.80 213.62 214.19 214.54 214.66
Q =    .036   .037   .038   .041   .045   .034   .023   .012   .000

T = 16.00 S  TAU = .000
H = 200.99 203.31 205.62 207.91 210.16 212.38 214.53 216.63 218.64
Q =   -.046  -.045  -.043  -.040  -.035  -.028  -.020  -.011   .000
```

11-11 Transients Caused by Pumps

Power failure to the pump motors usually produces the most critical conditions in a pipeline. After a power failure, pump discharge and speed reduce rapidly. Usually, flow reduces to zero and then reverses although the pump may still be rotating in the positive direction. This reverse flow causes rapid deceleration of the pump, and the pump speed reverses as well if no protective devices, such as a ratchet, are installed to prevent the reverse rotation. The

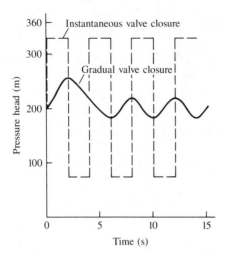

Figure 11-17   Pressure head at valve following
valve closure

pump speed keeps on increasing in the reverse direction until it reaches the
runaway speed. Because of increasing reverse speed, the reverse flow through
the pump is reduced, thereby increasing the pressure at the pump end.

Figure 11-18 shows the hydraulic grade line at different times during
the transient-state conditions produced by power failure. When the hydraulic
grade line falls below the pipeline, the liquid column may separate if the pressures
fall to the vapor pressure of the liquid. The subatmospheric pressure may collapse the pipe, or the high pressures generated by rejoining of the separated
liquid columns may burst the pipe. Therefore, if the analysis shows that the
transient-state hydraulic grade line falls below the centerline of the pipeline,

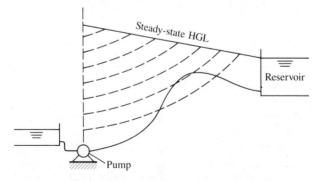

Figure 11-18   Transient-state hydraulic grade line
following power failure

either the pipe should be designed to withstand both the maximum and minimum pressures or control devices should be provided to prevent liquid-column separation.

11-12 Control Devices

Control devices are installed in a pipeline to keep the transient-state conditions within prescribed limits. These conditions may be maximum and minimum pressures, pump and turbine overspeed, water-level oscillations in a surge tank, and so forth. The main function of these devices is to reduce the rates of acceleration and deceleration of the liquid column in the pipeline.

Some common control devices are

Surge tanks
Air chambers
Valves
Flywheels

A surge tank is a standpipe connected to the pipeline. This tank stores excess liquid and provides it when the hydraulic grade line in the pipeline falls below the liquid level in the tank. Several types of surge tanks — simple, orifice, differential, closed, and one-way — are shown in Fig. 11-19.

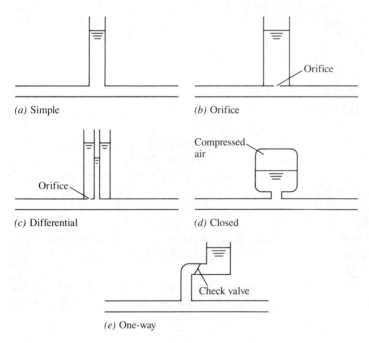

(a) Simple

(b) Orifice

(c) Differential

(d) Closed

(e) One-way

Figure 11-19 Types of surge tanks

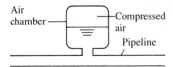

Figure 11-20   Air chamber

An air chamber has compressed air at the top (Fig. 11-20). The air acts as a cushion. When the pressure inside the pipeline falls, liquid flows out of the air chamber and the air expands. When the pressure rises, the liquid flows back into the chamber and the air is compressed. Thus, the air reduces the rates of acceleration and deceleration of the liquid column in the pipeline.

Valves are used to provide by-pass facilities so that the flow can be diverted to prevent sudden flow changes. The valve opening and closing rates are set by the pressure in the pipeline, or they are prespecified. Several different types of valves are available for transient control, such as pressure relief valves, pressure regulating valves, and safety valves. To prevent the pressure from falling too much below the atmospheric pressure, air valves are installed. These valves admit air into the pipeline when the pressure falls below the atmospheric pressure.

By increasing the inertia of the pump motor or turbogenerator set or by installing a flywheel, the transients may be kept within prescribed limits. Usually, an increase in the inertia of turbo-machine is used along with some other control methods.

11-13 Surge Tank Water-Level Oscillations

Figure 11-21 shows the schematic diagram of a typical hydroelectric power plant with an upstream surge tank. The oscillations of the water level in the tank following a flow change due to load changes on the turbo-generator set are usually very slow. Therefore, these oscillations may be analyzed using a lumped-system approach. In this section, we derive the governing equations for this system and list a number of methods for their solution.

By assuming the tunnel walls are rigid and the water is incompressible, and by neglecting the inertia of the water in the tank, the dynamic equation may be derived as follows.

Referring to the freebody diagram of a horizontal tunnel having constant cross-sectional area (Fig. 11-21), the following forces are acting on the water in the conduit:

$$F_1 = \gamma A_t (H_r - h_i - h_v) \tag{11-80}$$

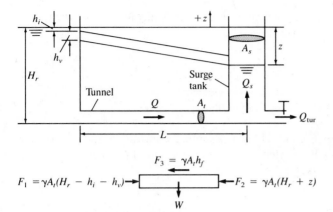

Freebody diagram

Figure 11-21   Surge-tank system

$$F_2 = \gamma A_t(H_r + z) \tag{11-81}$$

$$F_3 = \gamma A_t h_f \tag{11-82}$$

where A_t = cross-sectional area of the tunnel
    H_r = static head
    γ = specific weight of water
    h_v = velocity head
    h_i = intake losses
    h_f = friction and form losses in the tunnel
    z = water level in the tank above the reservoir level

The resultant force acting on the water in the tunnel is

$$F_r = F_1 - F_2 - F_3$$
$$= \gamma A_t(-z - h_v - h_i - h_f) \tag{11-83}$$

Now, rate of change of momentum of the water in the tunnel

$$= \frac{\gamma A_t L}{g} \frac{d}{dt}\left(\frac{Q}{A_t}\right) \tag{11-84}$$

where L is the length of the tunnel, and Q is the flow in the tunnel. Hence applying Newton's second law and simplifying the resulting equation, we can write

$$\frac{dQ}{dt} = \frac{gA_t}{L}(-z - h_v - h_i - h_f) \tag{11-85}$$

Let $h = h_v + h_i + h_f = cQ|Q|$, where c is a coefficient, and Q^2 is written as $Q|Q|$ to account for the reverse flow. Then,

$$\frac{dQ}{dt} = \frac{gA_t}{L}(-z - cQ|Q|) \tag{11-86}$$

This equation is called the *dynamic equation*.

Referring again to Fig. 11-21 and assuming constant mass density, the following equation may be written for the conservation of mass at the junction of the tunnel and the tank:

$$Q = Q_s + Q_{tur} \tag{11-87}$$

where Q_s is the flow into the tank, considered positive into the tank, and Q_{tur} is the turbine flow. Substituting $Q_s = A_s dz/dt$ into this equation and rearranging, we obtain

$$\frac{dz}{dt} = \frac{Q - Q_{tur}}{A_s} \tag{11-88}$$

where A_s = area of the surge tank.

This equation is called the *continuity equation*.

Equations (11-86) and (11-88) describe the water-level oscillations in the surge tank system shown in Fig. 11-21. Because of the presence of nonlinear terms (the expression for Q_{tur} may also be nonlinear), a closed form solution is not usually available. Therefore, the numerical methods we discussed in Chapter 10 may be used for the numerical integration of these equations.

PROBLEMS

11-1   Compute the pressure rise caused by instantaneous closure of a valve located at the downstream end of a 2-m diameter, 1962-m long pipeline conveying 10 m³/s of water. The walls of the pipe are 20 mm thick and made of steel.

11-2   Plot the variation of pressure at midlength of the pipeline of Prob. 11-1, assuming the system is frictionless and a constant-level upstream reservoir has water surface 100 m above the centerline of the pipeline.

11-3   A pipeline connects two constant-level reservoirs with a pump located at midlength of the pipeline. If the pump instantaneously stops pumping, plot the pressure variation at points A, B, C, and D. Assume there are no losses in the pipeline and the wave speed is 1000 m/s.

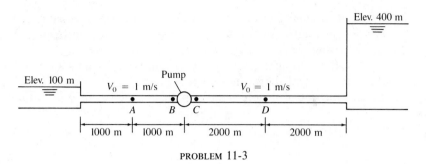

PROBLEM 11-3

11-4 The wave speed in a homogeneous, gas-liquid mixture may be approximated as

$$a = \sqrt{\frac{p^*}{\rho_l \alpha (1 - \alpha)}}$$

where p^* is absolute pressure, ρ_l is mass density of the liquid, void fraction $\alpha = \forall_g / \forall_m$, $\forall_g$ is volume of gas, and $\forall_m$ is volume of gas-liquid mixture. For standard temperature and pressure, compute the wave speed in an air-water mixture for different values of α, and plot a graph between a and α.

11-5 The inertia of the pump motor in this pipeline is very small so that the pump stops instantly upon power failure at time $t = 0$. A mechanical ratchet does not allow reverse pump rotation. The initial flow velocity in the pipeline is 1 m/s, and the wave speed is 1100 m/s. Determine the amount by which the pipeline would have to be lowered so that the hydraulic grade line at the summit does not fall below the centerline of the pipeline during the transient-state conditions. Assume the system is frictionless.

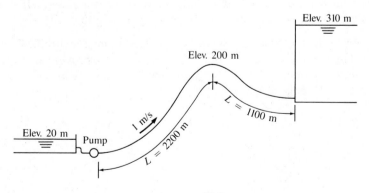

PROBLEM 11-5

11-6   Write a computer program to compute the transient-state conditions in a pipeline having a constant-level reservoir at the upstream end and a slowly closing valve at the downstream end as shown.

Use this computer program to compute the transient-state conditions in the piping system shown in the figure caused by closing the valve in 10 s. Assume the rate of closure is uniform, that is, the τ-t curve is a straight line.

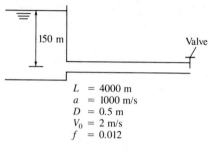

150 m

Valve

$L = 4000$ m
$a = 1000$ m/s
$D = 0.5$ m
$V_0 = 2$ m/s
$f = 0.012$

PROBLEM 11-6

11-7   Compute the time of flow establishment by instantaneously opening the downstream valve of the piping system of the figure in Prob. 11-6 by using the computer program of Prob. 11-6. Compare the results with those obtained from Eq. (11-4, page 577).

11-8   Develop the boundary conditions for an orifice located at the downstream end of a pipe.

11-9   Develop the boundary conditions for an opening or closing valve located at the junction of two pipes having the same diameter, wall material, and wall thickness.

11-10   Derive an expression for the pressure change in a pipeline caused by instantaneously closing a valve located at the downstream end of a frictionless pipe inclined downward at an angle with the horizontal. How would this expression be modified if the pipe were sloping upward?

11-11   By neglecting friction losses ($c = 0$) and assuming the turbine flow, Q_{tur}, is instantaneously changed from Q_0 to zero at $t = 0$, prove that the period, T, and the amplitude, Z, of the water-level oscillations in the surge tank of Fig. 11-21, page 605, are

$$T = 2\pi \sqrt{\frac{LA_s}{gA_t}}$$

$$Z = Q_0 \sqrt{\frac{L}{gA_sA_t}}$$

[*Hint*: Substitute $c = 0$ in Eq. (11-86) and eliminate Q from this equation and Eq. (11-88). Then, solve the resulting second-order differential equation in z subject to the following initial conditions at $t = 0$: $z = 0$, and $dz/dt = Q_0/A_s$. The second initial condition follows from the continuity equation at $t = 0$.]

REFERENCES

1. Bonin, C.C. "Water-Hammer Damage to Oigawa Power Station." *Jour. Engineering for Power*, Amer. Soc. Mech. Engineers (April 1960), pp. 111–19.
2. Chaudhry, M.H. *Applied Hydraulic Transients*, 2d ed. Van Nostrand Reinhold, New York, 1987.
3. Gray, C.A.M. "The Analysis of the Dissipation of Energy in Water Hammer." *Proc. Amer. Soc. of Civil Engineers*, 119 (1953), pp. 1176–94.
4. Jaeger, C. *Fluid Transients in Hydroelectric Practice*. Blackie & Sons, Glasgow, 1977.
5. Monge, G. "Graphical Integration." *Ann. des Ing. Sortis des Ecoles de Gand* (1789).
6. Roberson, J.A., and C.T. Crowe, *Engineering Fluid Mechanics*, 3d ed. Houghton Mifflin, Boston, 1985.
7. Serkiz, A.W. "Evaluation of Water Hammer Experience in Nuclear Power Plants." Report NUREG-0927, U.S. Nuclear Regulatory Commission, Washington, D.C. (May 1983).
8. Timoshenko, S. *Strength of Materials*, 2d ed., part 2. Van Nostrand Reinhold, New York, 1941.
9. Trenkle, C.J. "Failure of Riveted and Forge-Welded Penstock." *Jour. of Energy Division*, Amer. Soc. of Civil Engineers, 105 (January 1979), pp. 93–102.
10. Wylie, C.R. *Advanced Engineering Mathematics*, 3d ed. McGraw-Hill, New York, 1966.
11. Wylie, E.B., and V.L. Streeter. *Fluid Transients*. FEB Press, Ann Arbor, Mich., 1983.

12

Surge waves in Amenagement d'Oraison (Courtesy
Eletricitie de France. Photo by: Louis Chagnon)

Unsteady
Free-Surface
Flows

We discussed steady open-channel flows in Chapter 4 and unsteady closed-conduit flows in Chapter 11. In this chapter, we discuss unsteady free-surface flows. These flows occur in natural and manmade channels and are produced by changes in the water levels or changes in the inflow or outflow rates. Typical examples of these flows are floods in streams and rivers, surges in power canals, tidal flows in estuaries, unsteady flows in irrigation canals and channels, and storm runoff in sewers.

In this chapter, commonly used terms are first defined. Causes which produce these flows are discussed. The governing equations are derived, and numerical methods for their solution are presented.

12-1   Definitions

If the flow velocity at a point varies with time, the flow is called *unsteady flow*. Depending on the rate of variation of depth, unsteady flows may be classified as *gradually varied* (flood waves) or *rapidly varied* (surges). In the case of rapidly varied flows, steep fronts or discontinuities may occur in the flow depth. These discontinuities are referred to as shocks or bores.

A *wave* is a temporal or spatial variation of flow depth or rate of discharge. The wave velocity with respect to the flow it is traveling in is referred to as the *celerity*, c. In one-dimensional flows, the absolute wave velocity, V_w, is given by the equation

$$V_w = V \pm c \tag{12-1}$$

where V = flow velocity.

Considering the direction of downstream flow positive, the plus sign is used if the wave is traveling in the downstream direction, and the negative sign is used if the wave is traveling in the upstream direction.

The *wave length* is the distance between two consecutive crests. If the wave length is more than 20 times the undisturbed flow depth, the wave is called *shallow-water wave*; if the wave length is less than twice the flow depth, the wave is called *deep-water wave*. The ratio of the wave length and the flow depth defines whether the wave is shallow water or deep water. That is, a ripple in shallow water can be a deep-water wave, and a tide in the ocean can be a shallow-water wave. This distinction between shallow- and deep-water waves is important because each type has certain characteristics that dictate selection of the governing equations. In this chapter, we discuss only shallow-water waves.

Waves may be classified as positive or negative. A *positive wave* occurs when the flow depth is greater than the undisturbed flow depth, and a *negative wave* occurs when the flow depth is smaller than the undisturbed flow depth.

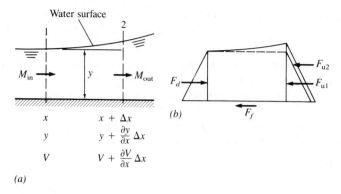

Figure 12-1   Definition sketches

12-2 Governing Equations

The continuity and dynamic equations describe the unsteady free-surface flows. These equations may be derived by making the following assumptions (3, 4, 7, 8):

1. The pressure distribution is hydrostatic. This is usually true if the flow surface does not have a sharp curvature.
2. The channel bottom slope is small so that the flow depth measured vertically is almost the same as the flow depth normal to the channel bottom and $\sin \theta \approx \tan \theta \approx \theta$, where θ is the angle between the channel bottom and horizontal datum.
3. The velocity distribution at a channel cross section is uniform.
4. The channel is prismatic, that is, the bottom slope and cross section remain unchanged with distance.
5. The friction losses in unsteady flow may be computed using the empirical formulas for steady-state flows.

Let us consider a segment of water between two cross sections located at distance x and $x + \Delta x$ (Fig. 12-1). Consider the downstream flow direction positive, and measure the flow depth vertically. Let the flow depth and the flow velocity at distance x be y and V, respectively. Then their values at distance $x + \Delta x$ will be $y + (\partial y/\partial x)\Delta x$ and $V + (\partial V/\partial x)\Delta x$.

Continuity Equation

To derive the continuity equation, we apply the law of conservation of mass to the segment of water between sections 1 and 2, as shown in Fig. 12-1a. Referring to this figure, the mass inflow into the segment of water during time Δt, is expressed by:

$$M_{\text{in}} = \frac{\gamma}{g} AV \Delta t \tag{12-2a}$$

where γ = specific weight of water. The mass outflow during time Δt, is:

$$M_{\text{out}} = \frac{\gamma}{g}\left(A + \frac{\partial A}{\partial x}\Delta x\right)\left(V + \frac{\partial V}{\partial x}\Delta x\right)\Delta t \tag{12-2b}$$

Hence, the net mass inflow into the segment of water

$$= M_{\text{in}} - M_{\text{out}}$$

$$= -\frac{\gamma}{g}\left(V\frac{\partial A}{\partial x} + A\frac{\partial V}{\partial x}\right)\Delta x\,\Delta t \tag{12-3}$$

where higher-order terms have been neglected.

Now, we can also write an expression for the increase in the mass of segment during time interval Δt as

$$= \frac{\gamma}{g}\frac{\partial A}{\partial t}\Delta x\,\Delta t \tag{12-4}$$

Equating the net mass inflow into the segment of water (given by Eq. 12-3) to the increase in its mass during time interval Δt (given by Eq. 12-4) and dividing throughout by $(\gamma/g)\Delta x\,\Delta t$, we obtain

$$\frac{\partial A}{\partial t} + V\frac{\partial A}{\partial x} + A\frac{\partial V}{\partial x} = 0 \tag{12-5}$$

By combining the second and the third terms of Eq. (12-5) we obtain

$$\frac{\partial A}{\partial t} + \frac{\partial(AV)}{\partial x} = 0 \tag{12-6}$$

This equation is referred to as the *continuity equation in the conservation form.*

We can express the variation of A with respect to x and t as

$$\frac{\partial A}{\partial x} = \frac{dA}{dy}\frac{\partial y}{\partial x} = B\frac{\partial y}{\partial x} \tag{12-7}$$

and $\quad\dfrac{\partial A}{\partial t} = \dfrac{dA}{dy}\dfrac{\partial y}{\partial t} = B\dfrac{\partial y}{\partial t} \tag{12-8}$

where B = top water surface width.

Substituting Eqs. (12-7) and (12-8) into Eq. (12-5) and simplifying, we obtain

$$\frac{\partial y}{\partial t} + D \frac{\partial V}{\partial x} + V \frac{\partial y}{\partial x} = 0 \tag{12-9}$$

where D = the hydraulic depth defined as $D = A/B$.

Dynamic Equation

To derive the dynamic equation, we apply the law of conservation of momentum to the segment of water between sections 1 and 2. To do this, we equate the rate of change of momentum of the segment of water to the resultant of the external forces acting on the segment.

Four forces act on the segment of water, as shown in Fig. 12-1b. The pressure force acting on the downstream face has been divided into two parts. Expressions for these forces are as follows:

Pressure force acting on the upstream face is

$$F_d = \gamma A \bar{z} \tag{12-10}$$

where $\bar{z}$ = the depth of the centroid below the water surface.

Similarly, $F_{u1} = \gamma A \bar{z}$. Neglecting the higher-order terms (these will correspond to the small shaded triangle shown in Fig. 12-1b), the second part of the pressure force acting on the downstream face is

$$F_{u2} = \gamma A \frac{\partial y}{\partial x} \Delta x \tag{12-11}$$

If S_f is the slope of the energy grade line, then the force due to friction, F_f, may be written as

$$F_f = \gamma A S_f \Delta x \tag{12-12}$$

This force will be acting in the upstream direction. S_f may be computed by using the Manning or Chezy formula. Now,

$$\text{Weight of segment of water} = \gamma A \, \Delta x \tag{12-13}$$

Since the slope of the channel bottom is assumed to be small, $S_0 = \sin \theta$. Therefore, the component of the weight of water, F_w, acting in the downstream direction is

$$F_w = \gamma A \, \Delta x S_0 \tag{12-14}$$

Hence, the resultant force, F_r, acting on the water segment is

$$F_r = F_d - F_{u1} - F_{u2} - F_f + F_w \tag{12-15}$$

Substituting the expressions for different forces into Eq. (12-15), we obtain

$$F_r = \gamma A \left(-\frac{\partial y}{\partial x} - S_f + S_0 \right) \Delta x \tag{12-16}$$

Now, the rate of momentum inflow into the segment of water is

$$M_i = \frac{\gamma}{g} A V^2 \tag{12-17}$$

Rate of momentum outflow is

$$M_o = \frac{\gamma}{g} \left[A V^2 + \frac{\partial}{\partial x} (A V^2) \Delta x \right] \tag{12-18}$$

Hence, the rate of net momentum influx

$$= M_i - M_o$$

$$= -\frac{\gamma}{g} \frac{\partial}{\partial x} (A V^2) \Delta x \tag{12-19}$$

And time rate of increase of momentum of the segment of water may be written as

$$= \frac{\partial}{\partial t} \left[\frac{\gamma}{g} A V \Delta x \right] \tag{12-20}$$

Now, the time rate of increase of momentum

= Rate of net influx of momentum + Resultant force
  acting on the control volume

Substituting into this equation expressions for various terms from Eqs. (12-16), (12-19), and (12-20), dividing throughout by $(\gamma/g) \Delta x$, and simplifying, we obtain

$$\frac{\partial}{\partial t} (A V) + \frac{\partial}{\partial x} (A V^2) + g A \frac{\partial y}{\partial x} = g A (S_0 - S_f) \tag{12-21}$$

By expanding the terms on the left-hand side of Eq. (12-21), dividing by A, and rearranging the resulting equation, we obtain the following.

$$g\frac{\partial y}{\partial x} + V\frac{\partial V}{\partial x} + \frac{\partial V}{\partial t} + \frac{V}{A}\left[\frac{\partial A}{\partial t} + V\frac{\partial A}{\partial x} + A\frac{\partial V}{\partial x}\right] = g(S_0 - S_f) \quad (12\text{-}22)$$

According to Eq. (12-5), the sum of the terms within the brackets is zero. Hence Eq. (12-22) may be written as

$$g\frac{\partial y}{\partial x} + \frac{\partial V}{\partial t} + V\frac{\partial V}{\partial x} = g(S_0 - S_f) \quad (12\text{-}23)$$

This is referred to as the *dynamic equation*. By rearranging the terms of this equation, we can write the equation to indicate the significance of each term for a particular type of flow (7):

$$S_f = S_0 - \frac{\partial y}{\partial x} - \frac{V}{g}\frac{\partial V}{\partial x} - \frac{1}{g}\frac{\partial V}{\partial t} \quad (12\text{-}24)$$

Steady, uniform $\longrightarrow$

Steady, nonuniform $\longrightarrow$

Unsteady, nonuniform $\longrightarrow$

12-3 Methods of Solution

Equations (12-9) and (12-23) are a set of hyperbolic, partial differential equations. Because of the presence of nonlinear terms, a closed-form solution of these equations is not available except for very simplified cases. Therefore, numerical methods are used for their solution (1, 2, 6, 10). There are several different types of these methods, notably, *method of characteristics*, *finite-difference methods*, *finite-element method*, and *spectral method*. The method of characteristics and various finite-difference methods have been used more extensively than the other methods for the analysis of unsteady free-surface flows; therefore, we discuss only these methods.

12-4 Method of Characteristics

This method, popular in the 1960s for the analysis of unsteady flows in open channels, has been replaced by various finite-difference schemes. It is, however, still being used in the explicit finite-difference schemes to simulate the boundary nodes (3). Details of this method follow.

Multiplying Eq. (12-9) by an unknown multiplier, λ, and adding to Eq. (12-23), we obtain

$$\left[\frac{\partial V}{\partial t} + (V + \lambda D)\frac{\partial V}{\partial x}\right] + \lambda\left[\frac{\partial y}{\partial t} + \left(V + \frac{g}{\lambda}\right)\frac{\partial y}{\partial x}\right] = g(S_0 - S_f) \quad (12\text{-}25)$$

Now, if we define the unknown multiplier so that

$$V + \lambda D = \frac{dx}{dt} = V + \frac{g}{\lambda} \tag{12-26a}$$

Since $D = A/B$, it follows from this equation that

$$\lambda = \pm \sqrt{\frac{gB}{A}} \tag{12-26b}$$

The celerity of a shallow-water wave in free-surface flows is given by the expression

$$c = \sqrt{\frac{gA}{B}} \tag{12-27}$$

Therefore, by defining $\lambda = g/c$, we can write Eq. (12-25) as

$$\frac{dx}{dt} = V + c \tag{12-28}$$

Using Eq. 12-28 and expressions for the total derivatives of V and y, we can write Eq. (12-25) as

$$\frac{dV}{dt} + \frac{g}{c} \frac{dy}{dt} = g(S_0 - S_f) \tag{12-29}$$

Similarly, by defining $\lambda = -g/c$, Eq. (12-25) may be written as

$$\frac{dx}{dt} = V - c \tag{12-30}$$

Then, using Eqs. (12-25) and (12-30), Eq. (12-24) becomes

$$\frac{dV}{dt} - \frac{g}{c} \frac{dy}{dt} = g(S_0 - S_f) \tag{12-31}$$

Equation (12-29) is valid if Eq. (12-28) is satisfied, and Eq. (12-31) is valid if Eq. (12-30) is satisfied. Equations (12-28) and (12-30) plot as characteristic curves (Fig. 12-2) in the x–t plane. Referring to this figure and noting the above conditions for the validity of Eqs. (12-29) and (12-31), we may say that Eq. (12-29) is valid along the positive characteristic curve, C^+, and Eq. (12-31) is valid along the negative characteristic curve, C^-.

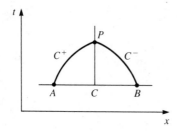

Figure 12-2   Characteristic curves

Multiplying Eqs. (12-29) and (12-31) by dt and integrating along the characteristic curves AP and BP, we obtain

$$\int_A^P dV + \int_A^P \frac{g}{c}\, dy = \int_A^P g(S_0 - S_f)\, dt \tag{12-32}$$

and

$$\int_B^P dV - \int_B^P \frac{g}{c}\, dy = \int_B^P g(S_0 - S_f)\, dt \tag{12-33}$$

In the preceding derivation of Eqs. (12-32) and (12-33), we have not made any approximation whatsoever. However, approximations become necessary to integrate the various terms, as we discuss in the following paragraph.

To determine the integrals of the second term on the left-hand side and the terms on the right-hand side of Eqs. (12-32) and (12-33), the variation of V and y along the characteristic curves should be known. However, V and y are the unknowns we want to compute. Therefore, we cannot directly evaluate the integrals of these terms, and some approximation has to be made for their evaluation. To do this, we may use the values of c and S_f computed by using the values of V and y at the known time level and assume these computed values of c and S_f remain unchanged from A to P and from B to P (Fig. 12-2). Then, we may write Eqs. (12-32) and (12-33) as

$$V_P - V_A + \left(\frac{g}{c}\right)_A (y_P - y_A) = g(S_0 - S_f)_A\, \Delta t \tag{12-34}$$

and

$$V_P - V_B - \left(\frac{g}{c}\right)_B (y_P - y_B) = g(S_0 - S_f)_B\, \Delta t \tag{12-35}$$

where subscripts P, A, and B refer to the grid points in the x–t plane. By combining the known quantities, Eqs. (12-34) and (12-35) may be written as

$$V_P = C_p - K_A y_P \tag{12-36}$$

and

$$V_P = C_n + K_B y_P \tag{12-37}$$

where $C_p = V_A + K_A y_A + g(S_0 - S_f)_A \Delta t$ $\hspace{2cm}$ (12-38)

$\hspace{1.5cm}$ $C_n = V_B - K_B y_B + g(S_0 - S_f)_B \Delta t$ $\hspace{1.8cm}$ (12-39)

$\hspace{1.5cm}$ $K = \dfrac{g}{c}$ $\hspace{4cm}$ (12-40)

12-5 Finite-Difference Methods

In the finite-difference methods, the channel is divided into a number of reaches, usually having equal length, Δx. The ends of each reach are called *computational nodes* or *grid points*. If the channel is divided into N reaches and the first node (upstream end) is numbered as 1, then the last node (downstream end) will be $N + 1$. The nodes at the upstream and downstream ends are called *boundary nodes*, and the remaining nodes are called *interior nodes*. Computations are performed at discrete times. The difference between two consecutive times is called *computational time interval* or *computational time step*. Thus, the x–t plane is divided into a grid (Fig. 12-3), usually referred to as the *computational grid* or *lattice*.

We replace the partial derivatives of the governing equations, Eqs. (12-9) and (12-23), with finite-difference approximations and then solve the resulting algebraic equations at each grid point or computational node. Referring to Fig. 12-3, let us say we know the flow velocity and flow depth at all grid points at time t_0, and we want to determine their values at time $t_0 + \Delta t$. The known values may be the initial conditions from which the unsteady flow conditions begin, or they may be computed during the previous time interval.

We will use a subscript to denote the grid point in the x direction and a superscript to denote the grid point in the t direction. For example, V_i^k refers to the flow velocity at the ith section and at the kth time level. We use superscript k for the time level at which flow conditions are known (referred to as the known time level) and $k + 1$ for the time level at which flow conditions are unknown (referred to as the unknown time level).

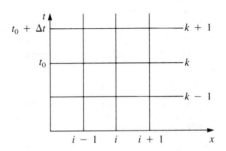

Figure 12-3 Computational grid

Depending on the type of finite-difference approximation, two different schemes are produced. If the finite-difference approximations for the spatial derivatives—that is, partial derivative with respect to x—are in terms of the quantities at the known time level, the resulting equations can be directly solved for each computational node one at a time. These methods are referred to as the *explicit methods*. In the *implicit methods*, the finite-difference approximation for the spatial derivatives are in terms of the unknown variables, and the algebraic equations for the entire system have to be solved simultaneously.

Details of each class of these methods are presented in the following sections.

12-6 Explicit Methods

Several explicit finite-difference methods have been used for unsteady free-surface flows. We will present details of one of these methods, the *Lax diffusive scheme* (3, 6, 8). It is simple to program and yields satisfactory results even when flows have bores.

In this scheme, the partial derivatives of the governing equations are replaced as follows:

$$\frac{\partial y}{\partial t} = \frac{y_i^{k+1} - y_i^*}{\Delta t} \tag{12-41}$$

$$\frac{\partial V}{\partial t} = \frac{V_i^{k+1} - V_i^*}{\Delta t} \tag{12-42}$$

$$\frac{\partial y}{\partial x} = \frac{y_{i+1}^k - y_{i-1}^k}{2\,\Delta x} \tag{12-43}$$

$$\frac{\partial V}{\partial x} = \frac{V_{i+1}^k - V_{i-1}^k}{2\,\Delta x} \tag{12-44}$$

where $y_i^* = 0.5(y_{i-1}^k + y_{i+1}^k)$ $\qquad\qquad$ (12-45)

$\qquad\quad V_i^* = 0.5(V_{i-1}^k + V_{i+1}^k)$ $\qquad\qquad$ (12-46)

Replacing the partial derivatives of the governing equations (Eqs. 12-9 and 12-23) by these finite-difference approximations and the coefficient D and the slope S_f by $D^* = 0.5(D_{i-1}^k + D_{i+1}^k)$ and $S_f^* = 0.5(S_{fi-1}^k + S_{fi+1}^k)$ and simplifying, we obtain

$$y_i^{k+1} = y_i^* - 0.5rD_i^*(V_{i+1}^k - V_{i-1}^k) - 0.5rV_i^*(y_{i+1}^k - y_{i-1}^k) \tag{12-47}$$

$$V_i^{k+1} = V_i^* - 0.5rg(y_{i+1}^k - y_{i-1}^k) - 0.5rV_i^*(V_{i+1}^k - V_{i-1}^k)$$
$$+ g\,\Delta t(S_0 - S_f^*) \tag{12-48}$$

where $r = \Delta t/\Delta x$.

Equations (12-47) and (12-48) yield the values at the interior nodes only, that is, at $i = 2, 3, \ldots, N$. Boundary nodes need special treatment, as discussed in the following paragraphs.

Boundary Conditions

The characteristic equations along with the conditions imposed by the boundaries determine the *conditions at the boundary nodes*. The conditions imposed by the boundary may be in the form of specifying the variation of discharge or depth with respect to time, or a relationship between the discharge and depth. For example, for a constant-level reservoir at the upstream end (node 1), we specify

$$y_1^{k+1} = y_{\text{res}} \tag{12-49}$$

where entrance and the velocity head at node 1 are assumed to be small and neglected, and y_{res} is the flow depth in the reservoir above the channel bottom at node 1.

At the upstream end, the negative characteristic equation, Eq. (12-37), and at the downstream end, the positive characteristic equation, Eq. (12-36), are used.

Stability

It is usually necessary in the explicit finite-difference schemes that the ratio of Δx and Δt satisfies a condition for stability. A scheme is said to be *stable* if an error introduced in the solution does not grow as the computations progress in time, and the scheme is said to be *unstable* if the error is amplified with time. In the case of an unstable scheme, the error is amplified very rapidly and masks the true solution in a few time intervals.

For the Lax scheme to be stable, the computational time step and the grid spacing must satisfy the following condition, referred to as Courant's stability condition:

$$\Delta t \leqslant \frac{\Delta x}{|V| + c} \tag{12-50}$$

12-7 Implicit Finite-Difference Methods

In the implicit methods, the spatial derivatives are replaced by the finite-difference approximations in terms of the variables at the unknown time level. Depending on the finite-difference approximations and the coefficients used, several different formulations are possible. Of these schemes, the *Preissmann scheme* has been used extensively for the analysis of unsteady free-surface flows.

In the Preissmann scheme, the partial derivatives and other variables are approximated as follows:

$$\frac{\partial f}{\partial t} = \frac{[(f_i^{k+1} + f_{i+1}^{k+1}) - (f_i^k + f_{i+1}^k)]}{2\,\Delta t} \tag{12-51}$$

$$\frac{\partial f}{\partial x} = \frac{\alpha[f_{i+1}^{k+1} - f_i^{k+1}]}{\Delta x} + \frac{(1-\alpha)[f_{i+1}^k - f_i^k]}{\Delta x} \tag{12-52}$$

$$f(x, t) = \frac{\alpha(f_{i+1}^{k+1} + f_i^{k+1})}{2} + \frac{(1-\alpha)(f_{i+1}^k + f_i^k)}{2} \tag{12-53}$$

where α = the weighting coefficient, $0.5 < \alpha \leqslant 1$.

Substituting these equations into Eqs. (12-9) and (12-23) and simplifying the resulting equations, we obtain

$$y_i^{k+1} + y_{i+1}^{k+1} + \frac{\Delta t}{\Delta x}\left[\alpha(D_{i+1}^{k+1} + D_i^{k+1}) + (1-\alpha)(D_{i+1}^k + D_i^k)\right]$$

$$\cdot \left[\alpha(V_{i+1}^{k+1} - V_i^{k+1}) + (1-\alpha)(V_{i+1}^k + V_i^k)\right] + \frac{\Delta t}{\Delta x}$$

$$\cdot \left[\alpha(V_{i+1}^{k+1} + V_i^{k+1}) + (1-\alpha)(V_{i+1}^k + V_i^k)\right]$$

$$\cdot \left[\alpha(y_{i+1}^{k+1} - y_i^{k+1}) + (1-\alpha)(y_{i+1}^k - y_i^k)\right]$$

$$= y_i^k + y_{i+1}^k \tag{12-54}$$

$$V_i^{k+1} + V_{i+1}^{k+1} + 2g\frac{\Delta t}{\Delta x}\left[\alpha(y_{i+1}^{k+1} - y_i^{k+1}) + (1-\alpha)(y_{i+1}^k - y_i^k)\right]$$

$$+ \frac{\Delta t}{\Delta x}\left[\alpha(V_{i+1}^{k+1} + V_i^{k+1}) + (1-\alpha)(V_{i+1}^k + V_i^k)\right]$$

$$\cdot \left[\alpha(V_{i+1}^{k+1} - V_i^{k+1}) + (1-\alpha)(V_{i+1}^k - V_i^k)\right]$$

$$= (V_i^k + V_{i+1}^k) + 2g\,\Delta t S_0 - g\,\Delta t$$

$$\cdot \left[\alpha(S_{fi+1}^{k+1} + S_{fi}^{k+1}) + (1-\alpha)(S_{fi+1}^k + S_{fi}^k)\right] \tag{12-55}$$

There are four unknowns in Eqs. (12-54) and (12-55). By writing similar equations for grid points $i = 1, 2, \ldots, N$, we will have a total of $2N$ equations. We cannot write Eqs. (12-54) and (12-55) for node $N + 1$, since we do not have node $N + 2$. However, there are two unknowns, V and y, per node. Therefore, we have $2(N + 1)$ unknowns, and to obtain a unique solution, we need two more equations. These equations are provided by the boundary conditions.

Boundary Conditions

Unlike the explicit methods, equations representing the conditions imposed by the boundary are directly included in the system of equations and not in combination with the characteristic equations. For example, for a constant-level reservoir at the upstream end, y will be constant at all times if the velocity head and entrance losses are neglected. Similarly, other conditions may be specified at the upstream and downstream ends.

The resulting system of equations are nonlinear algebraic equations. These may be solved by using the Newton-Raphson method.

Stability

The implicit methods are usually unconditionally stable. This means there is no restriction on the size of the grid spacing Δx and Δt for the stability of the numerical scheme. However, accuracy dictates that the computational time step nearly equal to that given by the Courant condition be used in the computations.

EXAMPLE 12-1 Analyze unsteady flow in a 1000-m long trapezoidal channel having bottom width of 20 m, side slope of 2 H to 1 V, the bottom slope, S_0 of 0.0001, and Manning's n of 0.013. Initial flow and flow depth in the channel are 110 m^3/s and 3.069 m, respectively. Unsteady flow is produced by instantaneously closing a downstream gate at $t = 0$.

SOLUTION   Figure 12-4 lists the computer program based on the Lax scheme. To ensure that Courant's condition is always satisfied at each node, the time interval required to satisfy this condition was determined at the end of each time step. If necessary, computations are repeated with a reduced value of the time interval; the value of the interval is increased for the next cycle of computations if it is found to be too small as compared to that required for stability.

Figure 12-4, pages 623–627, shows the program input and output, and Fig. 12-5, page 627, shows the computed water levels.   ∎

Figure 12-4   Program listing: Computation of unsteady, free-surface flows
Program Listing

```
c
c       COMPUTATION OF UNSTEADY, FREE-SURFACE FLOWS BY LAX'S
c        DIFFUSIVE SCHEME
c
c       ******************* NOTATION ***********************
c
c       A = FLOW AREA;
c       B = TOP WATER-SURFACE WIDTH;
c       BO = CHANNEL-BOTTOM WIDTH;
c       P = WETTED PERIMETER;
c       Q = DISCHARGE;                              (continued)
```

```
C        S = CHANNEL-SIDE SLOPE, S HORIZONTAL : 1 VERTICAL;
C        SO = CHANNEL-BOTTOM SLOPE;
C        X = DISTANCE ALONG CHANNEL BOTTOM, POSITIVE IN THE
C           DOWNSTREAM DIRECTION;
C        Y = FLOW DEPTH;
C        YU = FLOW DEPTH AT UPSTREAM END.
C
         REAL L, MN,MN2
         DIMENSION Y(100),YP(100),V(100),VP(100)
         BT(YY)=BO+2.*S*YY
         AR(YY)=YY*(BO+S*YY)
         WP(YY)=BO+2.*YY*SQRT(1.+S*S)
         G=9.81
         READ (5,*) N,IPRINT,QO,YD,MN,BO,SO,S,L,TMAX
         WRITE(6,10) N,QO,YD,MN,BO,SO,S,L
10       FORMAT(5X,'N =',I3,' QO =',F8.3,' M3/S',' YD =',F6.2,' M',
        1 2X, ' MN =',F6.3,' BO =',F6.2,' M'/5X,' SO =',F6.4,'S =',F8.4,
        2  ' L =',F8.2,' M')
C
C        STEADY-STATE CONDITIONS
C
         MN2=MN*MN
         NN=N+1
         A=AR(YD)
         VO=QO/A
         DO 30 I = 1,NN
         Y(I) = YD
         V(I)=VO
30       CONTINUE
         YU= Y(1)
         B=BT(YD)
         C=SQRT(G*A/B)
         DX=L/N
         DT=DX/(VO+C)
         T=0.0
35       K = 0
         WRITE(6,40) T
40       FORMAT(/5X,'T =',F8.3,' S')
         WRITE(6,50) (Y(I),I=1,NN)
50       FORMAT(6X,'Y =',(12F10.2))
         WRITE (6,60) (V(I),I=1,NN)
60       FORMAT(6X,' V=',(12F10.3))
70       T=T+DT
         K=K+1
         R=0.5*DT/DX
         IF (T.GT.TMAX) GO TO 160
C
C        UPSTREAM END
C
         YP(1) = YU
         AB=AR(Y(2))
         BB=BT(Y(2))
         CB=SQRT(G*BB/AB)
         RB=AB/WP(Y(2))
         SFB=(MN2*V(2)*V(2))/(RB**1.333)
         CN=V(2)-CB*Y(2)+G*(SO-SFB)*DT
         VP(1)=CN+CB*YP(1)
C
C        DOWNSTREAM END
C
         VP(NN)=0.
         AA=AR(Y(N))
         BA=BT(Y(N))
         CA=SQRT(G*BA/AA)
         RA=AA/WP(Y(N))
         SFA=(MN2*V(N)*V(N))/(RA**1.333)
         CP=V(N)+CA*Y(N)+G*(SO-SFA)*DT
         YP(NN)=CP/CA
C
C        INTERIOR NODES
C
         DO 80 I=2,N
         I1=I-1
         IP1=I+1
         AA=AR(Y(I1))
         PA=WP(Y(I1))
         RA=AA/PA
         SFA=(MN2*V(I1)*V(I1))/(RA**1.333)
         BA=BT(Y(I1))
         AB=AR(Y(IP1))
```

```
              BB=BT(Y(IP1))
              P=WP(Y(IP1))
              RB=AB/P
              SFB=(MN2*V(IP1)*V(IP1))/(RB**1.333)
              DM=0.5*(AA/BA + AB/BB)
              SFM=0.5*(SFA+SFB)
              VM=0.5*(V(I1)+V(IP1))
              YM=0.5*(Y(I1)+Y(IP1))
              VP(I)=VM-R*G*(Y(IP1)-Y(I1)) - R*VM*(V(IP1)-V(I1))
       1         + G*DT*(SO-SFM)
              YP(I)=YM-R*DM*(V(IP1)-V(I1))-R*VM*(Y(IP1)-Y(I1))
   80     CONTINUE
   C
   C      CHECK FOR STABILITY
   C
              DO 100 I=1,NN
              A=AR(YP(I))
              B=BT(YP(I))
              C=SQRT(G*A/B)
              DTN=DX/(ABS(VP(I))+C)
              IF (DTN.LE.DT) GO TO 110
              IF (DT.LT.0.75*DTN) DTNEW=1.15*DT
              IF (DTN.GE.0.75*DT) DTNEW=DT
  100     CONTINUE
              GO TO 120
   C
   C      REDUCE DT FOR STABILITY AND RE-CALCULATE
   C
  110     T=T-DT
              DT=.9*DTN
              K=K-1
              GO TO 70
  120     DT=DTNEW
              DO 130 I=1,NN
              V(I)=VP(I)
              Y(I)=YP(I)
  130     CONTINUE
              IF (K.EQ.IPRINT) GO TO 35
              GO TO 70
  160     STOP
              END
```

Input Data

```
10,1,110.,3.069,0.013,20.,0.0001,2.,1000.,400.
```

Program Output

```
N = 10 QO = 110.000 M3/S YD = 3.07 M   MN = .013 BO = 20.00 M
SO = .0001S = 2.0000 L = 1000.00 M


T =   .000 S
Y =    3.07    3.07    3.07    3.07    3.07    3.07    3.07    3.07    3.07    3.07    3.07
V=    1.371   1.371   1.371   1.371   1.371   1.371   1.371   1.371   1.371   1.371   1.371


T = 15.850 S
Y =    3.07    3.07    3.07    3.07    3.07    3.07    3.07    3.07    3.07    3.07    3.76
V=    1.371   1.371   1.371   1.371   1.371   1.371   1.371   1.371   1.371   1.371   .000


T = 31.701 S
Y =    3.07    3.07    3.07    3.07    3.07    3.07    3.07    3.07    3.07    3.67    3.76
V=    1.371   1.371   1.371   1.371   1.371   1.371   1.371   1.371   1.371   .231    .000


T = 47.551 S
Y =    3.07    3.07    3.07    3.07    3.07    3.07    3.07    3.07    3.58    3.67    3.81
V=    1.371   1.371   1.371   1.371   1.371   1.371   1.371   1.371   .412    .231    .000


T = 63.401 S
Y =    3.07    3.07    3.07    3.07    3.07    3.07    3.07    3.49    3.58    3.78    3.81
V=    1.371   1.371   1.371   1.371   1.371   1.371   1.371   .573    .412    .049    .000


T = 79.252 S
Y =    3.07    3.07    3.07    3.07    3.07    3.07    3.41    3.49    3.75    3.78    3.82
V=    1.371   1.371   1.371   1.371   1.371   1.371   .714    .573    .110    .049    .000
```

(continued)

Figure 12-4
Program Output (continued)

```
T =  95.102 S
Y =     3.07    3.07    3.07    3.07    3.07    3.35    3.41    3.70    3.75    3.81    3.82
V=      1.371   1.371   1.371   1.371   1.371   .836    .714    .185    .110    .017    .000

T = 110.952 S
Y =     3.07    3.07    3.07    3.07    3.29    3.35    3.66    3.70    3.79    3.81    3.83
V=      1.371   1.371   1.371   1.371   .939    .836    .272    .185    .038    .017    .000

T = 126.802 S
Y =     3.07    3.07    3.07    3.25    3.29    3.60    3.66    3.78    3.79    3.82    3.83
V=      1.371   1.371   1.371   1.024   .939    .368    .272    .065    .038    .011    .000

T = 142.653 S
Y =     3.07    3.07    3.21    3.25    3.55    3.60    3.76    3.78    3.81    3.82    3.83
V=      1.371   1.371   1.095   1.024   .467    .368    .101    .065    .022    .011    .000

T = 158.503 S
Y =     3.07    3.18    3.21    3.49    3.55    3.73    3.76    3.80    3.81    3.83    3.83
V=      1.371   1.152   1.095   .568    .467    .145    .101    .035    .022    .009    .000

T = 174.353 S
Y =     3.07    3.18    3.44    3.49    3.70    3.73    3.79    3.80    3.82    3.83    3.84
V=      .936    1.152   .667    .568    .197    .145    .051    .035    .019    .009    .000

T = 190.204 S
Y =     3.07    3.29    3.44    3.67    3.70    3.78    3.79    3.81    3.82    3.84    3.84
V=      .936    .540    .667    .258    .197    .071    .051    .028    .019    .009    .000

T = 206.054 S
Y =     3.07    3.29    3.53    3.67    3.77    3.78    3.81    3.81    3.83    3.84    3.85
V=      .134    .540    .127    .258    .094    .071    .038    .028    .018    .009    .000

T = 221.904 S
Y =     3.07    3.29    3.53    3.65    3.77    3.80    3.81    3.82    3.83    3.84    3.85
V=      .134    -.210   .127    -.060   .094    .050    .038    .026    .018    .009    .000

T = 237.755 S
Y =     3.07    3.29    3.44    3.65    3.70    3.80    3.82    3.82    3.84    3.84    3.86
V=      -.629   -.210   -.395   -.060   -.104   .050    .035    .026    .017    .009    .000

T = 253.605 S
Y =     3.07    3.22    3.44    3.51    3.70    3.73    3.82    3.83    3.84    3.85    3.86
V=      -.629   -.781   -.395   -.428   -.104   -.109   .035    .026    .017    .009    .000

T = 269.455 S
Y =     3.07    3.22    3.31    3.51    3.55    3.73    3.75    3.83    3.85    3.85    3.87
V=      -1.070  -.781   -.800   -.428   -.414   -.109   -.108   .026    .017    .009    .000

T = 285.306 S
Y =     3.07    3.15    3.31    3.36    3.55    3.58    3.75    3.77    3.85    3.86    3.87
V=      -1.070  -1.091  -.800   -.766   -.414   -.392   -.108   -.106   .017    .009    .000

T = 301.156 S
Y =     3.07    3.15    3.20    3.36    3.40    3.58    3.61    3.77    3.79    3.86    3.87
V=      -1.244  -1.091  -1.057  -.766   -.725   -.392   -.371   -.106   -.104   .009    .000

T = 317.006 S
Y =     3.07    3.11    3.20    3.24    3.40    3.44    3.61    3.64    3.79    3.81    3.87
V=      -1.244  -1.232  -1.057  -1.012  -.725   -.686   -.371   -.352   -.104   -.103   .000

T = 332.857 S
Y =     3.07    3.11    3.14    3.24    3.28    3.44    3.47    3.64    3.67    3.81    3.76
V=      -1.309  -1.232  -1.202  -1.012  -.967   -.686   -.651   -.352   -.335   -.103   .000

T = 348.707 S
Y =     3.07    3.09    3.14    3.18    3.28    3.32    3.47    3.51    3.67    3.64    3.76
V=      -1.309  -1.301  -1.202  -1.167  -.967   -.925   -.651   -.619   -.335   -.217   .000

T = 364.557 S
Y =     3.07    3.09    3.12    3.18    3.21    3.32    3.36    3.51    3.49    3.64    3.53
V=      -1.346  -1.301  -1.285  -1.167  -1.132  -.925   -.886   -.619   -.490   -.217   .000
```

T = 380.408 S										
Y =	3.07	3.09	3.12	3.14	3.21	3.25	3.36	3.35	3.49	3.40	3.53
V=	-1.346	-1.345	-1.285	-1.264	-1.132	-1.097	-.886	-.752	-.490	-.253	.000

T = 394.657 S										
Y =	3.07	3.09	3.11	3.15	3.17	3.25	3.24	3.35	3.28	3.41	3.27
V=	-1.378	-1.342	-1.336	-1.259	-1.235	-1.088	-.962	-.745	-.511	-.252	.000

12-8 Comparison of Explicit and Implicit Methods

The explicit and implicit finite-difference methods have their advantages and disadvantages and none of them is suitable for universal applications. We will discuss the advantages and disadvantages of these methods in this section so that they may be taken into consideration during their selection for a particular application.

The *implicit methods* are usually unconditionally stable whereas the Courant condition has to be satisfied for the explicit methods. This condition restricts the size of the computational time interval, thereby making the method uneconomical for the analysis of unsteady flows in a large system for long durations.

The *explicit methods* are easier to program and debug than the implicit methods. Therefore if the available time for the development of a computer program is short, then explicit methods should be selected.

The computer storage requirements for the implicit methods are usually more than those for the explicit methods. If a large system has to be analyzed and the computer memory is limited, then there may be no choice except to use the explicit method.

It might be necessary to use small time steps if the flow variables to be analyzed have sharp peaks. In such a case explicit methods are superior because the computational effort per time step is more for the implicit methods than that for the explicit methods.

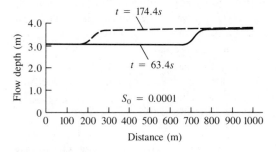

Figure 12-5 Flow depth along channel

12-9 Approximate Methods for Flood Routing

The term *flood routing* refers to the computation of the speed, height, and discharge of a flood wave as it propagates in waterways, such as rivers, lakes, reservoirs, and streams. For the routing of a flood wave, two numerical methods (presented in the preceding sections) that solve the complete continuity and momentum equations may be used. This is referred to as *hydraulic routing*. The mathematical models based on the numerical solution of complete equations are called *dynamic models*, and the waves computed by these models are called *dynamic waves*. In several situations, the relative effects of some terms of the governing equations are very small as compared with the other terms and may be neglected. For example, Henderson (7) gave the following values for different terms of Eq. (12-24), page 616, for a very fast-rising flood in a river in a steep, alluvial country:

Term	S_0	$\dfrac{\partial y}{\partial x}$	$\dfrac{V}{g}\dfrac{\partial V}{\partial x}$	$\dfrac{1}{g}\dfrac{\partial V}{\partial t}$
Value (ft/mi)	26	0.5	0.125–0.25	0.05

Therefore, the inertial terms in this case may be neglected without introducing large errors. Approximate procedures in which the continuity equation is solved simultaneously with a simplified form of the dynamic equation are called *hydrologic routing*. There are several of these procedures depending on the terms retained in the simplified form of the dynamic equation and the computational procedure used for solving the governing equations. For example, in a kinematic wave model, the steady uniform flow equation is used for the momentum equation; whereas, an additional term representing the slope of the water surface is included in the diffusion models. By approximating the complex relationships between storage capacity of a reach, inflow, outflow, stage, and so on, several procedures known as coefficient methods have been developed. Of these procedures, one developed by the U.S. Army Corps of Engineers on the Muskingum River in 1934 has been widely used. We discussed details of this method in Sect. 4-5, page 223, and we consider it again in this section to show how a rigorous derivation compares with the empirical approach.

Simplified models are easy to use and provide satisfactory results if the assumptions they are based on are satisfied. Besides simplicity, these methods do not require detailed information about the channel geometry. We give a brief description of some of these methods in the following paragraphs.*

---

* For a detailed description of these methods, see Chow (4), Cunge (5), Henderson (7), Mahmood (8), Ponce (9), and Weinmann (11).

A general form of the resistance formulas, such as Chezy or Manning, may be written as

$$Q = C' A R^m \sqrt{S_f} \qquad (12\text{-}56)$$

where C' is an empirical constant, R is the hydraulic radius, m is an empirical exponent, and S_f is the slope of the energy grade line. In Manning's formula (in SI units), $C' = 1/n$ and $m = 2/3$; whereas, in Chezy's formula, $C' = C$, and $m = 1/2$. For steady-uniform flows, Q is the normal discharge, Q_n, and $S_f = S_0$. Hence, for steady-uniform flows, Eq. (12-56) becomes

$$Q_n = C' A R^m \sqrt{S_0} \qquad (12\text{-}57)$$

Eliminating $C' A R^m$ from Eqs. (12-56) and (12-57), we obtain

$$Q = \frac{Q_n}{\sqrt{S_0}} \sqrt{S_f} \qquad (12\text{-}58)$$

Now, substitution of expression for S_f from Eq. (12-24) into this equation yields

$$Q = Q_n \sqrt{1 - \frac{1}{S_0} \frac{\partial y}{\partial x} - \frac{V}{gS_0} \frac{\partial V}{\partial x} - \frac{1}{gS_0} \frac{\partial V}{\partial t}} \qquad (12\text{-}59)$$

Kinematic $\longrightarrow$ |

Diffusion $\longrightarrow$ |

Dynamic $\longrightarrow$ |

In this equation, terms that are included in different types of models are marked.

Figure 12-6 shows a plot of typical relationship between Q and y, as given by Eq. (12-59). For steady-uniform flows, there is a unique, single-valued relationship between Q and y, shown by the dashed line in this figure. Simply, this

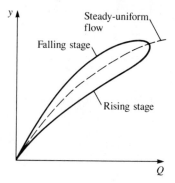

Figure 12-6　Looped rating curve

means only one value of normal discharge is possible for a given y. Closely examining Eq. (12-59) shows that the discharge for a specified y at a channel section during a rising stage will be more than the normal discharge for that y because $\partial y/\partial x$ and $\partial V/\partial x$ are negative during a rising stage, and the last term is usually small. Following a similar reasoning, discharge for a specified flow depth during a falling stage is less than the corresponding normal discharge, Q_n. This difference between the flows corresponding to a given flow depth during rising and falling stages is due to the effects of the above three terms (Kinematic, Diffusion, and Dynamic). In other words, the larger the hysteresis effect between the rising and the falling stages at a channel section, the more important it will be to include these terms in the analysis.

We will now briefly discuss three commonly used approximate methods for flood routing.

Kinematic Models

As we mentioned, kinematic models are based on the solution of the continuity equation and the steady-uniform equation for the dynamic equation. The waves propagated using these models are called *kinematic waves*, and routing is called *kinematic routing*. To study the properties of kinematic waves, we consider the solution of the governing equations.

The continuity equation, Eq. (12-6), page 613, may be written as

$$\frac{\partial A}{\partial t} + \frac{\partial Q}{\partial x} = 0 \tag{12-60}$$

where Q = discharge

For the momentum equation, we use the equation describing steady-uniform flows, $Q = Q_n$. Since Q_n is a single-valued function of flow depth y, and the flow area is a function of flow depth, we can write

$$Q = Q(A)$$

or $A = A(Q)$ \tag{12-61}

Because both A and Q are functions of x and t, we can write the following equation by applying the chain rule:

$$\frac{\partial A}{\partial t} = \frac{\partial A}{\partial Q} \frac{\partial Q}{\partial t}$$

$$= \frac{\partial Q}{\partial t} \frac{dA}{dQ} \tag{12-62}$$

Substituting Eq. (12-62) into Eq. (12-60) and multiplying throughout by $\dfrac{dQ}{dA}$ yield

$$\frac{\partial Q}{\partial t} + \frac{dQ}{dA}\frac{\partial Q}{\partial x} = 0 \qquad (12\text{-}63)$$

or

$$\frac{\partial Q}{\partial t} + c\frac{\partial Q}{\partial x} = 0 \qquad (12\text{-}64)$$

where $c = dQ/dA$

From the following discussion, we will see that c describes the speed of a kinematic wave and is called kinematic wave speed.

Eq. (12-64) is a first-order partial differential equation with Q as the dependent variable and x and t as the independent variables. For a general solution of Eq. (12-64), let us try the function

$$Q = f(x - ct) \qquad (12\text{-}65)$$

and assume that the partial derivatives of f with respect to x and t exist and that c is a constant. Then,

$$\frac{\partial f}{\partial x} = f'(x - ct)$$

$$\frac{\partial f}{\partial t} = -cf'(x - ct) \qquad (12\text{-}66)$$

Substituting Eqs. (12-66) into Eq. (12-64) shows that the function $Q = f(x - ct)$ is a general solution of Eq. (12-64). This is called D'Alembert's solution (12); it is useful in understanding the characteristics of the kinematic waves.

At time $t = 0$, the general solution of Eq. (12-64), $Q = f(x - ct)$, defines the curve $Q = f(x)$. This curve describes the initial condition (Fig. 12-7, page 632). At any later time t_1, the solution defines the curve $Q = f(x - ct_1)$. These two curves are identical in shape except that the curve representing the solution at time t_1 is translated to the right a distance equal to ct_1. That is, the entire curve moves in the positive x direction without distortion a distance equal to ct_1 during time t_1. Therefore, the velocity of the wave is $ct_1/t_1 = c$.

Thus we can now summarize the following three *properties of kinematic routing*:

1. Kinematic waves travel only in the positive x direction.
2. The wave shape does not change, and there is no attenuation of the wave height.
3. The wave speed, $c = dQ/dA$

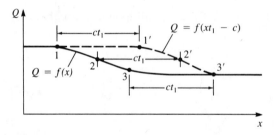

Figure 12-7   Propogation of kinematic wave

Kinematic wave models are based on an analytical solution or on a numerical solution of Eqs. (12-64) and (12-65). The value of c may be determined from the observed flood hydrographs, or it may be computed from the slope of the rating curve for a given section, that is, $c = dQ/dA = (1/B) dQ/dy$, where $B =$ the width of the water surface at depth y.

Some attenuation of the flood waves has reportedly been computed by certain kinematic models. This attenuation either is caused by dispersion introduced by the numerical scheme or is due to the value of Courant number, $C_N = c \Delta t/\Delta x$, being different from 1 in the computations.

Diffusion Routing

In the diffusion routing, the following simplified equation is used for the momentum equation:

$$S_f = S_0 - \frac{\partial y}{\partial x} \tag{12-67}$$

Replacing S_f in terms of conveyance factor, K, this equation may be written as

$$\frac{Q^2}{K^2} = S_0 - \frac{\partial y}{\partial x} \tag{12-68}$$

Differentiating Eq. (12-68) with respect to t yields

$$\frac{2Q}{K^2} \frac{\partial Q}{\partial t} - \frac{2Q^2}{K^3} \frac{\partial K}{\partial t} = -\frac{\partial^2 y}{\partial t \, \partial x} \tag{12-69}$$

By differentiating Eq. (12-60) with respect to x and noting that

$$\frac{\partial A}{\partial x} = \frac{\partial A}{\partial y} \frac{\partial y}{\partial x} = B \frac{\partial y}{\partial x}$$

we obtain $B\dfrac{\partial^2 y}{\partial x\, \partial t} + \dfrac{\partial^2 Q}{\partial x^2} = 0$ (12-70)

Dividing this equation by B and subtracting from Eq. (12-69) yields

$$\frac{2Q}{K^2}\frac{\partial Q}{\partial t} - \frac{2Q^2}{K^3}\frac{\partial K}{\partial t} = \frac{1}{B}\frac{\partial^2 Q}{\partial x^2}$$ (12-71)

Now, $\dfrac{\partial K}{\partial t} = \dfrac{\partial K}{\partial A}\dfrac{\partial A}{\partial t}$ (12-72)

Substitution for $\partial A/\partial t$ from Eq. (12-60) into this equation yields

$$\frac{\partial K}{\partial t} = \frac{dK}{dA}\left(-\frac{\partial Q}{\partial x}\right)$$ (12-73)

To determine the expression for dK/dA, assume $Q = K\sqrt{S_0}$ instead of $Q = K\sqrt{S_f}$. Then, $dQ/dA = dK/dA\sqrt{S_0}$. Substituting this relationship into Eq. (12-73), we obtain

$$\frac{\partial K}{\partial t} = \frac{1}{\sqrt{S_0}}\frac{dQ}{dA}\left(-\frac{\partial Q}{\partial x}\right)$$ (12-74)

Eliminating $\partial K/\partial t$ from Eqs. (12.71) and (12.74) gives

$$\frac{2BQ}{K^2}\frac{\partial Q}{\partial t} + \frac{2BQ^2}{K^3\sqrt{S_0}}\frac{dQ}{dA}\frac{\partial Q}{\partial x} = \frac{\partial^2 Q}{\partial x^2}$$

Multiplying throughout by $K^2/(2BQ)$, noting that $Q = K\sqrt{S_0}$, and simplifying the resulting equation, we obtain

$$\frac{\partial Q}{\partial t} + \frac{dQ}{dA}\frac{\partial Q}{\partial x} = \frac{Q}{2BS_0}\frac{\partial^2 Q}{\partial x^2}$$ (12-75)

Comparing this equation with the equation describing a kinematic wave (Eq. 12-64) shows that, except for the term on the right-hand side of Eq. (12-75), both equations are identical. The diffusion wave travels at the same speed, $c = dQ/dA$, as the kinematic wave. However, the additional term results in the diffusion of a wave. We can rewrite Eq. (12-75) compactly as

$$\frac{\partial Q}{\partial t} + c\frac{\partial Q}{\partial x} = D\frac{\partial^2 Q}{\partial x^2}$$ (12-76)

where $D = Q/(2BS_0)$.

D is called the diffusion coefficient that simulates the attenuation of the wave as it propagates in the channel. If D and C are determined from the observed flood hydrographs in a channel, they account for the effects of channel storage and other factors on the movement of flood waves.

Muskingum-Cunge Method

In Sec. 4-5, page 223, we presented details of the Muskingum method in which the coefficients for the method were determined from the observed flood records. In this section, we show that this method is actually a particular finite-difference approximation of the kinematic wave equations and present expressions for the Muskingum coefficients in terms of the physical properties of the channel.

Referring to Fig. 12-8 the spatial grid points i and $i + 1$ refer to the ends of a channel reach, and the temporal grid points k and $k + 1$ refer to the beginning and the end of the routing interval. Let us use the following finite-difference approximations for the partial derivatives:

$$\frac{\partial Q}{\partial x} = \frac{(Q_{i+1}^k + Q_{i+1}^{k+1}) - (Q_i^k + Q_i^{k+1})}{2\,\Delta x} \tag{12-77}$$

$$\frac{\partial Q}{\partial t} = \frac{\alpha(Q_i^{k+1} - Q_i^k) + (1 - \alpha)(Q_{i+1}^{k+1} - Q_{i+1}^k)}{\Delta t} \tag{12-78}$$

where α is the weighting coefficient used for the temporal partial derivative. By substituting Eqs. (12-77) and (12-78) into Eq. (12-64) and simplifying the resulting equation, we obtain

$$[(1 - \alpha) + 0.5cr]Q_{i+1}^{k+1} + (\alpha - 0.5cr)Q_i^{k+1}$$
$$+ (-\alpha - 0.5cr)Q_i^k + [-(1 - \alpha) + 0.5cr]Q_{i+1}^k = 0 \tag{12-79}$$

where $r = \Delta t/\Delta x$

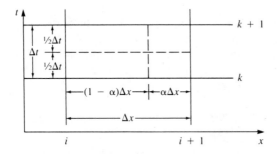

Figure 12-8   Definition sketch

Therefore, in terms of the notation of Fig. 12-8 and the terminology of Sec. 4-4, $Q_{i+1}^{k+1} = O_{i+1}$, $Q_{i+1}^{k} = O_i$, $Q_i^{k} = I_i$, and $Q_i^{k+1} = I_{i+1}$. Then we can write Eq. (12-79) as

$$Q_{i+1}^{k+1} = C_0 Q_i^{k+1} + C_1 Q_i^{k} + C_2 Q_{i+1}^{k} \qquad (12\text{-}80)$$

where $C_0 = \dfrac{0.5\,\Delta t - \alpha\,\Delta x/c}{0.5\,\Delta t + (1-\alpha)\,\Delta x/c}$

$C_1 = \dfrac{0.5\,\Delta t + (\alpha\,\Delta x/c)}{0.5\,\Delta t + (1-\alpha)\,\Delta x/c}$

$C_2 = \dfrac{-0.5\,\Delta t + (1-\alpha)\,\Delta x/c}{0.5\,\Delta t + (1-\alpha)\,\Delta x/c} \qquad (12\text{-}81)$

Comparing Eqs. (12-81) with Eqs. (4-58) through (4-60) shows that the expressions for the Muskingum coefficients are identical if we assume $\alpha = X$ and $\Delta x/c = K$. Thus, K is the travel time for the flood wave in the reach. By expanding Eq. (12-80) in a Taylor series and comparing it with the diffusion equation, Eq. (12-76), the following expression may be obtained for X in terms of the physical properties of the channel:

$$X = \frac{1 - Q_0}{2 S_0 B c\, \Delta x} \qquad (12\text{-}82)$$

We can show that the Muskingum routing becomes equivalent to kinematic routing for certain values of α and $c\,\Delta t/\Delta x$. For example, let us substitute $\alpha = 0.5$ and $c\,\Delta t/\Delta x = 1$ into Eqs. (12-81). This process gives $C_0 = 0$, $C_1 = 1$, and $C_2 = 0$. Substituting these values of Muskingum coefficients into Eq. (12-80) yields

$$Q_{i+1}^{k+1} = Q_i^{k} \qquad (12\text{-}83)$$

This equation shows us that the flood wave is not attenuated as it propagates through the channel reach under consideration; that is, we get the same result that we get for kinematic routing.

PROBLEMS

12-1   Develop the following boundary conditions for the Lax scheme:
   a. Constant-level upstream reservoir
   b. Constant-level downstream reservoir
   c. Rating curve for the downstream end
   Assume the entrance losses at the reservoir are small, and neglect the velocity head.

12-2   Write a computer program to analyze unsteady flows in a channel 5000 m long having a channel bottom slope of 0.0001 and Manning's $n = 0.025$. The channel cross section is trapezoidal in shape with bottom width of 10 m and side slopes of 1 vertical to 1.5 horizontal. The initial flow conditions are uniform at a flow depth of 3 m. There is a constant-level reservoir at the downstream end, and the flow velocity at the upstream end is instantaneously reduced to zero at $t = 0$. Use the Lax scheme.

12-3   Compute the celerity of a wave in a circular tunnel having a flow depth of 6 m and the tunnel diameter of 8 m.

12-4   A 10-m wide rectangular channel has a flow depth of 5 m and flow velocity of 4 m/s. Determine the absolute velocity of a surge wave produced by instantaneously closing a control gate located at the
a. Downstream end
b. Upstream end
c. Midlength of the channel

12-5   Develop a computer program for the analysis of unsteady flow in a channel having a constant-level reservoir at the upstream end and a slowly closing gate at the downstream end. Use the Preissmann scheme. Run this program for the channel of Prob. 12-2 assuming there is a constant-level reservoir at the upstream end and a control gate at the downstream end. Assume the downstream gate is closed in 45 s.

12-6   Run the programs of Probs. 12-2 and 12-5 using different values of time interval, and compare the computed results.

12-7   Using the computer program of Prob. 12-2, show that the computations become unstable if the Courant's stability condition is not satisfied. (*Hint*: Use a computational time interval larger than that required by the Courant condition.)

12-8   Lax scheme becomes unstable if the time derivative is replaced by $\partial f/\partial t = [f_i^{k+1} - f_i^k]/\Delta t$ instead of that given by Eqs. (12-41) and (12-42), page 620, (f stands for both y and V) even if the Courant condition is satisfied. By modifying the program of Prob. 12-2, prove that this is the case.

REFERENCES

1. Abbott, M.B. *Computational Hydraulics: Elements of the Theory of Free-Surface Flows.* Pitman, London, 1979.
2. Anderson, D.A., J.C. Tannehill, and R.H. Pletcher. *Computational Fluid Mechanics and Heat Transfer.* McGraw-Hill, New York, 1984.
3. Chaudhry, M.H. *Applied Hydraulic Transients,* 2d ed. Van Nostrand Reinhold, New York, 1987.
4. Chow, V.T. *Open Channel Hydraulics.* McGraw-Hill, New York, 1959.

5. Cunge, J.A. "On the Subject of a Flood Propagation Method (Muskingum Method)." *Jour. of Hydraulic Research*, International Assoc. for Hydraulic Research, 7, no. 2 (1969), pp. 205–30.
6. Cunge, J., F.M. Holly, and A. Verwey. *Practical Aspects of Computational River Hydraulics*. Pitman, London, 1980.
7. Henderson, F.M. *Open Channel Flow*. Macmillan, New York, 1966.
8. Mahmood, K., and V. Yevjevich (eds.). *Unsteady Flow in Open Channels*. Water Resources Publications, Fort Collins, Colo., 1975.
9. Ponce, V.M., R.M. Li, and D.B. Simons. "Applicability of Kinematic and Diffusion Models." *Jour. Hydraulics Div.*, Amer. Soc. of Civil Engineers, 104 (March 1978), pp. 353–60.
10. Stoker, J.J. *Water Waves*. Interscience Publishers, New York, 1965.
11. Weinmann, P.E., and E.M. Laurenson. "Approximate Flood Routing Methods: A Review." *Jour. Hydraulics Div.*, Amer. Soc. of Civil Engineers, 105 (December 1979), pp. 1521–36.
12. Wylie, C.R. *Advanced Engineering Mathematics*, 3d ed. McGraw-Hill, New York, 1966.

Appendix

NOMENCLATURE AND DIMENSIONS

Symbol	Dimensions	Description
A	L^2	Area
AF	L^3	Acre-feet
a	L/T^2	Acceleration
B	L	Aquifer thickness
B	L^4/T^3	Buoyancy flux for round plume
B	L^3/T^3	Buoyancy flux for plain plume
B_c	L	Conduit diameter
B_d	L	Trench width
b	L	Linear measure
b_0	L	Width of two-dimensional jet
C	...	Runoff coefficient
C	...	Discharge coefficient
C	$L^{1/2}/T$	Chezy coefficient
C	...	Resistance coefficient
C_c	...	Contraction coefficient
C_D	...	Discharge coefficient
C_D	...	Drag coefficient
C_H	...	Head coefficient
C_h	...	Hazen-Williams friction coefficient
CN	...	Curve number in runoff calculation
C_P	...	Power coefficient
C_P	...	Pressure coefficient
C_p	$L^2/T^2\theta$	Specific heat
C_p	...	Constant in synthetic unit hydrograph
C_Q	...	Discharge coefficient
C_t	...	Constant in synthetic unit hydrograph
C_V	...	Velocity coefficient
$°C$	θ	Temperature, centigrade
c	L/T	Wave celerity
c	L	Chord length
c_f	...	Shear stress coefficient
D	L	Depth in stilling basin
D	L	Diameter
D	T	Duration
d	L	Depth
d	L	Gate opening

NOMENCLATURE AND DIMENSIONS (continued)

Symbol	Dimensions	Description
d_c	L	Critical depth
E	L	Specific energy
E	L	Total head on weir
E	F/L^2	Elastic modulus
E	L	Evaporation, depth of water
e	. . .	Efficiency
e	F/L^2	Vapor pressure
F	L	Fetch
F	F	Force
F	. . .	Functional relation
°F	θ	Temperature, Fahrenheit
F_D	F	Drag force
F_L	F	Lift force
Fr	. . .	Froude number
f	. . .	Resistance coefficient (friction factor)
f	T^{-1}	Frequency
f	. . .	Coefficient of friction
f	L/T	Infiltration rate
G_0	L	Gate opening
G_0	L^3	Volume of subsurface flow
g	L/T^2	Acceleration due to gravity
H	L	Head (on weir, spillway, etc.)
H	L	Height of fill
H_0	L	Design head
h	L	Piezometric head
h	L	Depth
h_a	L	Velocity head
h_f	L	Head loss due to friction
h_L	L	Head loss
h_p	L	Head supplied by pump
h_t	L	Head supplied to turbine
h_v	L	Velocity head
I	L^3/T	Inflow discharge rate
I	L^4	Moment of inertia
I_a	L	Initial abstraction
i	. . .	Index number
$\mathbf{i}$	. . .	Unit vector in x direction
$\mathbf{j}$	. . .	Unit vector in y direction
j	. . .	Index number
K	F/L	Spring constant
K	. . .	A constant

(continued)

NOMENCLATURE AND DIMENSIONS (continued)

Symbol	Dimensions	Description
K	L/T	Hydraulic conductivity
K	...	Head-loss coefficient
K	T	Constant in Muskingum flood routing method
K	...	Flow coefficient in discharge equation
K	...	Heat transfer coefficient
K_f	...	Infiltration constant
k	...	Time constant
k	...	Maximizing factor of precipitation
k_s	L	Equivalent sand roughness
L	L	Linear measure
M	FT^2/L	Mass
M	...	Mean volume
M	FL	Moment
M	...	Order in series of events
M	L	Depth of snowmelt
M	L^3	Moisture volume in soil
m	L	Length, meter
N	T^{-1}	Angular speed in rpm
N	...	Number of events
N_s	$L^{3/4}/T^{3/2}$	Specific speed
n	...	Porosity
n	...	Number (years, events, etc.)
n	T^{-1}	Angular speed in rps
n	...	Manning's resistance coefficient
n_s	...	Specific speed
O	L^3/T	Outflow discharge
P	LF/T	Power
P	L	Wetted perimeter
P	L	Height of weir or dam
P	L	Depth of precipitation
P	L	Rainfall excess
p	F/L^2	Pressure
P	...	Probability
p_a	F/L^2	Atmospheric pressure
p_v	F/L^2	Vapor pressure
Q	L^3/T	Discharge
q	L^2/T	Discharge per unit width
q	L/T	Discharge per square mile
R	L^3	Volume of runoff
R	L	Hydraulic radius
R	L	Runup
R	...	Skew

NOMENCLATURE AND DIMENSIONS (continued)

Symbol	Dimensions	Description
R	F	Reaction force
R	L	Rainfall depth
Re	...	Reynolds number
r	L	Radial linear measure
r_0	L	Pipe radius
S	...	Slope
S	L	Potential maximum soil retention
S	...	Dilution
S	F	Strength of pipe
S	...	Storage coefficient
S	...	Standard deviation
S	L^3	Storage volume
S_f	...	Friction slope
S_o	...	Channel slope
S_G	L^3	Ground water storage
S_r	...	Specific retention
S_s	L^3	Surface water storage
S_y	...	Specific yield
s	T	Time, second
sfm	L^3	Second-foot-month
T	T	Recurrence interval
T	T	Wave period
T	L	Top width
T	L^2/T	Transmissivity
T	FL	Torque
T	F	Thrust force
T	T	Time
T_r	...	Time ratio
T_w	θ	Temperature, wet bulb Fahrenheit
t	L	Thickness
t	T	Time
t_c	T	Time of concentration
t_p	T	Time to peak of hydrograph
t_r	T	Rainfall duration
U	L/T	Velocity
u	L/T	Velocity
u	...	Argument of the well function
u'	L/T	Turbulence intensity in x direction
u_*	L/T	Shear velocity
$\forall$	L^3	Volume
V	L/T	Velocity

(continued)

NOMENCLATURE AND DIMENSIONS (continued)

Symbol	Dimensions	Description
V_d	L^3	Drainage volume
V_e	F	Earthquake force
V_t	L^3	Total volume
V_v	L^3	Voids volume
V_w	L^3	Water volume
V	L^3	Volume
v	$\ldots$	Velocity
v'	L/T	Turbulence intensity in y direction
W	T	Width of synthetic unit hydrograph
W	F	Weight
W	F	Load
W	FL/T	Power, watts
$W(u)$	$\ldots$	Well function
w	L/T	Velocity in z direction
X	$\ldots$	Weighting factor in flood routing
x	L	Linear measure
y	L	Linear measure
y_c	L	Critical depth
y_n	L	Normal depth
z	L	Linear measure
z	L	Elevation

Greek Letters

α	$\ldots$	Angular measure
α	$\ldots$	Kinetic energy correction factor
α	θ^{-1}	Coefficient of thermal expansion
α	$\ldots$	Angle of attack
β	$\ldots$	Angular measure
γ	F/L^3	Specific weight
Δ	$\ldots$	Increment
δ	$\ldots$	Increment
ϵ	$\ldots$	Strain
η	$\ldots$	Efficiency
θ	$\ldots$	Angular measure
λ	$\ldots$	Earthquake intensity
λ	T^{-1}	Multiplier
μ	FT/L^2	Viscosity, dynamic
ν	L^2/T	Viscosity, kinematic
τ	F/L^2	Shear stress
π	$\ldots$	3.1416
ρ	FT^2/L^4	Mass density
σ	F/L^2	Normal stress
σ	$\ldots$	Cavitation index
ϕ	$\ldots$	Speed ratio
ω	T^{-1}	Angular speed

Figure A-1 Centroids and Moments of Inertia of Plane Area

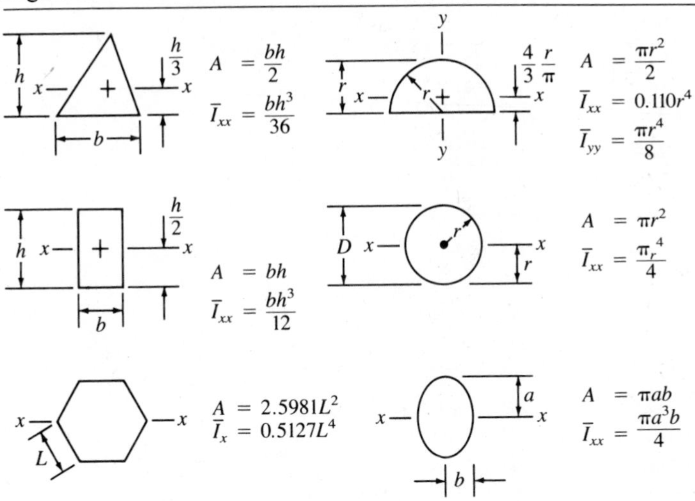

Triangle:
$$A = \frac{bh}{2}$$
$$\bar{I}_{xx} = \frac{bh^3}{36}$$

Semicircle:
$$\frac{4}{3}\frac{r}{\pi} \qquad A = \frac{\pi r^2}{2}$$
$$\bar{I}_{xx} = 0.110r^4$$
$$\bar{I}_{yy} = \frac{\pi r^4}{8}$$

Rectangle:
$$A = bh$$
$$\bar{I}_{xx} = \frac{bh^3}{12}$$

Circle:
$$A = \pi r^2$$
$$\bar{I}_{xx} = \frac{\pi_r^4}{4}$$

Hexagon:
$$A = 2.5981L^2$$
$$\bar{I}_x = 0.5127L^4$$

Ellipse:
$$A = \pi ab$$
$$\bar{I}_{xx} = \frac{\pi a^3 b}{4}$$

Table A-1 Conversion Factors from
Traditional to SI Units

Multiply number of	by	to obtain
in.	25.4	mm
in.	0.0254	m
ft	0.3048	m
yard	0.9144	m
mile	1,609.0	m
ft^2	0.0929	m^2
$in.^2$	6.452 E − 4	m^2
yd^2	0.8361	m^2
mi^2	2.590 E + 6	m^2
acre	4,047.0	m^2
ft^3	0.02832	m^3
yd^3	0.7646	m^3
U.S. gallon	3.785 E − 3	m^3
acre ft	1,233.0	m^3
ft/s	0.3048	m/s
mi/hr	0.447	m/s
ft^3/s	0.02832	m^3/s
gpm	6.309 E − 5	m^3/s
lbf	4.448	N
ton (2000 lbf)	8896.0	N
ft–lbf	1.356	$N \cdot m$
slug	14.59	kg
$slug/ft^3$	515.4	kg/m^3
lbf/ft^2	47.88	N/m^2
$lbf/in.^2$	6,895.0	N/m^2
lbf/ft^3	157.1	N/m^3
ft^2/s	0.0929	m^2/s
$lbf–s/ft^2$	47.88	$N \cdot s/m^2$
hp	0.747	kW

Table A-2 Commonly Used Equivalent Units in Hydraulic Engineering

Volume

Unit	cu. inch	liter	U.S. gallon	cu. foot	cu. yard	cu meter	acre-foot	sec-foot-day
				Equivalent[a][b]				
cubic inch	1	0.016 39	0.004 329	578.7 E − 6	21.43 E − 6	16.39 E − 6	13.29 E − 9	6.698 E − 9
liter	61.02	1	0.264 2	0.035 31	0.001 308	0.001	810.6 E − 9	408.7 E − 9
U.S. gallon	231.0	3.785	1	0.133 7	0.004 951	0.003 785	3.068 E − 6	1.547 E − 6
cubic foot	1728	28.32	7.481	1	0.037 04	0.028 32	22.96 E − 6	11.57 E − 6
cubic yard	46,660	764.6	202.0	27	1	0.764 6	619.8 E − 6	312.5 E − 6
meter³	61,020	1000	264.2	35.31	1.308	1	810.6 E − 6	408.7 E − 6
acre-foot	75.27 E + 6	1,233,000	325,900	43 560	1 613	1 233	1	0.504 2
sec-ft-da	149.3 E + 6	2,447,000	646,400	86 400	3 200	2 447	1.983	1

Discharge (Flow Rate, Volume/Time)

Unit	gallon/min	liter/sec	acre-foot/day	foot³/sec	million gal/day	meter³/sec
			Equivalent[a][b]			
gallon/minute	1	0.063 09	0.004 419	0.002 228	0.001 440	63.09 E − 6
liter/second	15.85	1	0.070 05	0.035 31	0.022 82	0.001
acre-foot/day	226.3	14.28	1	0.504 2	0.325 9	0.014 28
feet³/second	448.8	28.32	1.983	1	0.646 3	0.028 32
million gallons/day	694.4	43.81	3.069	1.547	1	0.043 81
meter³/second	15,850	1000	70.04	35.31	22.82	1

(continued)

Table A-2 (continued)

Velocity

Unit	Equivalent[a][b]				
	foot/day	kilometer/hour	foot/sec	mile/hour	meter/sec
foot/day	1	12.70 E − 6	11.57 E − 6	7.891 E − 6	3.528 E − 6
kilometer/hour	78,740	1	0.911 3	0.621 4	0.277 8
foot/second	86,400	1.097	1	0.681 8	0.304 8
mile/hour	126,700	1.609	1.467	1	0.447 0
meter/second	283,500	3.600	3.281	2.237	1

(a) Equivalent values are shown to 4 significant figures.
(b) Multiply the numerical amount of the given unit by the equivalent value shown (per single amount of given unit) to obtain the numerical amount of the equivalent unit
(e.g.: 5 inches × 0.025 40 m/inch = 0.127 0 m).

SOURCE: SI System of Units: Pamphlet prepared for the Universities Council on Water Resources by Peter C. Klingeman, 1976

Table A-3 Physical Properties of Water at Atmospheric Pressure–S.I. System of Units

Temperature	Density	Specific Weight	Dynamic Viscosity	Kinematic Viscosity	Vapor Pressure	Surface Tension[1]	Bulk Modulus
	kg/m^3	N/m^3	$N \cdot s/m^2$	m^2/s	N/m^2 abs.	N/m	GN/m^2
0°C	1000	9810	1.79×10^{-3}	1.79×10^{-6}	611	0.0756	1.99
5°C	1000	9810	1.51×10^{-3}	1.51×10^{-6}	872	0.0749	2.05
10°C	1000	9810	1.31×10^{-3}	1.31×10^{-6}	1230	0.0742	2.11
15°C	999	9800	1.14×10^{-3}	1.14×10^{-6}	1700	0.0735	2.16
20°C	998	9790	1.00×10^{-3}	1.00×10^{-6}	2340	0.0728	2.20
25°C	997	9781	8.91×10^{-4}	8.94×10^{-7}	3170	0.0720	2.23
30°C	996	9771	7.97×10^{-4}	8.00×10^{-7}	4250	0.0712	2.25
35°C	994	9751	7.20×10^{-4}	7.24×10^{-7}	5630	0.0704	2.27
40°C	992	9732	6.53×10^{-4}	6.58×10^{-7}	7380	0.0696	2.28
50°C	988	9693	5.47×10^{-4}	5.53×10^{-7}	12,300	0.0679	
60°C	983	9643	4.66×10^{-4}	4.74×10^{-7}	20,000	0.0662	
70°C	978	9594	4.04×10^{-4}	4.13×10^{-7}	31,200	0.0644	
80°C	972	9535	3.54×10^{-4}	3.64×10^{-7}	47,400	0.0626	
90°C	965	9467	3.15×10^{-4}	3.26×10^{-7}	70,100	0.0607	
100°C	958	9398	2.82×10^{-4}	2.94×10^{-7}	101,300	0.0589	

[1] Surface tension of water in contact with air

Table A-4 Physical Properties of Water at Atmospheric Pressure–Traditional System of Units

Temperature	Density slugs/ft³	Specific Weight lbf/ft³	Dynamic Viscosity lbf-s/ft²	Kinematic Viscosity ft²/s	Vapor Pressure psia	Surface Tension lbf/ft	Bulk Modulus psi
40°F	1.94	62.43	3.23×10^{-5}	1.66×10^{-5}	0.122	0.00514	297,000
50°F	1.94	62.40	2.73×10^{-5}	1.41×10^{-5}	0.178	0.00508	306,000
60°F	1.94	62.37	2.36×10^{-5}	1.22×10^{-5}	0.256	0.00504	313,000
70°F	1.94	62.30	2.05×10^{-5}	1.06×10^{-5}	0.363	0.00498	320,000
80°F	1.93	62.22	1.80×10^{-5}	0.930×10^{-5}	0.506	0.00492	324,000
100°F	1.93	62.00	1.42×10^{-5}	0.739×10^{-5}	0.949	0.00479	330,000
120°F	1.92	61.72	1.17×10^{-5}	0.609×10^{-5}	1.69	0.00466	332,000
140°F	1.91	61.38	0.981×10^{-5}	0.514×10^{-5}	2.89	0.00452	
160°F	1.90	61.00	0.838×10^{-5}	0.442×10^{-5}	4.74	0.00439	
180°F	1.88	60.58	0.726×10^{-5}	0.385×10^{-5}	7.51	0.00427	
200°F	1.87	60.12	0.637×10^{-5}	0.341×10^{-5}	11.53	0.00414	
212°F	1.86	59.83	0.593×10^{-5}	0.319×10^{-5}	14.70	0.00408	

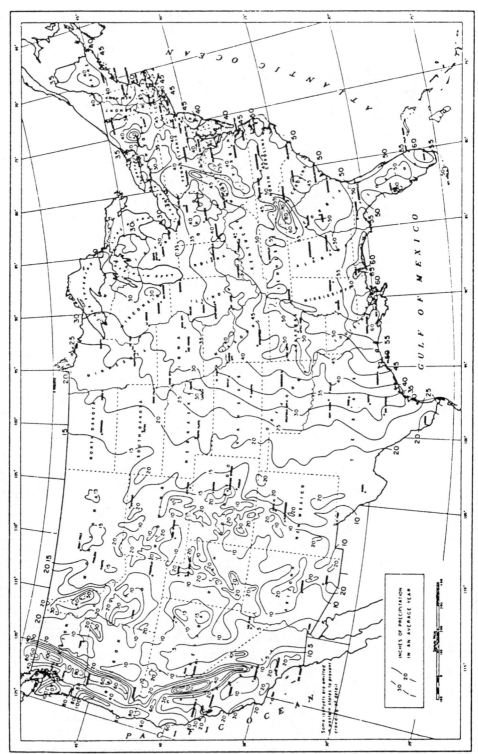

Enlarged Fig. 2-8, page 39, Average annual precipitation in the U.S. (31)

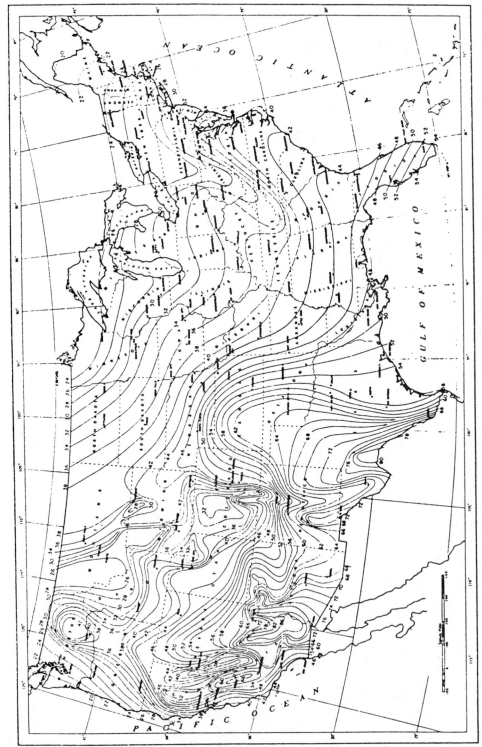

Enlarged Fig. 2-25, page 64, Average annual lake evaporation in the U.S. in inches (6)

Index